Student Solutions Manual and Study Guide for

APPLIED CALCULUS

Third Edition

by Berkey

FRED WRIGHT
Iowa State University

SAUNDERS COLLEGE PUBLISHING
Harcourt Brace College Publishers
Fort Worth Philadelphia San Diego New York
Orlando Austin San Antonio Toronto
Montreal London Sydney Tokyo

Printed in the United States of America.

Fred B. Wright: Student Solutions Manual and Study Guide to accompany APPLIED CALCULUS, 3/E
by Berkey

ISBN 0-03-076178-6

456 021 987654321

Contents

Preface

This *Student Solutions Manual and Study Guide* is a supplement for the third edition of Berkey's *Applied Calculus*. Each chapter is divided into three parts. The first contains review material; the second, solutions to selected exercises; and the third, supplementary practice problems. This combination of review and practice material makes this supplement a useful tool for both learning and studying calculus.

Each chapter begins with a summary of the main concepts including key formulas and definitions to aid you in your mastery of the material. The second part of each chapter contains solutions for every odd-numbered exercise in the text (both section exercises and review exercises). The third part has 13-20 supplementary problems modeled on the exercises in the chapter for further practice. Many of these problems have multiple parts, so there are actually over 525 supplementary practice problems in this supplement. The answers to these practice problems are at the end of the chapter.

If you have any comments or suggestions about this *Student Solutions Manual and Study Guide*, please address your correspondence to: Mathematics Editor, Saunders College Publishing, The Public Ledger Building, Suite 560, 620 Chestnut Street, Philadelphia, PA 19106.

CHAPTER 1

Summary of Chapter 1

1. Real numbers identified as points on the number line.
2. Inequalities and their properties.
3. Intervals and their notations.
4. The Cartesian coordinate plane and graphs of equations involving the real variables x and y.
5. Vertical lines; non-vertical lines, their slopes, and their equations.
6. Distance in a plane, and circles and their equations.
7. The function concept, domain of a function, notations for functions, graphs of functions and the vertical line property.
8. Power functions, quadratic functions, polynomials, rational functions.
9. Finding zeroes of functions by factoring, quadratic formula, equality of function values.
10. The algebra of functions, composite functions, split functions.
11. Revenue, cost, profit; demand, supply, equilibrium price.

Selected Solutions to Exercise Set 1.1

1. Any real number
3. Any real number
5. Integers
7. Integers
9. Any real number
11. d
13. a
15. b
17. g
19. f
21. True
23. Two, $m = -1$ or $m = 1$
25. $6x - 2 \leq 16, \quad 6x \leq 18, \quad x \leq 3, \quad (-\infty, 3]$
27. $4 - 2x \leq 12, \quad -2x \leq 8, \quad x \geq -4, \quad [-4, \infty)$
29. $6 + x \leq 2x - 5, \quad -x \leq -11, \quad x \geq 11, \quad [11, \infty)$
31. $5 - 3x < 7 - 2x, \quad -x < 2, \quad x > -2, \quad (-2, \infty)$
33. $x(6 + x) \geq 3 + x^2, \quad 6x + x^2 \geq 3 + x^2, \quad 6x \geq 3, \quad x \geq \frac{1}{2}, \quad [\frac{1}{2}, \infty)$
35. $200P + 250B \leq 10,000$
37. Let x = number of guests. $12x + 50 \leq 300, \quad 12x \leq 250, \quad x \leq 20$
39. $1296 < 54w < 1620, \quad 24 < w < 30$
41. a. $1100 + 0.02(20,000)(x) > 0.03(20,000)(x)$, $1100 + 400x > 600x$,
    $-200x > -1100$, $x < 6$

    b. $.03(20,000)(x) > 1100 + .02(20,000)(x), \quad 600x > 1100 + 400x,$
    $200x > 1100, \quad x \geq 6$
43. Suppose 11 oranges weight 3 ounces each. Then the twelth orange could weigh at most 15 ounces.

Selected Solutions to Exercise Set 1.2

1.

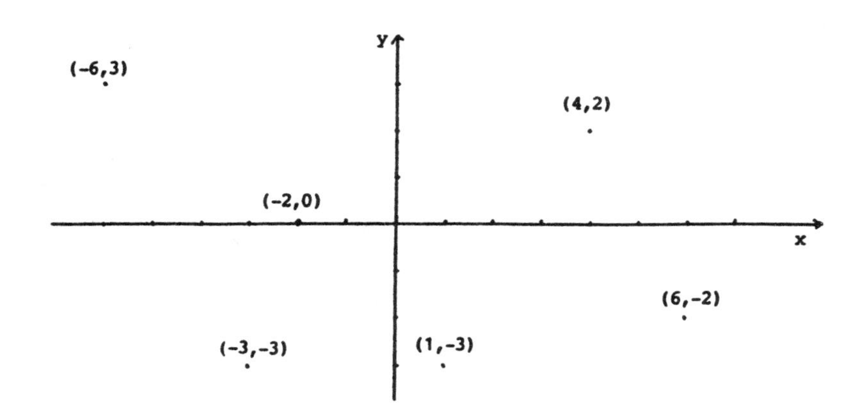

3. $3x - y = 5, \quad -y = 5 - 3x, \; y = 3x - 5$

x	-3	-2	-1	0	1	2	3
y	-14	-11	-8	-5	-2	1	4

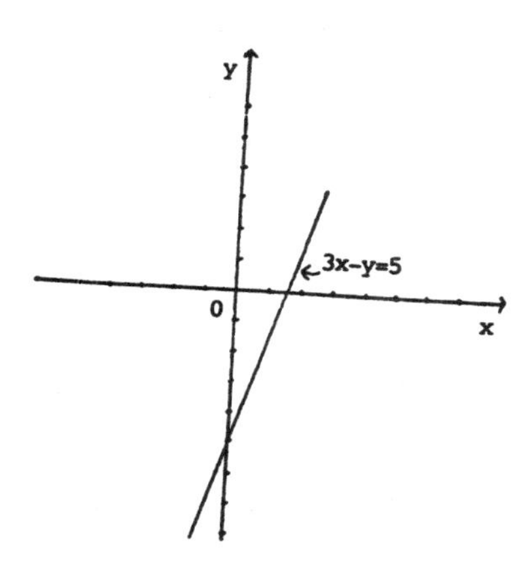

5. $y = \sqrt{x} + 2$

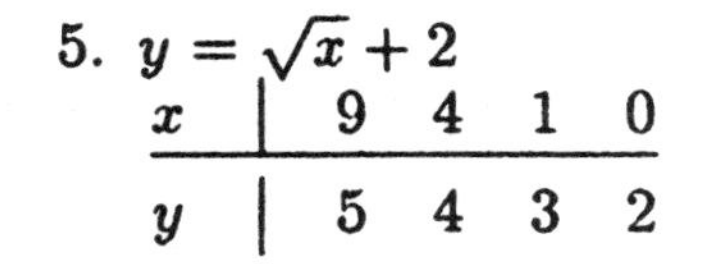

x	9	4	1	0
y	5	4	3	2

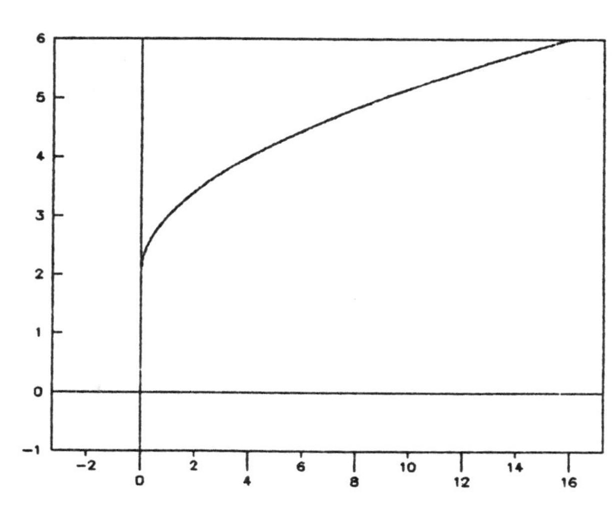

7. $x - y = 1, \quad -y = 1 - x, \quad y = x - 1$

x	-2	-1	0	1	2	3
y	-3	-2	-1	0	1	2

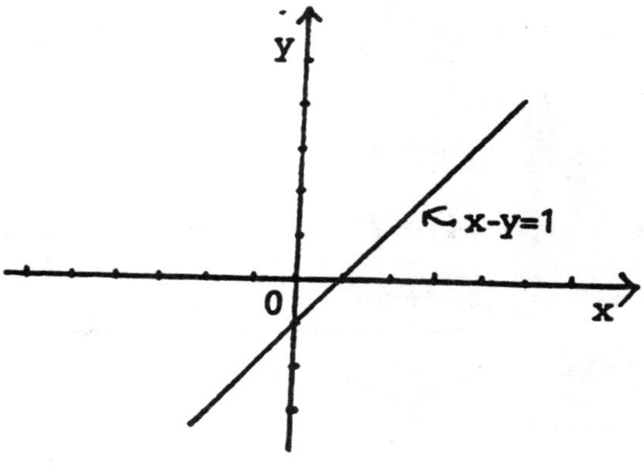

9. $y + \sqrt{x} = 4, \quad y = 4 - \sqrt{x}$

x	4	1	0
y	2	3	4

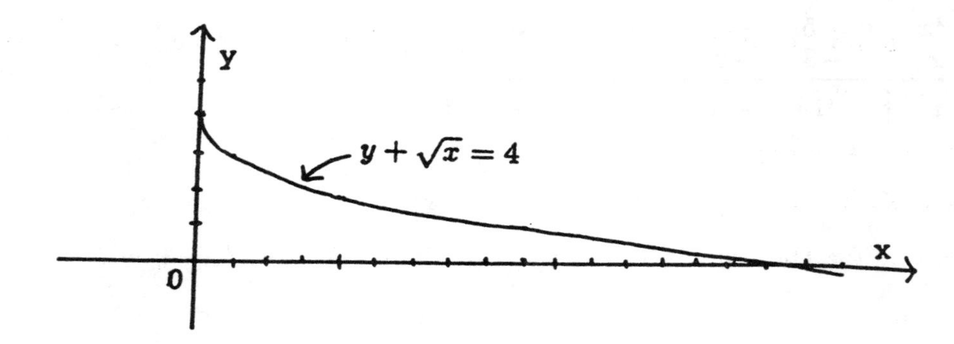

11. $y = \frac{1}{x-2}$

x	3	5	0	-5	1
y	1	$\frac{1}{3}$	$-\frac{1}{2}$	$-\frac{1}{7}$	-1

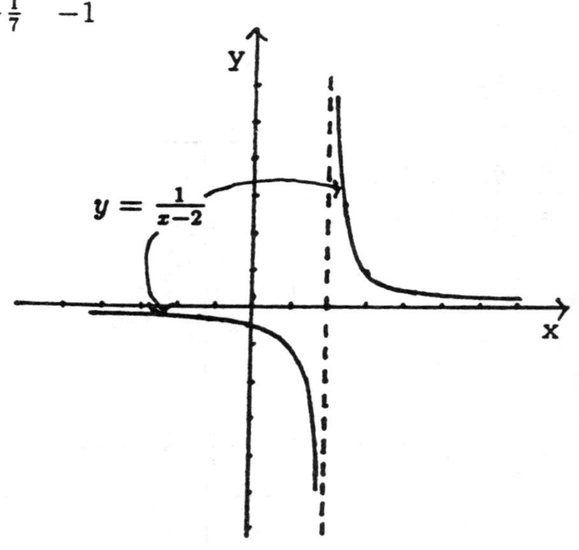

13. False, a vertical line does not have slope.

15. $\frac{a-4}{-2-2} = 3, \quad \frac{a-4}{-4} = 3, \quad a - 4 = -12, \quad a = -8$

17. $y = 4x - 2$

19. $y = -5$

21. $y - 1 = (-4)\cdot(x - 4), \quad y - 1 = -4x + 16, \quad y = -4x + 17$

23. $m = \frac{6-0}{0-(-3)} = \frac{6}{3} = 2$

$y = 2x + 6$

25. $y = -3$

27. $m = \frac{-4-2}{-1-0} = \frac{-6}{-1} = 6$

$y - 2 = 6(x - 0), \quad y - 2 = 6x, \quad y = 6x + 2$

29. Find the slope of the line $3x - 5y = 15$. $\quad -5y = -3x + 15, \quad y = \frac{3}{5}x - 3$.

The slope of this line is $m_1 = \frac{3}{5}$. The slope of the requested line is $m_2 = m_1 = \frac{3}{5}$.

Equation for requested line: $y - (-2) = \frac{3}{5}(x - 5), \quad 5y + 10 = 3x - 15,$

$3x - 5y = 25$.

31. Slope $m_1 = \frac{9-5}{-1-(-2)} = \frac{4}{1} = 4$.

$m_1 m_2 = -1, \quad 4 \cdot m_2 = -1, \quad m_2 = -\frac{1}{4}$

Equation for the requested line: $y - (-1) = -\frac{1}{4}(x - 4), \quad y + 1 = -\frac{1}{4}x + 1,$

$y = -\frac{1}{4}x$

33. $3x - 5y = 6, \quad -5y = 6 - 3x, \quad y = \frac{3}{5}x - \frac{6}{5},$

$m = \frac{3}{5}, \quad b = -\frac{6}{5}$

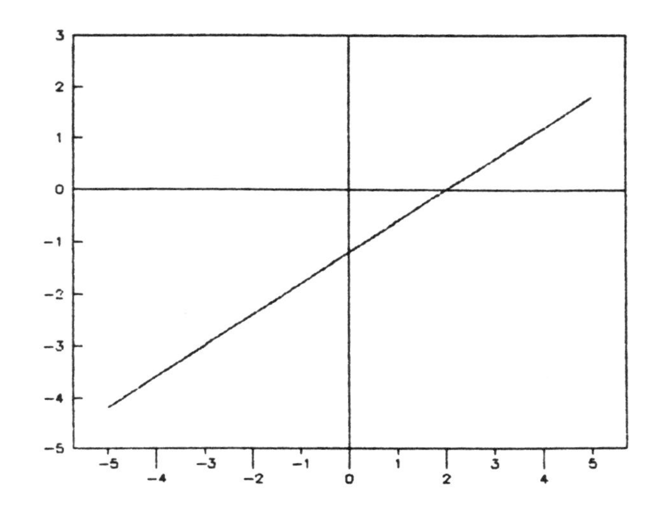

35. $2x = 10 - 3y$, $-3y = 2x - 10$, $y = -\frac{2}{3}x + \frac{10}{3}$,

$m = -\frac{2}{3}$, $b = \frac{10}{3}$

37. $y - 2x = 9$, $y = 2x + 9$

$m = 2$, $b = 9$

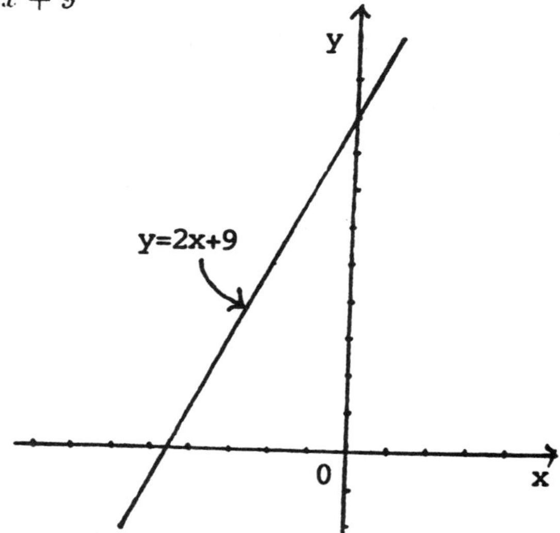

39. $y = x$

$m = 1$, $b = 0$

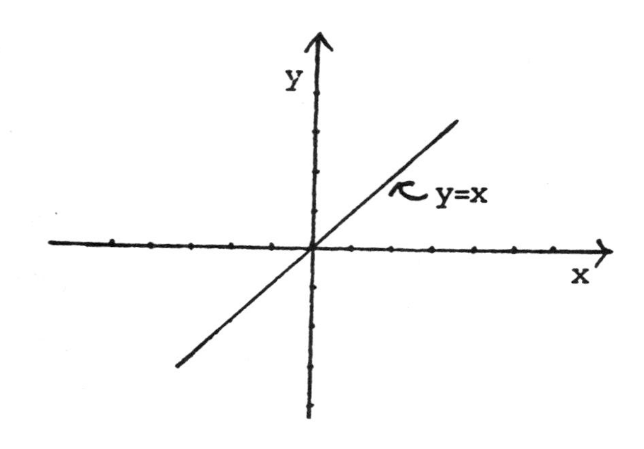

41. $9x - 5y = 45, \quad -5y = 45 - 9x, \quad y = \frac{9}{5}x - 9.$

$m = \frac{9}{5}, \quad b = -9.$

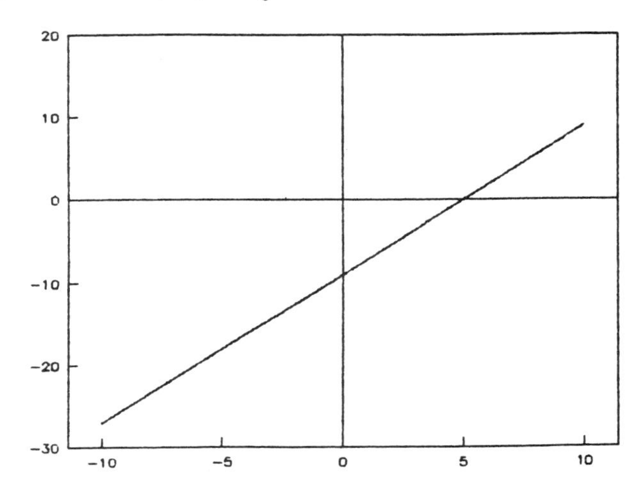

43. a. $m = \frac{600-200}{100-20} = \frac{400}{80} = 5$

$200 = 5(20) + b, \quad 200 = 100 + b, \quad b = 100$

b. $C = 5(150) + 100, \quad C = \$850$

c. $100/week

45. $t =$ the number of years after 1967

Slope $m = \frac{293-100}{15} = \frac{193}{15} = 12.87.$

When $t = 0, \quad C = 100.$

a. $C = 12.87t + 100$

b. $C = 12.87(22) + 100 = 383.14$

47. Let $t =$ the number of years after 1988, let y denote the corresponding industrial production for Germany.

Slope $m = \frac{120-105}{3} = 5$

$y - 105 = 5(t - 0) \implies y = 5t + 105.$

a. $t = 6 \implies y = 5(6) + 105 = 135.$

b. $t = 9 \implies y = 5(9) + 105 = 150.$

49. In this problem, $C = 20,000 dollars$, N $= 4$ years, and $S = 2,000$ dollars.

a. $V = C - t\left(\frac{C-S}{N}\right) = 20,000 - t\left(\frac{20,000-2,000}{4}\right) = 20,000 - 4,500t.$

b. $t = 1.5 \implies V = 20,000 - 4,500(1.5) = 13,250$ dollars.

c. When $V = 10,000$,

$$10,000 = 20,000 - 4,500t \implies 4,500t = 10,000$$
$$\implies t = \frac{20}{9} \text{ years.}$$

d. The annual rate of depreciation is 4,500 dollars per year.

51. a. $V = 12,000 - 200t$.

b. $A = 6,000 + 400t$.

c. $V = A \implies 12,000 - 200t = 6,000 + 400t \implies 600t = 6,000 \implies t = 10$ months.

d. The selling price will be $(12,000 - 200(10)) = 10,000$ dollars.

53. Let D denote the demand in reservations per month for the tour, let p denote the price in dollars.

a. The slope $m = \frac{40-200}{3,600-2,000} = -.1$

$D - 200 = -.1(p - 2000) \implies D = -0.1p + 400$.

b. If $p = 3000$, then $D = -0.1(3,000) + 400 = 100$ reservations per month.

c. When $D = 0$

$$0 = -0.1p + 400 \implies p = 4,000 \text{ dollars.}$$

d. The vertical intercept for this linear model is the value of D when $p = 0$. This value of D is 400 reservations per month, which is an upper bound for the number of reservations possible.

Selected Solutions to Exercise Set 1.3

1. $D = \sqrt{(0-2)^2 + (2-(-1))^2}$, $D = \sqrt{4+9}$, $D = \sqrt{13}$
3. $D = \sqrt{(1-0)^2 + (-9-2)^2}$, $D = \sqrt{1+121}$, $D = \sqrt{122}$
5. $D = \sqrt{(-1-1)^2 + (-1-1)^2}$, $D = \sqrt{4+4}$, $D = \sqrt{8} = 2\sqrt{2}$
7. $D = \sqrt{(-3-(-4))^2 + (-5-2)^2}$, $D = \sqrt{1+49}$, $D = \sqrt{50} = 5\sqrt{2}$

9. First we verify that the distance d_1 between (x_1, y_1) and $\left(\frac{x_1+x_2}{2}, \frac{y_1+y_2}{2}\right)$ equals the distance d_2 between (x_2, y_2) and $\left(\frac{x_1+x_2}{2}, \frac{y_1+y_2}{2}\right)$.

$$d_1 = \sqrt{\left(\frac{x_1+x_2}{2} - x_1\right)^2 + \left(\frac{y_1+y_2}{2} - y_1\right)^2} = \sqrt{\left(\frac{x_2-x_1}{2}\right)^2 + \left(\frac{y_2-y_1}{2}\right)^2}$$

$$d_2 = \sqrt{\left(\frac{x_1+x_2}{2} - x_2\right)^2 + \left(\frac{y_1+y_2}{2} - y_2\right)^2} = \sqrt{\left(\frac{x_1-x_2}{2}\right)^2 + \left(\frac{y_1-y_2}{2}\right)^2}$$

Thus $d_1 = d_2$.

Now we show that the point $\left(\frac{x_1+x_2}{2}, \frac{y_1+y_2}{2}\right)$ is on the line joining (x_1, y_1) and (x_2, y_2).

Case 1: $x_2 \neq x_1$. The line joining (x_1, y_1) and (x_2, y_2) has the equation

$$y - y_1 = \left(\frac{y_2-y_1}{x_2-x_1}\right)(x - x_1).$$

$$\frac{y_1+y_2}{2} - y_1 \overset{?}{=} \left(\frac{y_2-y_1}{x_2-x_1}\right)\left(\frac{x_1+x_2}{2} - x_1\right)$$

$$\frac{y_2-y_1}{2} \overset{?}{=} \left(\frac{y_2-y_1}{x_2-x_1}\right)\frac{(x_2-x_1)}{2} \qquad \text{true.}$$

Case 2: $x_2 = x_1$. The line joining (x_1, y_1) and (x_2, y_2) has the equation

$$x = x_1.$$

$$\frac{x_1+x_2}{2} \overset{?}{=} x_1$$

$$\frac{x_1+x_1}{2} \overset{?}{=} x_1 \qquad \text{true.}$$

Therefore the point $\left(\frac{x_1+x_2}{2}, \frac{y_1+y_2}{2}\right)$ is the midpoint of the line segment joining the points (x_1, y_2) and (x_2, y_2).

11. $x^2 + y^2 = 3^2 = 9$.

13. $(x-4)^2+(y-3)^2 = 2^2$, $\quad x^2-8x+16+y^2-6y+9 = 4$, $\quad x^2-8x+y^2-6y+21 = 0$

15. $(x-(-6))^2 + (y-(-4))^2 = 10^2$, $\quad x^2+12x+36+y^2+8y+16 = 100$,

$x^2 + 12x + y^2 + 8y - 48 = 0$

17. $x^2 - 10x = (x^2 - 10x + (-5)^2) - (-5)^2 = (x-5)^2 - 25$

19. $8x - x^2 = -(x^2 - 8x) = -\left[(x^2 - 8x + (-4)^2) - (-4)^2\right] = -(x-4)^2 + 16$

21. $3-2t-2t^2 = -2(t^2+t)+3 = -2\left[\left(t^2+t+\left(\frac{1}{2}\right)^2\right) - \left(\frac{1}{2}\right)^2\right]+3 = -2\left(t+\frac{1}{2}\right)^2+\frac{7}{2}$

23. $(x^2 - 2x) + (y^2 + 6y) - 12 = 0$

$(x^2 - 2x) = (x^2 - 2x + 1) - 1 = (x-1)^2 - 1$

$(y^2 + 6y) = (y^2 + 6y + 9) - 9 = (y+3)^2 - 9$

$\left[(x-1)^2 - 1\right] + \left[(y+3)^2 - 9\right] - 12 = 0, \quad (x-1)^2 + (y+3)^2 = 22$

The circle has the center $(1, -3)$ and radius $r = \sqrt{22}$.

25. $(x^2 + 14x) + (y^2 - 10y) + 70 = 0$

$(x^2 + 14x) = (x^2 + 14x + 49) - 49 = (x+7)^2 - 49$

$(y^2 - 10y) = (y^2 - 10y + 25) - 25 = (y-5)^2 - 25$

$\left[(x+7)^2 - 49\right] + \left[(y-5)^2 - 25\right] + 70 = 0, \quad (x+7)^2 + (y-5)^2 = 4$

The circle has the center $(-7, 5)$ and radius $r = \sqrt{4} = 2$.

27. $(x^2 - 2x) + (y^2 - 6y) + 3 = 0$

$(x^2 - 2x) = (x^2 - 2x + 1) - 1 = (x-1)^2 - 1$

$(y^2 - 6y) = (y^2 - 6y + 9) - 9 = (y-3)^2 - 9$

$\left[(x-1)^2 - 1\right] + \left[(y-3)^2 - 9\right] + 3 = 0, \quad (x-1)^2 + (y-3)^2 = 7$

The circle has the center $(1, 3)$ and radius $r = \sqrt{7}$.

29. $(x^2 - 2x) + (y^2 - 4y) + 1 = 0$

$(x^2 - 2x) = (x^2 - 2x + 1) - 1 = (x-1)^2 - 1$

$(y^2 - 4y) = (y^2 - 4y + 4) - 4 = (y-2)^2 - 4$

$\left[(x-1)^2 - 1\right] + \left[(y-2)^2 - 4\right] + 1 = 0, \quad (x-1)^2 + (y-2)^2 = 4$

The circle has the center $(1, 2)$ and radius $r = \sqrt{4} = 2$.

The distance between $(1, 2)$ and $(2, 1)$ is $\sqrt{(2-1)^2 + (1-2)^2} = \sqrt{1+1} = \sqrt{2}$ which is smaller than the radius $r = \sqrt{4}$. So point $(2, 1)$ lies inside the circle.

31. We assume that the points (x, y) located within 5 units of the tree are the solutions of the inequality.

$$(x-2)^2 + (y-4)^2 < 5^2.$$

The point $(0,0)$ satisfies this inequality since

$$(0-2)^2 + (0-4)^2 = 4 + 16 < 25.$$

Selected Solutions to Exercise Set 1.4

1. Function, because each package has a unique postage charge.
3. Not necessarily a function, because a professor might have more than one child.
5. Function, because each day has a unique average temperature.
7. Not necessarily a function, because a course might have more than one textbook.
9. a. $f(0) = 7 - 3(0) = 7$

   b. $f(3) = 7 - 3(3) = -2$

   c. $f(a) = 7 - 3a$

   d. $f(x+h) = 7 - 3(x+h)$

   e. The domain of f is the set of all real numbers, and the range of f is also the set of all real numbers.
11. a. $f(0) = \frac{0+3}{0-0+2} = \frac{3}{2}$

   b. $f(-3) = \frac{-3+3}{9+3+2} = \frac{0}{14} = 0$

   c. $f(-1) = \frac{(-1)+3}{(-1)^2-1+2} = \frac{2}{4} = \frac{1}{2}$

   d. $f(x+h) = \frac{(x+h)+3}{(x+h)^2-(x+h)+2}$
13. Not a function
15. Not a function
17. Function
19. Function

21. $f(x) = \frac{1}{x^2-7}$. The domain is the set of all real numbers x except $x = \sqrt{7}$ and $x = -\sqrt{7}$, because division by zero is not allowed.

23. $f(s) = \frac{s^2-1}{s+1} = s - 1$. The domain is the set of all real numbers s except $s = -1$, because division by zero is not allowed in the original formula for $f(s)$.

25. $g(x) = \frac{1}{\sqrt{1-x^2}}$. Because square roots of negative numbers are not allowed and division by zero also is not allowed, the domain is the set of all real numbers x such that $(1 - x^2) > 0$, $x^2 < 1$, $-1 < x < 1$. The domain is the open interval $(-1, 1)$.

27. $f(x) = \frac{\sqrt{x}}{x^3-x}$. The domain is the set of all positive real numbers x except $x = 1$, because division by zero is not allowed, and square roots of negative numbers are not allowed.

29. $f(x) = x^2 + 6x + 7 = (x^2 + 6x + (3)^2) - 9 + 7 = (x + 3)^2 - 2$.
The range is the interval $[-2, \infty)$.

31. $C(x) = 2x + 0.5$ dollars

33. $C(x) = 70x + 1000$ dollars

35. a) $D(5) = 800 - 40(5) = 800 - 200 = 600$

b) $D(0)$ gives the greatest demand. (At $p = 0$, $D(p)$ is greatest.)

c) $D(p) = 0$ for $800 - 40p = 0$, $40p = 800$, $p = 20$.

37. Let p = daily room rent in dollars, and let y = the corresponding number of rooms rented per day.

Let $R(p)$ = daily revenue in dollars from room rentals.

If $p < 50$, then $y = 400$, $R(p) = p \cdot y = 400p$ dollars.

Suppose $p \geq 50$. Note that y is linearly related to p, so $y = mp + b$. $m = -10$, because for each dollar increase in the daily rate above 50 dollars, 10 rooms will remain vacant. Because $y = 400$ when $p = 50$, $400 = (-10)(50) + b \implies b = 900$. Therefore $y = -10p + 900$. Thus $R(p) = p \cdot y = p(-10p + 900) = -10p^2 + 900p$.

39. a.

n	0	5	10	15	20
$W(n)$	20	28.28	34.64	40	44.72

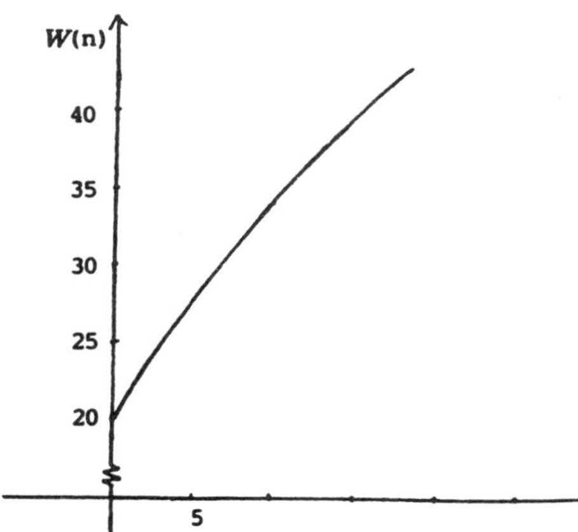

b. $W(0) = \sqrt{400} = 20$ words.

c. There is no maximum speed since $W(n)$ continually increases as n increases.

41. a. $P(x) = 0.005x + 100$.

b. When $x = 150,000$, $P(x) = 0.005(150,000) + 100 = \$850$.

43. t = number of years since 1958.

L = carbon dioxide level at the South Pole in parts per million.

$L(t) = mt + b$.

a. When $t = 0$, $L = 315$;

when $t = 8$, $L = 322$;

when $t = 16$, $L = 329$.

$m = \frac{322-315}{8} = \frac{7}{8}$

$b = 315$

$L(t) = \frac{7}{8}t + 315$

b. When $t = 40$, $L = \frac{7}{8}(40) + 315 = 350$ parts per million.

Selected Solutions to Exercise Set 1.5

1. $27^{2/3} = \left(27^{1/3}\right)^2 = 3^2 = 9$

3. $4^{-3/2} = \frac{1}{4^{3/2}} = \frac{1}{(4^{1/2})^3} = \frac{1}{2^3} = \frac{1}{8}$

5. $\frac{2(3x)^4}{6x^2} = \frac{2(81x^4)}{6x^2} = 27x^2$

7. $\frac{(3x^2y^{-1})^3}{\sqrt{xy}} = \frac{27x^6y^{-3}}{x^{1/2}\cdot y^{1/2}} = \frac{27x^{6-\frac{1}{2}}}{y^{\frac{1}{2}+3}} = \frac{27x^{11/2}}{y^{7/2}}.$

9. $\frac{6(xy)^3}{x^{2/3}y^{-3/2}} = \frac{6x^3y^3}{x^{2/3}y^{-3/2}} = 6x^{3-\frac{2}{3}}y^{3-(-\frac{3}{2})} = 6x^{7/3}y^{9/2}$

11. True

13. $f(x) = \frac{1}{2}x^2 - 3$

x	$f(x)$
0	−3.0
±1	2.5
±2	−1.0
±3	1.5
±4	5.0

The function is symmetric about the y-axis.

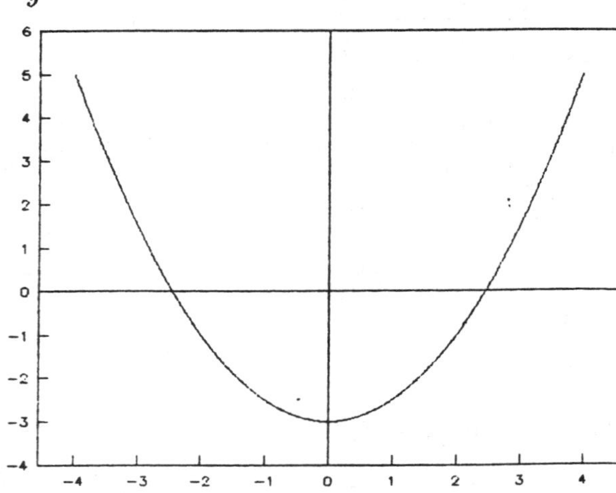

15. $f(x) = 3x^{2/3}$

x	$f(x)$
0	0
±1	3
±4	7.56
±8	12

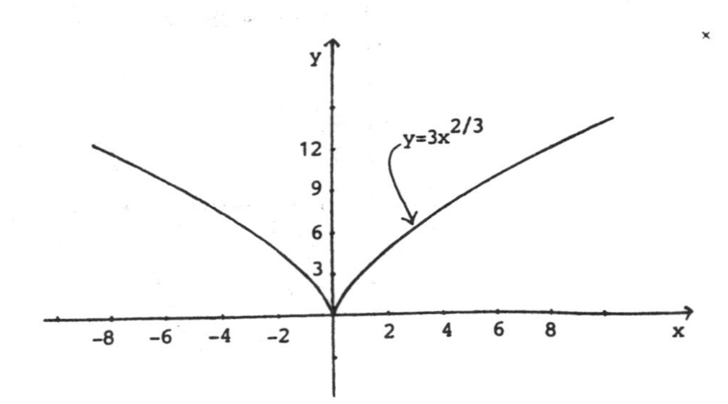

17. $f(x) = -4x^{1/3}$

x	$f(x)$
0	0
±1	∓4
±8	∓8

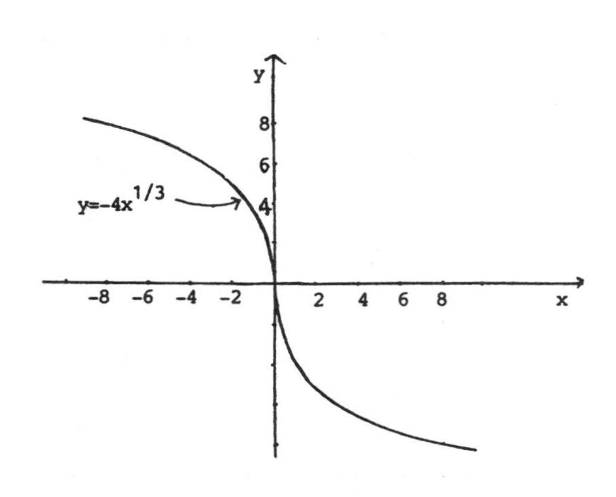

19. $y = -x^{-2/3}$

x	y
$\pm.1$	-4.64
$\pm.5$	-1.59
±1	-1
±4	$-.4$
±8	$-.25$

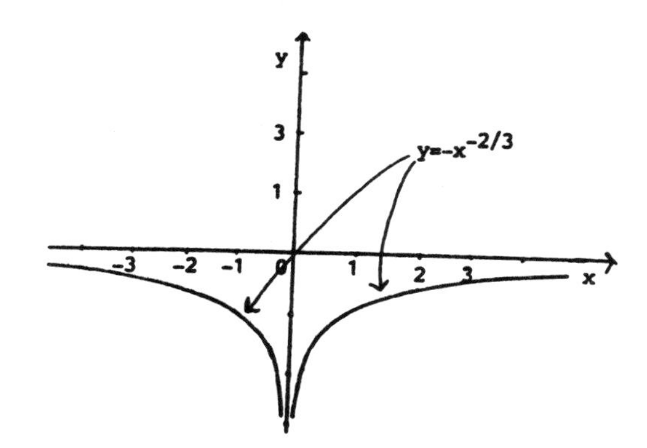

21. $f(x) = (x-3)^2 - 4$

x	$f(x)$
-1	12
0	5
1	0
2	-3
3	-4
4	-3
5	0
6	5

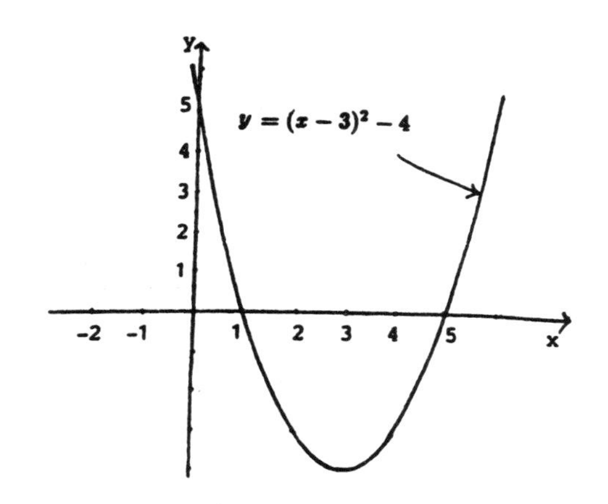

23. $f(x) = x^2 - 2x + 3$

x	$f(x)$
-1	6
0	3
1	2
2	3
3	6

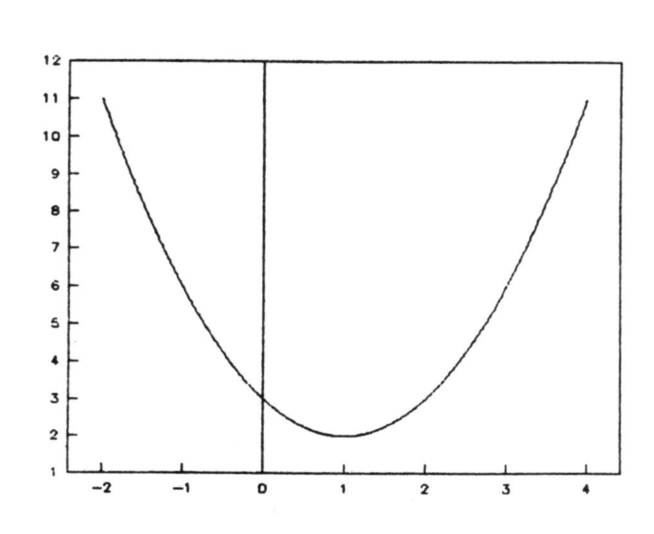

25. $f(x) = -2x^2 - 4x + 1$

x	$f(x)$
-3	-5
-2	1
-1	3
0	1
1	-5

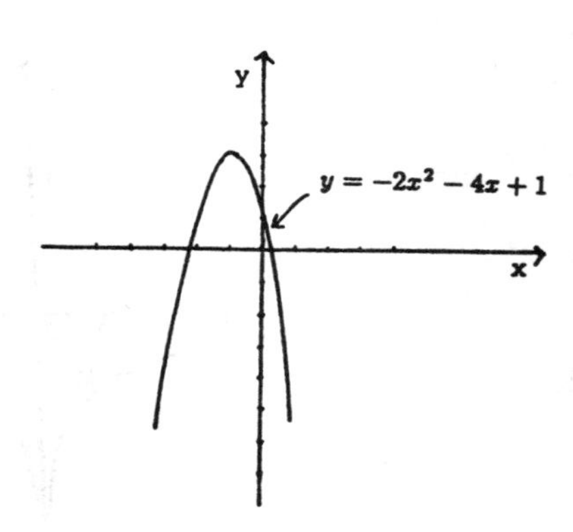

27. $f(x) = x^3 + 4$

x	$f(x)$
-2	-4
-1	3
0	4
1	5
2	12

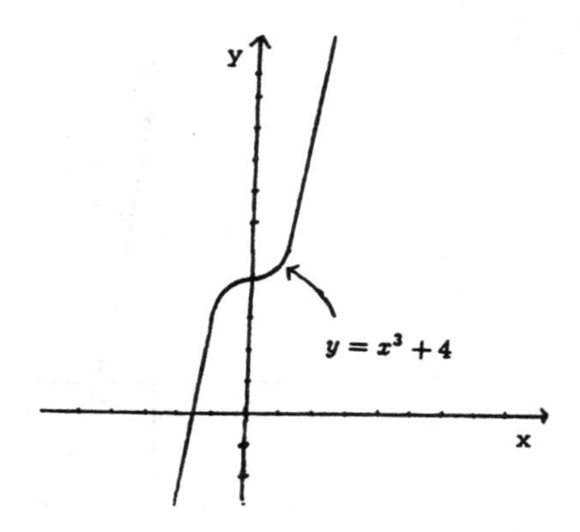

29. $f(x) = 4 - x^4$

x	$f(x)$
-2	-12
-1	3
0	4
1	3
2	-12

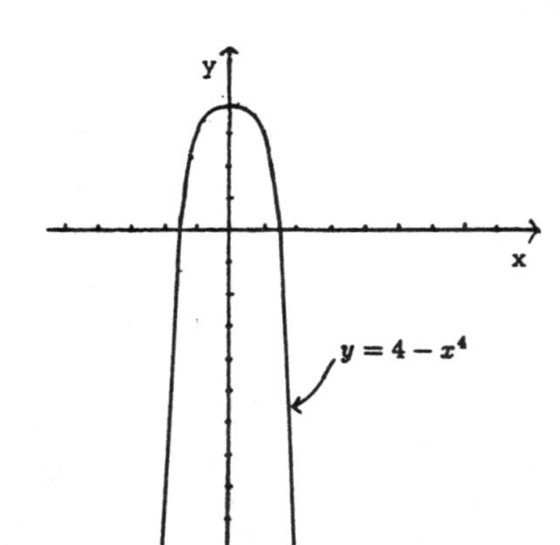

31. $f(x) = x^3 - x^2 - 2x + 2$

x	$f(x)$
-2	-6
-1	2
$-\frac{1}{2}$	$\frac{21}{8}$
0	2
1	0
$\frac{3}{2}$	$\frac{1}{8}$
2	2

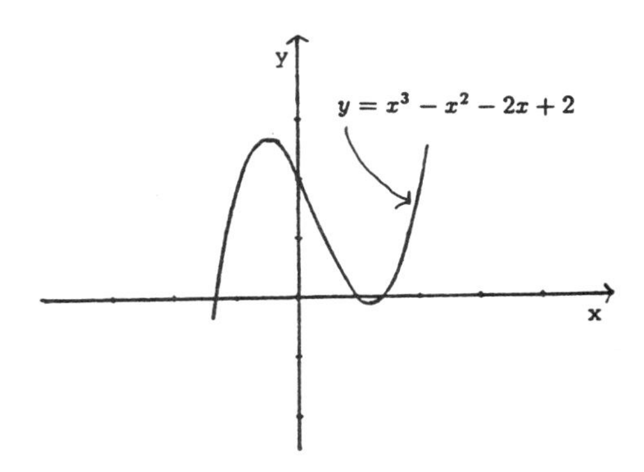

33. $f(x) = \frac{1}{x-3}$

x	$f(x)$
0	$-\frac{1}{3}$
1	$-\frac{1}{2}$
2	-1
4	1
5	$\frac{1}{2}$
6	$\frac{1}{3}$

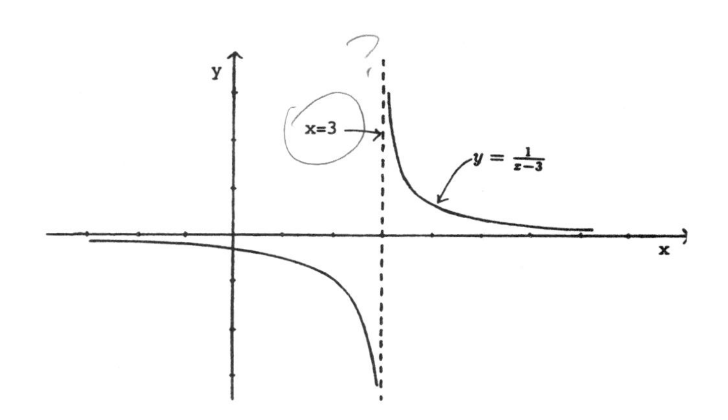

35. $f(x) = \frac{1-x}{1+x}$

x	$f(x)$
-10.0	$-\frac{11}{9} = -1.22$
$-\ 4.0$	$-\frac{5}{3}$
$-\ 3.0$	-2
$-\ 2.0$	-3
$-\ 1.5$	-5

x	$f(x)$
-0.5	3
0	1
1.0	0
2.0	$-\frac{1}{3}$
3.0	$-\frac{1}{2}$
10.0	$-\frac{9}{11} = -0.82$

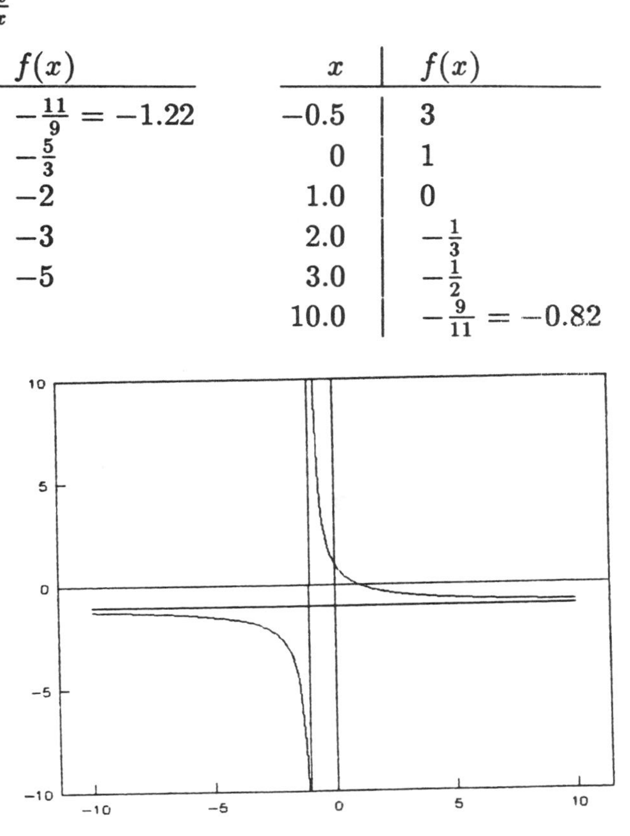

37. $f(x) = \frac{x^3}{1+x}$

x	$f(x)$
-4	$\frac{64}{3}$
-3	$\frac{27}{2}$
-2	8
$-\frac{3}{2}$	$\frac{27}{4}$
$-\frac{5}{4}$	$\frac{125}{16}$
$-\frac{1}{2}$	$-\frac{1}{4}$
0	0
1	$\frac{1}{2}$
2	$\frac{8}{3}$

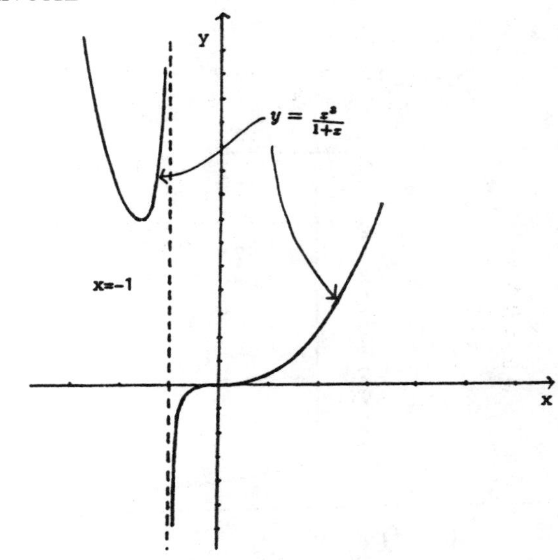

39. Let w = width in cm, ℓ = length in cm.
$w \cdot \ell = 80$. $w = \frac{80}{\ell}$.
The domain of this function is the set of all positive real numbers ℓ.

41. a.

x	0	1	2	3	4	14
$U_1(x)$	1.4	1.7	2	2.2	2.4	4

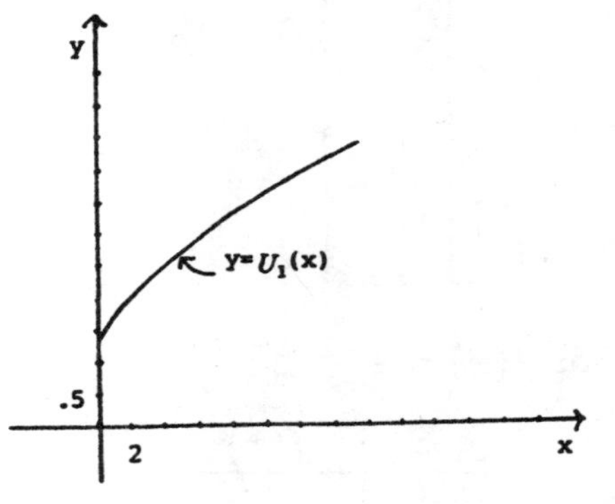

b.

x	0	1	2	3	4	5	6
$U_2(x)$	0	5	8	9	8	5	0

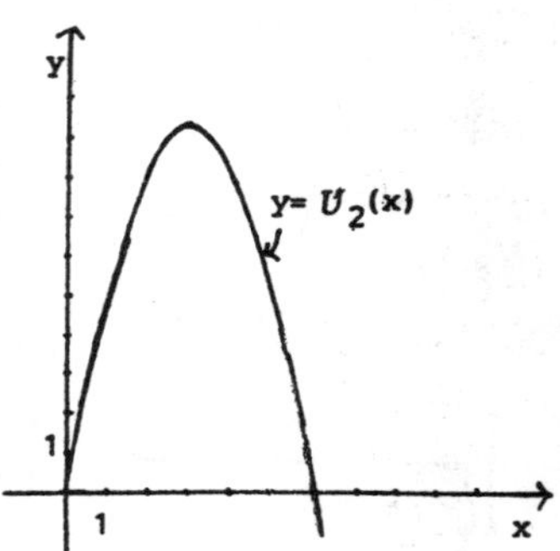

c. $U_2(x)$ would be more appropriate because after eating a certain number of slices of pizza, your satisfaction goes down, not up.

d. $U_1(x)$ because the more money you win the more satisfaction you get.

43. $T(x) = 30\left(1 + \frac{2}{\sqrt{x+1}}\right)$

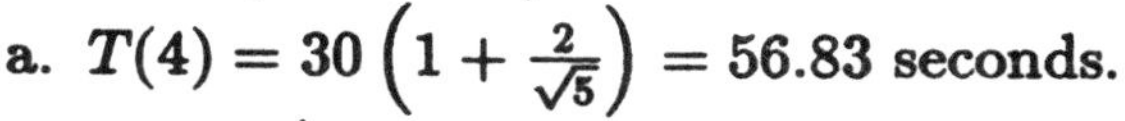

a. $T(4) = 30\left(1 + \frac{2}{\sqrt{5}}\right) = 56.83$ seconds.

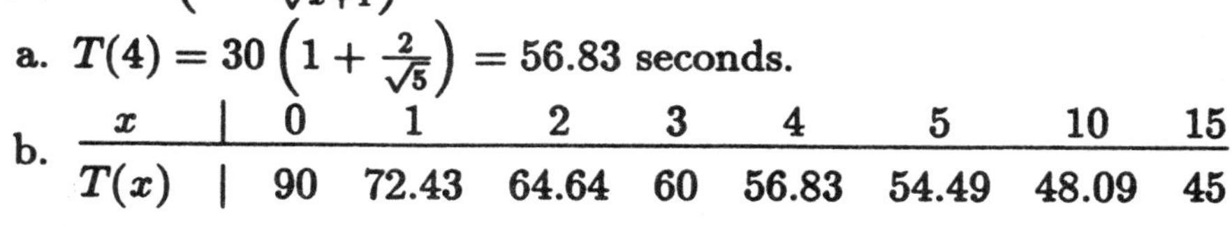

b.

x	0	1	2	3	4	5	10	15
$T(x)$	90	72.43	64.64	60	56.83	54.49	48.09	45

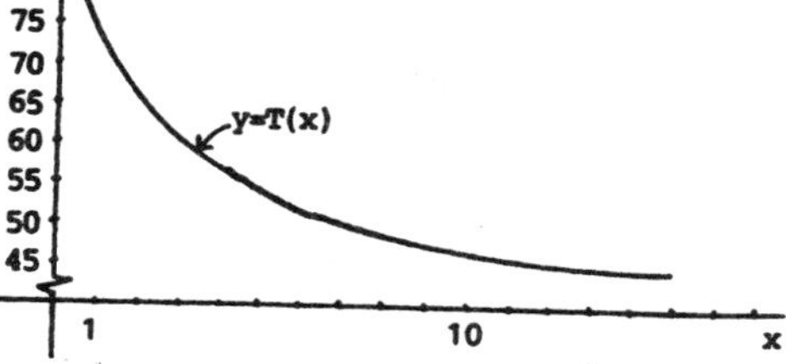

45. $C(x) = 400 + 40x + 0.2x^2$

x	0	40	80	120	160	200	240	280	320
$C(x)$	400	2320	4880	8080	11920	16400	21520	27280	33680

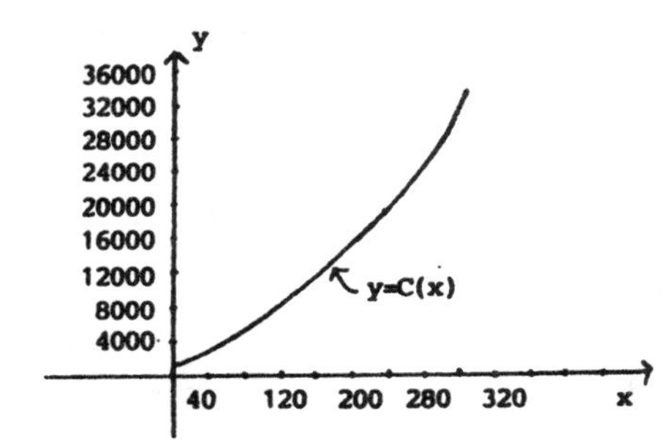

Fixed costs $= C(0) = 400$.

47. a. $C(5) = 500 + 10\left(\frac{5-5}{5}\right)^3 = 500$

$C(30) = 500 + 10\left(\frac{30-5}{5}\right)^3 = 1,750$

$C(40) = 500 + 10\left(\frac{40-5}{5}\right)^3 = 3,930$

$C(50) = 500 + 10\left(\frac{50-5}{5}\right)^3 = 7,790$

b. $R(x) = px = 100x$

c. $P(x) = R(x) - C(x) = 100x - 500 - 10\left(\frac{x-5}{5}\right)^3$

d. $P(t) = 100t - 500 - 10\left(\frac{t-5}{5}\right)^3$

$P(30) = 100(30) - 1750 = 1,250.$

$P(40) = 100(40) - 3,930 = 70.$

$P(50) = 100(50) - 7,790 = -2790.$

e. The profits are not largest at the largest production level because the cost $C(x)$ keeps increasing as x increases.

Selected Solutions to Exercise Set 1.6

1. $f(x) = x^2 - 4 = (x+2)(x-2)$
   $f(x) = 0$ for $x = -2$, $x = 2$.
3. $f(x) = x^2 + x - 6 = (x+3)(x-2)$
   $f(x) = 0$ for $x = -3$, $x = 2$.
5. $f(x) = x^2 - 2x - 35 = (x-7)(x+5)$
   $f(x) = 0$ for $x = 7$, $x = -5$.
7. $f(x) = \frac{\sqrt{x-2}}{x+1}$
   $f(x) = 0$ for $x - 2 = 0 \implies x = 2$.
9. $f(x) = x^{3/2} - 3x^{1/2} = x^{1/2}(x-3) = 0$ for $x = 0$, $x = 3$.
11. $f(x) = x^2 + x - 1 = 0$ for $x = \frac{-1\pm\sqrt{1^2-4(1)(-1)}}{2(1)} = \frac{-1\pm\sqrt{5}}{2}$.
13. $f(x) = 2x^2 + 3x - 1 = 0$ for $x = \frac{-3\pm\sqrt{3^2-4(2)(-1)}}{2(2)} = \frac{-3\pm\sqrt{17}}{4}$
15. $f(x) = 3 + 2x - x^2 = 0$ for $x = \frac{-2\pm\sqrt{2^2-4(-1)(3)}}{2(-1)} = \frac{-2\pm4}{-2} = 3$ or -1.
17. $f(x) = x - 3x^2 + 4 = 0$ for $x = \frac{-1\pm\sqrt{(1)^2-4(-3)(4)}}{2(-3)}$
    $= \frac{-1\pm\sqrt{49}}{-6} = \frac{-1\pm7}{-6} = \frac{4}{3}$ or -1.
19. $f(x) = g(x)$ for $2x + 1 = 7 - x$, $3x = 6$, $x = 2$.
21. $f(x) = g(x)$ for $x^2 - 2 = 2x + 1 \implies x^2 - 2x - 3 = 0 \implies (x-3)(x+1) = 0$
    $\implies x = 3$ or $x = -1$.
23. $f(x) = g(x)$ for $\sqrt{x} = 4x - 3 \implies x = 16x^2 - 24x + 9 \implies 16x^2 - 25x + 9 = 0$
    $\implies x = \frac{25\pm\sqrt{(25)^2-4\cdot(16)\cdot(9)}}{2\cdot(16)} = \frac{25\pm7}{32} \implies x = 1$ or $x = \frac{9}{16}$. $x = \frac{9}{16}$ does not work in the original equation.
25. $f(x) = g(x)$ for $x^3 + 2 = x + 2$, $x^3 - x = 0$, $x(x+1)(x-1) = 0$
    $\Rightarrow x = 0$, $x = -1$, $x = 1$.
27. $f(x) = g(x)$ for $3x^2 - 12x = 5x - 20$, $3x^2 - 17x + 20 = 0$,
    $(3x-5)(x-4) = 0 \Rightarrow x = \frac{5}{3}$, $x = 4$.
29. The demand is greater than the supply.
31. $L(x) = S(x)$ for $100\sqrt{x} = \frac{5}{3}x$, $10,000x = \frac{25}{9}x^2$,
    $25x^2 - 90,000x = 0$, $25x(x - 3600) = 0$, $x = \$3600$.
33. a. $C(x) = 500 + 20x + x^2$, $R(x) = 80x$,
       $P(x) = R(x) - C(x) = 80x - (500 + 20x + x^2) = -x^2 + 60x - 500$
       Break-even production level: $-x^2 + 60x - 500 = 0$,
       $(-x+50)(x-10) = 0 \Rightarrow x = 10$, $x = 50$.
    b. $P(x) > 0$ for $10 < x < 50$.
35. $f(t) = g(t)$ for $\frac{25}{t} = \frac{50}{1+(t-2)^2}$, $50t = 25\left(1 + (t-2)^2\right)$,
    $50t = 25\left(1 + (t^2 - 4t + 4)\right)$, $50t = 25t^2 - 100t + 125$,
    $25t^2 - 150t + 125 = 0$, $25(t^2 - 6t + 5) = 0$,
    $25(t-1)(t-5) = 0 \Rightarrow t = 1$, $t = 5$ hours.

37. Selling price = average cost when $70 = \frac{500+10x+x^2}{x}$
$500 + 10x + x^2 = 70x \implies x^2 - 60x + 500 = 0 \implies (x-50)(x-10) = 0$
$x = 50$, $x = 10$.
39. $C(x) = 2,000 + 80\sqrt{x}$.
The average cost $c(x) = \frac{C(x)}{x} = \frac{2000}{x} + \frac{80}{\sqrt{x}}$.
$c(x) = 1 \implies 1 = \frac{2000}{x} + \frac{80}{\sqrt{x}} \implies x - 80(\sqrt{x}) - 2000 = 0$
$\implies (\sqrt{x})^2 - 80\sqrt{x} - 2000 = 0 \implies (\sqrt{x} - 100)\cdot(\sqrt{x} + 20) = 0 \implies \sqrt{x} = 100 \implies x = 10,000$.

Selected Solutions to Exercise Set 1.7

1. $h_1(x) = (f+g)(x) = f(x) + g(x) = (x-7) + (x^2+2) = x^2 + x - 5$.
3. $h_1(5) = 5^2 + 5 - 5 = 25$.
5. $h_2(-4) = \frac{-4-7}{(-4)^2+2} = -\frac{11}{18}$.
7. $h_1(x) = (f-2g)(x) = f(x) - 2g(x) = \frac{1}{x+2} - 2\sqrt{3+x}$.
9. $h_2(-3) = ((-3)+2)\sqrt{3+(-3)} = 0$.
11. $h_1(1) = \frac{1}{1+2} - 2\sqrt{3+1} = \frac{1}{3} - 4 = -\frac{11}{3}$.
13. $(f+g)(x) = 4 - x$.
Since $(f+g)(x) = f(x) + g(x) = (x^2 - x) + g(x)$, $(x^2 - x) + g(x) = 4 - x$, $g(x) = (4-x) - (x^2 - x)$, $g(x) = -x^2 + 4$.
15. $(f \circ u)(x) = f(u(x)) = f(\sqrt{x}) = 3\sqrt{x} + 1$.
17. $(g \circ f)(x) = g(f(x)) = g(3x+1) = (3x+1)^3$.
19. $(h^2 \circ u)(x) = [h(u(x))]^2 = \left[\frac{1}{\sqrt{x}+1}\right]^2 = \frac{1}{x+2\sqrt{x}+1}$.
21. $(f \circ g \circ u)(x) = f(g(u(x))) = f(g(\sqrt{x})) = f\left((\sqrt{x})^3\right) = 3(\sqrt{x})^3 + 1 = 3x^{3/2} + 1$.
23. Since $u(x) = x - 3$, $f(u(x)) = f(x-3) = x^3 + 5$.
So let $f(x) = (x+3)^3 + 5$.
25. $f(g(u(x))) = 4x^2 - 8x + 8$.
Since $f(g(u(x))) = f\left((u(x)-2)^2\right) = (u(x)-2)^2 + 4$,
then $(u(x)-2)^2 + 4 = 4x^2 - 8x + 8$,
$(u(x)-2)^2 = 4x^2 - 8x + 4 = 4(x-1)^2$.
Take $u(x) - 2 = 2(x-1) \Rightarrow u(x) = 2(x-1) + 2 = 2x$.
27. a.
29. b.
31. d.
33. a.
35. d.

37. $f(x) = \begin{cases} \sqrt{4-x}, & x \le 0 \\ 2 - x^2, & x > 0. \end{cases}$

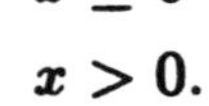

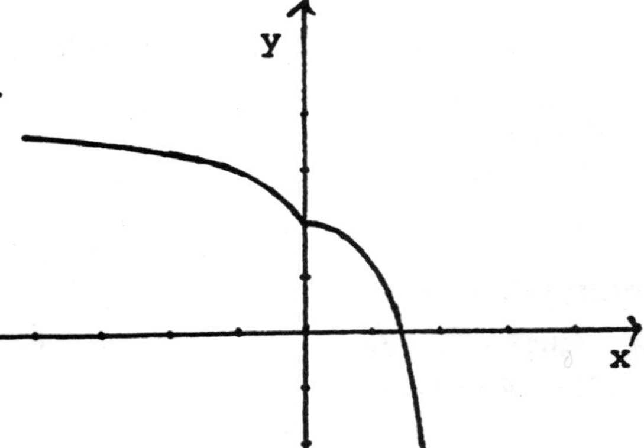

39. $f(x) = x - |x|$.

$$f(x) = \begin{cases} 2x, & x < 0 \\ 0, & x \ge 0 \end{cases}$$

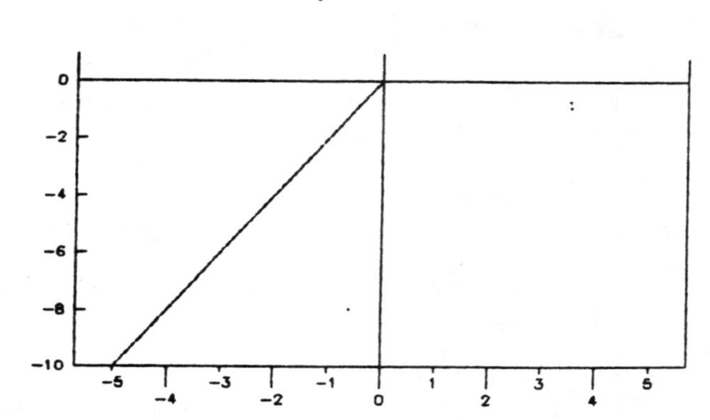

41. $f(x) = |x^2 - x - 2|$

$$f(x) = \begin{cases} x^2 - x - 2, & x < -1 \text{ or } x > 2 \\ -x^2 + x + 2, & -1 \le x \le 2 \end{cases}$$

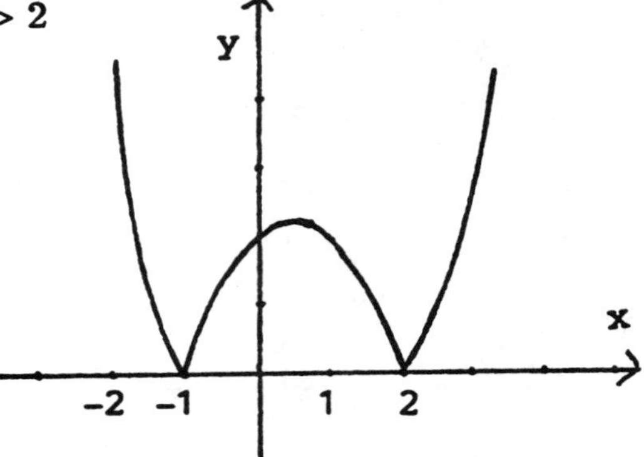

43. a. $p(x) = \begin{cases} 500, & 0 < x \le 5 \\ 500 - 10(x - 5) = 550 - 10x, & 5 < x \le 25 \\ 300, & 25 < x \end{cases}$

b.

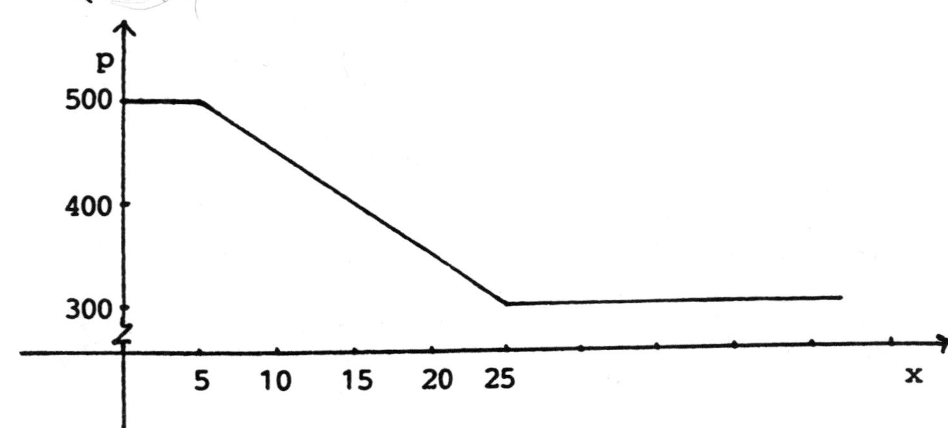

45. $(f \circ r)(c) = f(r(c)) = f(10 + 0.2c + 0.01c^2) = \sqrt{10 + 0.2c + 0.01c^2 + 5}$.

47. a. $C(x) = 200x + 75$ dollars.

b. $R(x) = (1 + .4) \cdot (200x + 75) = 280x + 105$ dollars.

c. The retail price per lawnmower $= \frac{R(x)}{x} = p(x) = \frac{280x+105}{x} = 280 + \frac{105}{x}$ dollars.

Selected Solutions to Chapter 1 – Review Exercises

1. a. $(-6, 3]$ b. $(-\infty, 4)$ c. $[2, \infty)$

3. $3 - x \geq 4 + x$, $-2x \geq 1$, $x \leq -\frac{1}{2}$.

5. $2 - x \leq 4 - 4x$, $-x \leq 2 - 4x$, $3x \leq 2$, $x \leq \frac{2}{3}$.

7.

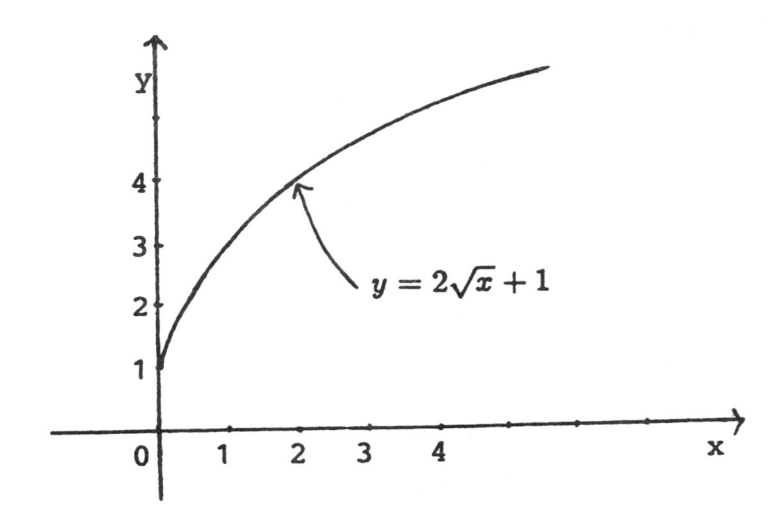

9.

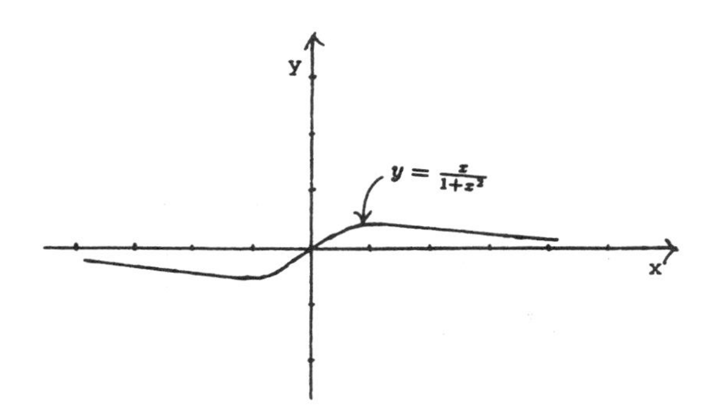

11. $x - 6y = 24$, $y = \frac{1}{6}x - 4$. The slope m_1 of this line is $\frac{1}{6}$. The required slope is $m_1 = \frac{1}{6}$.

13. $y = 5$.

15. The equation of the line is $x = 7$.
Since $(b, 4)$ lies on the line, b must be 7.
17. $-2y = 6 - x$ $\quad y-\text{int} = -3$
$y = -3 + \frac{x}{2}$
$m = \frac{1}{2}$

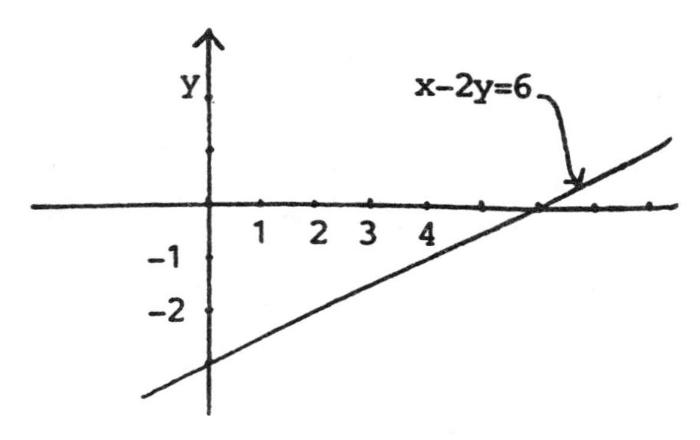

19. $2y = -2x + 8$
$y = -x + 4$
$m = -1, \quad b = 4.$

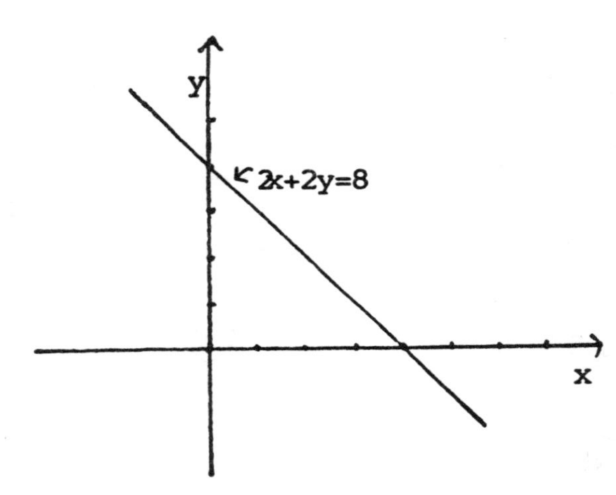

21. Domain is the set of all x except 0.
23. $f(x) = \sqrt{8 - x^3}$
$8 - x^3 = 0$ for $x = 2$.
$8 - x^3 > 0$ for $x < 2$.
Domain is the set of all x such that $x \leq 2$.
25. a. $f(6) = \frac{\sqrt{6-2}}{6+3} = \frac{2}{9}$.
b. $f(11) = \frac{\sqrt{11-2}}{11+3} = \frac{3}{14}$.
27. $16^{3/4} = (16^{1/4})^3 = 2^3 = 8$.
29. $\frac{(x^2y^4)^2}{xy^2} = \frac{x^4y^8}{xy^2} = x^3y^6$.

31.

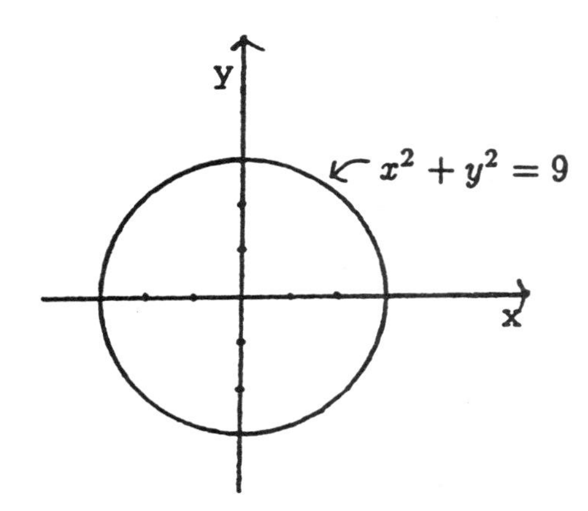

33. $x^2 + 2x + y^2 = 7 + 2y \implies x^2 + 2x + y^2 - 2y = 7$ *implies* $(x^2 + 2x + 1) + (y^2 - 2y + 1) = 7 + 1 + 1 \implies (x+1)^2 + (y-1)^2 = 3^2$. The center is $(-1, 1)$, and the radius is 3.

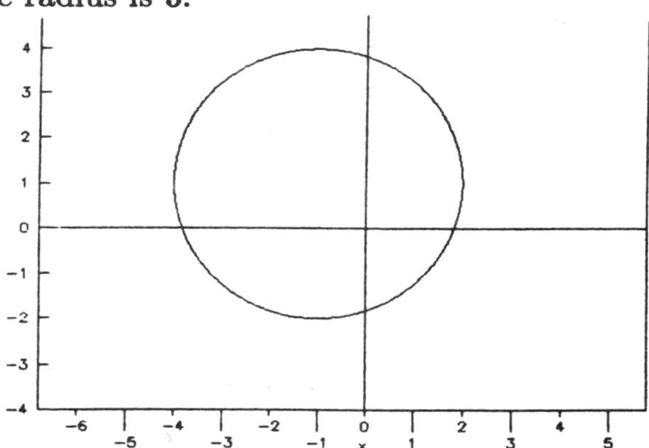

35. $x^2 + 4x + y^2 + 2y = 14$, $(x^2 + 4x + 2^2) - 4 + (y^2 + 2y + 1^2) - 1 = 14$, $(x+2)^2 + (y+1)^2 = 19$. The center is $(-2, -1)$, and the radius is $\sqrt{19} = 4.36$.

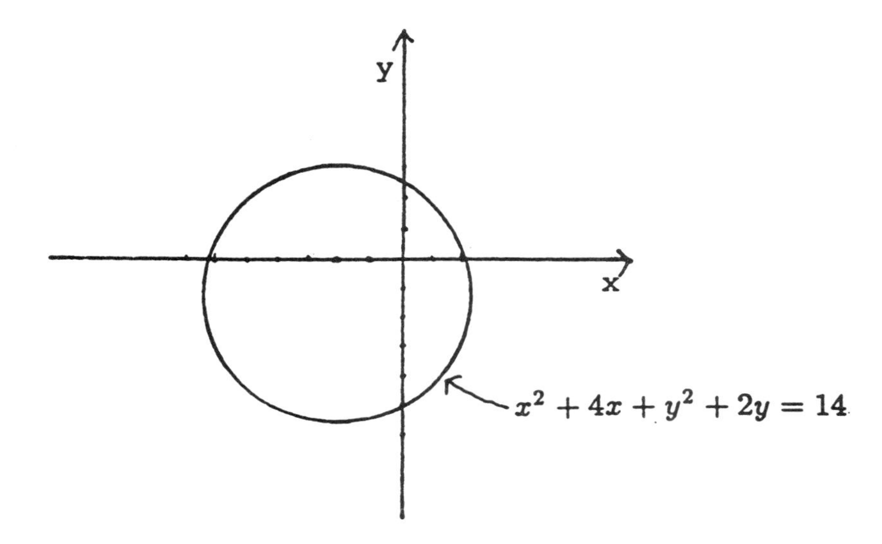

37. $f(x) = x^2 - 3x + 2 = (x-2)(x-1)$
$f(x) = 0$ for $x = 2$, $x = 1$.
39. $f(x) = 0$ for $x - 3 = 0 \Rightarrow x = 3$.
41. $f(x) = \frac{x^2+9}{x-2}$ $\quad f(x)$ has no zeros.
43. $f(x) = g(x)$, $x - 6 = 2x + 2$, $x = -8$.
45. $f(x) = g(x)$, $2 - x^2 = 4x^2 - 3$, $5x^2 = 5$, $x^2 = 1 \Rightarrow x = 1$, $x = -1$.
47. $h_1(x) = (x^3 + 2) \cdot (x + 4) = x^4 + 4x^3 + 2x + 8$.
49. $h_1(0) = (0^3 + 2) \cdot (0 + 4) = 8$.
51. $(g \circ f)(x) = g(f(x)) = g(x^2 - 7) = \sqrt{(x^2 - 7) + 3} = \sqrt{(x^2 - 4)}$.
53. $(u \circ f)(x) = u(f(x)) = u(x^2 - 7) = 4(x^2 - 7) = 4x^2 - 28$.
55. $f(x) = |x + 3| = \begin{cases} -x - 3, & x \le -3 \\ x + 3, & x \ge -3 \end{cases}$

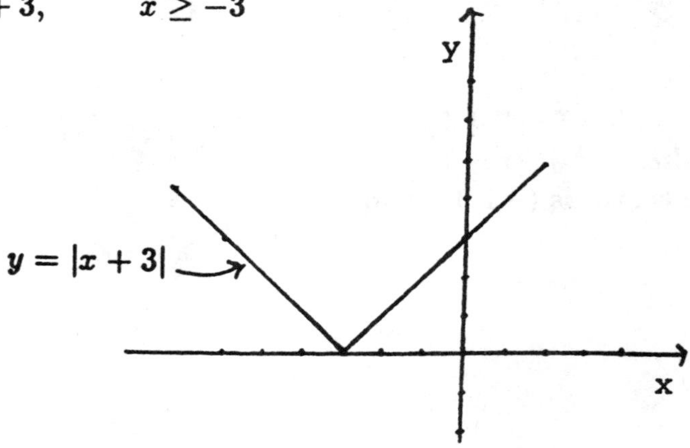

57. $f(x) = \begin{cases} x - 4, & x < 2 \\ -\frac{1}{2}x^2, & x \ge 2 \end{cases}$

x	$f(x)$
−1	−5
0	−4
1	−3
2	−2
37	$-\frac{9}{2}$
4	−8

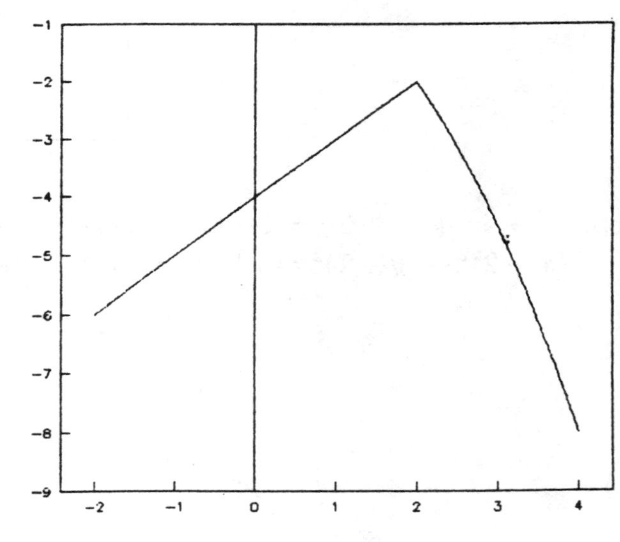

59. $D(p) = S(p)$. $\quad 2000 - 6p = 4p \Rightarrow 10p = 2000$, $\quad p = 200$.
The equilibrium price p_0 is \$200.
61. Figures b and c are graphs of functions.

63. False.
65. Let $x =$ the number of people served.
Let $C(x) =$ the corresponding total cost of the meal.
For $x \le 50$, $C(x) = 20x$.
For $x > 50$, $C(x) = 20 \cdot 50 + 15(x - 50) = 1000 + 15x - 750 = 250 + 15x$.
67. Let x be the number of years after 1989, and let P_2 be the corresponding cotton production in India in millions of bales.
$P_2 = m_2 x + b_2$.
$m_2 = \frac{9.5-10.7}{2} = -.6$.
$b_2 = 10.7$. $\quad P_2 = -.6x + 10.7$
When $x = 7$, $\quad P_2 = -.6(7) + 10.7 = 6.5$ million bales.
69. Let x by the number of years after 1940, and let T be the average daily temperature in degrees Fahrenheit in this certain city.
$T = mx + b$.
$m = \frac{59.2-58}{40} = .03$.
$b = 58$. $\quad T = .03x + 58$.
When $x = 60$, $\quad t = .03(60) + 58 = 59.8°$F.
71. $D = (1 - r)T$.
a. Since $r = 0.3$ and $T = 30{,}000$, $D = (1 - 0.3) \cdot 30{,}000 = \$21{,}000$.
b. Since $D = 16{,}000$ and $T = 20{,}000$, $16{,}000 = (1 - r) \cdot 20{,}000$.
$0.8 = (1 - r)$. $\therefore r = 0.2 = 20\%$.
c. Since $D = 30{,}000$ and $r = 0.4$, $30{,}000 = (1 - 0.4) \cdot T$.
$30{,}000 = 0.6T$. $\therefore T = \$50{,}000$.
73. $V(n) = -\frac{V_0 - V_s}{N} \cdot n + V_0$.
Since $V_0 = 30{,}000$, $V_s = 5{,}000$, and $N = 5$,

$$V(n) = -\frac{25{,}000}{5} \cdot n + 30{,}000 = -5{,}000 \cdot n + 30{,}000$$

For $n = 3$, $V(3) = -5{,}000 \cdot 3 + 30{,}000 = 15{,}000$.
75. $Y = (400 - 2x) \cdot x$.
77. $C(x) = C_0 + 16x + .02x^2$ and $R(x) = 24x$.
From $C(x) = R(x)$, we get $C_0 + 16x + 0.02x^2 = 24x$; $0.02x^2 - 8x + C_0 = 0$.
Since $C(x) = R(x)$ at $x = 100$, $\quad 0.02 \cdot 100^2 - 8 \cdot 100 + C_0 = 0 \quad \therefore C_0 = \$600$.
79. $V(n) = (100{,}000)\left(1 - \frac{n}{20}\right)$.
(a) $V(5) = (100{,}000)\left(1 - \frac{5}{20}\right) = (100{,}000)\left(\frac{3}{4}\right) = 75{,}000$.
(b) $V(n) = (100{,}000)\left(1 - \frac{n}{30}\right)$.
(c) $V(5) = (100{,}000)\left(1 - \frac{5}{30}\right) = (100{,}000)\left(\frac{5}{6}\right) = 83{,}333.33$.
81. Production level of sugar follows a linear function of times in U.S.A., Eastern Europe, South Africa and Australia.
83. Let t be the number of years after 1982, let C_1 be the avaerage manufacturing costs for small cars in Japan, and let C_2 be the average manufacturing costs for small cars in the United States.
a. $C_1 = m_1 t + b_1$.
$m_1 = \frac{7200-4200}{10} = 300$.

$b_1 = 4200.$
$C_1 = 300t + 4200.$
$C_2 = m_2 t + b_2.$
$m_2 = \frac{7400-6400}{10} = 100.$
$b_2 = 6400.$
$C_2 = 100t + 6400.$

b. The linear model for Japan has a larger slope, so the average manufacturing costs for small cars are increasing more rapidly in Japan.

c. $C_1 = C_2 \implies 300t + 4200 = 100t + 6400 \implies 200t = 2200 \implies t = 11.$ Thus 11 years after 1982, that is in 1993, the average manufacturing costs for small cars will be the same in Japan and the United States.

d. When $t = 15$, $C_1 = 300(15){+}4200 = \$8700$, and $C_2 = 100(15){+}6400 = \$7,900$.

Practice Problems for Chapter 1

1. a. Solve the inequality $3(x + 2) \geq 4 - x$, expressing the solution set as an interval.
   b. Solve the inequality $2(x^2 + 5x + 6) < 3(x + 3)$, expressing the solution set as an interval.
2. Let $f(x) = \frac{1}{\sqrt{4x+3}}$.
   a. Find the domain of f.
   b. Find $f(0)$ and $f\left(\frac{11}{2}\right)$.
   c. Sketch the graph of f.
3. Let x be the number of items of a certain product produced and sold per week, and let p be the corresponding price per item. Suppose the variable costs are \$50 per item, and suppose the weekly fixed costs are \$10,000. Suppose $p = 100 - .01x$.
   a. Find the weekly profit function $P(x)$.
   b. Find the values of x for which $P(x) > 0$.
   c. Find the value of x for which $P(x)$ is the largest, and determine the corresponding price and total profit for this level of production.
   d. Find the profit function if the government imposes an excise tax of \$10 per item on the manufacturer.
4. Let L be the straight line which passes through the points $(1, 3)$ and $(-2, 9)$.
   a. Find the slope of L.
   b. Find the y-intercept of L.
   c. Find an equation for L.
   d. Find an equation of the straight line which has x-intercept 4 and is perpendicular to L.

5. Suppose a moving company buys a van for \$25,000. Suppose the van depreciates in value at the rate of \$400 per month (linear depreciation). Suppose the useful lifetime of the van is 5 years.
   a. Find the value y of the van as a function of the number x of months the van has been used.
   b. Sketch the graph of the linear function in part (a).
   c. Find the value of the van at the end of its useful lifetime (scrap values).
6. a. Find the distance between the points (3,5) and (-1,8).
   b. Find an equation of the circle with center at the mid-point of the line segment joining the points in part (a) and passing through these points.
7. a. Find the center and radius of the circle whose equation is
   $x^2 + y^2 + 6x - 8y + 21 = 0$.
   b. Draw the graph of the circle in part (a).
8. Let $f(x) = 2x - 3$ and $g(x) = \sqrt{x^2 + 16}$.
   (a) Find $(f \circ g)(-3)$ and $(g \circ f)(-3)$.
   (b) Find equations for the composite functions $(f \circ g)(x)$ and $(g \circ f)(x)$.
9. a. Sketch the graph of the function

$$f(x) = \begin{cases} x^2 - 4, & x \le -2 \\ x + 2, & x > -2. \end{cases}$$

   b. Sketch the graph of the function

$$f(x) = |-x^2 + 4x + 5|.$$

10. Let x be the price in dollars per item of a certain product. Suppose the corresponding demand D at price x is given by the formula $D = -x^2 + 3x + 66$, and suppose the corresponding supply S at price x is given by the formula $S = 2x^2 + 21x - 15$. Find the equilibrium price.
11. Let L be the straight line determined by the equation $3x + 4y = 12$.
    a. Find the slope, the y-intercept, and the x-intercept of L.
    b. Sketch the graph of L.
    c. Find the values of x such that x and y are both nonnegative.
12. Sketch the graph of each of the following functions.
    a. $f(x) = x^3 - x^2 - 4x + 4$.
    b. $f(x) = x^3 - 3x^2 + 3x - 9$.
    c. $f(x) = \frac{2x-3}{x+1}$.
    d. $f(x) = \frac{2x-1}{x^2+1}$.
13. A certain manufacturer of calculators produces x calculators per day at the total daily cost in dollars of $C(x) = \frac{25x \cdot (x+150)}{x+100} + 500$.
    a. Sketch the graph of $y = C(x)$ by plotting points. From your graph decide when $C(x)$ is increasing as x increases and when $C(x)$ is decreasing as x increases.

b. Sketch the graph of the average cost function $y = c(x) = \frac{C(x)}{x}$ by plotting points. From your graph decide when $c(x)$ is increasing as x increases and when $c(x)$ is decreasing as x increases.
c. Find the value of x such that $C(x) = 1000$ dollars.
d. Find the value of x such that $c(x) = 50$ dollars per calculator.

14. As in Problem 38 in Exercise Set 1.6, let the total weekly costs in dollars of producing x jackets be given by the total cost function

$$C(x) = 400 + 10x + \frac{1}{2}x^2.$$

a. Graph the average cost function $y = c(x) = \frac{C(x)}{x}$ for $0 < x \leq 100$ by plotting points.
b. Explain how it follows from your graph that there is a positive real number x_0 such that $c(x_0) < c(x)$ for all positive real numbers x different from x_0.
c. By finding $c(x)$ for $x = 28.2, 28.25, 28.3, 28.35, 28.4$ show that this minimum value of $c(x)$ occurs for $x_0 \approx 28.3$ and that this minimum value is approximately 38.28.
d. Let k be a positive real number such that k is greater than the minimum average cost. Explain how it follows from your graph that there exist real numbers x_1 and x_2 such that $0 < x_1 < x_0$, $x_2 > x_0$, $c(x_1) = k = c(x_2)$, $c(x) < k$ for $x_1 < x < x_2$, and $c(x) > k$ for either $0 < x < x_1$ or $x > x_2$.

15. Consider again the total weekly costs function

$$C(x) = 400 + 10x + \frac{1}{2}x^2$$

in Problem 38 in Exercise Set 1.6.

a. Let k be a positive real number. By proceeding as in the solution of Problem 38, part a in Exercise Set 1.6, show that $c(x) = k$ implies that $x = (k - 10) \pm \sqrt{k^2 - 20k - 700}$. Show that $(k^2 - 20k - 700) = 0$ for $k = 10 + 20\sqrt{2}$, $(k^2 - 20k - 700) < 0$ for $0 < k < 10 + 20\sqrt{2}$, and $(k^2 - 20k - 700) > 0$ for $k > 10 + 20\sqrt{2}$. Explain why the minimum average cost is $(10 + 20\sqrt{2})$ and this occurs when $x = 20\sqrt{2}$.
b. Let k be a positive real number such that $k > 10 + 20\sqrt{2}$. Show that the production levels for which the average cost is equal to k are given by

$$x_1 = (k - 10) - \sqrt{k^2 - 20k - 700}$$

and

$$x_2 = (k - 10) + \sqrt{k^2 - 20k - 700}.$$

By considering the quadratic polynomial $2x\cdot[c(x)-k]$, show that $c(x)<k$ for $x_1<x<x_2$ and that $c(x)>k$ for either $0<x<x_1$ or $x>x_2$.

16. Suppose the total weekly cost in dollars of producing x boxes of candy bars is given by

$$C(x)=\frac{1}{10}x^2+5x+2250.$$

Suppose the selling price in dollars per box is given by

$$p(x)=50-.02x.$$

a. Find the weekly profit function $P(x)$.

b. Find the break-even points (that is, the break-even levels of production) for the weekly operation.

c. Find the production level x for which the weekly profit $P(x)$ is a maximum, and find this maximum profit.

d. Proceed as in Problem 15 to show that the minimum average cost is 35 dollars per box of candy bars and that this minimum occurs when the production level is 150 boxes of candy bars per week.

17. Tree A is located at the point $(3,4)$ on a landscape plot, and Tree B is located at the point (8,6) on this landscape plot. A study indicates that 40% of the nuts due to Tree A will fall within 4 units of that tree and that 35% of the squirrels born in Tree B will live within 6 units of that tree.

a. Draw a sketch showing the regions described above.

b. Indicate the overlap of these two regions.

Solutions to Practice Problems for Chapter 1

1. a. $3(x+2)\geq 4-x, \qquad 3x+6\geq 4-x$
$4x\geq -2 \implies x\geq -1/2.\ \therefore \left[-\frac{1}{2},\infty\right).$

b. $2(x^2+5x+6)<3(x+3)$
$2x^2+10x+12<3x+9$
$2x^2+7x+3<0$
$(2x+1)(x+3)<0 \implies -3<x<-\frac{1}{2}.\ \therefore \left(-3,-\frac{1}{2}\right).$

2. a. $4x+3>0 \implies x>-\frac{3}{4}.$

b. $f(0)=\frac{1}{\sqrt{3}}$ and $f\left(\frac{11}{2}\right)=\frac{1}{5}.$

c.

x	$-\frac{1}{2}$	0	$\frac{3}{2}$	2
$f(x)$	1	$\frac{1}{\sqrt{3}} \doteq 0.58$	$\frac{1}{3} \doteq 0.33$	$\frac{1}{\sqrt{11}} \doteq 0.30$

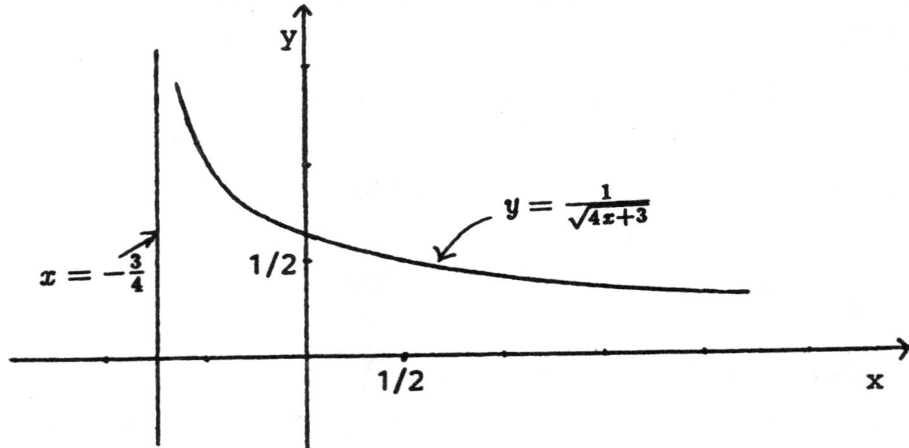

3. $R(x) = x(100 - .01x) = -.01x^2 + 100x.$
   $C(x) = 50x + 10,000.$
   a. $P(x) = R(x) - C(x) = -.01x^2 + 50x - 10,000.$
   b. $P(x) > 0 \implies -.01x^2 + 50x - 10,000 > 0$
   $x^2 - 5000x + 1,000,000 < 0.$
   From $x^2 - 5000x + 1,000,000 = 0$, we have

$$x = 2500 \pm \sqrt{6,250,000 - 1,000,000} = 2500 \pm \sqrt{5,250,000}$$
$$\doteq 2500 \pm 2291.3 = 4791.3 \quad \text{or} \quad 208.7.$$
$$\therefore 209 \leq x \leq 4791.$$

   c. $P(x) = -.01(x^2 - 5000x + 1,000,000)$
   $= -.01\left(x^2 - 5000x + (2500)^2 - (2500)^2 + 1,000,000\right)$
   $= -.01\{(x - 2500)^2 - 5,250,000\}$
   $= -.01(x - 2500)^2 + 52,500.$
   Maximum value of $P(x)$ occurs when $x = 2500$.
   The price $p = 100 - .01 \cdot 2500 = 100 - 25 = 75$ and the profit is \$52,500 for $x = 2500$.
   d. Profit function $= (-.01x^2 + 50x - 10,000) - 10x$
   $= -.01x^2 + 40x - 10,000.$
4. a. The slope of $L = \frac{3-9}{1-(-2)} = \frac{-6}{3} = -2$.
   b. Since the slope of L is -2 and L passes through the point (-2,9), L passes through (0,5). Hence the y- intercept of L is 5.
   c. $y = -2x + 5$.
   d. Since the straight line is perpendicular to L, its slope is $\frac{1}{2}$. Thus we can assume that its equation is of the form $y = \frac{1}{2}x + b$. Since it has

x-intercept 4, it passes through the point (4,0). So $0 = 2 + b \implies b = -2$.
$\therefore y = \frac{1}{2}x - 2$.

5. a. $y = 25,000 - 400x$.

b.

x	0	20	40	60
y	25,000	17,000	9,000	1,000

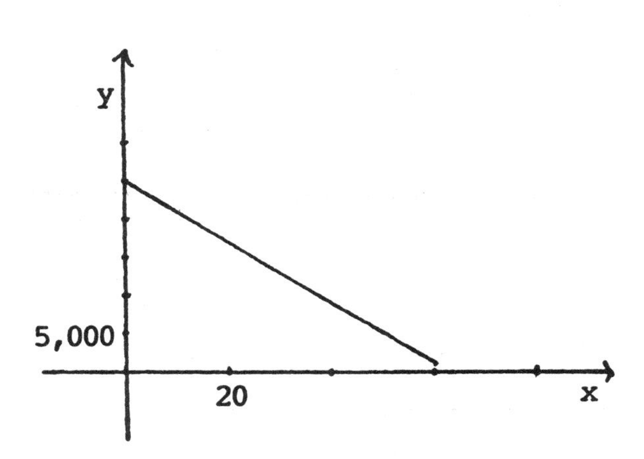

c. $y = 25,000 - 400 \cdot 60 = \$1,000$.

6. a. $\sqrt{(3-(-1))^2 + (5-8)^2} = \sqrt{25} = 5$.

b. The center is $\left(\frac{3-1}{2}, \frac{5+8}{2}\right) = (1, 6.5)$.
The radius is $\frac{5}{2} = 2.5$
$\implies$ The equation of the circle is

$$(x-1)^2 + (y-6.5)^2 = 2.5^2 = 6.25.$$

7. a. $x^2 + y^2 + 6x - 8y + 21 = 0$.
$x^2 + 6x + y^2 - 8y + 21 = 0$.
$x^2 + 6x + 9 - 9 + y^2 - 8y + 16 - 16 + 21 = 0$.
$(x+3)^2 - 9 + (y-4)^2 + 5 = 0$.
$(x+3)^2 + (y-4)^2 = 4 = 2^2$.
$\therefore$ The center is (-3,4) and radius is 2.

b.

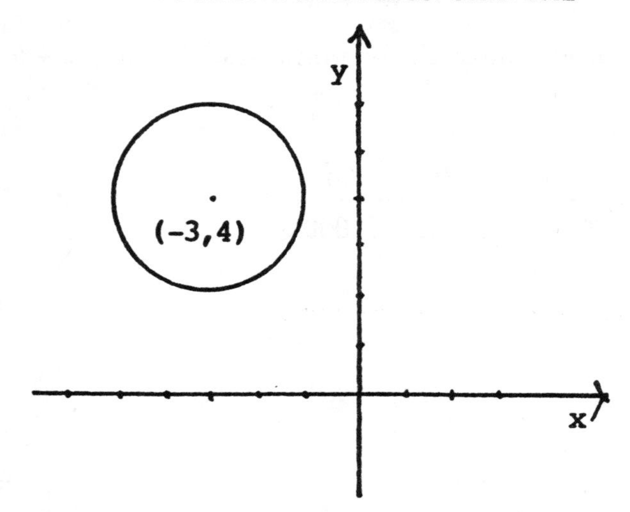

8. a. $(f \circ g)(-3) = f(g(-3)) = f(5) = 2 \cdot 5 - 3 = 7.$
$(g \circ f)(-3) = g(f(-3)) = g(-9) = \sqrt{81+16} = \sqrt{97}.$
b. $(f \circ g)(x) = f(g(x)) = f\left(\sqrt{x^2+16}\right) = 2\sqrt{x^2+16} - 3.$
$(g \circ f)(x) = g(f(x)) = g(2x-3) = \sqrt{(2x-3)^2+16} = \sqrt{4x^2-12x+25}.$

9. a. $f(x) = \begin{cases} x^2 - 4, & x \leq -2 \\ x + 2, & x > -2. \end{cases}$

x	-4	-3	-2	-1	0	1	2
y	12	5	0	1	2	3	4

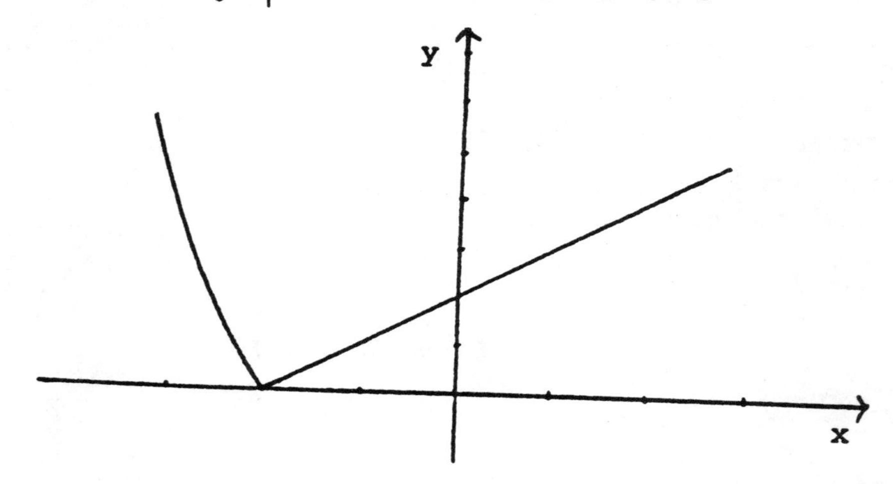

b. $f(x) = |-x^2 + 4x + 5|.$
Since $-x^2 + 4x + 5 = -(x-5)(x+1)$, $f(x) = 0$ for $x = 5$ or -1.

x	-2	-1	0	1	2	3	4	5	6
$f(x)$	7	0	5	8	9	8	5	0	7

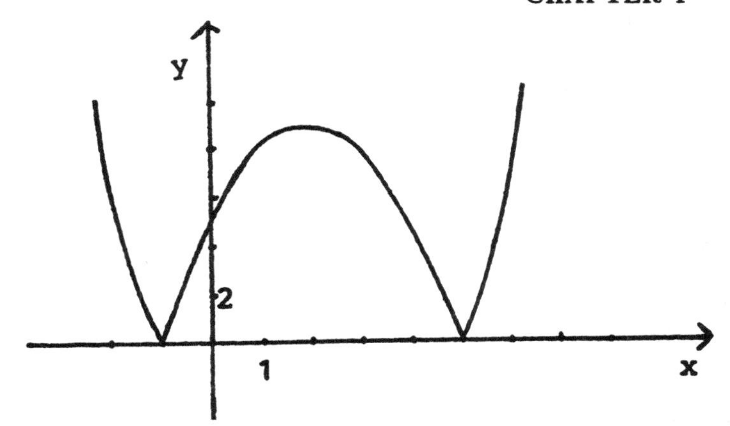

10. From $S(x) = D(x)$,
$2x^2 + 21x - 15 = -x^2 + 3x + 66$
$3x^2 + 18x - 81 = 0.$
$x^2 + 6x - 27 = 0$
$(x+9)(x-3) = 0$
$\therefore x = 3$ or -9.
The equilibrium price is \$3.

11. $3x + 4y = 12 \implies 4y = -3x + 12 \implies y = -\frac{3}{4}x + 3.$

a. Slope: $-\frac{3}{4}$ y-intercept: 3 x-intercept: 4

b.

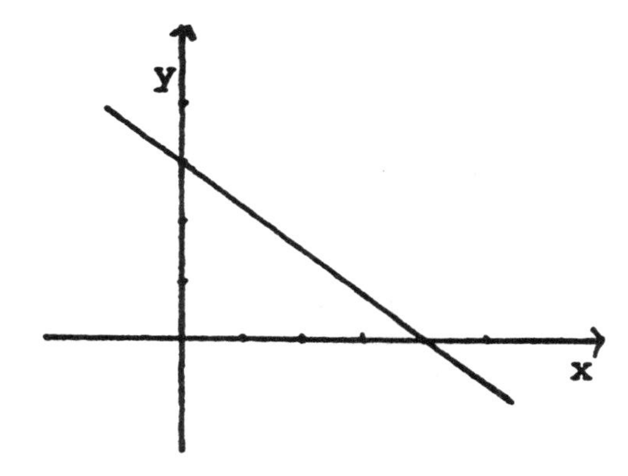

c. $0 \le x \le 4.$

12. a. $f(x) = x^3 - x^2 - 4x + 4$
$= (x-1)(x^2-4) = (x-1)(x+2)(x-2).$
$f(x) = 0$ for $x = 1, -2, 2.$

x	-2.5	-2	-1	0	1	$\frac{3}{2}$	2	3
$f(x)$	-7.875	0	6	4	0	$-.875$	0	10

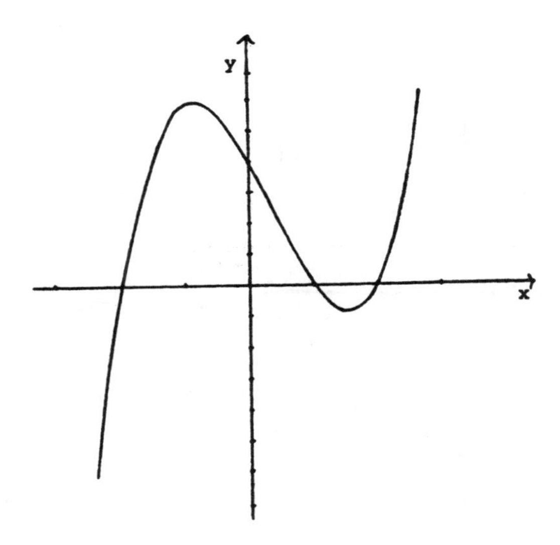

b. $f(x) = x^3 - 3x^2 + 3x - 9 = (x - 3)(x^2 + 3)$.
$f(x) = 0$ for $x = 3$.

x	-1	0	1	2	3	4
$f(x)$	-16	-9	-8	-7	0	19

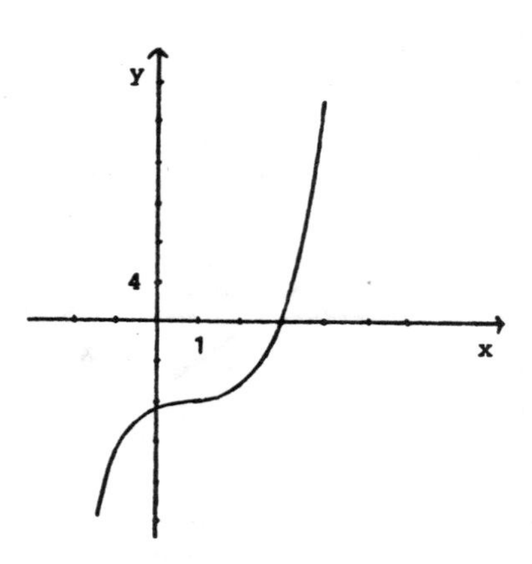

c. $f(x) = \frac{2x-3}{x+1} = \frac{2x+2-5}{x+1} = 2 - \frac{5}{x+1}$.
$f(x)$ is not defined at $x = -1$.

x	-4	-3	-2	-1	0	1	2	3
$f(x)$	$\frac{11}{3}$	$\frac{9}{2}$	7	Und.	-3	$-\frac{1}{2}$	$\frac{1}{3}$	$\frac{3}{4}$

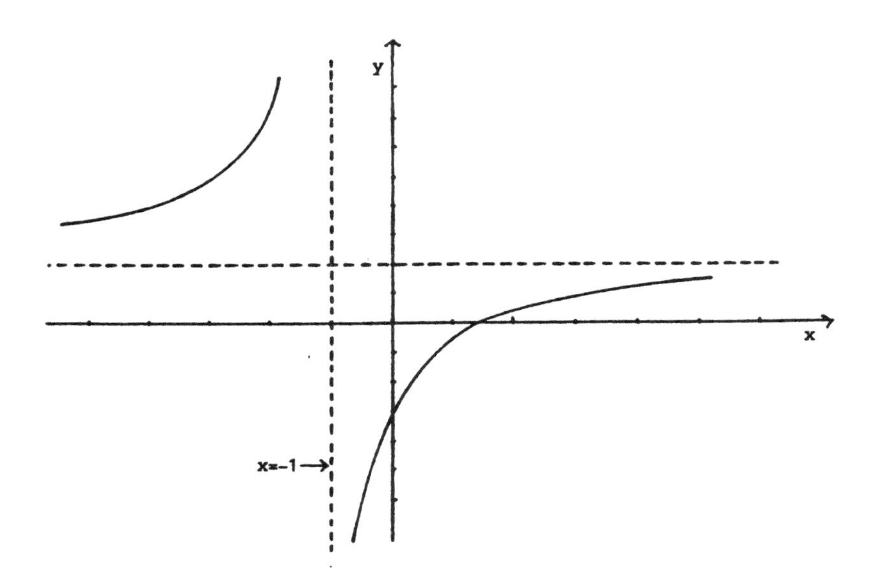

d. $f(x) = \frac{2x-1}{x^2+1}$.
$f(x) = 0$ for $x = \frac{1}{2}$.

x	-2	-1	0	$\frac{1}{2}$	1	2	3
$f(x)$	-1	$-\frac{3}{2}$	-1	0	$\frac{1}{2}$	$\frac{3}{5}$	$\frac{1}{2}$

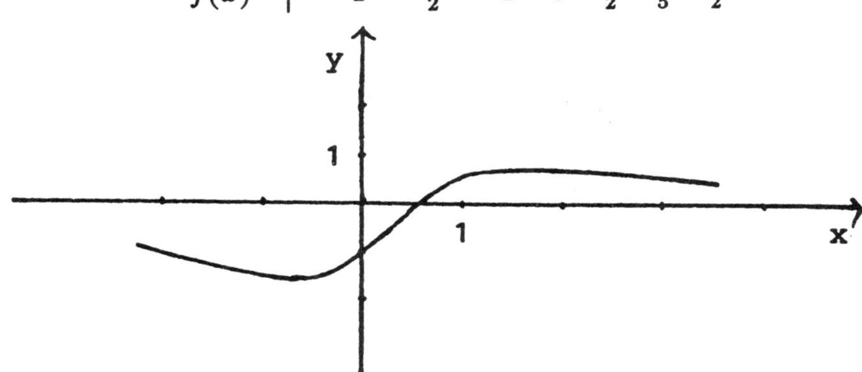

13. a.

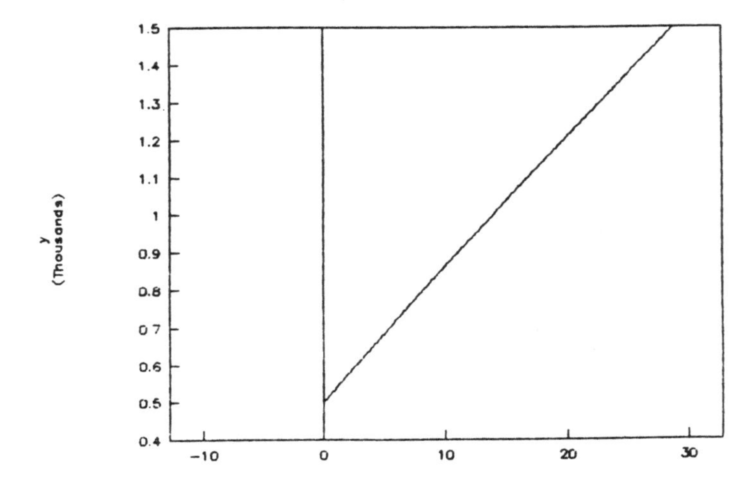

$C(x)$ keeps increasing as x increases.

b.

$c(x)$ keeps decraseing as x increases.

c. $C(x) = 1000 \implies \frac{25x\cdot(x+150)}{x+100} + 500 = 1000 \implies 25x \cdot (x+150) = 500 \cdot (x+100)$

$\implies 25x^2 + 3250x - 50,000 = 0 \implies x = \frac{-3250 \pm \sqrt{(3250)^2 - 4\cdot 25\cdot(-50,000)}}{2\cdot 25}$
$= \frac{-3250 \pm \sqrt{15,562,500}}{50} \implies x = 13.9.$

d. $c(x) = 50 \implies \frac{25\cdot(x+150)}{x+100} + \frac{500}{x} = 50 \implies 25\cdot(x+150)\cdot x + 500\cdot(x+100)$
$= 50 \cdot x \cdot (x+100) \implies x^2 + 30x - 2000 = 0 \implies$
$x = \frac{-30 \pm \sqrt{(30)^2 - 4\cdot 1\cdot(-2000)}}{2\cdot 1} = \frac{-30 \pm \sqrt{8900}}{2} \implies x = 32.17.$

14. a.

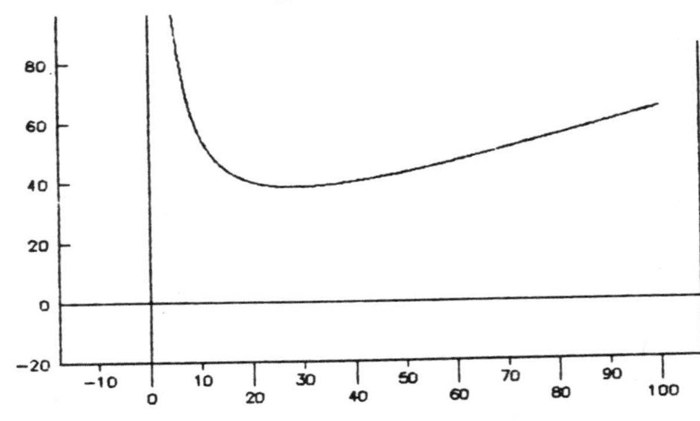

b. From the graph it can be seen that first $c(x)$ decreases as x increases, and then $c(x)$ increases as x increases. Therefore there is a positive real number x_0 such that $c(x_0) < c(x)$ for all positive real numbers x different from x_0.

c. $c(28.2) = 38.28439$, $c(28.25) = 38.28429$, $c(28.3) = 38.28427$, $c(28.35) = 38.28434$, $c(28.4) = 38.28450$. From this it follows that the minimum value of $c(x)$ occurs for $x_0 \approx 28.3$ and that the minimum value is approximately 38.28427.

d. From the graph of $y = c(x)$ it can be seen that $c(x)$ becomes very large as x becomes very close to 0, and that $c(x)$ also becomes very large as

x becomes very large. Moreover, $c(x)$ decreases as x increases for $0 < x < x_0$, and $c(x)$ increases as x increases for $x > x_0$. Therefore if k is any positive real number greater than the minimum average cost, then there exist real numbers x_1 and x_2 such that $0 < x_1 < x_0$, $x_2 > x_0$, $c(x_1) = k = c(x_2)$, $c(x) < k$ for $x_1 < x < x_2$, and $c(x) > k$ for either $0 < x < x_1$ or $x > x_2$.

15. a. $c(x) = k \implies \frac{400}{x} + 10 + \frac{1}{2}x = k \implies \frac{1}{2}x + (10 - k) + \frac{400}{x} = 0 \implies$
$x^2 + (20 - 2k)x + 800 = 0 \implies x = \frac{(2k-20)\pm\sqrt{(20-2k)^2-4\cdot 1\cdot 800}}{2\cdot 1}$
$= \frac{(2k-20)\pm\sqrt{4k^2-80k-2800}}{2} = \frac{(2k-20)\pm 2\cdot\sqrt{k^2-20k-700}}{2} = (k-10)\pm\sqrt{k^2 - 20k - 700}.$

$(k^2 - 20k - 700) = 0 \implies k = \frac{20\pm\sqrt{(20)^2-4\cdot 1\cdot(-700)}}{2\cdot 1}$
$= \frac{20\pm\sqrt{3200}}{2} = \frac{20\pm 40\sqrt{2}}{2} \implies$ ~~$k = 10 - 20\sqrt{2}$ or~~ $k = 10 + 20\sqrt{2}$.
Because the graph of $y = k^2 - 20k - 700$ is a parabola opening upward, $(k^2 - 20k - 700) < 0$ for $0 < k < 10 + 20\sqrt{2}$, and $(k^2 - 20k - 700) > 0$ for $k > 10 + 20\sqrt{2}$.
Therefore the minimum average cost is given by $k_0 = 10 + 20\sqrt{2}$, and the minimum average cost occurs when $x = 20\sqrt{2}$.

b. Note first that

$$(k^2 - 20k - 700) < (k - 10)^2 \implies \sqrt{k^2 - 20k - 700} < k - 10.$$

Therefore the production levels for which the average cost is equal to k are given by

$$x_1 = (k - 10) - \sqrt{k^2 - 20k - 700}$$

and

$$x_2 = (k - 10) + \sqrt{k^2 - 20k - 700}.$$

Observe that x_1 and x_2 are the zeros of $2x \cdot [c(x) - k] = x^2 + (20 - 2k)x + 800$, and the graph of $y = x^2 + (20 - 2k)x + 800$ is a parabola opening upward. Thus:

$$x_1 < x < x_2 \implies 2x \cdot [c(x) - k] < 0 \implies c(x) < k$$
$$0 < x < x_1 \text{ or } x > x_2 \implies 2x \cdot [c(x) - k] > 0 \implies c(x) > k.$$

(1) 16. a. The revenue $R(x) = x \cdot p(x) = x \cdot (50 - .02x) = -.02x^2 + 50x$.
The weekly profit function $P(x)$ is given by

$$P(x) = R(x) - C(x) = (-.02x^2 + 50x) - (.1x^2 + 5x + 2250)$$
$$= -.12x^2 + 45x - 2250.$$

b. The break-even points are the production levels x for which $P(x) = 0$.

$$P(x) = 0 \implies x = \frac{-45 \pm \sqrt{(45)^2 - 4 \cdot (-.12) \cdot (-2250)}}{2 \cdot (-.12)} = \frac{45 \pm \sqrt{945}}{.24}.$$

Thus the break-even points are $x_1 = 59.4$ and $x_2 = 315.6$.

c. $P(x) = -.12 \cdot \left(x^2 - \frac{45}{.12}x + \left(\frac{1}{2} \cdot \left(-\frac{45}{.12}\right)\right)^2\right) + \frac{1}{4} \cdot \frac{(45)^2}{.12} - 2250$
$\quad = -.12 \cdot \left(x - \frac{45}{.24}\right)^2 + 1968.75.$

Thus, the maximum profit occurs when $x = \frac{45}{.24} = 187.5$, and this maximum profit is \$1968.75. Note that the graph of the quadratic function $P(x) = -.12x^2 + 45x - 2250$ is a parabola opening downward, so the value of x for which $P(x)$ is the largest is halfway between the roots x_1 and x_2 of the quadratic equation $P(x) = 0$.

d. The average cost function $c(x) = \frac{C(x)}{x} = .1x + 5 + \frac{2250}{x}$.
Let k be a positive real number.

$$c(x) = k \implies .1x + (5 - k) + \frac{2250}{x} = 0 \implies x^2 + (50 - 10k)x + 22{,}500 = 0$$

$$\implies x = \frac{(10k - 50) \pm \sqrt{(50 - 10k)^2 - 4 \cdot 1 \cdot 22{,}500}}{2 \cdot 1} \implies$$

$$x = \frac{(10k - 50) \pm \sqrt{100k^2 - 1000k - 87500}}{2} \implies$$

$$x = \frac{(10k - 50) \pm 10 \cdot \sqrt{k^2 - 10k - 875}}{2} = (5k - 25) \pm 5 \cdot \sqrt{k^2 - 10k - 875}.$$

$(k^2 - 10k - 875) = (k - 35)(k + 25) = 0 \implies k = 35.$

For $0 < k < 35$, $(k^2 - 10k - 875) < 0$.

For $k > 35$, $(k^2 - 10k - 875) > 0$.

Therefore, the minimum average cost is 35, and this occurs when the production level x is 150.

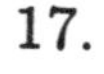

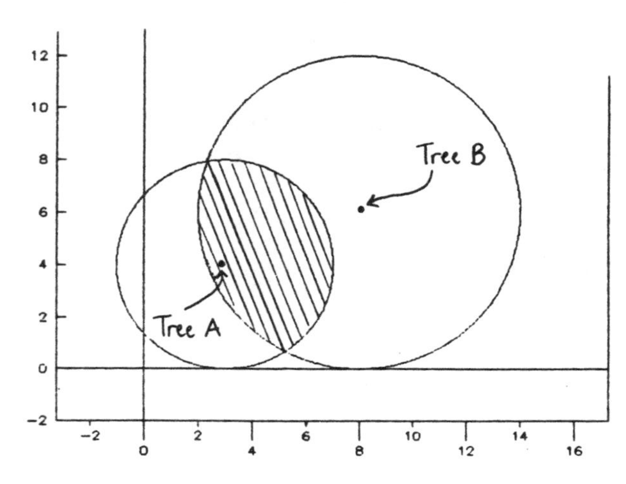

CHAPTER 2

SUMMARY OF CHAPTER 2

1. The tangent line to a curve at a point as the limiting position of secant lines through this point, the slope of this tangent line as a limit.

2. $L = \lim_{x \to a} f(x)$, where a and L are real numbers, properties of such limits.

3. One-sided limits and continuity.

4. The derivative function $f'(x)$ of a function f defined by the limit $f'(x) = \lim_{h \to 0} \frac{f(x+h)-f(x)}{h}$; $f'(x_0)$ as the slope of the tangent line to the graph of $y = f(x)$ at the point $P = (x_0, f(x_0))$; other notations for derivatives.

5. Rules for calculating derivatives: the power rule, the sum rule, the constant multiple rule, the product rule, the quotient rule.

6. Average velocity, instantaneous velocity as a derivative.
7. Business applications of the derivative: revenue and demand; marginal cost, marginal revenue, marginal profit; total utility.
8. The chain rule, the general power rule.

9. Limits at infinity.

Selected Solutions to Exercise Set 2.1

1. The tangent line appears to go through points $(1,8)$ and $(0,10)$, so the slope $m_{\tan}$ is $\frac{10-8}{0-1} = -2$.

3. The tangent line appears to go through points $(4,2)$ and $(0,1)$, so the slope $m_{\tan}$ is $\frac{1-2}{0-4} = \frac{1}{4}$.

5. The tangent line appears to go through points $(1,\frac{1}{2})$ and $(0,1)$, so the slope $m_{\tan}$ is $\frac{1-\frac{1}{2}}{0-1} = -\frac{1}{2}$.

7. $m = \lim_{h\to 0} \frac{f(1+h)-f(1)}{h} = \lim_{h\to 0} \frac{(1+h)^2-1^2}{h} = \lim_{h\to 0} \frac{1+2h+h^2-1}{h}$
$= \lim_{h\to 0} \frac{2h+h^2}{h} = \lim_{h\to 0} \frac{h(2+h)}{h} = \lim_{h\to 0}(2+h) = 2.$

9. $m = \lim_{h\to 0} \frac{f(1+h)-f(1)}{h} = \lim_{h\to 0} \frac{2-(1+h)^2-(2-1^2)}{h} = \lim_{h\to 0} \frac{2-1-2h-h^2-1}{h}$
$= \lim_{h\to 0} \frac{-h(2+h)}{h} = \lim_{h\to 0} -(2+h) = -2.$

11. $m = \lim_{h\to 0} \frac{f(1+h)-f(1)}{h} = \lim_{h\to 0} \frac{(1+h)^3-1^3}{h} = \lim_{h\to 0} \frac{1+3h+3h^2+h^3-1^3}{h}$
$= \lim_{h\to 0} \frac{h(3+3h+h^2)}{h} = \lim_{h\to 0}(3+3h+h^2) = 3.$

13. $m = \lim_{h\to 0} \frac{f(-1+h)-f(-1)}{h} = \lim_{h\to 0} \frac{(-1+h)^3+(-1+h)-(-2)}{h} = \lim_{h\to 0} \frac{-1+3h-3h^2+h^3-1+h+2}{h}$
$= \lim_{h\to 0} \frac{4h-3h^2+h^3}{h} = \lim_{h\to 0} \frac{h(4-3h+h^2)}{h} = \lim_{h\to 0}(4-3h+h^2) = 4.$

15. $m = \lim_{h\to 0} \frac{f(1+h)-f(1)}{h} = \lim_{h\to 0} \frac{\left((1+h)^4+(1+h)\right)-(1^4+1)}{h}$
$= \lim_{h\to 0} \frac{(1+4h+6h^2+4h^3+h^4+1+h)-(1+1)}{h}$
$= \lim_{h\to 0} \frac{h(5+6h+4h^2+h^3)}{h} = \lim_{h\to 0}(5+6h+4h^2+h^3) = 5.$

17. $m = \lim_{h\to 0} \frac{f(-2+h)-f(-2)}{h} = \lim_{h\to 0} \frac{\frac{1}{(-2+h)+3}-\frac{1}{(-2+3)}}{h} = \lim_{h\to 0} \frac{\frac{1}{1+h}-1}{h}$
$= \lim_{h\to 0} \frac{1}{h}\left\{\frac{1}{1+h}-1\right\} = \lim_{h\to 0} \frac{1}{h}\left\{\frac{1}{1+h}-\frac{1+h}{1+h}\right\} = \lim_{h\to 0} \frac{1}{h}\left\{\frac{1-1-h}{1+h}\right\}$
$= \lim_{h\to 0} \frac{1}{h}\left(\frac{-h}{1+h}\right) = \lim_{h\to 0} \frac{h}{h}\left(\frac{-1}{1+h}\right) = -1.$

19. $m = \lim_{h\to 0} \frac{f(3+h)-f(3)}{h} = \lim_{h\to 0} \frac{\sqrt{(3+h)+1}-\sqrt{3+1}}{h}$
$= \lim_{h\to 0} \frac{\sqrt{4+h}-2}{h} = \lim_{h\to 0} \frac{\sqrt{4+h}-2}{h} \cdot \frac{\sqrt{4+h}+2}{\sqrt{4+h}+2}$
$= \lim_{h\to 0} \frac{\left(\sqrt{4+h}\right)^2-2^2}{h\left(\sqrt{4+h}+2\right)} = \lim_{h\to 0} \frac{(4+h)-4}{h\left(\sqrt{4+h}+2\right)} = \lim_{h\to 0} \frac{1}{\sqrt{4+h}+2} = \frac{1}{4}.$

21. $m = \lim_{h\to 0} \frac{f(4+h)-f(4)}{h} = \lim_{h\to 0} \frac{(4+h)^{3/2}-4^{3/2}}{h}$

$= \lim_{h\to 0} \frac{\sqrt{(4+h)^3}-8}{h} = \lim_{h\to 0} \frac{\sqrt{(4+h)^3}-8}{h} \cdot \frac{\sqrt{(4+h)^3}+8}{\sqrt{(4+h)^3}+8}$

$= \lim_{h\to 0} \frac{(4+h)^3-64}{h\left(\sqrt{(4+h)^3}+8\right)} = \lim_{h\to 0} \frac{(64+48h+12h^2+h^3)-64}{h\left(\sqrt{(4+h)^3}+8\right)}$

$= \lim_{h\to 0} \frac{h(48+12h+h^2)}{h\left(\sqrt{(4+h)^3}+8\right)} = \lim_{h\to 0} \frac{48+12h+h^2}{\sqrt{(4+h)^3}+8} = \frac{48}{16} = 3.$

23. $m = \lim_{h\to 0} \frac{f(-2+h)-f(-2)}{h} = \lim_{h\to 0} \frac{9-(-2+h)^2-\left(9-(-2)^2\right)}{h} = \lim_{h\to 0} \frac{9-4+4h-h^2-5}{h}$

$= \lim_{h\to 0} \frac{h(4-h)}{h} = \lim_{h\to 0}(4-h) = 4.$

An equation for the line is $y-5 = 4(x+2)$, $y-5 = 4x+8$, $y = 4x+13$.

25. slope $= \lim_{h\to 0} \frac{f(x+h)-f(x)}{x} = \lim_{h\to 0} \frac{(m(x+h)+b)-(mx+b)}{h} = \lim_{h\to 0} \frac{mh}{h} = m.$

27. $m = \lim_{h\to 0} \frac{D(6+h)-D(6)}{h} = \lim_{h\to 0} \frac{\frac{10}{(6+h)+4}-\frac{10}{6+4}}{h} = \lim_{h\to 0} \frac{\frac{10}{10+h}-\frac{10}{10}}{h}$

$= \lim_{h\to 0} \frac{1}{h}\left(\frac{10}{10+h} - \frac{10+h}{10+h}\right) = \lim_{h\to 0} \frac{1}{h}\left(\frac{10-10-h}{10+h}\right) = \lim_{h\to 0} \frac{h}{h}\left(\frac{-1}{10+h}\right) = -\frac{1}{10}.$

29. $f(x) = -x+3$ for $x < 3$, $f(x) = x-3$ for $x \geq 3$.

a. for $a < 3$, the slope is -1.

b. for $a > 3$, the slope is 1.

31. $m = \lim_{h\to 0} \frac{f(1+h)-f(1)}{h} = \lim_{h\to 0} \frac{a+b(1+h)+c(1+h)^2-\left(a+b(1)+c(1)^2\right)}{h}$

$= \lim_{h\to 0} \frac{a+b+bh+c+2ch+ch^2-a-b-c}{h} = \lim_{h\to 0} \frac{h(b+2c+ch)}{h}$

$= \lim_{h\to 0}(b+2c+ch) = b+2c.$

33. a. The tangent line to the curve labeled Type I is almost horizontal for quite a while and then becomes almost vertical. The slope of the tangent line to the curve labeled Type III is negative and rather large in absolute value for quite some time, and then this slope decreases in absolute value and approaches zero as t becomes large.

b. The anwers in part a. distinguish one curve from the other in the sense that in Type I a large percentage of the population lives for quite a few years whereas in Type III a large number of the population dies rather soon.

c. As the time variable t increases, the slope of the tangent line to the curve labeled Type I decreases in absolute value and eventually approaches zero, and the cloper of the tangent line to the curve labeled Type II is constant.

35. The cost function is given by $C(x) = 2,000 + 20x + 2x^2$.

a. The average cost function is given by

$c(x) = \frac{C(x)}{x} = \frac{2,000}{x} + 20 + 2x.$

b. The slope of the tangent line to the graph of $y = c(x)$ at the point where $x = 10$ is

$$m = \lim_{h\to 0} \frac{c(10+h)-c(10)}{h}$$
$$= \lim_{h\to 0} \frac{\frac{2,000}{10+h}+20+2(10+h)-240}{h}$$
$$= \lim_{h\to 0} \frac{2,000-220(10+h)+2(10+h)^2}{h(10+h)}$$
$$= \lim_{h\to 0} \frac{2,000-2200-220h+200+40h+2h^2}{h(10+h)}$$
$$= \lim_{h\to 0} \frac{h(-180+2h)}{h(10+h)} = \lim_{h\to 0} \frac{-180+2h}{10+h} = \frac{-180}{10} = -18.$$

Because this slope is negative, the average cost per racket is decreasing at this level of production.

c. The slope here is given by

$$\lim_{h\to 0} \frac{c(100+h)-c(100)}{h}$$
$$= \lim_{h\to 0} \frac{\frac{2,000}{100+h}+20+2(100+h)-240}{h}$$
$$= \lim_{h\to 0} \frac{2,000-220(100+h)+2(100+h)^2}{h(100+h)}$$
$$= \lim_{h\to 0} \frac{2,000-22,000-220h+20,000+400h+2h^2}{h(100+h)}$$
$$= \lim_{h\to 0} \frac{h(180+2h)}{h(100+h)} = \lim_{h\to 0} \frac{180+2h}{(100+h)} = 1.8.$$

Because this slope is positive, the average cost per racket is increasing at this level of production.

d. The slope of the tangent line to the graph of $y = C(x)$ at the point $(x_0, C(x_0))$ is given by

$$\lim_{h\to 0} \frac{C(x_0+h)-C(x_0)}{h}$$
$$= \lim_{h\to 0} \frac{\left(2{,}000+20(x_0+h)+2(x_0+h)^2\right)-(2{,}000+20x_0+2x_0^2)}{h}$$
$$= \lim_{h\to 0} \frac{20h+2(x_0^2+2hx_0+h^2)-2x_0^2}{h}$$
$$= \lim_{h\to 0} \frac{h(20+4x_0+2h)}{h} = \lim_{h\to 0}(20+4x_0+2h)$$
$$= 20+4x_0.$$

The number x_0 for which this slope is zero is given by $20 + 4x_0 = 0$

$\Longrightarrow x_0 = -5.$

Selected Solutions to Exercise Set 2.2

1. $\lim_{x\to 3}(2x-6) = 2\cdot 3 - 6 = 0.$
3. $\lim_{x\to 1}(x^2-4x) = 1-4 = -3.$
5. $\lim_{x\to -3} \frac{x^2-9}{x+3} = \lim_{x\to -3} \frac{(x-3)(x+3)}{(x+3)} = \lim_{x\to -3}(x-3) = -6.$
7. $\lim_{x\to 4} \frac{x-4}{x^2-x-12} = \lim_{x\to 4} \frac{x-4}{(x-4)(x+3)} = \lim_{x\to 4} \frac{1}{x+3} = \frac{1}{7}.$
9. $\lim_{x\to 1} \frac{x^3-1}{x-1} = \lim_{x\to 1} \frac{(x-1)(x^2+x+1)}{x-1} = \lim_{x\to 1}(x^2+x+1) = 3.$
11. $\lim_{h\to 0} \frac{(4+h)^2-16}{h} = \lim_{h\to 0} \frac{16+8h+h^2-16}{h} = \lim_{h\to 0}(8+h) = 8.$
13. $\lim_{x\to 4} \frac{x-4}{\sqrt{x}-2} = \lim_{x\to 4} \frac{\left(\sqrt{x}+2\right)\left(\sqrt{x}-2\right)}{\sqrt{x}-2} = \lim_{x\to 4}(\sqrt{x}+2) = 4.$
15. $\lim_{x\to -3} \frac{x^2-4x-21}{x+3} = \lim_{x\to -3} \frac{(x-7)(x+3)}{x+3} = \lim_{x\to -3}(x-7) = -10.$
17. $\lim_{x\to 1} \frac{1-x^4}{x^2-1} = \lim_{x\to 1} \frac{(1-x^2)(1+x^2)}{(x^2-1)} = \lim_{x\to 1}\left(-(1+x^2)\right) = -2.$
19. $\lim_{x\to 0} \frac{\sqrt{x+1}-1}{x} = \lim_{x\to 0} \frac{\sqrt{x+1}-1}{x} \cdot \frac{\sqrt{x+1}+1}{\sqrt{x+1}+1}$
    $= \lim_{x\to 0} \frac{(x+1)-1}{x\left(\sqrt{x+1}+1\right)} = \lim_{x\to 0} \frac{1}{\sqrt{x+1}+1} = \frac{1}{2}.$
21. $\lim_{x\to 1} \frac{1-\sqrt{x}}{x+1} = 0.$
23. $\lim_{x\to 3} \frac{\sqrt{x+1}-\sqrt{x}}{1-x} = \frac{\sqrt{3+1}-\sqrt{3}}{1-3} = \frac{2-\sqrt{3}}{-2} = -1 + \frac{\sqrt{3}}{2}.$
25. $\lim_{x\to 1} \frac{1-\sqrt{x}}{x^{3/2}-\sqrt{x}} = \lim_{x\to 1} \frac{1-\sqrt{x}}{\sqrt{x}(x-1)} = \lim_{x\to 1} \frac{1-\sqrt{x}}{\sqrt{x}\left(\sqrt{x}+1\right)\cdot\left(\sqrt{x}-1\right)}$
    $= \lim_{x\to 1} \frac{-1}{\sqrt{x}\left(\sqrt{x}+1\right)} = -\frac{1}{2}.$

Selected Solutions to Exercise Set 2.3

1. a. $f(1)$ doesn't exist.

   b. $\lim_{x\to 1^-} f(x) = 2.$

   c. $\lim_{x\to 1} f(x) = 2.$

   d. $\lim_{x\to 0} f(x) = 0.$

3. a. $\lim_{x\to 0^-} f(x) = 2.$

   b. $\lim_{x\to 0^+} f(x) = +\infty.$

   c. $\lim_{x\to -2} f(x) = 0.$

5. $\lim_{x\to 0^-} \frac{|x|}{x} = \lim_{x\to 0^-} \frac{-x}{x} = \lim_{x\to 0^-} (-1) = -1.$

7. $\lim_{x\to 0^+} \sqrt{\frac{x}{x+2}} = \sqrt{\frac{0}{0+2}} = 0.$

9. $\lim_{x\to 1^-} \frac{|x-2|}{x+2} = \lim_{x\to 1^-} \frac{2-x}{x+2} = \frac{2-1}{1+2} = \frac{1}{3}.$

11. $f(x) = \frac{1}{1-x^2}$ is discontinuous at $x = 1$ and $x = -1$ because $f(1)$ and $f(-1)$ don't exist.

13. $f(x) = \frac{x^2-x-6}{x+3}$ is discontinuous at $x = -3$ because $f(-3)$ doesn't exist.

15. $f(x) = \frac{x+3}{x^3+3x^2-x-3} = \frac{(x+3)}{(x-1)(x+3)(x+1)}$ is discontinuous at $x = 1, -3,$ and -1 since $f(1)$, $f(-1)$ and $f(-3)$ don't exist.

17. $f(x) = \begin{cases} 1-x, & x \le 2 \\ x-1, & x > 2 \end{cases}$ is discontinuous at $x = 2$ since $\lim_{x\to 2^+} f(x) = 1$ and $\lim_{x\to 2^-} f(x) = -1$, that is, $\lim_{x\to 2} f(x)$ doesn't exist.

19. $$f(x) = \begin{cases} x^2 - 2x - 15, & x < -3 \\ 9 - x^2, & -3 < x < 5 \\ x^2 - 2x - 15, & x > 5 \end{cases}$$

    The function f is continuous everywhere except for $x = -3$ and $x = 5$, because $f(-3)$ and $f(5)$ are not defined.

21. From $\lim_{x\to 2^+} f(x) = \lim_{x\to 2^-} f(x)$ we have $8 = 2^k$.
    Therefore $k = 3$.

23. From $\lim_{x\to 2^+} f(x) = \lim_{x\to 2^-} f(x)$ we have $2k+5=(2-k)(2+k)$. Then $2k+5=4-k^2 \Rightarrow k^2+2k+1=0 \Rightarrow k=-1$.

25. $p(q)$ is discontinuous at $q=50$ and 100 since $\lim_{q\to 50} p(q)$ and $\lim_{q\to 100} p(q)$ don't exist.

27. Let x be income in dollars. Then the tax function

$$T(x) \text{ is } \begin{cases} 2097+0.24(x-15,000), & 15,000 \le x < 18,200 \\ A+0.28\cdot(x-18,200), & 18,200 \le x \le 23,500. \end{cases}$$

For $T(x)$ to be continuous at $x=18,200$, it is required that $\lim_{x\to 18,200} T(x)$ exists and equals $T(18,200)$.

$2097+0.24(18,200-15,000)=A$. This implies that $A=\$2865$.

29. From $\lim_{x\to 3^-} f(x) = \lim_{x\to 3^+} f(x)$, $9a+2=20$. Hence $a=2$.

31. Let $P(x)$ be the price in dollars for a movie ticket for a person x years old.

$$P(x) = \begin{cases} 2.50 & \text{for } 0 < x < 12 \\ 4.50 & \text{for } x \ge 12 \end{cases}$$

$\lim_{x\to 12^-} P(x) = 2.50 < P(12)$.

Selected Solutions to Exercise Set 2.4

1. $$f'(x) = \lim_{h\to 0} \frac{f(x+h)-f(x)}{h} = \lim_{h\to 0} \frac{[2(x+h)+5]-[2x+5]}{h} = \lim_{h\to 0} \frac{2x+2h+5-2x-5}{h} = 2.$$

3. $$f'(x) = \lim_{h\to 0} \frac{f(x+h)-f(x)}{h} = \lim_{h\to 0} \frac{[(x+h)^2+4]-[x^2+4]}{h} = \lim_{h\to 0} \frac{x^2+2hx+h^2+4-x^2-4}{h} = \lim_{h\to 0} \frac{h(2x+h)}{h} = 2x.$$

5. $$f'(x) = \lim_{h\to 0} \frac{f(x+h)-f(x)}{h} = \lim_{h\to 0} \frac{(x+h)^4-x^4}{h} = \lim_{h\to 0} \frac{x^4+4hx^3+6h^2x^2+4h^3x+h^4-x^4}{h} = \lim_{h\to 0} \frac{h(4x^3+6hx^2+4h^2x+h^3)}{h} = 4x^3.$$

7. $$f'(x) = \lim_{h\to 0} \frac{\frac{2}{(x+h)+3}-\frac{2}{x+3}}{h} = \lim_{h\to 0} \frac{\frac{2}{(x+h)+3}-\frac{2}{x+3}}{h} \cdot \frac{((x+h)+3)(x+3)}{((x+h)+3)(x+3)} = \lim_{h\to 0} \frac{2(x+3)-2((x+h)+3)}{h((x+h)+3)(x+3)} = \lim_{h\to 0} \frac{-2}{((x+h)+3)(x+3)} = \frac{-2}{(x+3)^2}.$$

9. $f'(x) = 3 \cdot 2x + 4 = 6x + 4.$

11. $\frac{dy}{dx} = 3 \cdot (4x^3) + 0 - 4 = 12x^3 - 4.$

13. $f'(x) = 3x^2 - 5x^4.$

15. $f'(x) = 3 \cdot 4x^3 - 6 = 12x^3 - 6.$

17. $\frac{dy}{dx} = 2 \cdot 5x^4 - 3 \cdot (-2)x^{-3} = 10x^4 + 6x^{-3}.$

19. $f'(x) = 2 \cdot \left(\frac{2}{3}\right) x^{\frac{2}{3}-1} + 3 \cdot \left(\frac{5}{3}\right) x^{\frac{5}{3}-1} = \frac{4}{3}x^{-\frac{1}{3}} + 5x^{\frac{2}{3}}.$

21. $\frac{dy}{dx} = 9 \cdot (-2)x^{-3} + 3\left(\frac{1}{2}\right) x^{-1/2} = -18x^{-3} + \frac{3}{2}x^{-1/2}.$

23. $y = 5x^9 - 10\sqrt{x} + 3x^{11/2} = 5x^9 - 10x^{1/2} + 3x^{11/2}$. $\frac{dy}{dx} = 5 \cdot 9x^8 - 10 \cdot \frac{1}{2}x^{-1/2} +$
$3 \cdot \frac{11}{2}x^{9/2} = 45x^8 - 5x^{-1/2} + \frac{33}{2}x^{9/2}.$

25. $f'(x) = 4 \cdot \left(-\frac{2}{3}\right) x^{-\frac{2}{3}-1} - 7\left(-\frac{1}{4}\right) x^{-\frac{1}{4}-1} = -\frac{8}{3}x^{-\frac{5}{3}} + \frac{7}{4}x^{-\frac{5}{4}}.$

27. $f'(x) = 12x^2 - 4x$. $f'(2) = 12(2)^2 - 4(2) = 48 - 8 = 40.$

29. $f'(x) = 12x^3 - 12x + 4$. $f'(-3) = 12(-3)^3 - 12(-3) + 4 = -284.$

31. $f'(x) = -3x^{-1/2} + \frac{10}{3}x^{-1/3}$. $f'(8) = -3(8)^{-1/2} + \frac{10}{3}(8)^{-1/3}$
$= -\frac{3}{2\sqrt{2}} + \frac{10}{3} \cdot \frac{1}{2} = -\frac{3}{2\sqrt{2}} + \frac{5}{3}.$

33. $f'(x) = 8x - 2$. At $(1,3)$, $f'(1) = 8(1) - 2 = 6.$

The line tangent to the graph of $f(x)$ at the point $(1,3)$ is $y - 3 = f'(1)(x-1)$, $y - 3 = 6(x-1)$, $6x - y - 3 = 0.$

35. $f'(x) = \frac{1}{2}x^{-1/2} + \frac{1}{2}x^{-3/2}$. The slope of the graph of $f(x)$ at $\left(4, \frac{3}{2}\right)$ is $f'(4) = \frac{1}{2}(4)^{-1/2} + \frac{1}{2}(4)^{-3/2} = \frac{1}{4} + \frac{1}{16} = \frac{5}{16}.$

37. $f'(x) = 2x + a.$

Since the slope of the tangent to the graph of $f(x)$ at $(x, f(x))$ equals $f'(x)$, then $m = 9 = f'(2) = 2(2) + a = 4 + a$, $a = 5.$

39. $f(x) = \sqrt{x+5}.$

$$f'(x) = \lim_{h\to 0} \frac{\sqrt{(x+h)+5} - \sqrt{x+5}}{h} = \lim_{h\to 0} \frac{\sqrt{(x+h)+5} - \sqrt{x+5}}{h} \cdot \frac{\sqrt{(x+h)+5} + \sqrt{x+5}}{\sqrt{(x+h)+5} + \sqrt{x+5}}$$
$$= \lim_{h\to 0} \frac{(x+h+5) - (x+5)}{h\left(\sqrt{x+h+5} + \sqrt{x+5}\right)} = \lim_{h\to 0} \frac{1}{\sqrt{x+h+5} + \sqrt{x+5}} = \frac{1}{2\sqrt{x+5}}.$$
$$f'(a) = \frac{1}{4} \implies \frac{1}{2\sqrt{a+5}} = \frac{1}{4} \implies 2\sqrt{a+5} = 4 \implies \sqrt{a+5} = 2$$
$$\implies a + 5 = 4 \implies a = -1.$$

41. The graph of $y = P(x) = 32x - 2x^2 - 120$ is a parabola opening downward.

a. $32x - 2x^2 - 120 = -2(x^2 - 16x + 60) = -2(x - 10)(x - 6) = 0$ for $x = 6$, $x = 10$.

Therefore $P(x) > 0$ for $6 < x < 10$.

b. The vertex of this parabola is the highest point of this parabola. The x-coordinate of the vertex is given by $x = -\frac{32}{2(-2)} = 8$. Therefore the profit $P(x)$ increases as the production level x increases for $0 < x < 8$.

c. The largest possible profit $P(x)$ occurs when the level of production $x = 8$, and this largest possible profit is $P(8) = 32(8) - 2(8^2) - 120 = 8$.

d. Note that $P'(x) = 32 - 4x = 4(8 - x)$.

$P'(x) = 0$ for $x = 8$.

For $0 < x < 8$, $P'(x) > 0$, so $P(x)$ increases as x increases.

Then for $x > 8$, $P'(x) < 0$, so $P(x)$ decreases as x increases.

43. $f(t) = t^{-3/2}$.

a. $f'(t) = -\frac{3}{2}t^{-5/2}$.

b. $f'(2) = -\frac{3}{2} \cdot 2^{-5/2} = -3 \cdot 2^{-7/2} = -\frac{3}{8\sqrt{2}} = -\frac{3\sqrt{2}}{1}6$.

c. $y - 2^{-3/2} = -\frac{3\sqrt{2}}{1}6(t - 2) \implies y = -\frac{3\sqrt{2}}{1}6t + \frac{5\sqrt{2}}{8}$.

45. a. By the definition of the derivative,

$$f'(x) = \lim_{h \to 0} \frac{f(x+h) - f(x)}{h} = \lim_{h \to 0} \frac{(x+h)^n - x^n}{h}.$$

b. $(x+h)^n = x^n + nx^{n-1}h + \frac{n(n-1)}{2} \cdot x^{n-2}h^2$
$+ \frac{n(n-1)(n-2)}{6}x^{n-3}h^3 + \frac{n(n-1)(n-2)(n-3)}{24}x^{n-4}h^4$
$+ \cdots + nxh^{n-1} + h^n$.

c. $f'(x) = \lim_{h \to 0} \frac{f(x+h) - f(x)}{h} = \lim_{h \to 0} \Big\{ nx^{n-1} + h\big(\frac{n(n-1)}{2}x^{n-2}$
$+ \frac{n(n-1)(n-2)}{6}x^{n-3}h + \frac{n(n-1)(n-2)(n-3)}{24}x^{n-4}h^2$
$+ \cdots + nxh^{n-3} + h^{n-2}\big) \Big\} = nx^{n-1}$.

Selected Solutions to Exercise Set 2.5

1. $s(t) = 7 + 3t^2, \quad t_1 = 0, \quad t_2 = 3$

   Average velocity $\quad \bar{v} = \frac{s(t_2)-s(t_1)}{t_2-t_1} = \frac{s(3)-s(0)}{3-0} = \frac{34-7}{3} = 9.$

3. $s(t) = \frac{10t}{1+t^2}, \quad t_0 = 0, \quad t_2 = 3$

   Average velocity $\quad \bar{v} = \frac{s(3)-s(0)}{3-0} = \frac{3-0}{3} = 1.$

5. $s(t) = 40 + 64t - 3t^2, \quad t_1 = 1, \quad t_2 = 2$

   Average velocity $\quad \bar{v} = \frac{s(2)-s(1)}{2-1} = \frac{156-101}{1} = 55.$

7. The velocity function $v(t) = s'(t) = 2t + 2.$

9. $v(t) = s'(t) = 3t^2 - 10t + 3.$

11. $v(t) = s'(t) = 3 \cdot \frac{1}{2}t^{-1/2} + 5 \cdot \frac{3}{2}t^{1/2} = \frac{3}{2}t^{-1/2} + \frac{15}{2}t^{1/2}.$

13. $v(t) = s'(t) = 2 \cdot \frac{1}{2}t^{-1/2} + 5 \cdot \left(-\frac{2}{3}\right)t^{-5/3} = t^{-1/2} - \frac{10}{3}t^{-5/3}.$

15. $s(t) = \frac{(t+6)(t-2)}{\sqrt{t}} = \frac{t^2+4t-12}{t^{1/2}} = \frac{t^2}{t^{1/2}} + 4\frac{t}{t^{1/2}} - \frac{12}{t^{1/2}}$

    $= t^{3/2} + 4t^{1/2} - 12t^{-1/2}.$

    $v(t) = s'(t) = \frac{3}{2}t^{1/2} + 4 \cdot \frac{1}{2}t^{-1/2} - 12\left(-\frac{1}{2}\right)t^{-3/2} = \frac{3}{2}t^{1/2} + 2t^{-1/2} + 6t^{-3/2}.$

17. $s(t) = (t^2 + 2)(t + 1) = t^3 + t^2 + 2t + 2.$

    $v(t) = s'(t) = 3t^2 + 2t + 2.$

    $v(3) = s'(3) = 3 \cdot 3^2 + 2 \cdot 3 + 2 = 35.$

19. $s(t) = 6t - t^2 = t(6 - t).$

    $v(t) = s'(t) = 6 - 2t.$

    a. The initial velocity $v(0) = 6 - 2(0) = 6.$

    b. The particle changes direction when $v(t) = 0.$

       $v(t) = 0 \implies t = 3.$

    c. The particle crosses the origin the second time when $t = 6.$

       $v(6) = 6 - 2(6) = -6.$

21. Let $y = s(t)$ denote the height of the water balloon t seconds after Carol dropped it. Let s_0 denote the initial height of the water balloon, and let v_0 denote the

initial velocity of the water baloon. We have the formula

$$s(t) = -16t^2 + v_0 t + s_0.$$

Since Carol dropped the water baloon, it follows that $v_0 = 0$. Therefore

$$s(t) = -16t^2 + s_0.$$

a. When the water balloon strikes the ground, $s(t) = 0 \implies 16t^2 = s_0 \implies t^2 = \frac{s_0}{16} \implies t = \frac{\sqrt{s_0}}{4}$. Since the water balloon strikes the ground with a velocity of -96 ft/sec, it follows that

$$-96 = -32 \cdot \frac{\sqrt{s_0}}{4} \implies \sqrt{s_0} = 12 \implies s_0 = 144.$$

Therefore Carol dropped the water balloon from a height of 144 feet.

b. The number of seconds the water balloon falls is $\frac{\sqrt{144}}{4} = 3$.

23. The position of the particle along the line after t seconds is given by the formula

$$s(t) = 6 + 5t - t^2.$$

The velocity of the particle after t seconds is given by the formula

$$v(t) = s'(t) = 5 - 2t.$$

Note that $v(t) = 0$ when $t = 2.5$.

When $0 < t < 2.5$, $v(t) > 0 \implies s(t)$ increases as t increases.

When $2.5 < t < 6$, $v(t) < 0 \implies s(t)$ decreases as t increases.

Note that $s(0) = 6$ and $s(6) = 0$.

Therefore the maximum distance of the particle from the position $s = 0$ in the time interval $[0, 6]$ is $s(2.5) = 12.25$.

Selected Solutions to Exercise Set 2.6

1. a. $R'(x) = 10.$

   b. $C'(x) = 6.$

   c. $P(x) = 10x - 50 - 6x = 4x - 50.$ $P'(x) = 4.$

3. a. $R'(x) = 100 - 4x.$

   b. $C'(x) = 20 + \frac{3}{2}x^{1/2}.$

   c. $P(x) = (100x - 2x^2) - (20x + x^{3/2} + 400) = -2x^2 - x^{3/2} + 80x - 400.$

   $P'(x) = -4x - \frac{3}{2}x^{1/2} + 80.$

5. a. $R'(x) = 40 + 25x^{-1/2}.$

   b. $C'(x) = 20 + \frac{2}{3}x^{-1/3}.$

   c. $P(x) = 40x + 50\sqrt{x} - 150 - 20x - x^{2/3} = 20x + 50\sqrt{x} - x^{2/3} - 150.$

   $P'(x) = 20 + 25x^{-1/2} - \frac{2}{3}x^{-1/3}.$

7. a. $R'(x) = 400 - \frac{20}{3}x^{-1/3}.$

   b. $C'(x) = -5000x^{-2}.$

   c. $P(x) = 400x - 10x^{2/3} - 5000x^{-1} - 40.$

   $P'(x) = 400 - \frac{20}{3}x^{-1/3} + 5000x^{-2}.$

9. a. $P'(x) = 160 - 2x.$ $P'(40) = 160 - 80 = 80.$

   b. $P'(x) = 160 - 2x > 0 \implies -2x > -160, \quad x < 80.$

11. a. $R(x) = x(600 - 3x) = 600x - 3x^2.$

    b. $P(x) = 600x - 3x^2 - 4000 - 150x - 0.5x^2 = -3.5x^2 + 450x - 4000.$

    c. $MC(x) = 150 + x$, $MR(x) = 600 - 6x.$

    d. $MC(x) = MR(x) \implies 150 + x = 600 - 6x, \quad 7x = 450, \, x \approx 64.$

13. a. $N(3) - N(2) = \{20(3) - (3)^2\} - \{20(2) - (2)^2\} = 15.$

    b. $N'(t) = 20 - 2t.$ $N'(2) = 20 - 4 = 16.$

15. a. $P'(t) = 1000t^{-1/2}.$

    b. No, the population of the city in 1980 does not affect its growth rate.

17. Let $R(x)$ = weekly revenue function.

    $R(x) = x \cdot p(x) = x \cdot (a - bx) = ax - bx^2$.

    $R'(x) = a - 2bx$ is a linear function with slope $m = -2b$.

19. $y(0) = A(0)^2 + B(0) + C = 40, \quad C = 40$.

    $y(5) = A(5)^2 + B(5) + 40 = 315, \quad 25A + 5B = 275, \quad 5A + B = 55$.

    $v(5) = 2A(5) + B = 105$, $10A + B = 105$, $B = 105 - 10A$.

    $5A + (105 - 10A) = 55 \implies A = 10$.

    $B = 105 - 10(10) = 5$.

21. a. $R(x) = x \cdot p(x) = x(1600 - 5x) = -5x^2 + 1600x$.

    b. $P(x) = R(x) - C(x) = (-5x^2 + 1600x) - (25x^2 + 400x + 10{,}000)$

    $= -30x^2 + 1200x - 10{,}000$.

    c. $MP(x) = P'(x) = -60x + 1200$.

    d. $MP(x) = 0 \implies x = 20$.

    e. $MP(x) = MR(x) - MC(x)$.

    $MP(20) = 0 \implies MR(20) = MC(20)$.

    f. For $0 < x < 20$, $MP(x) > 0 \implies P(x)$ increases as x increases.

    For $20 < x < 320$, $MP(x) < 0 \implies P(x)$ decreases as x increases.

    Therefore the maximum possible profit occurs when $x = 20$.

23. a. $R(x) = x \cdot p = x(10{,}000 - 5x) = -5x^2 + 10{,}000x$.

    b. $MR(x) = R'(x) = -10x + 10{,}000$.

    c. $MR(500) = 5{,}000$.

    $MR(1{,}000) = 0$.

    $MR(1{,}500) = -5{,}000$.

    d. For $0 < x < 1{,}000$, $MR(x) > 0 \implies R(x)$ increases as x increases.

    For $1{,}000 < x < 2{,}000$, $MR(x) < 0 \implies R(x)$ decreases as x increases.

    Therefore the revenues are the largest when the production level x is 1,000.

25. Let x_0 be a production level such that $MP(x_0) > 0$. Suppose it would be true that $MR(x_0) < 0$. Then it would follow that $MC(x_0) < 0$, which is not likely.

Selected Solutions to Exercise Set 2.7

1. $f'(x) = 1(x+1) + (x-1)1 = x + 1 + x - 1 = 2x.$

3. $f'(x) = (6x-8)(x^2+2) + (3x^2-8x)(2x) = 12x^3 - 24x^2 + 12x - 16.$

5. $f(x) = (x^3 - x)(x^3 - x).$

   $f'(x) = (3x^2-1)(x^3-x)+(x^3-x)(3x^2-1) = 3x^5-3x^3-x^3+x+3x^5-x^3-3x^3+x$

   $= 6x^5 - 8x^3 + 2x.$

7. $f(x) = \sqrt{x}\left(x^3 + x^{-2/3}\right).$

   $f'(x) = \frac{1}{2}x^{-1/2}\left(x^3 + x^{-2/3}\right) + x^{1/2}\left(3x^2 - \frac{2}{3}x^{-5/3}\right)$

   $= \frac{1}{2}x^{5/2} + \frac{1}{2}x^{-7/6} + 3x^{5/2} - \frac{2}{3}x^{-7/6}$

   $= \frac{7}{2}x^{5/2} - \frac{1}{6}x^{-7/6}.$

9. $f'(x) = (-6x^{-3} - x^{-2})(x-4) + (3x^{-2} + x^{-1})(1)$

   $= -6x^{-2} + 24x^{-3} - x^{-1} + 4x^{-2} + 3x^{-2} + x^{-1} = 24x^{-3} + x^{-2}.$

11. $f'(x) = \frac{(x-2)(1)-(x+2)(1)}{(x-2)^2} = \frac{x-2-x-2}{(x-2)^2} = -\frac{4}{(x-2)^2}.$

13. $g(x) = (8x+2)(x+1) \implies g'(x) = 8(x+1) + (8x+2)(1) = 8x + 8 + 8x + 2 =$

    $16x + 10.$

    $f'(x) = \frac{(x-3)(16x+10)-(8x+2)(x+1)(1)}{(x-3)^2}$

    $= \frac{16x^2+10x-48x-30-8x^2-8x-2x-2}{(x-3)^2} = \frac{8x^2-48x-32}{(x-3)^2}.$

15. $f'(x) = \frac{(x+2)(2x)-(x^2-4)(1)}{(x+2)^2} = \frac{2x^2+4x-x^2+4}{(x+2)^2}$

    $= \frac{x^2+4x+4}{x^2+4x+4} = 1.$

17. $h(x) = (1+x)(1+x) \implies h'(x) = 1(1+x)+(1+x)(1) = 1+x+1+x = 2+2x.$

    $f'(x) = \frac{(1+x)^2(-1)-(1-x)(2+2x)}{\{(1+x)^2\}^2} = \frac{-1-2x-x^2-2-2x+2x+2x^2}{\{(1+x)^2\}^2}$

    $= \frac{x^2-2x-3}{\{(1+x)^2\}^2} = \frac{x-3}{(1+x)^3}.$

19. $f'(x) = \frac{(x^{1/2}+1)\left(2x-\frac{2}{3}x^{-1/3}\right)-(x^2-x^{2/3})\left(\frac{1}{2}x^{-1/2}\right)}{\left(\sqrt{x}+1\right)^2}$

    $= \frac{2x^{3/2}-\frac{2}{3}x^{1/6}+2x-\frac{2}{3}x^{-1/3}-\frac{1}{2}x^{3/2}+\frac{1}{2}x^{1/6}}{\left(\sqrt{x}+1\right)^2}$

    $= \frac{\frac{3}{2}x^{3/2}+2x-\frac{1}{6}x^{1/6}-\frac{2}{3}x^{-1/3}}{\left(\sqrt{x}+1\right)^2}.$

21. $f(x) = x^{1/3}(x^2+2)$.

$f'(x) = \frac{1}{3}x^{-2/3}(x^2+2) + x^{1/3}(2x)$

$f'(x) = \left(\frac{1}{3}x^{4/3} + \frac{2}{3}x^{-2/3}\right) + (2x^{4/3})$

$= \frac{7}{3}x^{4/3} + \frac{2}{3}x^{-2/3}$.

23. $f(x) = g(x)h(x)$ where $g(x) = \frac{1}{x+1}$ and $h(x) = \frac{x-3}{x}$.

$g'(x) = \frac{(x+1)(0)-1(1)}{(x+1)^2} = \frac{-1}{(x+1)^2}$.

$h'(x) = \frac{x\cdot 1-(x-3)\cdot 1}{x^2} = \frac{3}{x^2}$.

$f'(x) = g'(x)\cdot h(x) + g(x)\cdot h'(x)$

$= \frac{-1}{(x+1)^2}\cdot\frac{(x-3)}{x} + \frac{1}{x+1}\cdot\frac{3}{x^2}$

$= \frac{-(x-3)\cdot x}{x^2(x+1)^2} + \frac{3(x+1)}{x^2(x+1)^2} = \frac{-x^2+6x+3}{x^2(x+1)^2}$.

25. $f(x) = g(x)h(x)$ where $g(x) = ax+b$ and $h(x) = cx^2+d$.

$g'(x) = a$.

$h'(x) = 2cx$.

$f'(x) = a(cx^2+d) + (ax+b)(2cx)$

$= acx^2 + ad + 2acx^2 + 2bcx$

$= 3acx^2 + 2bcx + ad$.

27. $f'(x) = \frac{(2x-4)3-3x(2)}{(2x-4)^2} = \frac{-12}{(2x-4)^2}$.

$f'(x) = -3 \implies \frac{-12}{(2x-4)^2} = -3, \quad -12 = -3(2x-4)^2$,

$4x^2 - 16x + 16 = 4 \implies 4x^2 - 16x + 12 = 0 \implies 4(x^2-4x+3) = 0 \implies$

$4(x-3)(x-1) = 0 \implies x = 3$ or $x = 1$. The points are $\left(3, \frac{9}{2}\right)$ and $\left(1, -\frac{3}{2}\right)$.

29. $\frac{d}{dx}(fgh) = \frac{d}{dx}\left(f(gh)\right) = f'(gh) + f(gh)' = f'(gh) + f(g'h+gh')$

$= (f')gh + f(g')h + fg(h')$.

31. $f'(x) = 1(1-x)(x^3-x) + (x-3)(-1)(x^3-x) + (x-3)(1-x)(3x^2-1)$.

$= -5x^4 + 16x^3 - 6x^2 - 8x + 3$.

33. $f'(x) = (1-2x)(x^3-x^{-2})(x^{-1}-x^{-3}) + (x-x^2)(3x^2+2x^{-3})(x^{-1}-x^{-3})$

$+(x-x^2)(x^3-x^{-2})(-x^{-2}+3x^{-4})$.

35. $R(x) = \frac{10000x^{3/2}}{2+x^{1/2}}$

a. $MR(x) = R'(x)$

$$= \left[(2 + x^{1/2})(15000x^{1/2}) - (10000x^{3/2})(\tfrac{1}{2}x^{-1/2})\right]/(2 + x^{1/2})^2$$

$$= \frac{30000x^{1/2}+15000x-5000x}{(2+x^{1/2})^2} = \frac{30000x^{1/2}+10000x}{(2+x^{1/2})^2}.$$

b. $MR(9) = \frac{30000(9)^{1/2}+10000(9)}{(2+9^{1/2})^2} = \frac{90000+90000}{25} = \frac{180000}{25} = 7200.$

37. $C(x) = 400 + 20x + x^2.$

a. $MC(x) = C'(x) = 20 + 2x.$

b. $c(x) = \frac{C(x)}{x} = \frac{400+20x+x^2}{x} = 400x^{-1} + 20 + x.$

c. $c'(x) = -400x^{-2} + 1.$

39. $W(t) = \frac{50t^2}{10+t^2}$

$W'(t) = \frac{(10+t^2)100t-(50t^2)(2t)}{(10+t^2)^2} = \frac{1000t+100t^3-100t^3}{(10+t^2)^2}$

$= \frac{1000t}{(10+t^2)^2}.$

41. $U(x) = \frac{x}{x^2+9}.$

a. $MU(x) = U'(x) = \frac{(x^2+9)(1)-x(2x)}{(x^2+9)^2} = \frac{9-x^2}{(x^2+9)^2}.$

b. $MU(x) = 0 \implies 9 - x^2 = 0 \implies x = 3.$

43. $P(t) = 10 + \frac{100t}{t^2+9}, \qquad t > 0.$

a. $P'(t) = 0 + \frac{(t^2+9)(100)-(100t)(2t)}{(t^2+9)^2}$

$= \frac{900-100t^2}{(t^2+9)^2}.$

b. $P'(0) = \frac{900}{(9)^2} = \frac{100}{9} > 0.$

Therefore the population increases initially.

c. $P'(t) = 0$ when $t = 3$.

For $t > 3$, $P'(t) < 0$, so the population is declining in size.

d. As t becomes very large, $\frac{100t}{t^2+9}$ becomes close to zero,

so $P(t)$ becomes close to 10. Therefore the eventual size of the population is 10.

Selected Solutions to Exercise Set 2.8

1. $f'(x) = 3(x+4)^2 \cdot \frac{d}{dx}(x+4) = 3(x+4)^2 \cdot 1 = 3(x+4)^2$.

3. $f(x) = (3x^2 - 2x)^5$.

   $f'(x) = 5(3x^2-2x)^4 \cdot \frac{d}{dx}(3x^2-2x) = 5(3x^2-2x)^4(6x-2) = 10(3x^2-2x)^4(3x-1)$.

5. $f(x) = \sqrt{x^4 - 2x^2} = (x^4 - 2x^2)^{1/2}$.

   $f'(x) = \frac{1}{2}(x^4 - 2x^2)^{-1/2} \cdot \frac{d}{dx}(x^4 - 2x^2) = \frac{1}{2}(x^4 - 2x^2)^{-1/2} \cdot (4x^3 - 4x)$

   $= 2x(x^4 - 2x^2)^{-1/2}(x^2 - 1)$.

7. $f(x) = \frac{x}{(x^2-9)^3}$.

   $f'(x) = \frac{(x^2-9)^3(1)-(x)\left(3(x^2-9)^2(2x)\right)}{(x^2-9)^6}$

   $= \frac{(x^2-9)^2\left((x^2-9)-6x^2\right)}{(x^2-9)^6} = \frac{-5x^2-9}{(x^2-9)^4}$.

9. $f'(x) = 4\left(3\sqrt{x}-2\right)^3 \cdot \frac{d}{dx}\left(3\sqrt{x}-2\right) = 4\left(3\sqrt{x}-2\right)^3 \cdot 3 \cdot \left(\frac{1}{2}x^{-1/2}\right)$

   $= 6x^{-1/2}\left(3\sqrt{x}-2\right)^3$.

11. $f'(x) = 4(x^{2/3} - x^{1/4})^3 \cdot \frac{d}{dx}(x^{2/3} - x^{1/4}) = 4(x^{2/3} - x^{1/4})^3 \cdot \left(\frac{2}{3}x^{-1/3} - \frac{1}{4}x^{-3/4}\right)$.

13. $f'(x) = 4\left(\frac{x-3}{x+3}\right)^3 \cdot \frac{d}{dx}\left(\frac{x-3}{x+3}\right)$

    $= 4\left(\frac{x-3}{x+3}\right)^3 \cdot \frac{(x+3)\frac{d}{dx}(x-3)-(x-3)\frac{d}{dx}(x+3)}{(x+3)^2}$

    $= 4\frac{(x-3)^3}{(x+3)^3} \cdot \frac{(x+3)\cdot 1-(x-3)\cdot 1}{(x+3)^2} = 4\frac{(x-3)^3}{(x+3)^3} \cdot \frac{6}{(x+3)^2} = \frac{24(x-3)^3}{(x+3)^5}$.

15. $f'(x) = 3\left(\frac{x+2}{\sqrt[3]{x}}\right)^2 \cdot \frac{d}{dx}\left(\frac{x+2}{\sqrt[3]{x}}\right) = 3\left(\frac{x+2}{\sqrt[3]{x}}\right)^2 \cdot \frac{\sqrt[3]{x}\cdot 1-(x+2)\cdot\frac{1}{3}x^{-2/3}}{\left(\sqrt[3]{x}\right)^2}$

    $= 3\frac{(x+2)^2}{\left(\sqrt[3]{x}\right)^2} \cdot \frac{1}{3} \cdot \frac{3\sqrt[3]{x}-(x+2)x^{-2/3}}{\left(\sqrt[3]{x}\right)^2} = \frac{(x+2)^2\left(3\sqrt[3]{x}-(x+2)x^{-2/3}\right)}{x\sqrt[3]{x}}$

    $= 2(x+2)^2(x^{-1} - x^{-2})$.

17. $f'(x) = -3(4x^{-1/4}+6)^{-4} \cdot \frac{d}{dx}(4x^{-1/4}+6)$

    $= -3\left(4x^{-1/4}+6\right)^{-4}\left(4\cdot\left(-\frac{1}{4}\right)x^{-5/4}\right)$

    $= 3x^{-5/4}\left(4x^{-1/4}+6\right)^{-4}$.

19. $f'(x) = \frac{(1-x)\frac{d}{dx}\left[(x^3+1)^3+1\right]-\left[(x^3+1)^3+1\right]\frac{d}{dx}(1-x)}{(1-x)^2}$

    $= \frac{(1-x)\left[3(x^3+1)^2(3x^2)\right]-\left[(x^3+1)^3+1\right](-1)}{(1-x)^2}$

    $= \frac{9x^2(1-x)(x^3+1)^2+(x^3+1)^3+1}{(1-x)^2}$.

21. $f(x) = \frac{1}{1+(x^2+2)^2} = \left[1+(x^2+2)^2\right]^{-1}$.

$f'(x) = (-1)\cdot\left[1+(x^2+2)^2\right]^{-2}\cdot(2(x^2+2)(2x))$

$= \frac{-4x^3-8x}{[1+(x^2+2)^2]^2}$.

23. $f'(x) = 3\left(\frac{x^{2/3}}{1+\sqrt{x}}\right)^2\cdot\frac{d}{dx}\left(\frac{x^{2/3}}{1+\sqrt{x}}\right) = 3\left(\frac{x^{2/3}}{1+\sqrt{x}}\right)^2\cdot\frac{(1+\sqrt{x})\frac{2}{3}x^{-1/3}-x^{2/3}\cdot\frac{1}{2}x^{-1/2}}{(1+\sqrt{x})^2}$

$= 3\frac{x^{4/3}}{(1+\sqrt{x})^2}\cdot\frac{\frac{2}{3}x^{-1/3}(1+\sqrt{x})-\frac{1}{2}x^{1/6}}{(1+\sqrt{x})^2} = \frac{3x^{4/3}\left[\frac{2}{3}x^{-1/3}(1+\sqrt{x})-\frac{1}{2}x^{1/6}\right]}{(1+\sqrt{x})^4}$

$= \frac{4x+x^{3/2}}{2(1+\sqrt{x})^4}$

25. $f'(x) = 3\left(\sqrt{x}-\frac{1}{\sqrt{x}}\right)^2\cdot\frac{d}{dx}\left(\sqrt{x}-\frac{1}{\sqrt{x}}\right) = 3\left(\sqrt{x}-\frac{1}{\sqrt{x}}\right)^2\cdot\frac{d}{dx}\left(x^{1/2}-x^{-1/2}\right)$

$= 3\left(\sqrt{x}-\frac{1}{\sqrt{x}}\right)^2\left(\frac{1}{2}x^{-1/2}+\frac{1}{2}x^{-3/2}\right) = \frac{3}{2}\left(\sqrt{x}-\frac{1}{\sqrt{x}}\right)^2\left(\frac{1}{\sqrt{x}}+\frac{1}{x\sqrt{x}}\right)$

$= \frac{3}{2}\left[x^{1/2}-x^{-1/2}-x^{-3/2}+x^{-5/2}\right]$

27. $f'(x) = \frac{\sqrt{x^3+6}\cdot 5(x-4)^4-(x-4)^5\cdot\frac{d}{dx}\sqrt{x^3+6}}{\left(\sqrt{x^3+6}\right)^2}$

$= \frac{\sqrt{x^3+6}\cdot 5(x-4)^4-(x-4)^5\cdot\frac{1}{2}\cdot\frac{1}{\sqrt{x^3+6}}\cdot(3x^2)}{\left(\sqrt{x^3+6}\right)^2} = \frac{5(x^3+6)(x-4)^4-\frac{3}{2}x^2(x-4)^5}{\left(\sqrt{x^3+6}\right)^3}$.

29. $f'(x) = \left(\frac{d}{dx}\left(x^{1/3}-4\right)^3\right)\cdot\left(x-\sqrt{x}\right)^{-2/3}+\left(x^{1/3}-4\right)^3\cdot\frac{d}{dx}\left(x-\sqrt{x}\right)^{-2/3}$

$= 3\left(x^{1/3}-4\right)^2\left(\frac{1}{3}x^{-2/3}\right)\left(x-\sqrt{x}\right)^{-2/3}+\left(x^{1/3}-4\right)^3$

$\cdot\left(-\frac{2}{3}\right)\left(x-\sqrt{x}\right)^{-5/3}\left(1-\frac{1}{2}x^{-1/2}\right)$

$= x^{-2/3}\left(x^{1/3}-4\right)^2\left(x-\sqrt{x}\right)^{-2/3}-\frac{1}{3}\left(x^{1/3}-4\right)^3\left(x-\sqrt{x}\right)^{-5/3}\left(2-x^{-1/2}\right)$.

31. $f'(x) = 3(1-x^2)^2(-2x) = -6x(1-x^2)^2$.

At $P = (1,0)$, $f'(1) = -6(1)(1-1^2)^2 = 0$.

The tangent line to the graph of $y = f(x)$ at P is $y-0 = f'(1)(x-1)$, that is, $y = 0$.

33. $f(x) = x\sqrt{1+x^3}$; $P = (2,6)$

$f'(x) = 1\sqrt{1+x^3}+x\left(\frac{1}{2}(1+x^3)^{-1/2}(3x^2)\right)$.

$f'(2) = 1\sqrt{1+(2)^3}+2\left(\frac{1}{2}(1+2^3)^{-1/2}\left(3(2)^2\right)\right)$

$= 3+\frac{1}{3}\cdot 12$

$= 7$.

The tangent line has the equation

$$y-6 = 7(x-2) \implies y = 7x-8.$$

35. $MC(x) = C'(x) = \frac{1}{2}(40 + 16x^2)^{-1/2} \cdot 16(2x) = 16x(40 + 16x^2)^{-1/2}$.

37. a. The marginal weekly profit

$$MP(x) = P'(x) = \frac{1}{3}(x^3 + 10x + 125)^{-2/3}(3x^2 + 10).$$

b. Use the fact that if $P'(x) > 0$ (or $P'(x) < 0$) for all x, $P(x)$ is always increasing (or decreasing) with x.

From part (a), $P'(x) = \frac{3x^2+10}{3\left(\sqrt[3]{(x^3+10x+125)}\right)^2} > 0$ for all $x \geq 0$.

Therefore the profit, $P(x)$, is always increasing with increasing sales x.

39. a. $N'(t) = \frac{d}{dt}\left(20 - 30(9+t^2)^{-1/2}\right) = -30 \cdot \left(-\frac{1}{2}\right)(9+t^2)^{-3/2} \cdot (2t)$

$N'(t) = 30t(9+t^2)^{-3/2} = \frac{30t}{\left(\sqrt{9+t^2}\right)^3}$.

b. Since $N'(t) = \frac{30t}{\left(\sqrt{9+t^2}\right)^3} > 0$ for $t > 0$, the number of items $N(t)$ a cashier can ring up per minute is always increasing with t.

Thus he or she will never stop improving.

41. The average monthly cost for the manufacturer $= \frac{C(x)}{x} = \frac{500+\sqrt{40+16x^2}}{x}$.

The marginal average monthly cost for the manufacturer $= \left(\frac{C(x)}{x}\right)'$

$$= \frac{\frac{1}{2}(40+16x^2)^{-1/2}(16 \cdot 2x) \cdot x - \left(500+\sqrt{40+16x^2}\right) \cdot 1}{x^2} = \frac{16}{\sqrt{40+16x^2}} - \frac{500+\sqrt{40+16x^2}}{x^2}.$$

$$= \frac{-\left(500\sqrt{40+16x^2}+40\right)}{x^2\sqrt{40+16x^2}}$$

Selected Solutions to the Exercise Set 2.9

1. $\lim\limits_{x\to\infty} \frac{3x^2+2}{10x^2-3x} = \lim\limits_{x\to\infty} \frac{3+\frac{2}{x^2}}{10-\frac{3}{x}} = \frac{3+0}{10-0} = \frac{3}{10}$.

3. $\lim\limits_{x\to\infty} \frac{x(4-x^3)}{3x^4+2x^2} = \lim\limits_{x\to\infty} \frac{4x-x^4}{3x^4+2x^2} = \lim\limits_{x\to\infty} \frac{\frac{4}{x^3}-1}{3+\frac{2}{x^2}} = -\frac{1}{3}$.

5. $\lim\limits_{x\to-\infty} \frac{2x^2+7}{3-4x^2} = \lim\limits_{x\to-\infty} \frac{2x^2+7}{3-4x^2} \cdot \frac{\frac{1}{x^2}}{\frac{1}{x^2}}$

$= \lim\limits_{x\to-\infty} \frac{2+\frac{7}{x^2}}{\frac{3}{x^2}-4} = \frac{2+0}{0-4} = -\frac{1}{2}$.

7. $\lim\limits_{x\to\infty} \frac{3x^2+7x}{1-x^4} = \lim\limits_{x\to\infty} \frac{\frac{3}{x^2}+\frac{7}{x^3}}{\frac{1}{x^4}-1} = \frac{0}{-1} = 0$.

9. $\lim\limits_{x\to+\infty} \frac{\sqrt{x-1}}{x^2} = \lim\limits_{x\to\infty} \sqrt{\frac{x-1}{x^4}} = \lim\limits_{x\to\infty} \sqrt{\frac{1}{x^3} - \frac{1}{x^4}} = 0$.

11. $\lim\limits_{x\to\infty} \frac{x+7-x^3}{10x^2+18} = \lim\limits_{x\to\infty} \frac{\frac{1}{x^2}+\frac{7}{x^3}-1}{\frac{10}{x}+\frac{18}{x^3}} = -\infty$.

13. $C(x) = 3,000 + 24x$.

a. The average cost per item $c(x) = \frac{C(x)}{x} = \frac{3,000}{x} + 24$.

b. $\lim_{x\to\infty} c(x) = 0 + 24 = 24$.

15. $P(x) = \frac{1,000+30x+\sqrt{x}}{10+2x}$.

$$\lim_{x\to\infty} P(x) = \lim_{x\to\infty} \frac{1,000+30x+\sqrt{x}}{10+2x} \cdot \frac{\frac{1}{x}}{\frac{1}{x}}$$
$$= \lim_{x\to\infty} \frac{\frac{1,000}{x}+30+\frac{1}{\sqrt{x}}}{\frac{10}{x}+2} = 15.$$

We assume from this positive limit that the company is profitable at high production levels.

17. a. $\lim_{t\to\infty} MP(t) = \lim_{t\to\infty} P'(t) = D$.

b. It is not necessarily true that the population function $P(t)$ will level off in the long-run. For example, suppose $P(t) = \sqrt{250,000t^2 + 5,000t + 10,000} + 110,000$. Then

$$P'(t) = \frac{1}{2} \cdot (250,000t^2 + 4,000t + 10,000)^{-1/2} \cdot (500,000t + 5,000)$$
$$= \frac{250,000t + 2,500}{\sqrt{250,000t^2 + 5,000t + 10,000}}.$$

Therefore,

$$\lim_{t\to\infty} P'(t) = \lim_{t\to\infty} \frac{250,000t + \frac{2,500}{t}}{\sqrt{250,000 + \frac{5,000}{t} + \frac{10,000}{t^2}}} = \frac{250,000}{500} = 500.$$

However, $\lim_{t\to\infty} P(t) = \infty$.

c. If $D > 0$, then in the long-run the population size will be increasing as time t increases at the rate approximately equal to D. If $D < 0$, then in the long-run the population size $P(t)$ will be decreasing as time t increases at the rate approximately equal to D.

Selected Solutions to the Review Exercises - Chapter 2

1. $\lim_{x\to 2}(3x-1) = 3\cdot 2 - 1 = 5.$
3. $\lim_{x\to 3}(x^4-4) = 3^4 - 4 = 77.$
5. $\lim_{x\to 3}\frac{x^2-9}{x-3} = \lim_{x\to 3}\frac{(x-3)(x+3)}{x-3} = \lim_{x\to 3}(x+3) = 6.$
7. $\lim_{x\to -2}\frac{x^2+x-2}{x+2} = \lim_{x\to -2}\frac{(x+2)(x-1)}{x+2} = \lim_{x\to -2}(x-1) = -2-1 = -3.$
9. $\lim_{x\to 1}\frac{x^2+2x-3}{x^2+x-2} = \lim_{x\to 1}\frac{(x-1)(x+3)}{(x-1)(x+2)} = \lim_{x\to 1}\frac{x+3}{x+2} = \frac{1+3}{1+2} = \frac{4}{3}.$
11. $\lim_{t\to -1}\sqrt{\frac{1-t^2}{1-t}} = \sqrt{\frac{1-(-1)^2}{1-(-1)}} = \sqrt{\frac{0}{2}} = 0.$
13. $\lim_{x\to 0}\frac{(2+x)^2-4}{x} = \lim_{x\to 0}\frac{4+4x+x^2-4}{x} = \lim_{x\to 0}\frac{x(4+x)}{x} = \lim_{x\to 0}(4+x) = 4+0 = 4.$
15. The denominator $x-2$ has only the zero $x=2$. Therefore $f(x)$ is discontinuous only at $x=2$.
17. The denominator of $f(x)$ is $x^2-x-6 = (x-3)(x+2)$; it has zeros $x=-2$ and $x=3$. Therefore $f(x)$ is discontinuous only at $x=-2$ and $x=3$.
19. $f(x) = \begin{cases} 2-x^2 & , \quad x \le 2 \\ -x & , \quad x > 2 \end{cases}$

    $\lim_{x\to 2^-} f(x) = \lim_{x\to 2^-}(2-x^2) = 2-2^2 = -2.$
    $\lim_{x\to 2+} f(x) = \lim_{x\to 2+}(-x) = -2.$
    $f(2) = 2-2^2 = -2.$

    Because $\lim_{x\to 2^-} f(x) = \lim_{x\to 2+} f(x) = f(2)$, the function $f(x)$ is continuous at $x=2$. Therefore $f(x)$ is continuous everywhere.
21. $f'(x) = 2x-1.$
23. $f(x) = \sqrt{x}+4.$

    $f'(x) = \frac{1}{2}x^{-1/2} = \frac{1}{2\sqrt{x}}.$
25. $f'(x) = \left(\frac{1}{x^2}\right)' = (x^{-2})' = -2x^{-3} = -\frac{2}{x^3}.$
27. $f'(x) = \left((x-2)^{-1}\right)' = (-1)(x-2)^{-2}\frac{d}{dx}(x-2) = -(x-2)^{-2}\cdot 1 = -(x-2)^{-2}.$
29. $f'(x) = m.$

31. $f(x) = \frac{1}{3x-7}$

$f'(x) = \frac{(3x-7)(0)-(1)(3)}{(3x-7)^2} = \frac{-3}{(3x-7)^2}.$

33. $f'(x) = \frac{(x+3)\cdot 1-(x-1)\cdot 1}{(x+3)^2} = \frac{4}{(x+3)^2}.$

35. $f(x) = (x^2+6)^{-1}.$

$f'(x) = (-1)(x^2+6)^{-2}\cdot \frac{d}{dx}(x^2+6) = -(x^2+6)^{-2}\cdot(2x) = -\frac{2x}{(x^2+6)^2}.$

37. $f(x) = (x^2+4x+4)^2.$

$f'(x) = 2(x^2+4x+4)\cdot\frac{d}{dx}(x^2+4x+4) = 2(x^2+4x+4)\cdot(2x+4).$

39. $f(x) = (x^2+3)^{-2/3}$

$f'(x) = -\frac{2}{3}(x^2+3)^{-5/3}\cdot\frac{d}{dx}(x^2+3) = -\frac{2}{3}\left(x^2+3\right)^{-5/3}\cdot(2x).$

41. $f'(x) = \left((1+x^2)^{-1/2}\right)' = -\frac{1}{2}(1+x^2)^{-3/2}\cdot\frac{d}{dx}(1+x^2) = -\frac{1}{2}(1+x^2)^{-3/2}\cdot(2x) =$
$-\frac{x}{(1+x^2)^{3/2}}.$

43. $f(x) = x^4\sqrt{1+x^2}.$

$f'(x) = (4x^3)(1+x^2)^{1/2} + (x^4)\left(\frac{1}{2}(1+x^2)^{-1/2}\cdot 2x\right)$

$= x^3(1+x^2)^{-1/2}\left(4(1+x^2)+x^2\right)$

$= x^3(1+x^2)^{-1/2}(4+5x^2).$

45. $f(x) = (1-x^2)^3(6+2x+5x^3)^{-3}.$

$f'(x) = \left(3(1-x^2)^2(-2x)\right)\cdot(6+2x+5x^3)^{-3}+(1-x^2)^3\left(-3(6+2x+5x^3)^{-4}(2+15x^2)\right)$

$= -6x(1-x^2)^2(6+2x+5x^3)^{-3} - 3(1-x^2)^3(6+2x+5x^3)^{-4}(2+15x^2).$

47. $D(p) = \frac{200}{40+p^2}.$

$D'(p) = \frac{(40+p^2)(0)-(200)(2p)}{(40+p^2)^2} = \frac{-400p}{(40+p^2)^2}.$

The slope of the demand curve at the point where $p = 10$ is

$$D'(10) = \frac{-400(10)}{(40+10^2)^2} = \frac{-4000}{19600} - \frac{-10}{49}.$$

49. The tangent line is $y-4 = f'(1)(x-1)$, $y-4 = -\frac{2}{3}(x-1)$

$2x+3y-14 = 0.$

51. a. $S(p_0) = D(p_0) \implies \frac{12}{p_0-1} = p_0 - 5$

$12 = (p_0 - 5)(p_0 - 1), \qquad 12 = p_0^2 - 6p_0 + 5$

$p_0^2 - 6p_0 - 7 = 0, \qquad (p_0 - 7)(p_0 + 1) = 0, \qquad p_0 = 7, \qquad p_0 = -1.$

But $p_0 > 5$ so the equilibrium price $p_0 = 7$.

b. $D'(p) = \left(12(p-1)^{-1}\right)' = 12(-1)(p-1)^{-2} = -12(p-1)^{-2}.$

The slope of $D(p)$ at $(7, D(7)) = D'(7) = -12(7-1)^{-2} = -\frac{12}{36} = -\frac{1}{3}.$

c. $S'(p) = 1$. The slope of $S(p)$ at $(7, S(7)) = S'(7) = 1$.

53. For all $p \geq 50$ such that $p \neq 200$ and $p \neq 400$, $D(p)$ is continuous.

For $p = 200$: $\lim_{p \to 200^-} D(p) = \lim_{p \to 200^-} (2p - 5) = 2 \cdot 200 - 5 = 395.$

$\lim_{p \to 200^+} D(p) = \lim_{p \to 200^+} (p + 400) = 200 + 400 = 600 \neq \lim_{p \to 200^-} D(p).$

$D(p)$ is discontinuous at $p = 200$.

For $p = 400$: $\lim_{p \to 400^-} D(p) = \lim_{p \to 400^-} (p + 400) = 400 + 400 = 800.$

$\lim_{p \to 400^+} D(p) = \lim_{p \to 400^+} \left(\frac{1}{2}p + 800\right) = \frac{1}{2}(400) + 800 = 1000 \neq \lim_{p \to 400^-} D(p).$

$D(p)$ is discontinuous at $p = 400$.

55. a. $MC(x) = C'(x) = 30 + \frac{1}{2} \cdot 2x = 30 + x.$

b. $MC(x) = 30 + x = 60 \implies x = 30.$

57. a. $MC(x) = C'(x) = 40 + \frac{1}{3} \cdot 3x^2 = x^2 + 40.$

b. $MC(x) = x^2 + 40 = 76, \quad x^2 = 36, \quad x = \pm 6.$

But x must not be negative. So for $x = 6$, $MC(x) = 76$.

59. a. $x = 500 - 2p, \quad 2p = 500 - x, \quad p = 250 - \frac{1}{2}x.$

b. $R(x) = xp(x) = x\left(250 - \frac{1}{2}x\right) = 250x - \frac{1}{2}x^2.$

c. $MR(x) = \frac{dR}{dx} = 250 - x.$

d. $MR(x) = 0 \implies 250 - x = 0, \quad x = 250.$

61. a. $c(x) = \frac{C(x)}{x} = \frac{400 + 50x}{x} = \frac{400}{x} + 50.$

b. $c'(x) = -\frac{400}{x^2}.$

63. $u(x) = \sqrt{x(x+24)} = \sqrt{x^2 + 24x}.$

$mu(x) = u'(x) = \frac{1}{2}(x^2 + 24x)^{-1/2} \cdot (2x + 24) = \frac{x+12}{\sqrt{x^2+24x}}.$

65. Because $u'(1) > u'(3)$, the employee earning 1,000 per month would appreciate more an increase of an equal amount in salary.

PRACTICE PROBLEMS FOR CHAPTER 2

1. Use the definition of the derivative to find $f'(x)$ for each of the following functions.

   (a) $f(x) = -2x^2 + 3x + 1$.

   (b) $f(x) = \frac{2x-1}{x+2}$.

   (c) $f(x) = \sqrt{2x+1}$.

2. (a) For the function in part (a) of Problem 1, find an equation of the tangent line to the graph at the point where $x = 2$.

   (b) For the function in part (b) of Problem 1, find an equation of the tangent line to the graph at the point $\left(1, \frac{1}{3}\right)$.

3. Find each of the following limits.

   (a) $\lim_{x\to 2} \frac{x^2+x-6}{x-2}$ (b) $\lim_{x\to -3} \frac{x+3}{x^3+27}$

   (c) $\lim_{x\to 0} \frac{\sqrt{x+1}-1}{x}$ (d) $\lim_{x\to\infty} \frac{2x^2+3x-4}{4x^2+1}$

   (e) $\lim_{x\to -\infty} \frac{3x+4}{4x^2-1}$ (f) $\lim_{x\to\infty} \frac{\sqrt{2x+1}}{\sqrt{4x-1}}$

4. (a) If $f(x) = \begin{cases} 9 - x^2 & \text{if } x \le 3 \\ 3x - 7 & \text{if } x > 3, \end{cases}$
   find $\lim_{x\to 3+} f(x)$, $\lim_{x\to 3^-} f(x)$. Is $f(x)$ continuous at $x = 3$?

   (b) If $f(x) = \frac{x+1}{1-x^2}$, find $\lim_{x\to -1+} f(x)$, $\lim_{x\to -1^-} f(x)$, $\lim_{x\to 1+} f(x)$, $\lim_{x\to 1^-} f(x)$.
   Where is $f(x)$ discontinuous?

5. Find the derivative of each of the following functions.

   (a) $f(x) = 2x^3 - 3x^2 + 6x + 7$.

   (b) $f(x) = 4x^{-2} - \frac{6}{x} + 11$.

   (c) $f(x) = 3x^{2/3} - \frac{4}{\sqrt[3]{x}} + 5$.

   (d) $f(x) = \sqrt[3]{2x^2+5}$.

6. If the revenue $R(x) = -.01x^2 + 1,000x$ and the cost $C(x) = 5x + 100$, find the marginal revenue, the marginal cost, the profit $P(x)$, and the marginal profit, where x is the number of items to be produced and sold.
7. An object is launched vertically upward from a height of 112 feet with an initial velocity of 96 feet/second. The height of the object t seconds later until it hits the ground is given by the formula $s = -16t^2 + 96t + 112$.
   (a) What is the maximum height attained by the object, and when does this occur?
   (b) When does the object strike the ground?
8. Find the derivative of each of the following functions.
   (a) $f(x) = (3x^2 + 1)^5 \cdot (x^3 - x^2 + 2x - 1)^{10}$.
   (b) $f(x) = \sqrt{2x + 1} \cdot (x^2 - 3x + 1)$.
   (c) $f(x) = \frac{3x+4}{2x-1}$.
   (d) $f(x) = \frac{1-3x}{(4x+1)^2}$.
   (e) $f(x) = \frac{2x+1}{x^2-1}$.
   (f) $f(x) = \sqrt{x + 1} \cdot \sqrt[3]{x^2 - 3} \cdot (x^3 - x + 1)$.
9. Suppose the weekly cost for producing x items is given by the formula $C(x) = x^3 + 2x^2 + 4x + 900$.
   (a) Find the average cost $c(x)$.
   (b) Find the marginal average cost $c'(x)$.
   (c) Verify that $c'(x) = 0$ implies $C'(x) = c(x)$.
10. Let $f(x) = \frac{1+x^2}{2x-1}$.
    (a) Find the derivative $f'(x)$.
    (b) Find an equation of the tangent line to the graph of f at the point (0,-1).
    (c) Find the values of x for which $f'(x) = 0$.

11. Suppose your company produces a certain brand of candy bar. Let p be the price in dollars you charge for a box of these candy bars, and let $x = D(p)$ be the corresponding number of boxes of these candy bars you are able to sell per week. Suppose this demand function $x = D(p)$ is a linear function. Suppose that when the price p is $20 you are able to sell 1500 boxes of these candy bars per week, and suppose that if you increase the price p to $30 you are then only able to sell 1000 boxes of these candy bars per week.

    a. Find the demand function $x = D(p)$.

    b. Find the price p as a function of the number of boxes of these candy bars sold per week.

    c. Find your weekly revenue in dollars from the sale of these candy bars as a function of the price p.

    d. Find the price p which wil maximize this weekly revenue. How many boxes of these candy bars will you sell per week at this price?

12. Consider again your company's production of candy bars described in Problem 11. Suppose your total weekly cost in dollars of producing x boxes of these candy bars is given by

$$C(x) = \frac{1}{10}x^2 + 5x + 2250.$$

Let $c(x) = \frac{C(x)}{x}$ be the average cost per box.

    a. Find the derivative function $c'(x)$.

    b. Find the value of x such that $c'(x) = 0$, and explain why the average cost per box is the smallest at this production level.

    c. Find the price p for a box of these candy bars when the average cost per box is the smallest.

13. Suppose a consumer's utility function associated with a certain product is given by

$$U(x) = \sqrt{-x^2 + 12x - 20}.$$

a. Find the domain of $U(x)$.

b. Find the marginal utility function $MU(x)$.

c. For values of x is $U(x)$ increasing as x increases?

SOLUTIONS TO PRACTICE PROBLEMS FOR CHAPTER 2

1. (a) $f(x) = -2x^2 + 3x + 1.$

$$f'(x) = \lim_{h\to 0} \frac{f(x+h)-f(x)}{h}$$
$$= \lim_{h\to 0} \frac{\left[-2(x+h)^2+3(x+h)+1\right]-(-2x^2+3x+1)}{h}$$
$$= \lim_{h\to 0} \frac{-2x^2-4xh-2h^2+3x+3h+1+2x^2-3x-1}{h}$$
$$= \lim_{h\to 0} \frac{-4xh-2h^2+3h}{h} = \lim_{h\to 0} \frac{h(-4x+3-2h)}{h}$$
$$= \lim_{h\to 0}(-4x + 3 - 2h) = -4x + 3.$$

(b) $f(x) = \frac{2x-1}{x+2}.$

$$f'(x) = \lim_{h\to 0} \frac{f(x+h)-f(x)}{h} = \lim_{h\to 0} \frac{\frac{2(x+h)-1}{(x+h)+2} - \frac{2x-1}{x+2}}{h}$$
$$= \lim_{h\to 0} \frac{[2(x+h)-1](x+2)-(2x-1)(x+h+2)}{h(x+2)(x+h+2)}$$
$$= \lim_{h\to 0} \frac{2x^2+2xh+4x+4h-x-2-2x^2-2xh+x+h-4x+2}{h(x+2)(x+h+2)}$$
$$= \lim_{h\to 0} \frac{5h}{h(x+2)(x+h+2)} = \lim_{h\to 0} \frac{5}{(x+2)(x+h+2)} = \frac{5}{(x+2)^2}.$$

(c) $f(x) = \sqrt{2x+1}.$

$$f'(x) = \lim_{h\to 0} \frac{f(x+h)-f(x)}{h} = \lim_{h\to 0} \frac{\sqrt{2(x+h)+1}-\sqrt{2x+1}}{h}$$
$$= \lim_{h\to 0} \frac{\left(\sqrt{2(x+h)+1}-\sqrt{2x+1}\right)\left(\sqrt{2(x+h)+1}+\sqrt{2x+1}\right)}{h\left(\sqrt{2(x+h)+1}+\sqrt{2x+1}\right)}$$
$$= \lim_{h\to 0} \frac{[2(x+h)+1]-(2x+1)}{h\left(\sqrt{2(x+h)+1}+\sqrt{2x+1}\right)}$$
$$= \lim_{h\to 0} \frac{2h}{h\left(\sqrt{2(x+h)+1}+\sqrt{2x+1}\right)} = \frac{1}{\sqrt{2x+1}}.$$

2. (a) When $x = 2$, $y = f(2) = -2(2)^2 + 3(2) + 1 = -1$,

and $f'(2) = -4(2) + 3 = -5$.

The tangent line to the graph at the point (2,-1) is

$$y - (-1) = f'(2)(x - 2) \implies y + 1 = -5(x - 2)$$
$$\implies y = -5x + 9.$$

(b) At the point $(1, \frac{1}{3})$, $f'(1) = \frac{5}{(1+2)^2} = \frac{5}{9}$.

The tangent line to the graph at $(1, \frac{1}{3})$ is

$$y - \frac{1}{3} = f'(1)(x - 1) \implies y - \frac{1}{3} = \frac{5}{9}(x - 1)$$
$$\implies y = \frac{5}{9}x - \frac{2}{9}.$$

3. (a) $\lim_{x\to 2} \frac{x^2+x-6}{x-2} = \lim_{x\to 2} \frac{(x-2)(x+3)}{(x-2)}$
$= \lim_{x\to 2}(x + 3) = 2 + 3 = 5.$

(b) $\lim_{x\to -3} \frac{x+3}{x^3+27} = \lim_{x\to -3} \frac{x+3}{(x+3)(x^2-3x+9)}$
$= \lim_{x\to -3} \frac{1}{(x^2-3x+9)} = \frac{1}{(-3)^2-3(-3)+9} = \frac{1}{27}.$

(c) $\lim_{x\to 0} \frac{\sqrt{x+1}-1}{x} = \lim_{x\to 0} \frac{(\sqrt{x+1}-1)(\sqrt{x+1}+1)}{x(\sqrt{x+1}+1)}$
$= \lim_{x\to 0} \frac{(x+1)-1}{x(\sqrt{x+1}+1)} = \lim_{x\to 0} \frac{1}{\sqrt{x+1}+1} = \frac{1}{2}.$

(d) $\lim_{x\to\infty} \frac{2x^2+3x-4}{4x^2+1} = \lim_{x\to\infty} \frac{(2x^2+3x-4)\frac{1}{x^2}}{(4x^2+1)\frac{1}{x^2}}$
$= \lim_{x\to\infty} \frac{2+\frac{3}{x}-\frac{4}{x^2}}{4+\frac{1}{x^2}} = \frac{2+0-0}{4+0} = \frac{1}{2}.$

(e) $\lim_{x\to -\infty} \frac{3x+4}{4x^2-1} = \lim_{x\to -\infty} \frac{(3x+4)\frac{1}{x^2}}{(4x^2-1)\frac{1}{x^2}} = \lim_{x\to -\infty} \frac{\frac{3}{x}+\frac{4}{x^2}}{4-\frac{1}{x^2}} = \frac{0}{4} = 0.$

(f) $\lim_{x\to\infty} \frac{\sqrt{2x+1}}{\sqrt{4x-1}} = \lim_{x\to\infty} \frac{\sqrt{2+\frac{1}{x}}}{\sqrt{4-\frac{1}{x}}} = \frac{\sqrt{2}}{\sqrt{4}} = \frac{\sqrt{2}}{2}.$

4. (a) $\lim_{x\to 3+} f(x) = \lim_{x\to 3+}(3x - 7) = 3(3) - 7 = 2.$

$\lim_{x\to 3^-} f(x) = \lim_{x\to 3^-}(9 - x^2) = 9 - 3^2 = 0.$

$f(x)$ is not continuous at $x = 3$.

(b) $f(x) = \frac{x+1}{1-x^2}$.

$\lim_{x\to -1+} f(x) = \lim_{x\to -1+} \frac{x+1}{(1+x)(1-x)} = \lim_{x\to -1+} \frac{1}{1-x} = \frac{1}{2}.$

$\lim\limits_{x\to -1^-} f(x) = \lim\limits_{x\to -1^-} \frac{1}{1-x} = \frac{1}{2}$.

$\lim\limits_{x\to 1^+} f(x) = \lim\limits_{x\to 1^+} \frac{1}{1-x} = -\infty$.

$\lim\limits_{x\to 1^-} f(x) = \lim\limits_{x\to 1^-} \frac{1}{1-x} = \infty$.

$f(x)$ is discontinuous at the points where $x = 1$ and $x = -1$.

5. (a) $f(x) = 2x^3 - 3x^2 + 6x + 7$.

   $f'(x) = 6x^2 - 6x + 6$.

   (b) $f(x) = 4x^{-2} - \frac{6}{x} + 11$.

   $f'(x) = -8x^{-3} + \frac{6}{x^2}$.

   (c) $f(x) = 3x^{2/3} - \frac{4}{\sqrt[3]{x}} + 5 = 3x^{2/3} - 4x^{-1/3} + 5$.

   $f'(x) = 2x^{-1/3} + \frac{4}{3}x^{-4/3}$.

   (d) $f(x) = \sqrt[3]{2x^2 + 5}$.

   $f'(x) = \frac{1}{3}(2x^2+5)^{-2/3}(4x) = \frac{4}{3}x(2x^2+5)^{-2/3}$.

6. $R(x) = -0.01x^2 + 1{,}000x, \quad C(x) = 5x + 100$.

   $R'(x) = -0.02x + 1{,}000. \quad C'(x) = 5$.

   $P(x) = R(x) - C(x) = -0.01x^2 + 1{,}000x - (5x + 100)$

   $= -0.01x^2 + 995x - 100$.

   $P'(x) = -0.02x + 995$.

7. (a) $s' = -32t + 96 = -32(t - 3)$.

   $s' = 0$ for $t = 3$ sec where $s = -16(3)^2 + 96(3) + 112 = 256$ ft.

   So the maximum height attained is 256 feet occuring when $t = 3$ seconds.

   (b) $s = -16t^2 + 96t + 112 = 0 \implies s = -16(t^2 - 6t - 7) = 0$

   $\implies (t+1)(t-7) = 0 \implies t = 7$ sec.

   So the object strikes the ground when $t = 7$ seconds.

8. (a) $f(x) = (3x^2+1)^5 \cdot (x^3 - x^2 + 2x - 1)^{10}$.

   $$f'(x) = 5(3x^2+1)^4(6x)(x^3 - x^2 + 2x - 1)^{10}$$
   $$+(3x^2+1)^5 \cdot 10(x^3 - x^2 + 2x - 1)^9(3x^2 - 2x + 2)$$
   $$= 30x(3x^2+1)^4(x^3 - x^2 + 2x - 1)^{10}$$
   $$+10(3x^2+1)^5(3x^2 - 2x + 2)(x^3 - x^2 + 2x + 1)^9.$$

(b) $f(x) = \sqrt{2x+1} \cdot (x^2 - 3x + 1)$.

$$f'(x) = \tfrac{1}{2}(2x+1)^{-1/2}(2)(x^2 - 3x + 1) + \sqrt{2x+1}(2x-3)$$
$$= (2x+1)^{-1/2}(x^2 - 3x + 1) + \sqrt{2x+1}(2x-3).$$

(c) $f(x) = \frac{3x+4}{2x-1}$.

$$f'(x) = \frac{3(2x-1)-(3x+4)\cdot 2}{(2x-1)^2} = \frac{6x-3-6x-8}{(2x-1)^2} = \frac{-11}{(2x-1)^2}.$$

(d) $f(x) = \frac{1-3x}{(4x+1)^2}$.

$$f'(x) = \frac{(-3)(4x+1)^2-(1-3x)2(4x+1)(4)}{(4x+1)^4}$$
$$= \frac{-3(4x+1)-8(1-3x)}{(4x+1)^3} = \frac{-12x-3-8+24x}{(4x+1)^3} = \frac{12x-11}{(4x+1)^3}.$$

(e) $f(x) = \frac{2x+1}{x^2-1}$.

$$f'(x) = \frac{2(x^2-1)-(2x+1)\cdot 2x}{(x^2-1)^2} = \frac{2x^2-2-4x^2-2x}{(x^2-1)^2} = \frac{-2(x^2+x+1)}{(x^2-1)^2}.$$

(f) $f(x) = \sqrt{x+1} \cdot \sqrt[3]{x^2-3} \cdot (x^3 - x + 1)$.

$$f'(x) = \tfrac{1}{2}(x+1)^{-1/2} \cdot \sqrt[3]{x^2-3} \cdot (x^3 - x + 1)$$
$$+\sqrt{x+1}\left[\tfrac{1}{3}(x^2-3)^{-2/3}(2x)(x^3-x+1) + \sqrt[3]{x^2-3} \cdot (3x^2-1)\right]$$
$$= \tfrac{1}{2}(x+1)^{-1/2} \cdot \sqrt[3]{x^2-3}(x^3-x+1) + \tfrac{2}{3}x\sqrt{x+1}(x^2-3)^{-2/3}(x^3-x+1)$$
$$+\sqrt{x+1} \cdot \sqrt[3]{x^2-3}(3x^2-1).$$

9. $C(x) = x^3 + 2x^2 + 4x + 900$.

(a) The average cost $c(x) = \frac{C(x)}{x} = x^2 + 2x + 4 + \frac{900}{x}$.

(b) $c'(x) = 2x + 2 - \frac{900}{x^2}$.

(c) $c'(x) = \frac{C'(x)\cdot x - C(x)\cdot 1}{x^2} = 0 \implies C'(x)\cdot x - C(x) = 0$

$\implies C'(x) = \frac{C(x)}{x} = c(x)$.

10. $f(x) = \frac{1+x^2}{2x-1}$.

(a) $f'(x) = \frac{2x(2x-1)-(1+x^2)\cdot 2}{(2x-1)^2} = \frac{4x^2-2x-2-2x^2}{(2x-1)^2} = \frac{2(x^2-x-1)}{(2x-1)^2}$.

(b) $f'(0) = \frac{2(0^2-0-1)}{(2\cdot 0-1)^2} = -2$.

The tangent line to the graph at the point (0,-1) is

$$y - (-1) = f'(0)(x-0) \implies y + 1 = -2x \implies y = -2x - 1.$$

(c) $f'(x) = \frac{2(x^2-x-1)}{(2x-1)^2} = 0 \implies x^2 - x - 1 = 0$

$\implies x = \frac{1 \pm \sqrt{1^2 - 4(1)(-1)}}{2} = \frac{1\pm\sqrt{5}}{2}$.

11. a. $x = mp + b$.

$m = \frac{1000-1500}{30-20} = -50$.

$1500 = (-50) \cdot 20 + b \implies b = 2500$.

$x = D(p) = -50p + 2500$.

b. $50p = 2500 - x \implies p = 50 - .02x$.

c. Weekly revenue $R(p) = p \cdot D(p) = -50p^2 + 2500p$.

d. $R'(p) = -100p + 2500$.

$R'(p) = 0$ for $p = 25$.

For $0 < p < 25$, $R'(p) > 0$, so $R(p)$ is increasing as p increases.

For $25 < p < 50$, $R'(p) < 0$, so $R(p)$ is decreasing as p increases.

The maximum weekly revenue occurs when the price per box is 25 dollars, and the number of boxes sold per week at this price is 1250.

12. a. $c(x) = \frac{C(x)}{x} = \frac{1}{10}x + 5 + \frac{2250}{x}$.

$c'(x) = \frac{1}{10} - \frac{2250}{x^2}$.

b. $c'(x) = 0 \implies x^2 = 22,500 \implies x = 150$ boxes of candy bars.

For $0 < x < 150$, $c'(x) < 0$, so $c(x)$ is decreasing as x increases.

For $150 < x < 2500$, $c'(x) > 0$, so $c(x)$ is increasing as x increases.

Therefore the minimum average cost per box occurs when the production level is 150 boxes of these candy bars per week.

c. The price for a box of these candy bars when the average cost per box is the smallest is $50 - .02(150) = 47$ dollars.

13. a. $-x^2 + 12x - 20 = (-x + 10)(x - 2) = 0$ for $x = 10$, $x = 2$.

$(-x^2 + 12x - 20) \geq 0$ for $2 \leq x \leq 10$.

The domain of $U(x)$ is the closed interval $[2, 10]$.

b. $MU(x) = U'(x) = \frac{1}{2}(-x^2 + 12x - 20)^{-1/2} \cdot (-2x + 12)$

$= \frac{-x+6}{\sqrt{-x^2+12x-20}}$.

c. $MU(x) > 0 \implies 0 < x < 6$.

Thus $U(x)$ increases as x increases for $0 < x < 6$.

CHAPTER 3

SUMMARY OF CHAPTER 3

1. Use of the derivative $f'(x)$ in studying increasing and decreasing functions.

2. Use of the derivative $f'(x)$ in studying relative maxima and minima.

3. The second derivative $f''(x)$, concavity, and the use of $f''(x)$ in studying concavity.

4. The second derivative test for extrema.

5. Vertical and horizontal asymptotes, oblique asymptotes.

6. Sketching the graph of $y = f(x)$ by using the following information:
   (1) Determining the domain of f.
   Examining the symmetry of the graph if evident.
   (2) Locating the zeros of f if convenient.
   (3) Finding the vertical asymptotes.
   (4) Finding the horizontal asymptotes.
   Trying to determine any oblique asymptotes.
   (5) Determining the open intervals on which f is increasing or decreasing, and determining where f has relative maxima and minima.
   (6) Determining the open intervals on which the graph of f is concave up or concave down, and locating the inflection points on the graph.
   (7) Calculating the values of $f(x)$ at a few convenient numbers and plotting the corresponding points.

7. Use of the derivative $f'(x)$ in finding the absolute maximum or the absolute minimum of a function on a closed interval.

8. The procedure described in Section 3.7 for solving applied max-min problems.

9. Applications for maximizing profits, minimizing average costs, inventory cost control, studying taxation problems, studying elasticity of demand.

10. Related rates.

11. Implicit differentiation.

12. Linear approximations and differentials.

12. The Mean Value Theorem.

Selected Solutions to Exercise Set 3.1

1. $f'(x) = 3 > 0$ for all x, so f is increasing on $(-\infty, \infty)$.

3. $f'(x) = -2x$, so $f'(x) = 0$ at $x = 0$.
   If $x < 0$, $f'(x) > 0$; and hence f is increasing on $(-\infty, 0)$.
   If $x > 0$, $f'(x) < 0$; and hence f is decreasing on $(0, \infty)$.

5. $f'(x) = 1 - 3x^2$, so $f'(x) = 0$ at $x = \sqrt{\frac{1}{3}}$ and $-\sqrt{\frac{1}{3}}$.

Interval I	Test Number	Sign of $f'(x)$	Conclusion
$\left(-\infty, -\sqrt{\frac{1}{3}}\right)$	$x_1 = -2$	$f'(-2) < 0$	decreasing
$\left(-\sqrt{\frac{1}{3}}, \sqrt{\frac{1}{3}}\right)$	$x_2 = 0$	$f'(0) > 0$	increasing
$\left(\sqrt{\frac{1}{3}}, \infty\right)$	$x_3 = 2$	$f'(2) < 0$	decreasing

7. $f'(x) = 3x^2 - 3 = 3(x+1)(x-1)$, so $f'(x) = 0$ at $x = 1$ and -1.

Interval I	Test Number	Sign of $f'(x)$	Conclusion
$(-\infty, -1)$	$x_1 = -3$	$f'(-3) > 0$	increasing
$(-1, 1)$	$x_2 = 0$	$f'(0) < 0$	decreasing
$(1, \infty)$	$x_3 = 2$	$f'(2) > 0$	increasing

9. $f(x) = x^3 - 12x^2 - 48x + 12$.
   $f'(x) = 3x^2 - 24x - 48 = 3(x^2 - 8x - 16)$.
   $f'(x) = 0$ for $x = \frac{8 \pm \sqrt{8^2 - 4 \cdot 1 \cdot (-16)}}{2} = \frac{8 \pm \sqrt{128}}{2} = \frac{8 \pm 8\sqrt{2}}{2} = 4 \pm 4\sqrt{2}$.

Interval I	Test Number	Sign of $f'(x)$	Conclusion
$(-\infty, 4 - 4\sqrt{2})$	$x_1 = -2$	$f'(-2) > 0$	increasing
$(4 - 4\sqrt{2}, 4 + 4\sqrt{2})$	$x_2 = 0$	$f'(0) < 0$	decreasing
$(4 + 4\sqrt{2}, \infty)$	$x_3 = 10$	$f'(10) > 0$	increasing

11. $f'(x) = 6x^2 - 6x - 36 = 6(x^2 - x - 6) = 6(x-3)(x+2) = 0$ for $x = 3$ and -2.

Interval I	Test Number	Sign of $f'(x)$	Conclusion
$(-\infty, -2)$	$x_1 = -3$	$f'(-3) > 0$	increasing
$(-2, 3)$	$x_2 = 0$	$f'(0) < 0$	decreasing
$(3, \infty)$	$x_3 = 4$	$f'(4) > 0$	increasing

13. $f'(x) = 12x^3 + 12x^2 - 24x = 12x(x^2 + x - 2) = 12x(x+2)(x-1) = 0$ for $x = 0$, -2, and 1.

Interval I	Test Number	Sign of $f'(x)$	Conclusion
$(-\infty, -2)$	$x_1 = -3$	$f'(-3) < 0$	decreasing
$(-2, 0)$	$x_2 = -1$	$f'(-1) > 0$	increasing
$(0, 1)$	$x_3 = \frac{1}{2}$	$f'\left(\frac{1}{2}\right) < 0$	decreasing
$(1, \infty)$	$x_4 = 2$	$f'(2) > 0$	increasing

15. $f'(x) = \frac{4}{3}x^{1/3} = 0$ for $x = 0$.
If $x < 0$, $f'(x) < 0$; so f is decreasing on $(-\infty, 0)$.
If $x > 0$, $f'(x) > 0$; so f is increasing on $(0, \infty)$.

17. $f(x) = |4 - x^2|$.

$$f(x) = \begin{cases} x^2 - 4, & -\infty < x \le -2 \\ 4 - x^2, & -2 < x \le 2 \\ x^2 - 4, & 2 < x < \infty. \end{cases}$$

If $x < -2$, $f'(x) = 2x < 0$, so f is decreasing.
If $-2 < x < 0$, $f'(x) = -2x > 0$, so f is increasing.
If $0 < x < 2$, $f'(x) = -2x < 0$, so f is decreasing.
If $x > 2$, $f'(x) = 2x > 0$, so f is increasing.

19. $f'(x) = \frac{1}{3}(8 - x^3)^{-2/3} \cdot (-3x^2) = \frac{-x^2}{(8-x^3)^{2/3}}$ fails to exist at $x = 2$, and $f'(x) = 0$ at $x = 0$.

Interval I	Test Number	Sign of $f'(x)$	Conclusion
$(-\infty, 0)$	$x_1 = -1$	$f'(-1) < 0$	decreasing
$(0, 2)$	$x_2 = 1$	$f'(1) < 0$	decreasing
$(2, \infty)$	$x_3 = 3$	$f'(3) < 0$	decreasing

21. $f'(x) = 6x^2 - 6ax$.
$f'(3) = 6 \cdot 3^2 - 6 \cdot a \cdot (3) = 0$, $54 - 18a = 0$.
$\therefore a = 3$.

23. The numbers are x and $50 - x$. Let $f(x) = x(50 - x) = 50x - x^2$.
Then $f'(x) = 50 - 2x = 0$ for $-2x = -50$, $x = 25$.
If $x < 25$, $f'(x) > 0$; so $f(x)$ is increasing on $(-\infty, 25)$.
If $x > 25$, $f'(x) < 0$; so $f(x)$ is decreasing on $(25, \infty)$.

25. $c(x) = \frac{C(x)}{x} = \frac{20+200x+0.01x^3}{x} = 0.01x^2 + 200 + \frac{20}{x}.$
$c'(x) = 0.02x - \frac{20}{x^2} = \frac{0.02x^3-20}{x^2} = \frac{0.02(x^3-1000)}{x^2}.$
$c'(10) = 0.$
a. If $x > 10$, $c'(x) > 0$, so $c(x)$ is increasing.
b. If $0 < x < 10$, $c'(x) < 0$, so $c(x)$ is decreasing.

27. $R(x) = 1500x - 60x^2 - x^3.$
$R'(x) = -3x^2 - 120x + 1500 = -3(x^2 + 40x - 500)$
$= -3(x+50)(x-10).$
$R'(x) = 0$ at $x = 10$ and -50.
Since $x > 0$, $R'(x) > 0$ on $(0,10)$ and $R'(x) < 0$ on $(10,\infty)$.
a. $(0,10)$.
b. $MR(x) = R'(x) = -3(x^2 + 40x - 500).$
$(MR)'(x) = -3(2x+40)$. $\therefore MR'(x) = 0$ at $x = -20$.
MR is decreasing on $(-20,\infty)$. Since $x > 0$, MR is decreasing on $(0,\infty)$.

29. $x = 1000 - 2p \implies p = 500 - \frac{x}{2}$. $C(x) = 400 + 2x$.
a. $R(x) = xp = x\left(500 - \frac{x}{2}\right) = 500x - \frac{1}{2}x^2.$
b. $R'(x) = 500 - x \implies R'(x) = 0$ at $x = 500$.
$\therefore R$ is increasing on $(0,500)$ since $R'(x) > 0$ on $(0,500)$.
c. $P(x) = R(x) - C(x) = 500x - \frac{1}{2}x^2 - (400 + 2x)$
$= -\frac{1}{2}x^2 + 498x - 400.$
d. $P'(x) = -x + 498$, so $P'(x) = 0$ at $x = 498$.
$\therefore P$ is increasing on $(0,498)$ since $P'(x) > 0$ there.
e. $(498,1000)$.

31. $P(t) = \frac{Kt}{100+t^2}$, $t > 0$.
$P'(t) = \frac{(100+t^2)\cdot K-(Kt)\cdot(2t)}{(100+t^2)^2} = \frac{K(100-t^2)}{(100+t^2)^2}.$
$P'(t) = 0$ for $t = 10$.
If $0 < t < 10$, $P'(t) > 0$, so $P(t)$ is increasing.
If $10 < t < 20$, $P'(t) < 0$, so $P(t)$ is decreasing.

33.

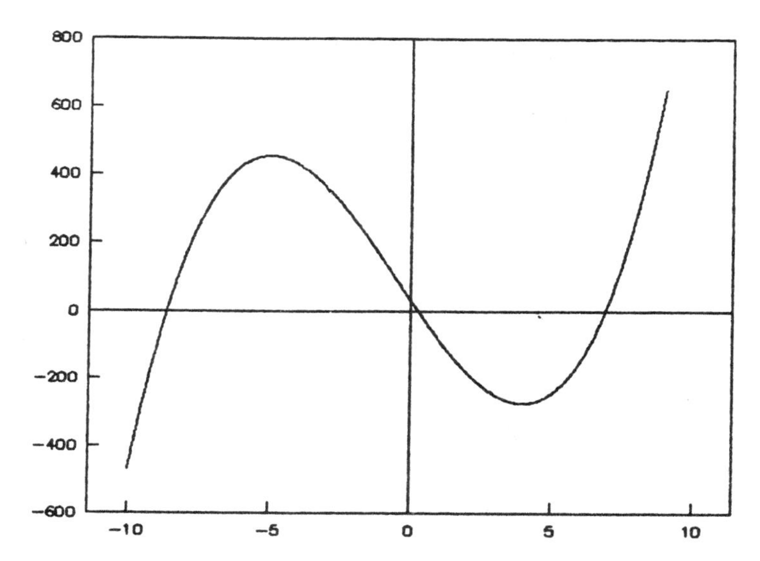

35. $c(x) = \frac{C(x)}{x} = \frac{500}{x} + 20 + 5x.$
$c'(x) = \frac{-500}{x^2} + 5.$
$c'(x) = 0$ for $x = 10$.
If $x < 10$, $c'(x) < 0$, so $c(x)$ is decreasing.
If $x > 10$, $c'(x) > 0$, so $c(x)$ is increasing.
From problem 34, we have that the total cost $C(x)$ keeps increasing as x increases.

Selected Solutions to Exercise Set 3.2

1. $fx) = x^2 - 2x.$
$f'(x) = 2x - 2.$
$f'(x) = 0$ for $x = 1$.
For $x < 1$, $f'(x) < 0$. For $x > 1$, $f'(x) > 0$.
$f(1)$ is a relative minimum.

x	-1	0	1	2	3
$f(x)$	3	0	-1	0	3

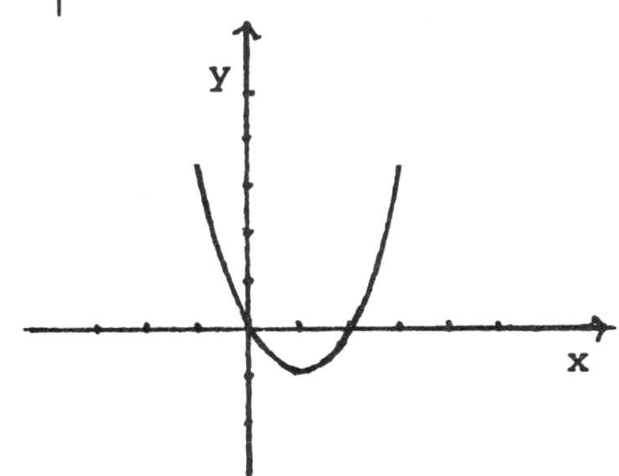

3. $f(x) = 3x^2 - 6x + 3$.

$f'(x) = 6x - 6$.

$f'(x) = 0$ for $x = 1$.

For $x < 1$, $f'(x) < 0$.

For $x > 1$, $f'(x) > 0$.

$f(1)$ is a relative minimum.

x	-1	0	1	2	3
$f(x)$	12	3	0	3	12

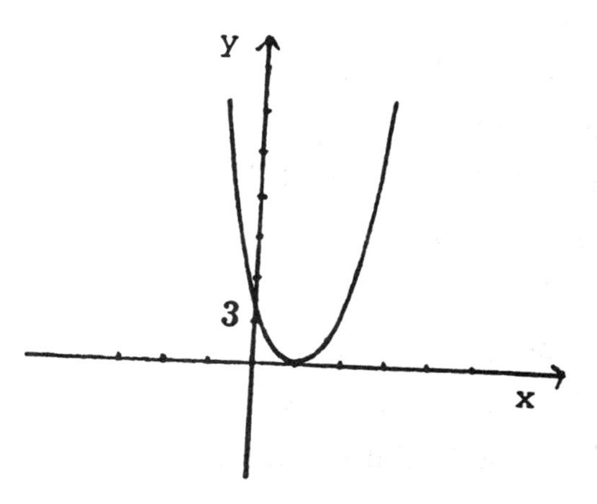

5. $f(x) = 9 - 4x - x^2$.

$f'(x) = -4 - 2x$.

$f'(x) = 0$ for $x = -2$.

For $x < -2$, $f'(x) > 0$.

For $x > -2$, $f'(x) < 0$.

$f(-2)$ is a relative maximum.

x	-4	-3	-2	-1	0
$f(x)$	9	12	13	12	9

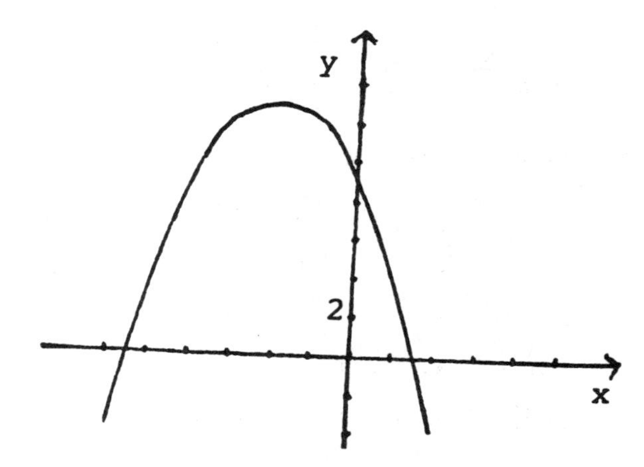

7. $f(x) = \sqrt{x^2 + 2}$.

$f'(x) = \frac{1}{2}(x^2 + 2)^{-1/2} \cdot (2x) = \frac{x}{\sqrt{x^2+2}}$.

$f'(x) = 0$ for $x = 0$.

For $x < 0$, $f'(x) < 0$, so $f(x)$ is decreasing.

For $x > 0$, $f'(x) > 0$, so $f(x)$ is increasing.

$f(0)$ is a relative minimum.

x	± 3	± 2	± 1	0
$f(x)$	$\sqrt{11} = 3.3166$	$\sqrt{6} = 2.4495$	$\sqrt{3} = 1.7321$	$\sqrt{2} = 1.4142$

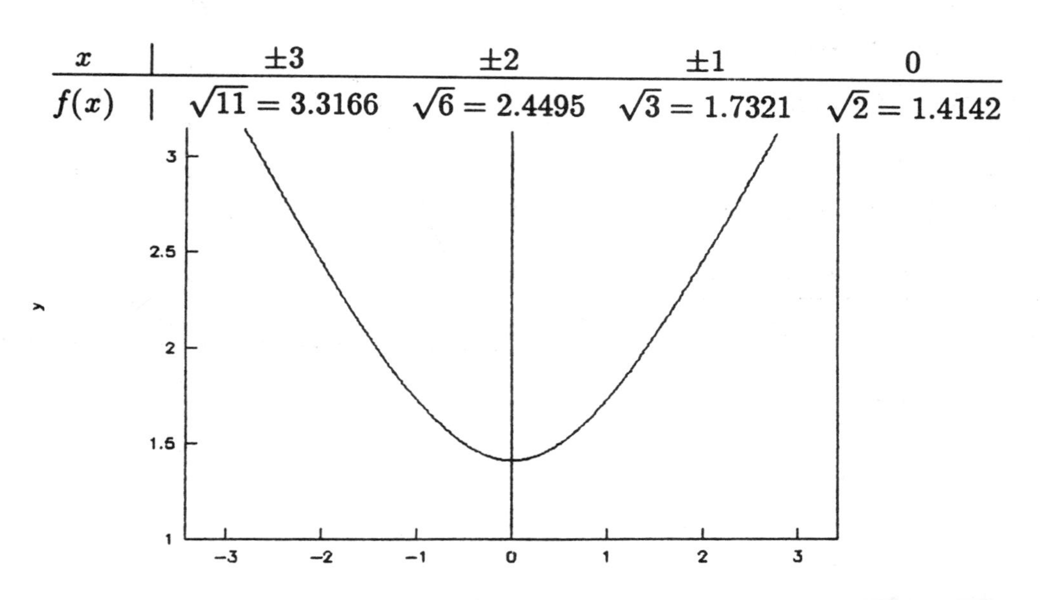

9. $f(x) = x^3 - 3x + 7$.

$f'(x) = 3x^2 - 3$.

$f'(x) = 0$ for $x = 1$ and $x = -1$

For $x < -1$, $f'(x) > 0$.

For $-1 < x < 1$, $f'(x) < 0$.

For $x > 1$, $f'(x) > 0$.

$f(-1)$ is a relative maximum and $f(1)$ is a relative minimum.

x	-2	-1	0	1	2
$f(x)$	5	9	7	5	9

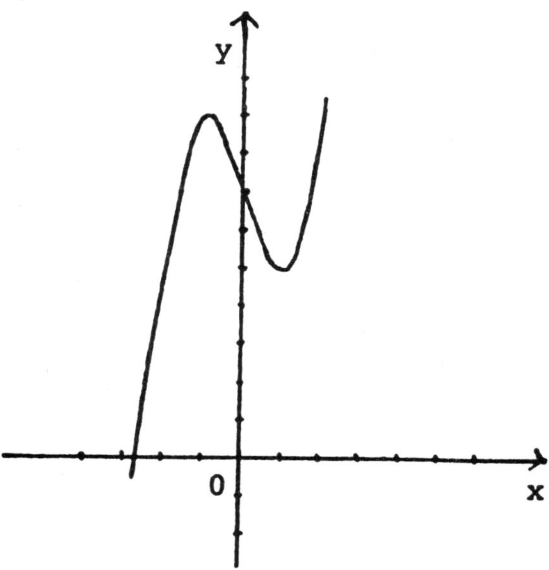

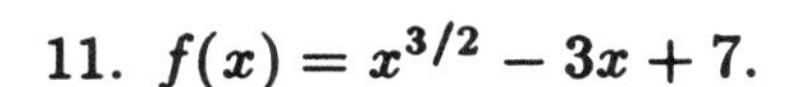

11. $f(x) = x^{3/2} - 3x + 7$.

$f'(x) = \frac{3}{2}x^{1/2} - 3$ for $x > 0$.

$f'(x) = 0$ for $x = 4$.

For $0 < x < 4$, $f'(x) < 0$.

For $x > 4$, $f'(x) > 0$.

$f(4)$ is a relative minimum.

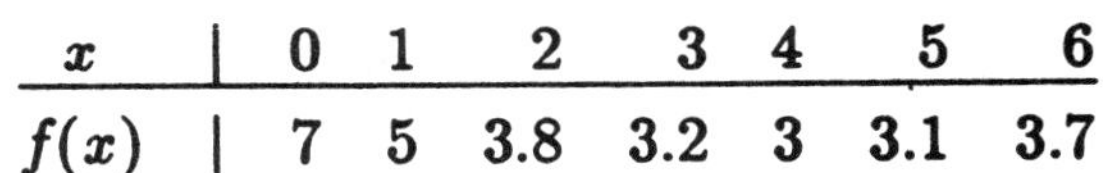
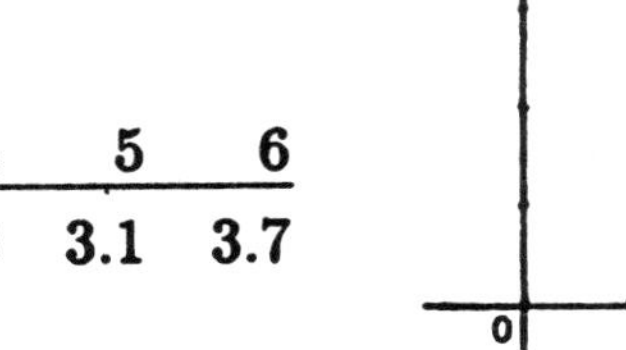

x	0	1	2	3	4	5	6
$f(x)$	7	5	3.8	3.2	3	3.1	3.7

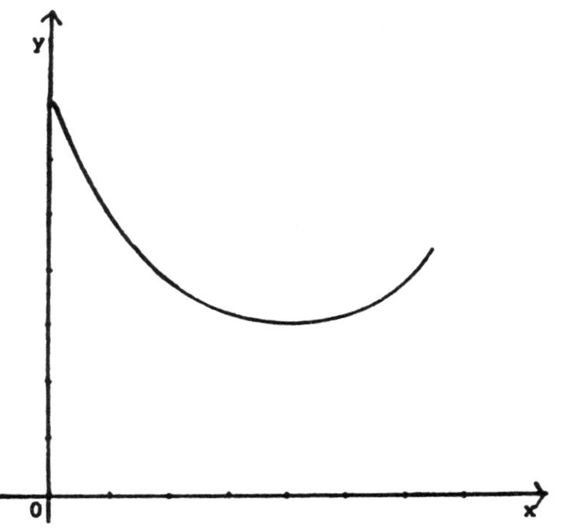

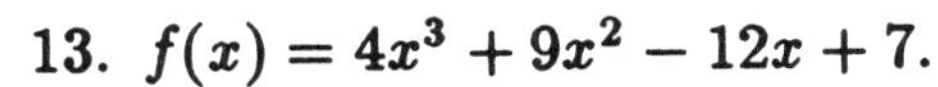

13. $f(x) = 4x^3 + 9x^2 - 12x + 7$.

$f'(x) = 12x^2 + 18x - 12 = 6(2x^2 + 3x - 2) = 6(2x - 1)(x + 2)$.

$f'(x) = 0$ for $x = -2$ and $x = \frac{1}{2}$.

For $x < -2$, $f'(x) > 0$.

For $-2 < x < \frac{1}{2}$, $f'(x) < 0$.

For $x > \frac{1}{2}$, $f'(x) > 0$.

$f(-2)$ is a relative maximum and $f\left(\frac{1}{2}\right)$ is a relative minimum.

x	-3	-2	-1	0	$\frac{1}{2}$	1
$f(x)$	16	35	24	7	3.75	8

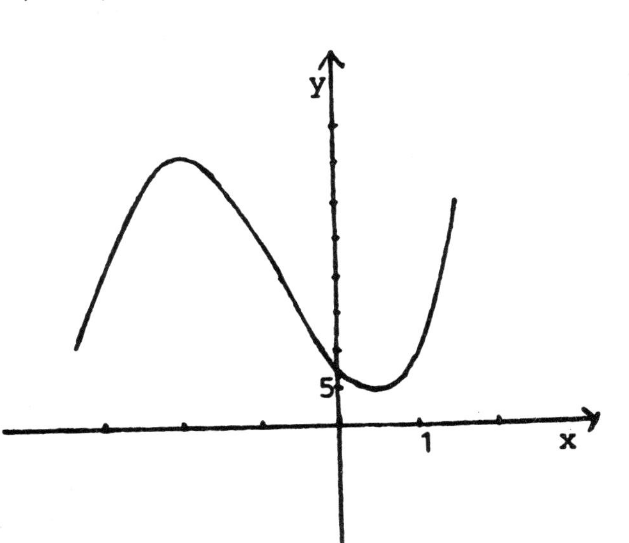

15. $f(x) = 3x^4 - 8x^3 - 18x^2 + 36$.

$f'(x) = 12x^3 - 24x^2 - 36x = 12x(x^2 - 2x - 3) = 12x\,((x-3)(x+1))$. $f'(x) = 0$ for all $x = -1$, $x = 0$, $x = 3$.

For $x < -1$, $f'(x) < 0$, so $f(x)$ is decreasing.

For $-1 < x < 0$, $f'(x) > 0$, so $f(x)$ is increasing.

For $0 < x < 3$, $f'(x) < 0$, so $f(x)$ is decreasing.

For $x > 3$, $f'(x) > 0$, so $f(x)$ is increasing.

$f(-1)$ is a relative minimum,

$f(0)$ is a relative maximum,

$f(3)$ is a relative minimum.

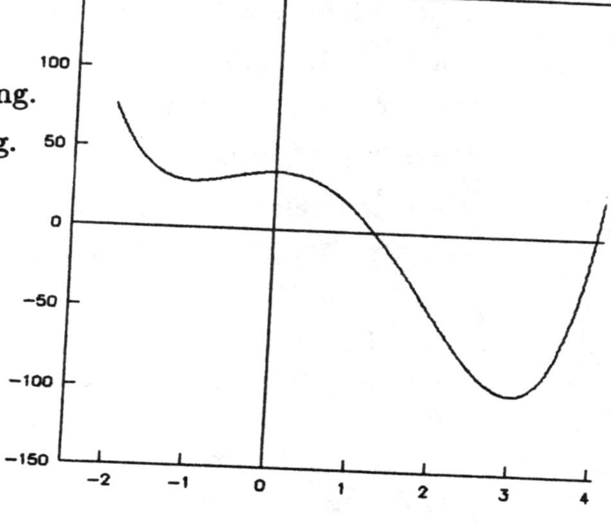

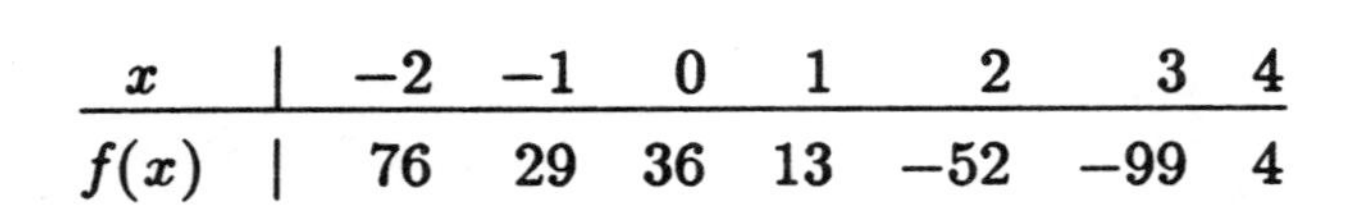

x	-2	-1	0	1	2	3	4
$f(x)$	76	29	36	13	-52	-99	4

17. $f(x) = 9x - x^{-1}$.

$f'(x) = 9 + x^{-2}$.

$f'(x)$ does not exist for $x = 0$.

$f'(x) > 0$ for $x \neq 0$, so $f(x)$ is increasing on $(-\infty, 0)$ and $(0, \infty)$.

x	-2	-1	$-\frac{1}{2}$	$-\frac{1}{4}$	$-\frac{1}{8}$
$f(x)$	-17.5	-8	$-\frac{5}{2}$	$\frac{7}{4}$	$\frac{55}{8}$
x	$\frac{1}{8}$	$\frac{1}{4}$	$\frac{1}{2}$	1	2
$f(x)$	$-\frac{55}{8}$	$-\frac{7}{4}$	$\frac{5}{2}$	8	17.5

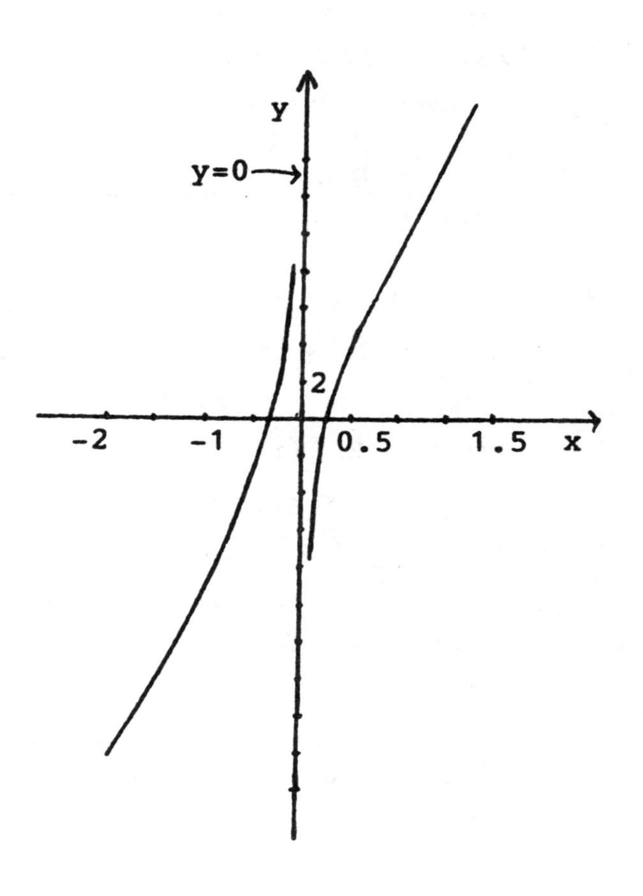

19. $f(x) = x^{5/3} - 5x^{2/3} + 3$.

$f'(x) = \frac{5}{3}x^{2/3} - \frac{10}{3}x^{-1/3} = \frac{5}{3}x^{-1/3}(x-2)$.

$f'(x) = 0$ for $x = 2$.

$f'(x)$ does not exist for $x = 0$.

For $x < 0$, $f'(x) > 0$.

For $0 < x < 2$, $f'(x) < 0$.

For $x > 2$, $f'(x) > 0$.

$f(0)$ is a relative maximum.

$f(2)$ is a relative minimum.

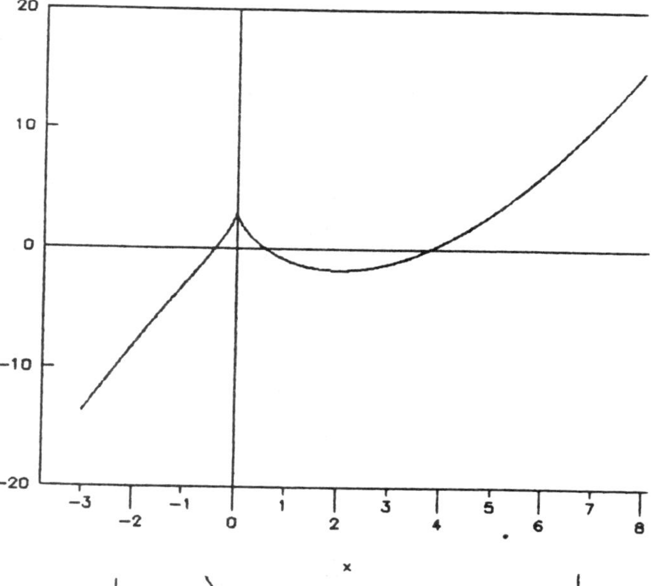

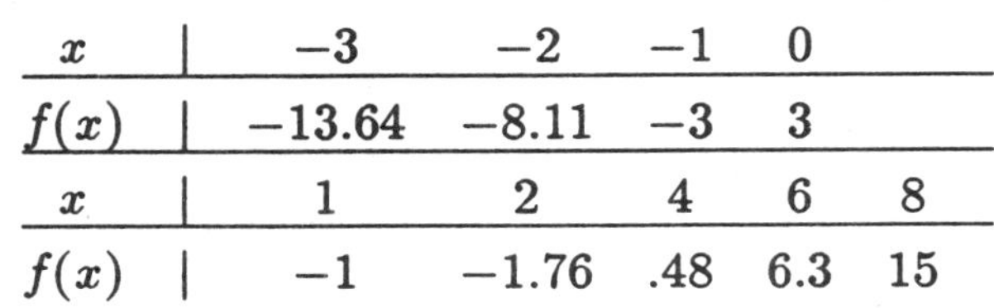

x	-3	-2	-1	0
$f(x)$	-13.64	-8.11	-3	3

x	1	2	4	6	8
$f(x)$	-1	-1.76	$.48$	6.3	15

21. $f(x) = \sqrt{x^2 + 4x + 6}$.

$f'(x) = \frac{1}{2}(x^2 + 4x + 6)^{-1/2} \cdot (2x + 4) = \frac{x+2}{\sqrt{x^2+4x+6}}$.

$f'(x) = 0$ for $x = -2$.

For $x < -2$, $f'(x) < 0$, so $f(x)$ is decreasing.

For $x > -2$, $f'(x) > 0$, so $f(x)$ is increasing.

$f(-2)$ is a relative minimum.

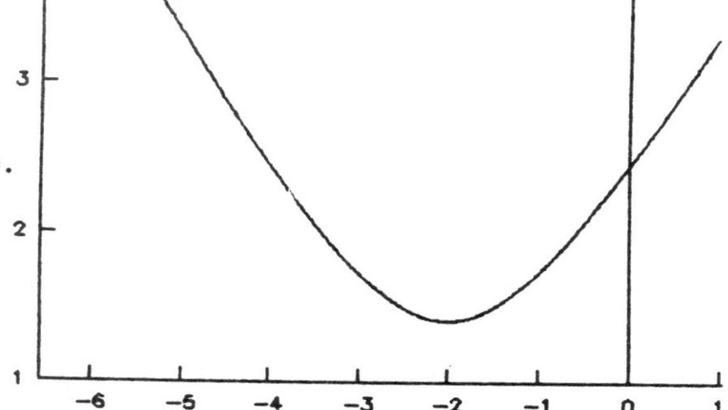

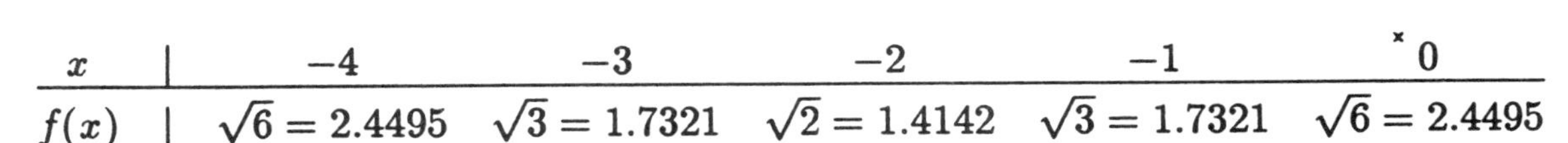

x	-4	-3	-2	-1	0
$f(x)$	$\sqrt{6} = 2.4495$	$\sqrt{3} = 1.7321$	$\sqrt{2} = 1.4142$	$\sqrt{3} = 1.7321$	$\sqrt{6} = 2.4495$

23. $P'(x) = -96 + 36x - 3x^2 = -3\{(x-4)(x-8)\} = 0$ for $x = 4$ and $x = 8$. For $0 < x < 4$, $P'(x) < 0$; for $4 < x < 8$, $P'(x) > 0$; for $8 < x < \infty$, $P'(x) < 0$. Thus, $P(4)$ is a relative minimum and $P(8)$ is a relative maximum.

25. $f'(x) = a$ for all x. If $a = 0$, $f(x) = b$ for all x.

If $a > 0$, $f'(x) > 0$ for all x, so $f(x)$ is always increasing.

If $a < 0$, $f'(x) < 0$ for all x, so $f(x)$ is always decreasing.

27. $f(x) = x^2 - ax + b$.

$f'(x) = 2x - a = 0$ for $x = 2$, so $a = 4$.

29. Let x be one of these numbers. Then the other number is $50 - x$.

Let $f(x)$ be the product of these two numbers.

$f(x) = x \cdot (50 - x) = -x^2 + 50x$.

$f'(x) = -2x + 50$.

$f'(x) = 0$ for $x = 25$.

For $0 < x < 25$, $f'(x) > 0$, so $f(x)$ is increasing.

For $25 < x < 50$, $f'(x) < 0$, so $f(x)$ is decreasing.

Therefore the product is a maximum when the two numbers are both 25.

31. $P(t) = \frac{Kt}{100+t^2}$, $t > 0$.

$P'(t) = K \cdot \frac{(100+t^2)(1)-(t)(2t)}{(100+t^2)^2} = K \cdot \frac{100-t^2}{(100+t^2)^2}$.

$P'(t) = 0$ for $t = 10$.

For $0 < t < 10$, $P'(t) > 0$, so $P(t)$ is increasing.

For $10 < t < 20$, $P'(t) < 0$, so $P(t)$ is decreasing.

The population has a maximum value when $t = 10$.

$P(10) = .05K$.

Selected Solutions to Exercise Set 3.3

1. $f(x) = x^2 + 6x + 9$.

$f'(x) = 2x + 6$. $\quad f''(x) = 2$.

3. $f(x) = x^3 - 4x^2 + 10x - 7$.

$f'(x) = 3x^2 - 8x + 10$. $\quad f''(x) = 6x - 8$.

5. $f(x) = 4x^4 - 3x^3 + 9x^2 - x + 6$.

$f'(x) = 16x^3 - 9x^2 + 18x - 1$. $\quad f''(x) = 48x^2 - 18x + 18$.

7. $f(x) = 9x^3 - x^6$.

$f'(x) = 27x^2 - 6x^5$. $\quad f''(x) = 54x - 30x^4$.

9. $f'(x) = 18x^5 - 24x^3$. $f''(x) = 90x^4 - 72x^2$.

11. $f(x) = 3x^{5/3} - 2x^{-2/3} + x^{-2}$.

$f'(x) = 5x^{2/3} + \frac{4}{3}x^{-5/3} - 2x^{-3}$. $\quad f''(x) = \frac{10}{3}x^{-1/3} - \frac{20}{9}x^{-8/3} + 6x^{-4}$.

13. $f'(x) = \left((x-1)^{-1}\right)' = -(x-1)^{-2} \cdot 1 = -\frac{1}{(x-1)^2}$.
$f''(x) = \left(-(x-1)^{-2}\right)' = 2(x-1)^{-3} = \frac{2}{(x-1)^3}$.

15. $f'(x) = \frac{1}{2}(x+2)^{-1/2}$. $f''(x) = \frac{1}{2}\left(-\frac{1}{2}\right)(x+2)^{-3/2} = -\frac{1}{4}(x+2)^{-3/2}$.

17. $f(x) = 2(x^3-3x)^3$.
$f'(x) = 2 \cdot 3(x^3-3x)^2(3x^2-3) = 18(x^3-3x)^2(x^2-1)$.

$$\begin{aligned} f''(x) &= 18\left[2(x^3-3x)\cdot(3x^2-3)\cdot(x^2-1) + (x^3-3x)^2\cdot(2x)\right] \\ &= 36(x^3-3x)\left[3(x^2-1)^2 + (x^3-3x)\cdot x\right] \\ &= 36(x^3-3x)\cdot\left[3x^4-6x^2+3+x^4-3x^2\right] \\ &= 36(x^3-3x)(4x^4-9x^2+3) \\ &= 36(4x^7-9x^5+3x^3-12x^5+27x^3-9x) \\ &= 36(4x^7-21x^5+30x^3-9x). \end{aligned}$$

19. $f(x) = \frac{2x+1}{9-3x}$.
$f'(x) = \frac{(9-3x)(2)-(2x+1)(-3)}{(9-3x)^2} = \frac{21}{(9-3x)^2}$.
$f''(x) = 21(-2)(9-3x)^{-3}(-3) = \frac{126}{(9-3x)^3}$.

21. $f'(x) = 1\cdot(x-1)^{2/3} + x\cdot\frac{2}{3}(x-1)^{-1/3}\cdot 1 = (x-1)^{2/3} + \frac{2}{3}x(x-1)^{-1/3}$.

$$\begin{aligned} f''(x) &= \frac{2}{3}(x-1)^{-1/3} + \frac{2}{3}\left[1\cdot(x-1)^{-1/3} + x\cdot\left(-\frac{1}{3}\right)(x-1)^{-4/3}\cdot 1\right] \\ &= \frac{4}{3}(x-1)^{-1/3} - \frac{2}{9}x(x-1)^{-4/3}. \end{aligned}$$

23. $f'(x) = 3(1-x^2)^2\cdot(-2x) = -6x(1-x^2)^2$.

$$\begin{aligned} f''(x) &= -6\left[1\cdot(1-x^2)^2 + x\cdot 2(1-x^2)(-2x)\right] \\ &= -6(1-x^2)^2 + 24x^2(1-x^2). \end{aligned}$$

25. $f(x) = x^3 - 3x + 2$.

$f'(x) = 3x^2 - 3$.

$f''(x) = 6x$.

$f''(x) = 0$ for $x = 0$.

For $x < 0$, $f''(x) < 0$, so the graph is concave down.

For $x > 0$, $f''(x) > 0$, so the graph is concave up.

The point $(0, 2)$ is an inflection point.

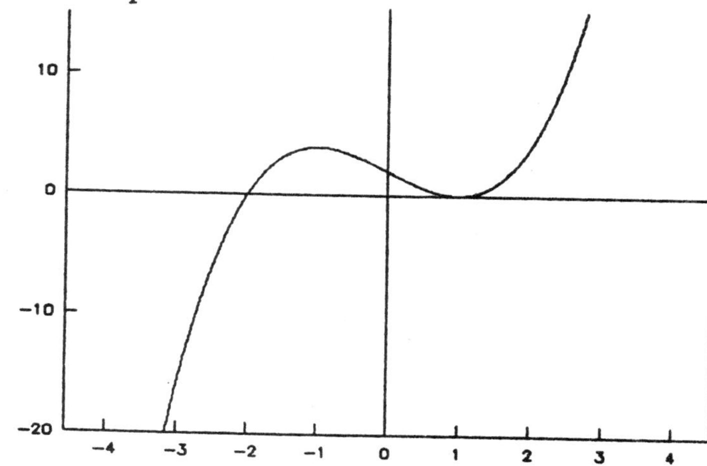

27. $f'(x) = 3x^2 - 18x + 12$. $f''(x) = 6x - 18 = 6(x - 3)$.
$f''(x) = 0$ for $x = 3$.

Interval	Test number t	$f''(t)$	Sign of $f''(x)$	Concavity of the graph of $f(x)$
$(-\infty, 3)$	0	-18	$-$	concave down
$(3, \infty)$	4	6	$+$	concave up

The point $(3, -24)$ is an inflection point.

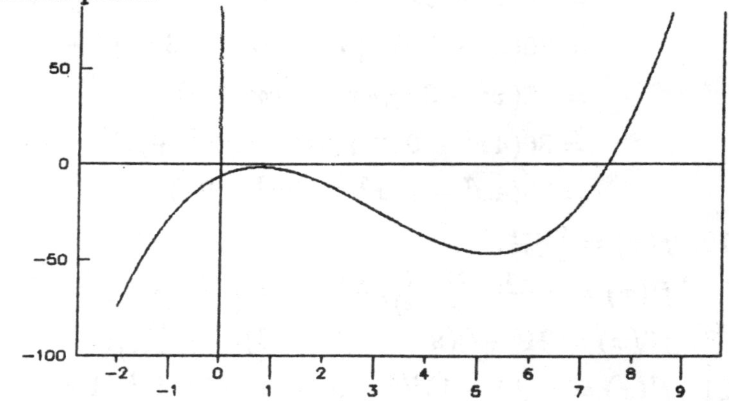

29. $f'(x) = -\frac{1}{x^2}$. $f''(x) = \frac{2}{x^3}$. $f''(x)$ is undefined for $x = 0$.

Intervals	Sign of $f''(x)$	Concavity of the graph of $f(x)$
$(-\infty, 0)$	$-$	concave down
$(0, \infty)$	$+$	concave up

The function $f(x)$ is not defined for $x = 0$. There is no inflection point.

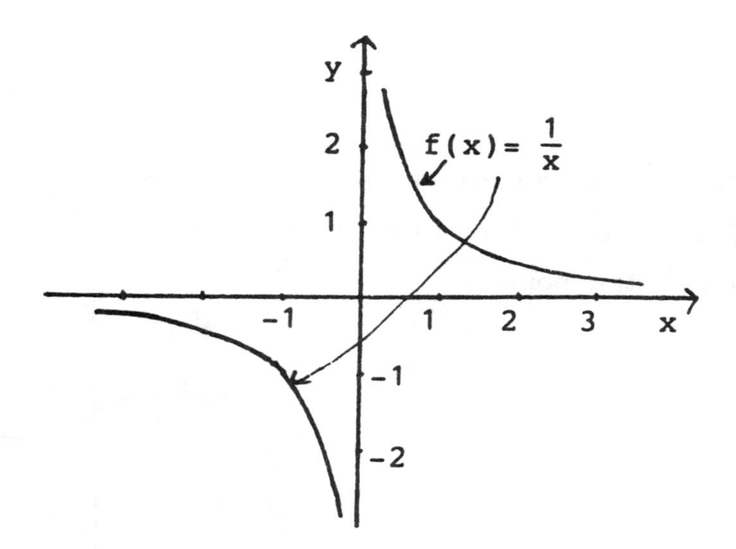

31. $f'(x) = \frac{(x+1)-x}{(x+1)^2} = \frac{1}{(x+1)^2}$. $f''(x) = -\frac{2}{(x+1)^3}$.

$f''(x)$ is undefined for $x = -1$.

On $(-\infty, -1)$, $f''(x) > 0$, the graph of $f(x)$ is concave up.

On $(-1, \infty)$, $f''(x) < 0$, the graph of $f(x)$ is concave down.

$f(x)$ is undefined for $x = -1$. Therefore, it has no inflection point.

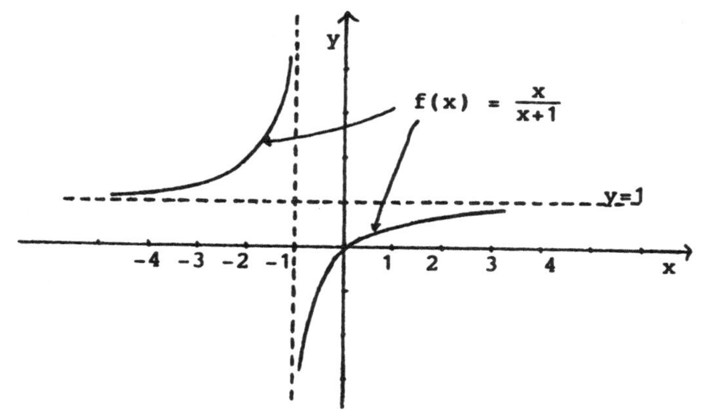

33. $f'(x) = 3(2x+1)^2 \cdot 2 = 6 \cdot (2x+1)^2$.

$f''(x) = 2 \cdot 6(2x+1) \cdot 2 = 48x + 24$.

$f''(x) = 0$ for $x = -\frac{1}{2}$.

Intervals	Sign of $f''(x)$	Concavity of the graph of $f(x)$
$(-\infty, -\frac{1}{2})$	$-$	concave down
$(-\frac{1}{2}, \infty)$	$+$	concave up

The point $(-\frac{1}{2}, 0)$ is an inflection point.

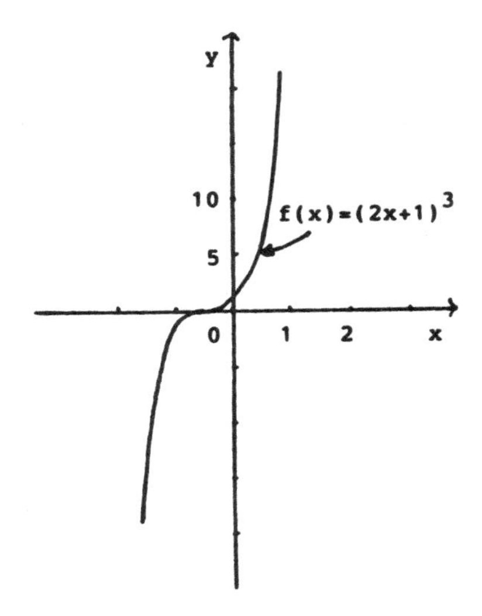

35. $f'(x) = 6x^2 - 6x + 18$. $f''(x) = 12x - 6$.
$f''(x) = 0$ for $x = \frac{1}{2}$.

Intervals	Sign of $f''(x)$	Concavity of the graph of $f(x)$
$(-\infty, \frac{1}{2})$	$-$	concave down
$(\frac{1}{2}, \infty)$	$+$	concave up

The point $(\frac{1}{2}, -\frac{7}{2})$ is an inflection point.

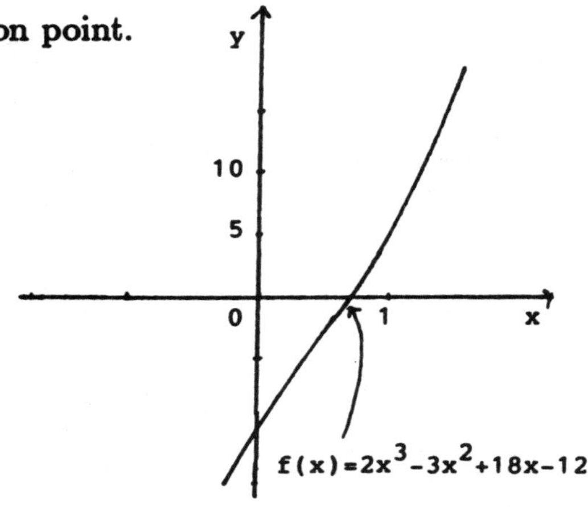

37. $f(x) = (x-2)^{1/3}$. $f'(x) = \frac{1}{3}(x-2)^{-2/3}$. $f''(x) = -\frac{2}{9}(x-2)^{-5/3}$.
$f''(x)$ is undefined for $x = 2$.
On $(-\infty, 2)$, $f''(x) > 0$, the graph of $f(x)$ is concave up.
On $(2, \infty)$, $f''(x) < 0$, the graph of $f(x)$ is concave down.
The point $(2, 0)$ is an inflection point.

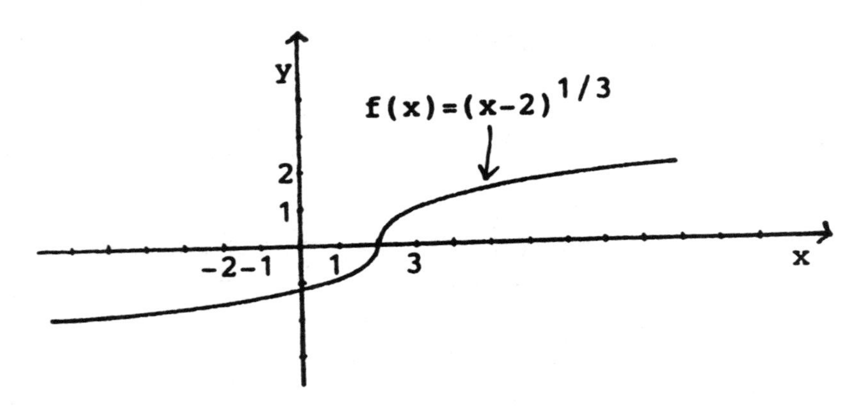

39. $f(x) = \frac{x}{1-x}$. $f'(x) = \frac{1\cdot(1-x)-x\cdot(-1)}{(1-x)^2} = \frac{1}{(1-x)^2}$. $f''(x) = \frac{2}{(1-x)^3}$.
$f''(x)$ is undefined for $x = 1$.
On $(-\infty, 1)$, $f''(x) > 0$, the graph of $f(x)$ is concave up.
On $(1, \infty)$, $f''(x) < 0$, the graph of $f(x)$ is concave down.
$f(x)$ is undefined for $x = 1$. Therefore, it has no inflection point.

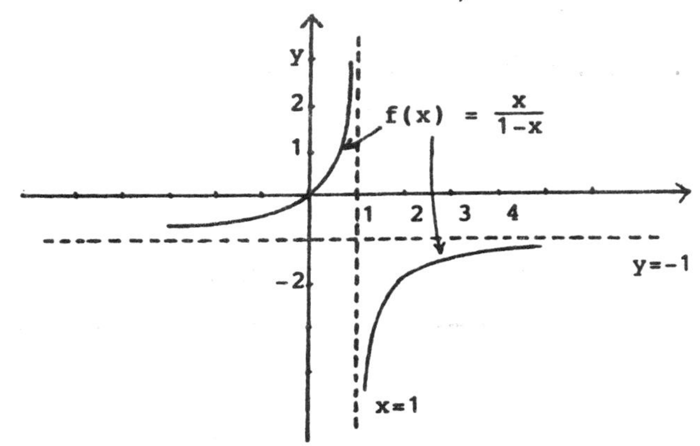

41. $f'(x) = 2x - \frac{2}{x^2}$. $f''(x) = 2 + \frac{4}{x^3}$.
$f'(1) = 2(1) - \frac{2}{1^2} = 0$ and $f''(1) = 2 + \frac{4}{1^3} = 6 > 0$.
So at $x = 1$, $f(x)$ has a relative minimum.

43. $f'(x) = \frac{x-1}{x+1}$. $f''(x) = \frac{1\cdot(x+1)-(x-1)\cdot 1}{(x+1)^2} = \frac{2}{(x+1)^2}$.
$f'(1) = \frac{1-1}{1+1} = 0$ and $f''(1) = \frac{2}{(1+1)^2} = \frac{1}{2} > 0$.
So $f(x)$ has a relative minimum at $x = 1$.

45. $f'(x) = (x^3 - 4x)^{4/3}$. $f''(x) = \frac{4}{3}(x^3 - 4x)^{1/3}(3x^2 - 4)$.
$f'(2) = (2^3 - 4\cdot 2)^{4/3} = 0$ and $f''(2) = \frac{4}{3}(2^3 - 4\cdot 2)^{1/3}(3\cdot 2^2 - 4) = 0$.
The test failed.

47. a. Since $R'(x)$ is decreasing for all $x > 0$, the graph of $y = R(x)$ is concave down on $(0, \infty)$.

b. Since $C'(x)$ is increasing for all $x > 0$, the graph of $y = C(x)$ is concave up on $(0, \infty)$.

49. a. For curve $U_1(x)$ marginal utility is increasing, because the graph of $U_1(x)$ is concave up.

b. The investor whose amount of annual return is $Y = U_1(x)$ knows how to make money work for him, because $U_1(x) > U_2(x)$ for $x > b$.

c. (i) At $x = a$, the investor whose amount of annual return is $Y = U_2(x)$ is more inclined to risk an additional dollar on his investments, because $U_2'(a) > U_1'(a)$.
At $x = b$, the investor whose amount of annual return is $Y = U_1(x)$ is more inclined to risk an additional dollar on his investments, because $U_1'(b) > U_2'(b)$.

Selected Solutions to Exercise Set 3.4

1. $f(x) = \frac{3}{x+5}$
   $\lim_{x\to\infty} f(x) = 0$ and $\lim_{x\to-\infty} f(x) = 0.$
   The line $y = 0$ is a horizontal asymptote.
3. $f(x) = \frac{3x^2}{1+x^2}.$
   $\lim_{x\to\pm\infty} f(x) = \lim_{x\to\pm\infty} \frac{3x^2}{1+x^2} \cdot \frac{\frac{1}{x^2}}{\frac{1}{x^2}} = \lim_{x\to\pm\infty} \frac{3}{\frac{1}{x^2}+1} = \frac{3}{0+1} = 3.$
   The line $y = 3$ is a horizontal asymptote.
5. $f(x) = \frac{x^3}{9-x^3}.$
   $\lim_{x\to\pm\infty} f(x) = \lim_{x\to\pm\infty} \frac{x^3}{9-x^3} \cdot \frac{\frac{1}{x^3}}{\frac{1}{x^3}} = \lim_{x\to\pm\infty} \frac{1}{\frac{9}{x^3}-1} = \frac{1}{0-1} = -1.$
   The line $y = -1$ is a horizontal asymptote.
7. $f(x) = \sqrt{\frac{1+7x^2}{x^2+3}}.$
   $\lim_{x\to\pm\infty} f(x) = \lim_{x\to\pm\infty} \sqrt{\frac{1+7x^2}{x^2+3} \cdot \frac{1/x^2}{1/x^2}} = \lim_{x\to\pm\infty} \sqrt{\frac{(1/x^2)+7}{1+(3/x^2)}} = \sqrt{7}.$
   The line $y = \sqrt{7}$ is a horizontal asymptote.
9. $f(x) = 6 - 4x^{-2/3}.$
   $\lim_{x\to\pm\infty} f(x) = 6 - 0 = 6.$
   The line $y = 6$ is a horizontal asymptote.
11. $x = 4$ is a vertical asymptote since $x - 4 = 0$ for $x = 4$.
13. $x = 2$ and $x = -2$ are vertical asymptotes since $x^2 - 4 = 0$ for both $x = 2$ and $x = -2$.
15. $x = -1$ and $x = -4$ are the vertical asymptotes since $x^2+5x+4 = (x+4)(x+1)$.
17. $f(x) = \frac{x+3}{x^2+x-6} = \frac{(x+3)}{(x+3)(x-2)} = \frac{1}{(x-2)}$ for $x \neq -3$, $x \neq 2$. Thus $x = 2$ is a vertical asymptote.

19. $f(x) = \frac{x^2+5x+6}{x^2-5x+6} = \frac{(x+2)(x+3)}{(x-2)(x-3)}$. Thus $x = 2$ and $x = 3$ are vertical asymptotes.

21. $f(x) = 3 + x + \frac{\sqrt{x}}{1+\sqrt{x}}$.

There is no vertical asymptote and there is no horizontal asymptote.

Note that $1 = \frac{1+\sqrt{x}}{1+\sqrt{x}} = \frac{1}{1+\sqrt{x}} + \frac{\sqrt{x}}{1+\sqrt{x}} \implies \frac{\sqrt{x}}{1+\sqrt{x}} = 1 - \frac{1}{1+\sqrt{x}}$.

Therefore $f(x) = 4 + x - \frac{1}{1+\sqrt{x}}$. Observe that $\frac{1}{1+\sqrt{x}} \to 0$ as $x \to \infty$.

Thus the line $y = 4 + x$ is an oblique asymptote.

23. $f(x) = \frac{2x^3+3x+5}{x^2+1}$.

For $x \neq 0$, $f(x) = \frac{2x^3+3x+5}{x^2+1} \cdot \frac{\frac{1}{x^2}}{\frac{1}{x^2}} = \frac{2x+\frac{3}{x}+\frac{5}{x^2}}{1+\frac{1}{x^2}}$.

$\lim_{x\to\infty} f(x) = \infty$ and $\lim_{x\to-\infty} f(x) = -\infty$.

There is no horizontal asmptote.

$$\begin{array}{r} 2x \\ x^2+1\overline{)2x^3+3x+5} \\ \underline{2x^3+2x} \\ x+5 \end{array}$$

Thus
$$f(x) = 2x + \frac{x+5}{x^2+1}$$

$$\lim_{x\to\pm\infty} \frac{x+5}{x^2+1} = \lim_{x\to\pm\infty} \frac{x+5}{x^2+1} \cdot \frac{\frac{1}{x^2}}{\frac{1}{x^2}} = \lim_{x\to\pm\infty} \frac{\frac{1}{x}+\frac{5}{x^2}}{1+\frac{1}{x^2}}$$
$$= \frac{0+0}{1+0} = 0.$$

The line $y = 2x$ is an oblique asymptote. There are no vertical asymptotes.

25. $f(x) = \frac{2x-2x^3+1}{1-x^2}$.

Because the numerator is a polynomial of higher degree than the polynomial in the denominator, there is no horizontal asymptote.

The lines $x = 1$ and $x = -1$ are vertical asymptotes.

$$\begin{array}{r} 2x \\ -x^2+1\overline{)-2x^3+2x+1} \\ \underline{-2x^3+2x} \\ 1 \end{array}$$

Therefore

$$f(x) = 2x + \frac{1}{-x^2+1}.$$

$$\lim_{x\to\pm\infty} \frac{1}{-x^2+1} = 0.$$

The line $y = 2x$ is an oblique asymptote.

27. It is possible for a function to have both an oblique asymptote and a horizontal asymptote. For example, let

$$f(x) = \begin{cases} \frac{3x^2}{1+x^2} & , \quad x \le -1 \\ \frac{x^2+x+3}{x+1} & , \quad x > -1. \end{cases}$$

As shown in Problem 3, $\lim_{x\to-\infty} f(x) = 3$, so the line $y = 3$ is a horizontal asymptote. As shown in Problem 24, we have that

$$f(x) = x + \frac{3}{x+1} \qquad \text{for} \quad x > -1.$$

So the line $y = x$ is an oblique asymptote.

29. a. $C(x) = 80x + 2500$.

b. $c(x) = \frac{C(x)}{x} = 80 + \frac{2500}{x}$.

c. The line $x = 0$ is a vertical asymptote.

$\lim_{x\to\infty} c(x) = 80 \implies$ the line $y = 80$ is a horizontal asymptote.

31. $C(x) = 2000 + 500x^{1/3} + Ax$.

$\lim_{x\to\infty} c(x) = 35 \implies \lim_{x\to\infty} \frac{2000+500x^{1/3}+Ax}{x} = 35$

$\implies \lim_{x\to\infty} \left(\frac{2000}{x} + \frac{500}{x^{2/3}} + A\right) = 35.$

$\therefore A = 35.$

33. $\lim_{x\to\infty} d(x) = \lim_{x\to\infty} \frac{72+96x^{1/3}}{10x^{1/5}+6x^{1/3}} = \lim_{x\to\infty} \frac{\frac{72}{x^{1/3}}+96}{\frac{10}{x^{2/15}}+6} = 16.$

Thus $y = 16$ is a horizontal asymptote.

Selected Solutions to Exercise Set 3.5

1. $f(x) = 9 - x^2$.
   $f(-x) = 9 - (-x)^2 = 9 - x^2 = f(x)$.
   Therefore f is an even function.
3. $f(x) = x - x^5$.
   $f(-x) = (-x) - (-x)^5 = -x + x^5 = -f(x)$.
   Therefore f is an odd function.
5. $f(x) = x(9 - x^2)$.
   $f(-x) = (-x) \cdot (9 - (-x)^2) = -x(9 - x^2) = -f(x)$.
   Therefore f is an odd function.
7. $f(x) = \frac{1+x^2}{9-x}$.
   $f(-1) = \frac{1+(-1)^2}{9-(-1)} = \frac{2}{10} = \frac{1}{5}$ and $f(1) = \frac{1+1^2}{9-1} = \frac{2}{8} = \frac{1}{4}$, so $f(-1)$ is neither $f(1)$ nor $-f(1)$.
   Therefore f is neither even nor odd.
9. $f(x) = \frac{1-x^{2/3}}{1+x^{4/3}}$.
   $f(-x) = \frac{1-(-x)^{2/3}}{1+(-x)^{4/3}} = \frac{1-x^{2/3}}{1+x^{4/3}} = f(x)$.
   Therefore f is an even function.
11. $f(x) = x^2 - 2x - 8$.
   (1) The domain of f is all real numbers since $f(x)$ is a polynomial.
   (2) f is neither even nor odd. (3) $f(x) = x^2 - 2x - 8 = (x-4)(x+2) = 0$ has two solutions $x = 4$ and $x = -2$.
   (4,5) No vertical asymptote and no horizontal asymptote since $f(x)$ is a polynomial. (6) $f'(x) = 2x - 2 = 0$ for $x = 1$.
   $f'(x) > 0$ for $x > 1$, so f is increasing on $(1, \infty)$.
   $f'(x) < 0$ for $x < 1$, so f is decreasing on $(-\infty, 1)$.
   Hence, $f(1) = -9$ is a relative minimum.
   (7) $f''(x) = 2 > 0$ for all real numbers, so f is concave up on $(-\infty, \infty)$.

(8)

x	-2	-1	0	1	2	3	4
$y = f(x)$	0	-5	-8	-9	-8	-5	0

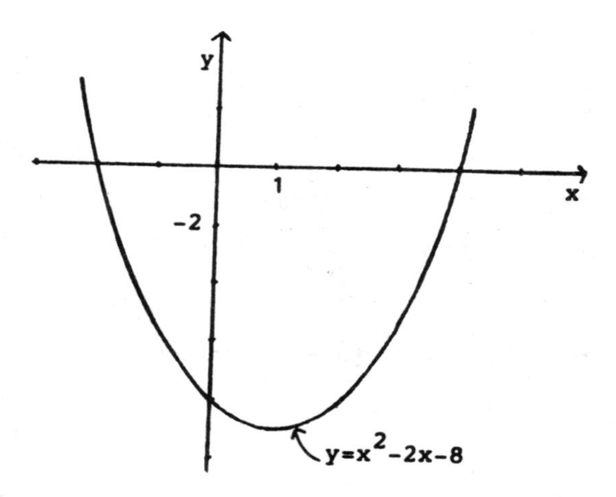

13. $f(x) = 2x^3 - 3x^2$.

(1) The domain of f is all real numbers since $f(x)$ is a polynomial.

(2) f is neither even nor odd.

(3) $f(x) = 2x^3 - 3x^2 = x^2(2x-3) = 0$ for $x = 0$, $x = \frac{3}{2}$.

(4,5) Since $f(x)$ is a polynomial, there are no asymptotes.

(6) $f'(x) = 6x^2 - 6x = 6x(x-1) = 0$ for $x = 0$ and $x = 1$.

Interval	Test number	$f'(x)$	Conclusion
$(-\infty, 0)$	$x = -1$	$f'(-1) > 0$	f increasing
$(0, 1)$	$x = \frac{1}{2}$	$f'\left(\frac{1}{2}\right) < 0$	f decreasing
$(1, \infty)$	$x = 2$	$f'(2) > 0$	f increasing

From these tests $f(0) = 0$ is a relative maximum for f and $f(1) = -1$ is a relative minimum for f.

(7) $f''(x) = 12x - 6 = 0$ for $x = \frac{1}{2}$.
Since $f''(x) > 0$ for $x > \frac{1}{2}$, f is concave up on $\left(\frac{1}{2}, \infty\right)$.
Since $f''(x) < 0$ for $x < \frac{1}{2}$, f is concave down on $\left(-\infty, \frac{1}{2}\right)$.
Thus $\left(\frac{1}{2}, -\frac{1}{2}\right)$ is an inflection point for f.

(8)

x	-1	0	$\frac{1}{2}$	1	2
$y = f(x)$	-5	0	$-\frac{1}{2}$	-1	4

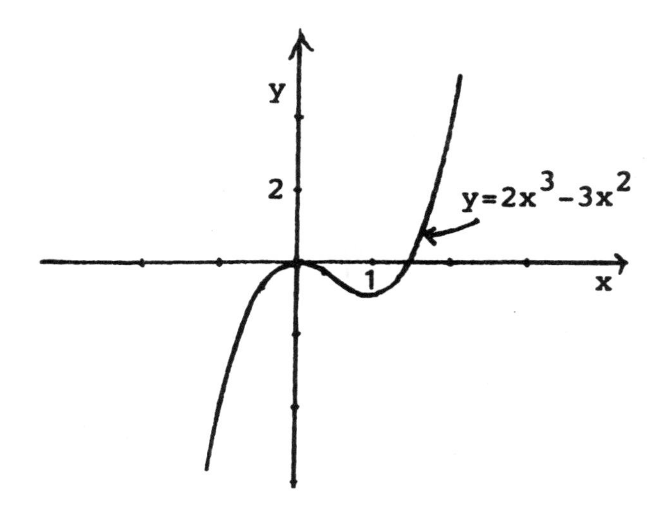

15. $f(x) = x^3 + x^2 - 8x + 8$.

(1) The domain of f is all real numbers since $f(x)$ is a polynomial.

(2) f is neither even nor odd.

(3) It is difficult to find the zeros of $f(x)$, so we skip this step. (4,5) Since $f(x)$ is a polynomial, there are no asymptotes. (6) $f'(x) = 3x^2 + 2x - 8 = (x+2)(3x-4) = 0$ for $x = \frac{4}{3}$ and $x = -2$.

Interval	Test number	$f'(x)$	Conclusion
$(-\infty, -2)$	$x = -3$	$f'(-3) > 0$	f increasing
$(-2, \frac{4}{3})$	$x = 0$	$f'(0) < 0$	f decreasing
$(\frac{4}{3}, \infty)$	$x = 3$	$f'(3) > 0$	f increasing

From these tests, $f(-2) = 20$ is a relative maximum for f and $f\left(\frac{4}{3}\right) = 1.48$ is a relative minimum for f.

(7) $f''(x) = 6x + 2 = 0$ for $x = -\frac{1}{3}$.
$f(x)$ is concave up on $\left(-\frac{1}{3}, \infty\right)$ since $f''(x) > 0$ on $\left(-\frac{1}{3}, \infty\right)$.
$f(x)$ is concave down on $\left(-\infty, -\frac{1}{3}\right)$ since $f''(x) < 0$ on $\left(-\infty, -\frac{1}{3}\right)$.
$\therefore \left(-\frac{1}{3}, 10.74\right)$ is an inflection point for f.

(8)

x	-3	-2	$-\frac{1}{3}$	0	$\frac{4}{3}$	2
$f(x)$	14	20	10.74	8	1.48	4

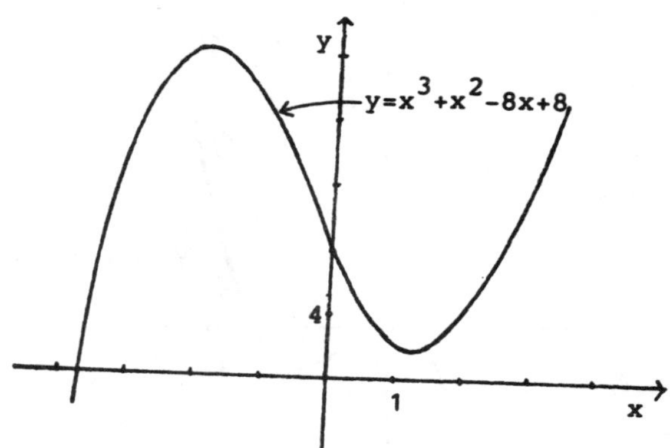

17. $y = f(x) = \frac{x+4}{x-4}$.

(1) The domain of f is all real numbers except $x = 4$ since $(x-4) = 0$.

(2) f is neither even nor odd.

(3) $f(x) = 0$ for $x + 4 = 0 \implies x = -4$.

(4) $x = 4$ is a vertical asymptote.

(5) $y = 1$ is a horizontal asymptote since $\lim_{x\to\pm\infty} \frac{x+4}{x-4} = 1$.

(6) $f'(x) = \frac{(x-4)-(x+4)}{(x-4)^2} = \frac{-8}{(x-4)^2} < 0$ for all real numbers except $x = 4$, so that f is decreasing on both $(-\infty, 4)$ and $(4, \infty)$.

(7) $f''(x) = \frac{8\cdot 2\cdot(x-4)}{(x-4)^4} = \frac{16}{(x-4)^3}$, from which $f''(x) < 0$ on $(-\infty, 4)$ and $f''(x) > 0$ on $(4, \infty)$. Hence $f(x)$ is concave up on $(4, \infty)$ and concave down on $(-\infty, 4)$. There is no inflection point.

(8)

x	-10	-4	0	1	3	3.5	4.5	5	10
$f(x)$	$\frac{3}{7}$	0	-1	$-\frac{5}{3}$	-7	-15	17	9	$\frac{7}{3}$

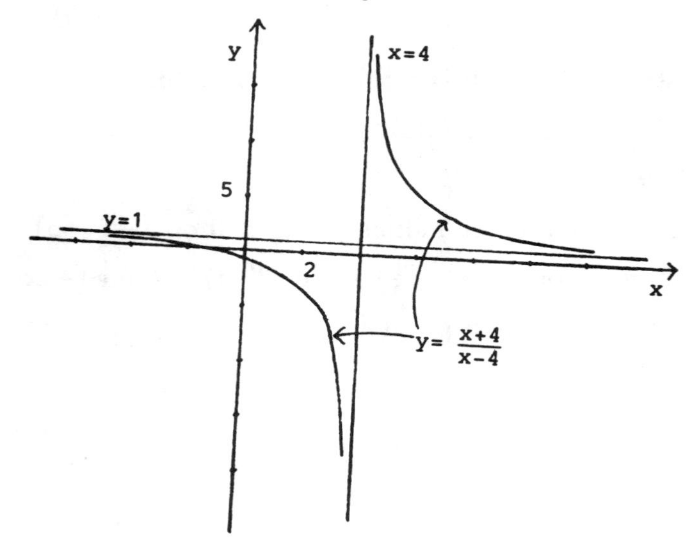

19. $f(x) = |4 - x^2|$.

This was discussed and drawn for Problem 38 in Exercise Set 3.3.

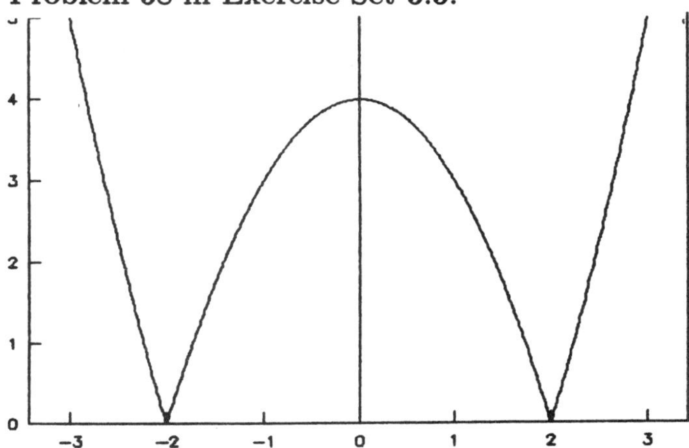

21. $y = \frac{1}{x(x-4)}$.

(1) The domain is all x except $x = 0$ and $x = 4$.

(2) The function f is neither even nor odd. The straight line $x = 2$ is an axis of symmetry for the graph of f.

(3) $y = \frac{1}{x(x-4)} \neq 0$ for all x in the domain of f.

(4) Vertical asymptotes are $x = 0$ and $x = 4$.

(5) Horizontal asymptote is $y = 0$ since $\lim_{x \to \pm\infty} \frac{1}{x(x-4)} = 0$.

(6) $y' = -1(x^2 - 4x)^{-2} \cdot (2x - 4) = (-2x + 4) \cdot (x^2 - 4x)^{-2}$.

$y' = 0$ for $x = 2$, and y' does not exist for $x = 0$ and $x = 4$.

Interval	Test Number	$f'(x)$	Conclusion
$(-\infty, 0)$	-1	$f'(-1) > 0$	increasing
$(0, 2)$	1	$f'(1) > 0$	increasing
$(2, 4)$	3	$f'(3) < 0$	decreasing
$(4, \infty)$	5	$f'(5) < 0$	decreasing

$f(2)$ is a relative maximum.

(7) $y'' = (-2)(x^2 - 4x)^{-2} + (-2x + 4)(-2)(x^2 - 4x)^{-3}(2x - 4)$

$= -2(x^2 - 4x)^{-3}\{(x^2 - 4x) + (-2x + 4)(2x - 4)\}$

$= -2(x^2 - 4x)^{-3}\{-3x^2 + 12x - 16\}$.

y'' has no zeros, and y'' does not exist for $x = 0$ and $x = 4$.

Interval	Test Number	$f''(x)$	Conclusion
$(-\infty, 0)$	-1	$f''(-1) > 0$	concave up
$(0, 4)$	1	$f''(1) < 0$	concave down
$(4, \infty)$	5	$f''(5) > 0$	concave up

There is no inflection point.

(8)

x	-5	-2	-1	1	2	3	5	7	10
y	$\frac{1}{45}$	$\frac{1}{12}$	$\frac{1}{5}$	$-\frac{1}{3}$	$-\frac{1}{4}$	$-\frac{1}{3}$	$\frac{1}{5}$	$\frac{1}{21}$	$\frac{1}{60}$

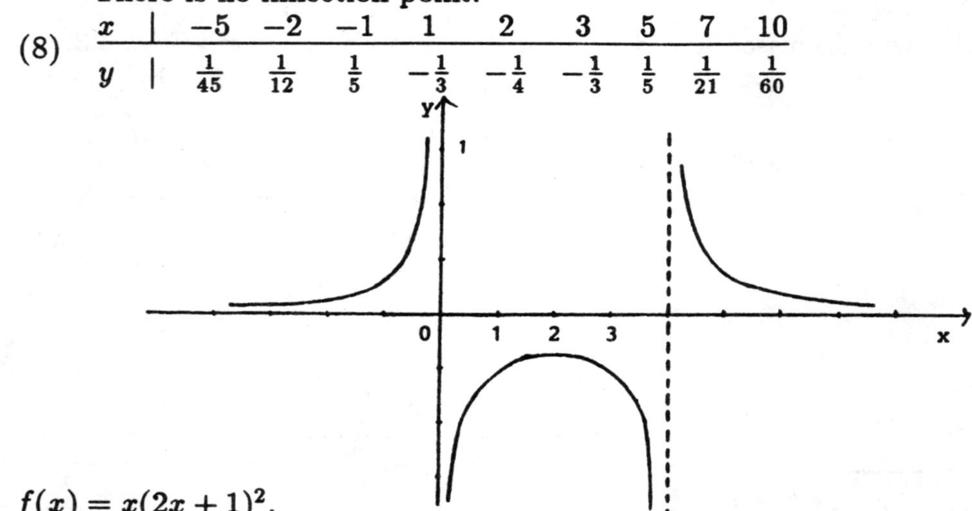

23. $f(x) = x(2x+1)^2$.

(1) The domain of f is all real numbers since $f(x)$ is a polynomial.

(2) f is neither even nor odd.

(3) $f(x) = 0$ for $x = 0$ and $x = -\frac{1}{2}$.

(4,5) There are no asymptotes.

(6) $f'(x) = (1)(2x+1)^2 + (x)(2(2x+1)(2)) = (2x+1)((2x+1)+(4x))$

$= (2x+1)(6x+1) = 12x^2 + 8x + 1$.

$f'(x) = 0$ for $x = -\frac{1}{6}$ and $x = -\frac{1}{2}$.

For $-\infty < x < -\frac{1}{2}$, $f'(x) > 0$, so $f(x)$ is increasing.

For $-\frac{1}{2} < x < -\frac{1}{6}$, $f'(x) < 0$, so $f(x)$ is decreasing.

For $-\frac{1}{6} < x < \infty$, $f'(x) > 0$, so $f(x)$ is increasing.

$f\left(-\frac{1}{2}\right) = 0$ is a relative maximum.

$f\left(-\frac{1}{6}\right) = -\frac{2}{27} = -.0741$ is a relative minimum.

(7) $f''(x) = 24x + 8$.

$f''(x) = 0$ for $x = -\frac{1}{3}$.

For $x < -\frac{1}{3}$, $f''(x) < 0$; the graph of f is concave down on the interval $\left(-\infty, -\frac{1}{3}\right)$.

For $x > -\frac{1}{3}$, $f''(x) > 0$; the graph of f is concave up on the interval $\left(-\frac{1}{3}, \infty\right)$.

The point $\left(-\frac{1}{3}, f\left(-\frac{1}{3}\right) = -\frac{1}{27}\right)$ is an inflection point.

(8)

x	-3	-2	-1	$-\frac{1}{2}$	$-\frac{1}{6}$	0	1	2
$y = f(x)$	-75	-18	-1	0	$-\frac{2}{27}$	0	9	50

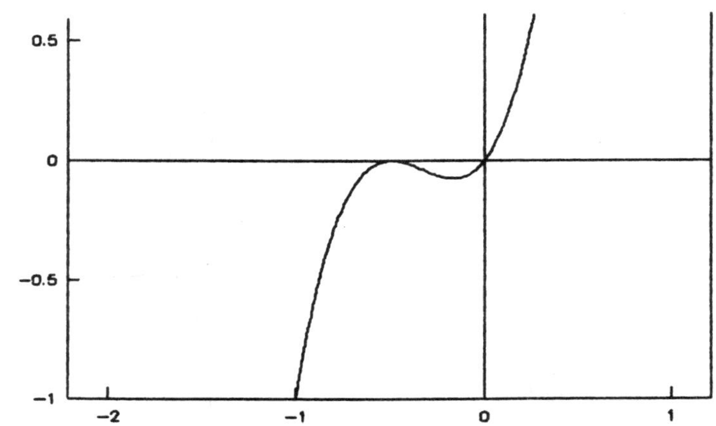

25. $y = \frac{1}{3}x^3 - x^2 - 3x + 4$.

(1) The domain is all x.

(2) f is neither even nor odd.

(3) It is difficult to determine for what x we have $y = 0$.

(4,5) There is no vertical asymptote and no horizontal asymptote.

(6) $y' = x^2 - 2x - 3 = (x-3)(x+1)$.

$y' = 0$ for $x = -1$ and $x = 3$.

Interval	Test Number	$f'(x)$	Conclusion
$(-\infty, -1)$	-2	$f'(-2) > 0$	increasing
$(-1, 3)$	0	$f'(0) < 0$	decreasing
$(3, \infty)$	4	$f'(4) > 0$	increasing

$f(-1)$ is a relative maximum and $f(3)$ is a relative minimum.

(7) $y'' = 2x - 2$. $y'' = 0$ for $x = 1$.

Interval	Test Number	$f''(x)$	Conclusion
$(-\infty, 1)$	0	$f''(0) < 0$	concave down
$(1, \infty)$	2	$f''(2) > 0$	concave up

$(1, \frac{1}{3})$ is the inflection point.

(8)

x	-3	-2	-1	0	1	2	3	4	5
y	-5	$\frac{10}{3}$	$\frac{17}{3}$	4	$\frac{1}{3}$	$-\frac{10}{3}$	-5	$-\frac{8}{3}$	$\frac{17}{3}$

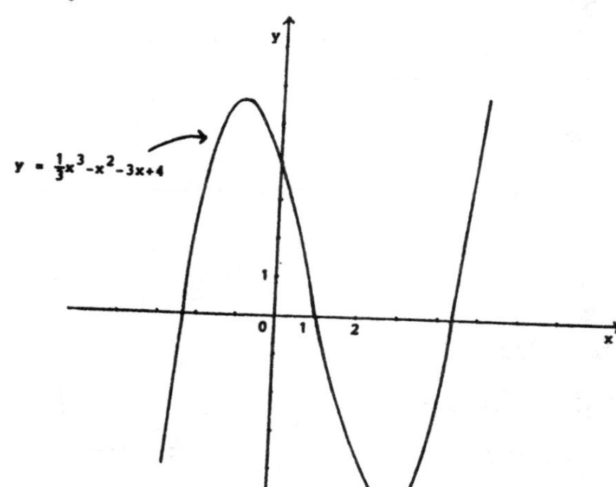

27. $y = f(x) = 4x^2(1 - x^2)$.

(1) The domain is all x.

(2) f is an even function. (3) $f(x) = 0$ for $x = 0$, $x = 1$, $x = -1$.

(4,5) There are no asymptotes.

(6) $f'(x) = 4x^2(-2x) + (1 - x^2)(8x) = -8x^3 + 8x - 8x^3$
$= -16x^3 + 8x = 8x(-2x^2 + 1)$.

$f'(x) = 0$ for $x = 0$, $x = \pm\sqrt{\frac{1}{2}}$.

Interval	Test Number	$f'(x)$	Conclusion
$\left(-\infty, -\sqrt{\frac{1}{2}}\right)$	-1	$f'(-1) > 0$	increasing
$\left(-\sqrt{\frac{1}{2}}, 0\right)$	$-\frac{1}{2}$	$f'\left(-\frac{1}{2}\right) < 0$	decreasing
$\left(0, \sqrt{\frac{1}{2}}\right)$	$\frac{1}{2}$	$f'\left(\frac{1}{2}\right) > 0$	increasing
$\left(\sqrt{\frac{1}{2}}, \infty\right)$	1	$f'(1) < 0$	decreasing

$f\left(-\sqrt{\frac{1}{2}}\right)$ and $f\left(\sqrt{\frac{1}{2}}\right)$ are relative maxima and $f(0)$ is a relative minimum.

(7) $f''(x) = -48x^2 + 8$. $f''(x) = 0$ for $x = \pm\sqrt{\frac{1}{6}}$.

Interval	Test Number	$f''(x)$	Conclusion
$\left(-\infty, -\sqrt{\frac{1}{6}}\right)$	$-\frac{1}{2}$	$f''\left(-\frac{1}{2}\right) < 0$	concave downward
$\left(-\sqrt{\frac{1}{6}}, \sqrt{\frac{1}{6}}\right)$	0	$f''(0) > 0$	concave upward
$\left(\sqrt{\frac{1}{6}}, \infty\right)$	$\frac{1}{2}$	$f''\left(\frac{1}{2}\right) < 0$	concave downward

$\left(-\sqrt{\frac{1}{6}}, \frac{5}{9}\right)$ and $\left(\sqrt{\frac{1}{6}}, \frac{5}{9}\right)$ are inflection points.

(8)

x	$\pm\frac{3}{2}$	± 1	$\pm\sqrt{\frac{1}{2}}$	0
$f(x)$	$-\frac{45}{4}$	0	1	0

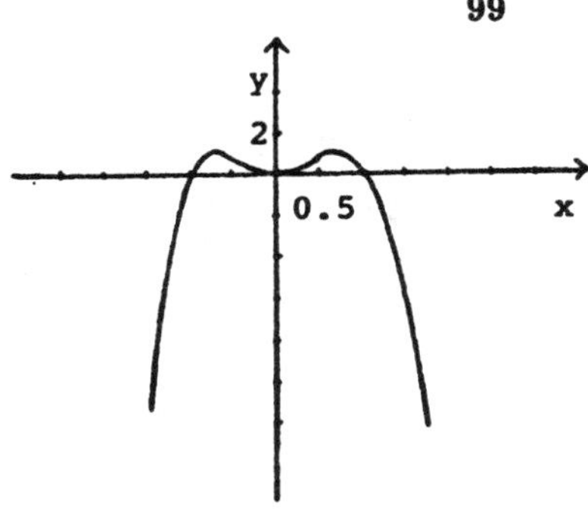

29. $f(x) = 16 - 20x^3 + 3x^5$.

(1) The domain is all x.

(2) This function f is neither even nor odd.

Note that

$$f(-x) - 16 = \left(16 - 20(-x)^3 + 3(-x)^5\right) - 16 = 20x^3 - 3x^5 = -f(x) + 16$$
$$= -\left[f(x) - 16\right].$$

Therefore the graph of f is symmetric with respect to the point (0,16).

(3) It is difficult to find the zeros of $f(x)$. (4,5) There are no asymptotes.

(6) $f'(x) = -60x^2 + 15x^4 = 15x^2(-4 + x^2)$.

$f'(x) = 0$ for $x = 0$, $x = 2$, $x = -2$.

Interval	Test Number	$f'(x)$	Conclusion
$(-\infty, -2)$	-3	$f'(-3) > 0$	increasing
$(-2, 0)$	-1	$f'(-1) < 0$	decreasing
$(0, 2)$	1	$f'(1) < 0$	decreasing
$(2, \infty)$	3	$f'(3) > 0$	increasing

$f(-2)$ is a relative maximum. $f(2)$ is a relative minimum.

(7) $f''(x) = -120x + 60x^3 = 60x(-2 + x^2)$.

$f''(x) = 0$ for $x = 0$, $x = \sqrt{2}$, $x = -\sqrt{2}$.

Interval	Test Number	$f''(x)$	Conclusion
$(-\infty, -\sqrt{2})$	-2	$f''(-2) < 0$	concave downward
$(-\sqrt{2}, 0)$	-1	$f''(-1) > 0$	concave upward
$(0, \sqrt{2})$	1	$f''(1) < 0$	concave downward
$(\sqrt{2}, \infty)$	2	$f''(2) > 0$	concave upward

The inflection points occur when $x = 0$, $x = \sqrt{2}$, $x = -\sqrt{2}$.

(8)

x	-3	-2	-1	0	1	2	3
$f(x)$	-173	80	33	16	-1	-48	205

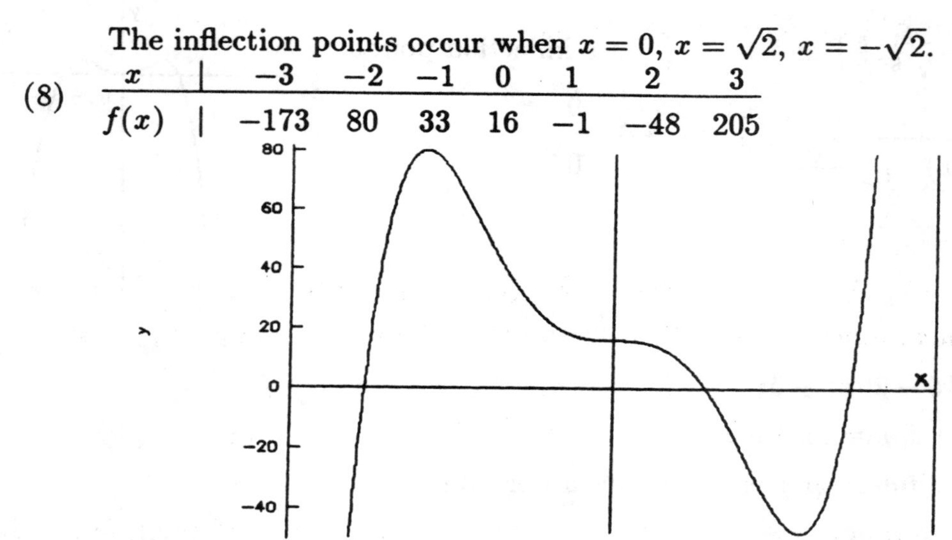

31. The function $f(x) = 9x - \frac{1}{x}$ in Problem 16 is an odd function, but the graph of f does not contain the origin.

33. Let $g(x) = x^2 + 1$ for all real numbers x. Let $h(x) = \sqrt{g(x)}$ for all real numbers x. For every real number x, $h(-x) = \sqrt{g(-x)} = \sqrt{g(x)} = h(x)$. Therefore h is an even function.

35. Let $g(x)$ and $h(x)$ be two odd functions defined, for simplicity, on the entire x-axis, and let $f(x) = g(x) \cdot h(x)$ for all real numbers x.
For any real number x, $f(-x) = g(-x) \cdot h(-x) = [-g(x)] \cdot [-h(x)] = g(x) \cdot h(x) = f(x)$. Therefore f is an even function.

37. Let $f(x) = (x-a)^2 + b$. For any real number h, $f(a+h) = f(a-h)$, so the graph of f is symmetric with respect to the vertical line $x = a$.

39. $f(x) = x^5 + 2$. Let $g(x) = f(x) - 2 = x^5$. The graph of g is symmetric with respect to the origin, so the graph of f is symmetric with respect to the point (0,2).

Selected Solutions to Exercise Set 3.6

1. $f(x) = 9 - x^2, \qquad x \text{ in } [-1, 3].$
   $f'(x) = -2x.$
   $f'(x) = 0$ for $x = 0$.
   $f(-1) = 9 - (-1)^2 = 8, \quad f(3) = 9 - (3)^2 = 0, \quad f(0) = 9.$
   The maximum value is $f(0) = 9$, and the minimum value is $f(3) = 0$.
3. $f'(x) = 2x - 2$. $f'(x) = 0$ for $x = 1$.
   $f(-1) = (-1)^2 - 2(-1) + 3 = 6$, $f(1) = 1^2 - 2 \cdot 1 + 3 = 2$,
   and $f(3) = 9 - 6 + 3 = 6$.
   The maximum value of f is $f(-1) = f(3) = 6$ and the minimum value is $f(1) = 2$.
5. $f(x) = 3 + x + x^2 + 2x^3, \qquad x \text{ in } [-2, 3].$
   $f'(x) = 1 + 2x + 6x^2.$
   $f'(x) > 0$ for all x.
   $f(-2) = -11, \qquad f(3) = 69.$
   The maximum value if $f(3) = 69$, and the minimum value is $f(-2) = -11$.
7. $f'(x) = 2x(x-1) + x^2 = x(3x - 2)$. $f'(x) = 0$ for $x = 0$ and $x = \frac{2}{3}$.
   $f(0) = 0$, $f\left(\frac{2}{3}\right) = \frac{4}{9}\left(-\frac{1}{3}\right) = -\frac{4}{27}$, and $f(3) = 9 \cdot 2 = 18$.
   The maximum value of f is $f(3) = 18$ and the minimum value is $f\left(\frac{2}{3}\right) = -\frac{4}{27}$.
9. $f'(x) = 3x^2 - 4x$. $f'(x) = 0$ for $x = 0$ and $x = \frac{4}{3}$.
   $f(-1) = (-1)^3 - 2(-1)^2 = -3$, $f(0) = 0$, $f\left(\frac{4}{3}\right) = \left(\frac{4}{3}\right)^3 - 2\left(\frac{4}{3}\right)^2 = -\frac{32}{27}$, $f(2) = 0$.
   The maximum value of f is $f(0) = f(2) = 0$ and the minimum value is $f(-1) = -3$.
11. $f'(x) = 1 - \frac{1}{x^2}$. $f'(x) = 0$ for $x = \pm 1$.
    $f\left(\frac{1}{2}\right) = \frac{5}{2}$, $f(1) = 2$, and $f(2) = \frac{5}{2}$.
    The maximum value of f is $f\left(\frac{1}{2}\right) = f(2) = \frac{5}{2}$ and the minimum value is $f(1) = 2$.
13. $f'(x) = \frac{2}{3x^{1/3}}$.
    The only critical number for f is $x = 0$.

$f(0) = 2$. There is no maximum value for $f(x)$ since $f(x)$ is increasing.

The minimum value is $f(0) = 2$.

15. $f'(x) = \frac{4-2x}{(x^2-4x)^2}$. $f'(x) = 0$ for $x = 2$.

$f(2) = -\frac{1}{4}$, $f(1) = -\frac{1}{3}$, and $f(3) = -\frac{1}{3}$.

The maximum value of f is $f(2) = -\frac{1}{4}$, and the minimum value is $f(1) = f(3) = -\frac{1}{3}$.

17. $f(x) = 2x^3 - 15x^2 + 24x + 10, \qquad x$ in $[0, 4]$.

$f'(x) = 6x^2 - 30x + 24 = 6(x^2 - 5x + 4) = 6(x-1)(x-4)$.

$f'(x) = 0$ for $x = 1, \quad x = 4$.

$f(0) = 10, \quad f(4) = -6, \quad f(1) = 21$.

The maximum value is $f(1) = 21$, and the minimum value is $f(4) = -6$.

19. $f'(x) = \left(\sqrt{x} - x^{3/2}\right)' = \frac{1}{2}x^{-1/2} - \frac{3}{2}x^{1/2} = \frac{1}{2}x^{-1/2}(1-3x)$.

$f'(x) = 0$ for $x = \frac{1}{3}$ and $f'(x)$ does not exist for $x = 0$.

$f(0) = 0$, $f\left(\frac{1}{3}\right) = \sqrt{\frac{1}{3}} \cdot \frac{2}{3} = \frac{2\sqrt{3}}{9}$, $f(4) = -6$.

The maximum value of f is $f\left(\frac{1}{3}\right) = \frac{2\sqrt{3}}{9}$ and the minimum value is $f(4) = -6$.

21. $P(t) = 100,000 + 48t^{3/2} - 4t^2$, t in $[0, \infty)$.

$P'(t) = 72t^{1/2} - 8t = 8t^{1/2}(9 - t^{1/2})$. $P'(t) = 0$ for $t = 0$, $t = 81$.

On $(0, 81)$, $P'(t) > 0$, $P(t)$ is increasing.

On $(81, \infty)$, $P'(t) < 0$, $P(t)$ is decreasing.

Therefore the maximum population occurs when $t = 81$ which is the year 2066.

23. $P'(x) = -96 + 36x - 3x^2 = -3(x-4)(x-8)$.

$P'(x) = 0$ for $x = 4$, $x = 8$.

$P(0) = 160$, $P(4) = 0$, $P(8) = 32$, $P(10) = 0$.

The maximum value of P is $P(0) = 160$ which occurs at $x = 0$, and the minimum value is $P(4) = P(10) = 0$ which occurs at $x = 4$ and $x = 10$.

25. Let $R(p)$ = revenue from the sales per week at price p.

$R(p) = xp = (400 - 4p)p = 400p - 4p^2$, p in $[0, 100]$.

$R'(p) = 400 - 8p$.

$R'(p) = 0$ for $p = 50$.

$R(0) = 0$, $R(50) = 10,000$, $R(100) = 0$.

The maximum revenue of 10,000 occurs when $p = 50$.

Selected Solutions to Exercise Set 3.7

1. $p(t) = \sqrt{24 + 10t - t^2}$, t in $[0, 10]$.

   $p'(t) = \frac{1}{2}(24 + 10t - t^2)^{-1/2}(10 - 2t) = \frac{5-t}{(24+10t-t^2)^{1/2}}$.

   $p'(t) = 0$ for $t = 5$.

   $p(0) = \sqrt{24}$, $p(5) = 7$, $p(10) = \sqrt{24}$.

   Therefore this percentage will be the largest when $t = 5$ which is the year 1990.

3. $R'(t) = 0.08 - 0.04t = 0.04(2 - t)$.

   $R'(t) = 0$ for $t = 2$.

   For $0 < t < 2$, $R'(t) > 0$, $R(t)$ is increasing.

   For $2 < t < \infty$, $R'(t) < 0$, $R(t)$ is decreasing.

   Therefore efficiency is maximum when $t = 2$ hours.

5. Let two nonnegative numbers be x and y.

   $x + y = 36$, so $y = 36 - x$.

   $S(x) = x + (36 - x)^2 = x + 1296 - 72x + x^2 = x^2 - 71x + 1296$.

   $S'(x) = 2x - 71 = 0$ for $x = 35.5$.

   $S(0) = 1296$, $S(35.5) = 35.75$, $S(36) = 36$.

   The maximum of $S(x)$ occurs when $x = 0$, so the numbers are 0 and 36.

7. Let N be the number of people in the population.

   Let $I(t)$ be the number of people infected after time t.

   Let $R(t)$ be the rate at which the disease is spreading after time t.

   There is a positive constant k such that

$$R(t) = k \cdot I(t) \cdot \{N - I(t)\}.$$

$$\begin{aligned} R'(t) &= k \cdot \{I'(t) \cdot [N - I(t)] + I(t) \cdot [-I'(t)]\} \\ &= k\{I'(t)N - 2 \cdot I'(t)I(t)\} \\ &= k \cdot I'(t) \cdot \{N - 2 \cdot I(t)\}. \end{aligned}$$

$$R'(t) = 0 \quad \text{for} \quad N - 2 \cdot I(t) = 0 \implies I(t) = \frac{N}{2}.$$

   The disease is spreading most rapidly when $I(t) = \frac{N}{2}$, because $R'(t) > 0$ when $I(t) < \frac{N}{2}$ and $R'(t) < 0$ when $I(t) > \frac{N}{2}$, and because $I(t)$ increases as t increases.

9. $f(x) = x^3 - 9x^2 + 7x - 6$.

   $f'(x) = 3x^2 - 18x + 7$

   $f''(x) = 6x - 18 = 0$ for $x = 3$.

   $f'(1) = -8$. $f'(3) = -20$, $f'(4) = -17$.

   The maximum slope if $f'(1) = -8$. The minimum slope is $f'(3) = -20$.

11. $R(p) = p \cdot x = p(2000 - 100p) = 2000p - 100p^2$.

    $R'(p) = 2000 - 200p = 0$ for $p = 10$.

    $R(0) = 0$, $R(10) = 10,000$, $R(20) = 0$.

    The maximum revenue occurs for $p = 10$.

13. Let p be the daily rental fee in dollars.

    Let x be the number of daily rentals of jackhammers per year.

    Let R be the revenue in dollars.

$$x = 500 - 10(p - 30) = 800 - 10p.$$

$$R = x \cdot p = (800 - 10p)p = 800p - 10p^2.$$

$$R' = 800 - 20p = 0 \quad \text{for} \quad p = 40.$$

$$R(30) = 15,000, \quad R(40) = 16,000, \quad R(80) = 0.$$

The maximum revenue is for $p = \$40$.

15. $\frac{6-x}{y} = \frac{6}{8} \implies 6y = 48 - 8x$, $y = 8 - \frac{4}{3}x$.

    The area of the rectangle $A(x) = x \cdot y = x\left(8 - \frac{4}{3}x\right) = 8x - \frac{4}{3}x^2$.

    $A'(x) = 8 - \frac{8}{3}x = 0$ for $x = 3$.

    $A(0) = 0$, $A(3) > 0$, $A(6) = 0$.

    The maximum area occurs when $x = 3$ cm and $y = 4$ cm.

17. (Area of window) $= w \cdot h + \frac{1}{2}r^2\pi$.

    $r = \frac{w}{2} \implies$ (Area of window) $= wh + \frac{1}{2}\pi\left(\frac{w}{2}\right)^2$

    $= wh + \frac{1}{8}\pi w^2$.

    $2h + w + \frac{1}{2}w\pi = 10 \implies 4h + (2 + \pi)w = 20$.

    $h = \frac{20-(2+\pi)w}{4}$.

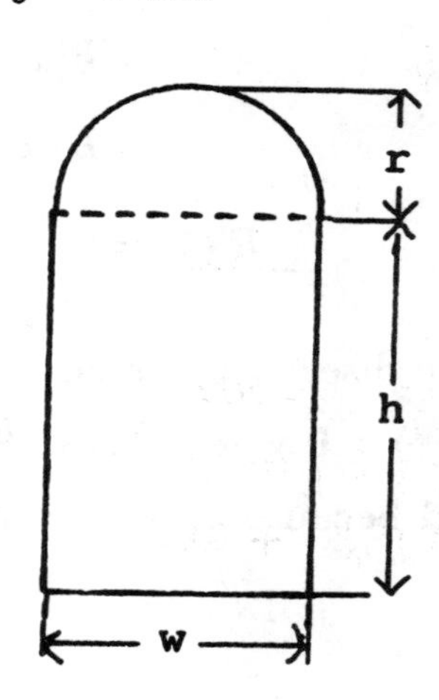

Therefore, (Area of window) $= w \cdot \left(\frac{20-(2+\pi)w}{4}\right) + \frac{1}{8}\pi w^2$

$$= 5w - \frac{(2+\pi)}{4}w^2 + \frac{1}{8}\pi w^2$$
$$= w^2\left(\frac{\pi}{8} - \frac{1}{2} - \frac{\pi}{4}\right) + 5w$$
$$= w^2\left(-\frac{1}{2} - \frac{\pi}{8}\right) + 5w = f(w).$$

$f'(w) = \left(-1 - \frac{\pi}{4}\right)w + 5 = 0$ for $w = \frac{5}{1+\frac{\pi}{4}} = \frac{20}{4+\pi}$.

$f'(w) > 0$ for $0 < w < \frac{20}{4+\pi}$, and $f'(w) < 0$ for $\frac{20}{4+\pi} < w < \frac{20}{2+\pi}$.

So maximum area occurs when $w = \frac{20}{(4+\pi)}$ and

$$h = \left(20 - (2+\pi)\cdot\frac{20}{(4+\pi)}\right)\cdot\frac{1}{4} = 5\left(1 - \frac{2+\pi}{4+\pi}\right) = \frac{10}{4+\pi} \text{ meters.}$$

19. $f(t) = \frac{6t}{t^2+2t+1}$, $t > 0$.

$f'(t) = \frac{6(t^2+2t+1)-6t(2t+2)}{(t^2+2t+1)^2} = \frac{6(t+1)-12t}{(t+1)^3} = \frac{-6t+6}{(t+1)^3} = \frac{-6(t-1)}{(t+1)^3}$.

So $f'(t) = 0$ for $t = 1$.

Interval	Test Number	$f'(t)$	Conclusion
$(0,1)$	$\frac{1}{2}$	$f'\left(\frac{1}{2}\right) > 0$	increasing
$(1,\infty)$	2	$f'(2) < 0$	decreasing

Thus, $f(t)$ has a maximum when $t = 1$ minute.

21. $U(x) = 2 - (x-8)^{2/3}$, $x \geq 0$.

$U'(x) = -\frac{2}{3}(x-8)^{-1/3} = \frac{-2}{3\sqrt[3]{x-8}}$.

$U'(x) > 0$ for $0 \leq x < 8$, and $U'(x) < 0$ for $x > 8$.

Hence $U(x)$ has a maximum when $x = 8$.

23.

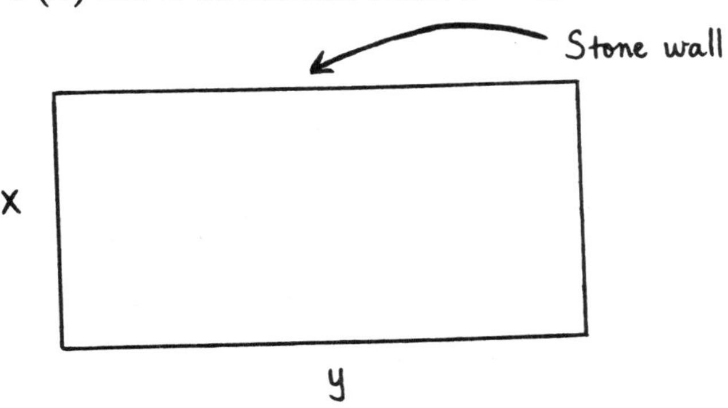

The cost $C = 50y + 2(20x) + 20y = 70y + 40x$.

$x \cdot y = 800 \implies y = \frac{800}{x}$.

$C = 70\left(\frac{800}{x}\right) + 40x = \frac{56000}{x} + 40x$.

$\frac{dC}{dx} = \frac{-56000}{x^2} + 40 \implies x^2 = 1400 \implies x = 10\sqrt{14}.$

For $0 < x < 10\sqrt{14}$, $\frac{dC}{dx} < 0$, so C decreases as x increases.

For $x > 10\sqrt{14}$, $\frac{dC}{dx} > 0$, so C increases as x increases.

Therefore the minimum cost C occurs when $x = 10\sqrt{14}$ and $y = \frac{800}{10\sqrt{14}} = \frac{800\sqrt{14}}{140} = \frac{40\sqrt{14}}{7}$.

25. Let P denote the annual profit for the bank.

$P = .10(50,000r) - (100,000r)r = -100,000r^2 + 5,000r.$

$\frac{dP}{dr} = -200,000r + 5,000.$

$\frac{dP}{dr} = 0$ for $r = .025$.

For $0 < r < .025$, $\frac{dP}{dr} > 0$, so P increases as r increases.

For $r > .025$, $\frac{dP}{dr} < 0$, so P decreases as r increases.

Therefore the profit on P is maximized when the interest rate paid on savings deposits is 2.5% per year.

27. The cost of enrolling n students in the summer program is

$$C(n) = 3n^2 + 2,000n + 7,500.$$

The average cost per child is

$$c(n) = \frac{C(n)}{n} = 3n + 2,000 + \frac{7,500}{n}.$$

$$c'(n) = 3 - \frac{7500}{n^2}.$$

$$c'(n) = 0 \text{ for } \frac{7500}{n^2} = 3 \implies n^2 = 2500 \implies n = 50.$$

For $0 < n < 50$, $c'(n) < 0$, so $c(n)$ decreases as n increases.

For $n > 50$, $c'(n) > 0$, so $c(n)$ increases as n increases.

Therefore the smallest average cost per child occurs when the number of children is 50.

29. The revenue R is given by

$$R = p \cdot x = 10x - .002x^2.$$

The profit P is given by

$$\begin{aligned} P = R - C &= (10x - .002x^2) - (2,000 + 2x) \\ &= -.002x^2 + 8x - 2,000. \\ \frac{dP}{dx} &= -.004x + 8. \\ \frac{dP}{dx} &= 0 \text{ for } x = 2,000. \end{aligned}$$

For $0 < x < 2,000$, $\frac{dP}{dx} > 0$, so P increases as x increases.
For $2,000 < x < 5,000$, $\frac{dP}{dx} < 0$, so P decreases as x increases.
Therefore the maximum profit occurs when 2,000 sets of cards are produced.

31. Let x be the number of additional trees per acre.

Let y be the yield per tree $\Longrightarrow$ $y = 300 - 5x$.

(Yield per acre) $= (x + 30)y = (x + 30)(300 - 5x) = f(x)$.

$f(x) = (30 + x)(300 - 5x) = 9000 + 150x - 5x^2$.

$f'(x) = 150 - 10x$. $f'(x) = 0$ for $x = 15$.

$f(0) = 9000$, $f(15) = 10,125$, $f(60) = 0$.

So, 15 additional trees per acre should be planted to maximize yield.

33. $v(\rho) = \frac{100}{1+\rho^2}$.

$\rho(t) = -10(t - 3)(t - 7)$, t in $[3, 7]$.

$\widehat{v}(t) = v(\rho(t))$.

$\widehat{v}'(t) = v'(\rho) \cdot \rho'(t) = \frac{-100(2\rho)}{(1+\rho^2)^2} \cdot [-10(t - 3 + t - 7)]$

$= \frac{-20000(t-3)(t-7)(2t-10)}{(1+100(t-3)^2(t-7)^2)^2}$.

$\widehat{v}'(t) = 0$ for $t = 3$, $t = 7$, and $t = 5$.

$\widehat{v}(3) = v(\rho(3)) = 100$, $\widehat{v}(5) = \frac{100}{1601} = .0625$, $\widehat{v}(7) = 100$.

The maximum velocity is 100 km/hr when $t = 3$ p.m. and $t = 7$ p.m., and the minimum velocity is 0.0625 km/hr when $t = 5$ p.m.

35. Let $y =$ the number of days it takes to complete the job.

Let $x =$ the number of overtime hours worked during each of the first $(y-1)$ days.

Let $z =$ the number of regular hours worked during the yth day.

Let $w =$ the number of overtime hours worked during the yth day.

$$8(y-1) + x(y-1) + z + w = 200.$$

$$\text{Cost } C = 80(y-1) + 15x(y-1) + 10z + 15w + 200(y-1).$$

$$\begin{aligned} 8(y-1) + x(y-1) + z + w &= 200 \\ (8+x)(y-1) &= 200 - z - w \\ y - 1 &= \frac{200-z-w}{8+x} \end{aligned}$$

$$\begin{aligned} C &= 80\left(\frac{200-z-w}{8+x}\right) + 15x\left(\frac{200-z-w}{8+x}\right) + 10z + 15w + 200\left(\frac{200-z-w}{8+x}\right) \\ &= (280+15x)\left(\frac{200-z-w}{8+x}\right) + 10z + 15w. \end{aligned}$$

Think of z and w as being given and that C then depends only on x. Then

$$\begin{aligned} \frac{d}{dx}C &= (200-z-w)\left(\frac{15(8+x) - 1(280+15x)}{(8+x)^2}\right) \\ &= (200-z-w)\frac{120 + 15x - 280 - 15x}{(8+x)^2} \\ &= (200-z-w)\left(\frac{-160}{(8+x)^2}\right). \end{aligned}$$

Since $z \leq 8$ and $w \leq 4$, it follows that $\frac{d}{dx}C < 0$ for $0 \leq x \leq 4$. Therefore the minimum value of C occurs when $x = 4$.

$$y - 1 = \frac{50}{3} - \frac{z+w}{12}.$$

For $(y-1)$ to be an integer, we must have $z = 8$ and $w = 0$. Thus, the minimum cost is obtained by putting in 17 days, and having the worker put in 4 hours of overtime during each of the first 16 days and no overtime on the 17th day.

37. The total cost of laying the cable as in the figure

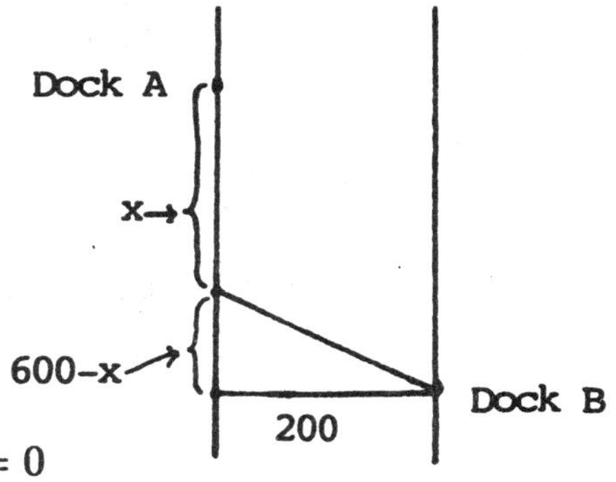

$$= 30x + 50 \cdot \sqrt{(600-x)^2 + 200^2}$$

$$= 30x + 50 \cdot \sqrt{x^2 - 1200x + 400,000} = f(x).$$

Dock A

x

600−x

200

Dock B

$$f'(x) = 30 + 25 \cdot \frac{2x-1200}{\sqrt{x^2-1200x+400,000}}$$

$$= \frac{30\sqrt{x^2-1200x+400,000}+50x-30,000}{\sqrt{x^2-1200x+400,000}}.$$

$f'(x) = 0$ for $30\sqrt{x^2 - 1200x + 400,000} + 50x - 30,000 = 0$

$\Longrightarrow 3\sqrt{x^2 - 1200x + 400,000} = 3000 - 5x \Longrightarrow 9(x^2 - 1200x + 400,000)$

$= 9,000,000 - 30,000x + 25x^2 \Longrightarrow 9x^2 - 10800x + 3,600,000$

$= 9,000,000 - 30000x + 25x^2 \Longrightarrow 16x^2 - 19,200x + 5,400,000 = 0$

$\Longrightarrow x^2 - 1200x + 337,500 = 0$

$$\Longrightarrow x = \frac{1200 \pm \sqrt{(1200)^2 - 4(337,500)}}{2} = \frac{1200 \pm \sqrt{1,440,000 - 1,350,000}}{2}$$

$$= \frac{1200 \pm \sqrt{90,000}}{2} = \frac{1200 \pm 300}{2} = \frac{900}{2} \text{ or } \frac{1500}{2}$$

$$= 450 \text{ or } 750.$$

$f'(x) < 0$ for $0 < x < 450$ and $f'(x) > 0$ for $450 < x < 600$. Hence the total cost of laying the cable is minimized when 450 m of cable are laid on the land.

39. Here $f(x) = k \cdot d(x) \cdot p(x) = k \cdot \frac{1}{x} \cdot \frac{1}{(x-10)^2}$,

where k is a constant and $1 \le x \le 9$.

$$f'(x) = -\frac{k(3x^2-40x+100)}{x^2(x-10)^4} = -\frac{k(3x-10)(x-10)}{x^2(x-10)^4} = 0 \text{ for } x = \tfrac{10}{3}.$$

$f'(x) < 0$ for $0 < x < \frac{10}{3}$ and $f'(x) > 0$ for $\frac{10}{3} < x < 9$.

Hence the survival rate is a minimum when the distance from the parent tree is $\frac{10}{3}$ m.

Selected Solutions to Exercise Set 3.8

1. $P(x) = \left(100x - \frac{1}{4}x^2\right) - (500 + 4x)$

   $= -\frac{1}{4}x^2 + 96x - 500,\ 0 \leq x \leq 200.$

   $P'(x) = -\frac{1}{2}x + 96 = 0$ for $x = 192$.

   Since $P'(x) > 0$ for $x < 192$ and $P'(x) < 0$ for $x > 192$, $P(x)$ has a maximum at $x = 192$.

3. $P(x) = 30x - (500 + 20x + x^2)$

   $= -x^2 + 10x - 500,\ 0 \leq x \leq 100.$

   $P'(x) = -2x + 10 = 0$ for $x = 5$.

   Since $P'(x) > 0$ for $x < 5$ and $P'(x) < 0$ for $x > 5$, $P(x)$ has a maximum at $x = 5$.

5. Maximum Profit in 1: $P(192) = 8716$

   Maximum Profit in 2: $P(50) = 250$

   Maximum Profit in 3: $P(5) = -475$.

   Maximum Profit in 4: $P(40) = 1480$.

   Companies in Exercises 1, 2, and 4 are profitable.

7. Let x be the number of times per year the store orders bulbs.

   Yearly cost $= 3000 \cdot \frac{1}{2x}(.48) + 20x + 150 = 720\frac{1}{x} + 20x + 150 = f(x)$.

   $f'(x) = -720\frac{1}{x^2} + 20$. $f'(x) = 0$ for $x = 6$.

   For $0 < x < 6$, $f'(x) < 0$; for $x > 6$, $f'(x) > 0$.

   The minimum value of $f(x)$ occurs for $x = 6$. Hence the store should order bulbs 6 times a year.

9. Let x be the number of times per year the merchant orders shirts.

   Yearly costs $= 2000 \cdot \frac{1}{2x} \cdot 3 + 50x + 400 = 3000 \cdot \frac{1}{x} + 50x + 400 = f(x)$.

   $f'(x) = -\frac{3000}{x^2} + 50 = 0$ for $x^2 = 60 \implies x = 7.746$.

   For $x < 7.746$, $f'(x) < 0$; for $x > 7.746$, $f'(x) > 0$.

   $f(x)$ is a minimum for $x = 7.746$.

   $f(7) = 1178.57$ and $f(8) = 1175$, so $f(8) < f(7)$.

   Thus the merchant should order shirts 8 times a year.

11. The profit $P(x) = R(x) - C(x) = (200x - 4x^2) - (900 + 40x) = -4x^2 + 160x - 900$.

    $P'(x) = -8x + 160 = 0$ for $x = 20$.

    $P(0) = -900$, $P(20) = 700$, $P(50) = -2900$.

    The maximum profit occurs when $x = 20$.

13. a. $R(x) = x \cdot p = x(600 - 3x) = -3x^2 + 600x$.

    b. $P(x) = R(x) - C(x)$

    $$= (-3x^2 + 600x) - (400 + 150x + 0.5x^2)$$

    $$= -3.5x^2 + 450x - 400.$$

    c. $C'(x) = 150 + x$. $R'(x) = -6x + 600$.

    d. $P'(x) = -7x + 450$.

    $P'(x) = 0$ for $x = \frac{450}{7} = 64.29$.

    For $0 < x < \frac{450}{7}$, $P'(x) > 0$; for $\frac{450}{7} < x < 100$, $P'(x) < 0$.

    $P(64) = 14{,}064$, $P(65) = 14{,}062.50$.

    Hence, the maximum profit occurs when the weekly production level is 64 dishwashers.

    e. $C'(64) = 214$. $R'(64) = 216$.

15. $c(x) = \frac{C(x)}{x}$. $c'(x) = \frac{x \cdot C'(x) - C(x)}{x^2}$.

    When the average cost is a minimum, $[x \cdot C'(x) - C(x)] = 0 \implies C'(x) = \frac{C(x)}{x}$, so marginal cost = average cost.

17. $P(x) = p \cdot x - C(x) - t \cdot x$ where x = number of items

$C(x)$ = total cost of producing these x items

$P(x)$ = profit from sales of these items

t = per item tax imposed on the manufacturer

Suppose C is differentiable and the range of C' contains the open interval $(0, p)$. Suppose $C'(x)$ is an increasing function of x. The output level which maximizes profit is given by $P'(x) = p - C'(x) - t = 0 \implies C'(x) = p - t$. Let t_1 and t_2 be per item taxes such that $t_2 > t_1$, and let x_1 and x_2 be the output levels for tax t_1 and tax t_2, respectively, which maximize profit. Then

$$C'(x_1) = p - t_1$$

$$C'(x_2) = p - t_2.$$

Thus, $C'(x_2) - C'(x_1) = t_1 - t_2 < 0 \implies C'(x_2) < C'(x_1) \implies x_2 < x_1$.

19. Now, $P(x) = -3.5x^2 + 485x - 400$.

$P'(x) = -7x + 485 = 0$ for $x = \frac{485}{7} = 69.3$.

$P(69) = 16{,}401.5, \quad P(70) = 16{,}400$.

Therefore the manufacturer should manufacture 69 dishwashers per week.

21. $E(p) = \frac{-pQ'(p)}{Q(p)} = \frac{-p \cdot (-\frac{5}{4})}{-\frac{5}{4}p + 40}$.

$E(20) = \frac{-20 \cdot (-\frac{5}{4})}{-\frac{5}{4} \cdot (20) + 40} = \frac{25}{15} = \frac{5}{3}$.

23. $E(p) = \frac{-pQ'(p)}{Q(p)} = \frac{-p \cdot (\frac{1}{2})(1000-4p)^{-1/2}(-4)}{(1000-4p)^{1/2}} = \frac{2p}{(1000-4p)}$.

$E(40) = \frac{2(40)}{1000 - 4 \cdot 40} = \frac{80}{840} = \frac{2}{21}$.

25. $E(p) = \frac{-p(-\frac{1}{2}p^{-3/2})}{p^{-1/2}} = \frac{1}{2}$.

27. $E(p) = \frac{-p(-20)}{(p+2)^3} \Big/ \frac{10}{(p+2)^2} = \frac{2p}{p+2}$. $\quad E(2) = 1$.

29. $E(p) = -p\left(\frac{-2}{p^3} \Big/ \frac{1}{p^2}\right) = 2$.

The curve is elastic for all $p > 0$.

31. $E(p) = \frac{-p\left((1+p)\cdot\frac{1}{2\sqrt{p}} - \sqrt{p}\right)}{(1+p)^2} \Big/ \frac{\sqrt{p}}{1+p}$

$= \frac{-p\left(\frac{1+p}{2\sqrt{p}} - \sqrt{p}\right)}{(1+p)^2} \cdot \frac{1+p}{\sqrt{p}}$

$= \frac{-\sqrt{p}\cdot\left(\frac{1+p}{2\sqrt{p}} - \sqrt{p}\right)}{(1+p)} = \frac{p - \frac{1+p}{2}}{1+p}$

$= \frac{2p-(1+p)}{2(1+p)} = \frac{p-1}{2(p+1)}.$

Then $E(p) < 1 \implies p - 1 < 2(p+1) \implies p > -3$. So the curve is inelastic for all $p > 1$.

33. $Q = 200 - p \implies p = 200 - Q$.

$R(Q) = pQ = 200Q - Q^2$.

$P(Q) = 200Q - Q^2 - (2Q + 100) = -Q^2 + 198Q - 100$.

$P'(Q) = -2Q + 198 = 0$ for $Q = 99$.

a. The profit is maximized when $Q = 99$.

The resulting price is $p = 200 - 99 = 101$.

b. $E(p) = \frac{-p\cdot Q'(p)}{Q(p)} = \frac{-p(-1)}{200-p}$. $E(101) = \frac{101}{99}$. The demand is elastic.

35. a. $x = Q(p) = \sqrt{1000 - 10p}$.

b. $E(p) = \frac{-pQ'(p)}{Q(p)} = \frac{-p\left(\frac{1}{2}\right)(1000-10p)^{-1/2}\cdot(-10)}{\sqrt{1000-10p}} = \frac{5p}{1000-10p}$.

$E(50) = \frac{5(50)}{1000-10(50)} = \frac{250}{500} = \frac{1}{2} < 1$.

Therefore the demand is inelastic when $p = 50$.

c. $E(70) = \frac{5(70)}{1000-10(70)} = \frac{350}{300} = \frac{7}{6} > 1$.

Therefore the demand is elastic when $p = 70$.

d. $E(p) = 1 \implies 5p = 1000 - 10p \implies 15p = 1000 \implies p = 66.67$.

Selected Solutions to Exercise Set 3.9

1. $x^2 + y^2 = 9$. $2x + 2y\frac{dy}{dx} = 0$, $2y\frac{dy}{dx} = -2x$,

$\frac{dy}{dx} = \frac{-2x}{2y} = \frac{-x}{y}$.

3. $xy^2 = 6.$ $\quad x \cdot 2y\frac{dy}{dx} + y^2 \cdot 1 = 0,$ $\quad 2xy\frac{dy}{dx} = -y^2,$

$\frac{dy}{dx} = \frac{-y^2}{2xy} = \frac{-y}{2x}.$

5. $x^2 + 2xy + y^2 = 8.$ $\quad 2x + 2x \cdot \frac{dy}{dx} + y \cdot 2 + 2y\frac{dy}{dx} = 0,$

$2x\frac{dy}{dx} + 2y\frac{dy}{dx} = -2x - 2y,$ $\quad (2x + 2y)\frac{dy}{dx} = -(2x + 2y),$

$\frac{dy}{dx} = \frac{-(2x+2y)}{(2x+2y)} = -1.$

7. $x^3 + 2xy^2 + x^2y + 2y^3 = 5.$

$3x^2 + 2y^2 + 2x\left(2y\frac{dy}{dx}\right) + 2xy + x^2\frac{dy}{dx} + 6y^2\frac{dy}{dx} = 0.$

$(4xy + x^2 + 6y^2)\frac{dy}{dx} = -3x^2 - 2y^2 - 2xy.$

$\frac{dy}{dx} = \frac{-3x^2-2y^2-2xy}{4xy+x^2+6y^2}.$

9. $x^3 + 2x^2y + 5xy^2 - y^2 = 10.$

$3x^2 + 4xy + 2x^2\frac{dy}{dx} + 5y^2 + 5x\left(2y\frac{dy}{dx}\right) - 2y\frac{dy}{dx} = 0.$

$(2x^2 + 10xy - 2y)\frac{dy}{dx} = -3x^2 - 4xy - 5y^2.$

$\frac{dy}{dx} = \frac{-3x^2-4xy-5y^2}{2x^2+10xy-2y}.$

11. $xy - y^4 + \sqrt{xy} = 10.$

$y + x\frac{dy}{dx} - 4y^3\frac{dy}{dx} + \frac{1}{2}(xy)^{-1/2} \cdot \left(y + x\frac{dy}{dx}\right) = 0.$

$\left(x - 4y^3 + \frac{1}{2}(xy)^{-1/2} \cdot (x)\right)\frac{dy}{dx} = -y - \frac{1}{2}(xy)^{-1/2} \cdot (y).$

$\frac{dy}{dx} = \frac{-y-\frac{1}{2}(xy)^{-1/2}(y)}{x-4y^3+\frac{1}{2}(xy)^{-1/2}(x)}.$

13. $\sqrt{x} - xy + x^{2/3}y^{4/3} = 30.$

$\frac{1}{2}x^{-1/2} - y - x\frac{dy}{dx} + \frac{2}{3}x^{-1/3}y^{4/3} + x^{2/3} \cdot \frac{4}{3}y^{1/3}\frac{dy}{dx} = 0.$

$\left(-x + \frac{4}{3}x^{2/3}y^{1/3}\right)\frac{dy}{dx} = -\frac{1}{2}x^{-1/2} + y - \frac{2}{3}x^{-1/3}y^{4/3}.$

$\frac{dy}{dx} = \frac{-\frac{1}{2}x^{-1/2}+y-\frac{2}{3}x^{-1/3}y^{4/3}}{-x+\frac{4}{3}x^{2/3}y^{1/3}}.$

15. $xy = 9.$ $\quad x\frac{dy}{dx} + y = 0,$ $\quad \frac{dy}{dx} = -\frac{y}{x}.$

At $(3,3)$, $\frac{dy}{dx} = \frac{-3}{3} = -1.$

17. $x^3 + y^3 = 16.$ $\quad 3x^2 + 3y^2\frac{dy}{dx} = 0,$ $\quad \frac{dy}{dx} = \frac{-3x^2}{3y^2} = \frac{-x^2}{y^2}.$

At $(2,2)$, $\frac{dy}{dx} = \frac{-4}{4} = -1.$

19. $\frac{x+y}{x-y} = 4.$ $\frac{(x-y)(1+\frac{dy}{dx})-(x+y)(1-\frac{dy}{dx})}{(x-y)^2} = 0,$

$2x\frac{dy}{dx} - 2y = 0,$ $\frac{dy}{dx} = \frac{2y}{2x} = \frac{y}{x}.$

At (5,3), $\frac{dy}{dx} = \frac{3}{5}.$

21. $\sqrt{x} + \sqrt{y} = 4.$ $\frac{1}{2}x^{-1/2} + \frac{1}{2}y^{-1/2}\frac{dy}{dx} = 0,$ $\frac{1}{2y^{1/2}} \cdot \frac{dy}{dx} = \frac{-1}{2x^{1/2}},$

$\frac{dy}{dx} = \frac{-2y^{1/2}}{2x^{1/2}} = \frac{-y^{1/2}}{x^{1/2}}.$

At (4,4), $\frac{dy}{dx} = \frac{-\sqrt{4}}{\sqrt{4}} = -1.$

Hence, -1 is the desired slope. The equation for the line tangent to the graph is $(y-4) = -1(x-4)$ where $y = -x + 8$.

23. $y^3 - x^2 = 7.$ $3y^2\frac{dy}{dx} - 2x = 0,$ $\frac{dy}{dx} = \frac{2x}{3y^2}.$

At (1,2), $\frac{dy}{dx} = \frac{2(1)}{3(2^2)} = \frac{1}{6}.$ Hence, $\frac{1}{6}$ is the desired slope.

25. $A = \pi r^2.$ $\frac{dA}{dt} = \pi \cdot \frac{dr^2}{dt} = \pi \cdot 2r\frac{dr}{dt} = 2\pi r\frac{dr}{dt}.$

When $r = 10$ and $\frac{dr}{dt} = 2$,

$\frac{dA}{dt} = 2\pi \cdot 10 \cdot 2 = 40\pi \cdot m^2/\text{sec}.$

27. $N(t) = I(t) + S(t),$ $\frac{dI}{dt} = 24,$ and $\frac{dS}{dt} = -20.$

$\frac{dN}{dt} = \frac{dI}{dt} + \frac{dS}{dt} = 24 + (-20) = 4$ persons/day.

29. $2x\frac{dx}{dt} + 5\frac{dx}{dt}p + 5x\frac{dp}{dt} + 2p\frac{dp}{dt} = 0.$

$2(20)\frac{dx}{dt} + 5(10)\frac{dx}{dt} + 5(20)(-1) + 2(10)(-1) = 0.$

$90\frac{dx}{dt} - 120 = 0 \implies \frac{dx}{dt} = \frac{4}{3}.$

31. $5\frac{dx}{dt} + 5\frac{dx}{dt}p + 5x\frac{dp}{dt} + 2p\frac{dp}{dt} = 0.$

$5\frac{dx}{dt} + 5(40)\frac{dx}{dt} + 5(20)(.20) + 2(40)(.20) = 0.$

$205\frac{dx}{dt} + 36 = 0.$

$\frac{dx}{dt} = \frac{-36}{205}.$

33. $2x\frac{dx}{dp} + 4\frac{dx}{dp}p + 4x + 4 = 0.$

$2(40)\frac{dx}{dp} + 4(100)\frac{dx}{dp} + 4(40) + 4 = 0.$

$480\frac{dx}{dp} + 164 = 0.$

$\frac{dx}{dp} = -\frac{164}{480} = -\frac{41}{120}.$ $E(p) = -\frac{p\frac{dx}{dp}}{x}.$

$E(100) = -\frac{100\left(-\frac{41}{120}\right)}{40} = \frac{4100}{4800} = \frac{41}{48}.$

Selected Solutions to Exercise Set 3.10

1. $f(9.2) \approx f(8) + f'(8)(9.2 - 8)$.

   $f(8) = 3(8^2) + 7 = 199$.

   $f'(x) = 6x \implies f'(8) = 48$.

   $f(9.2) \approx 199 + 48(1.2) = 256.6$.

3. $f(-5.4) \approx f(-6) + f'(-6)(-5.4 - (-6))$.

   $f(-6) = 2(-6)^2 + 3(-6) = 54$.

   $f'(x) = 4x + 3 \implies f'(-6) = 4(-6) + 3 = -21$.

   $f(-5.4) \approx 54 + (-21)(.6) = 41.4$.

5. $f(2.2) \approx f(2) + f'(2)(2.2 - 2)$.

   $f(2) = \sqrt{\frac{2+2}{2-1}} = 2$.

   $f'(x) = \frac{1}{2}\left(\frac{x+2}{x-1}\right)^{-1/2} \cdot \frac{(x-1)(1)-(x+2)(1)}{(x-1)^2} = \frac{1}{2}\left(\frac{x+2}{x-1}\right)^{-1/2} \cdot \frac{-3}{(x-1)^2}$.

   $f'(2) = \frac{1}{2}\left(\frac{2+2}{2-1}\right)^{-1/2} \frac{-3}{(2-1)^2} = \frac{1}{2} \cdot \frac{1}{2} \cdot (-3) = -\frac{3}{4}$.

   $f(2.2) \approx 2 + \left(-\frac{3}{4}\right)(.2) = 1.85$.

7. $f(120) \approx f(125) + f'(125)(120 - 125)$.

   $f(125) = 5 + 125^{2/3} = 5 + 25 = 30$.

   $f'(x) = \frac{2}{3}x^{-1/3} \implies f'(125) = \frac{2}{3}(125)^{-1/3} = \frac{2}{3}\left(\frac{1}{5}\right) = \frac{2}{15}$.

   $f(120) \approx 30 + \frac{2}{15}(-5) = \frac{88}{3}$.

9. Let $f(x) = \sqrt{x}$.

   $f'(x) = \frac{1}{2\sqrt{x}}$.

   Take $\sqrt{38.6} \approx f(36) + f'(36)(38.6 - 36) = 6 + \frac{1}{12}(2.6) = 6.2167$.

11. Let $f(x) = \frac{1}{\sqrt{x}}$.

    $f'(x) = -\frac{1}{2}x^{-3/2}$.

    Take $\frac{1}{\sqrt{17.2}} \approx f(16) + f'(16)(17.2 - 16) = \frac{1}{4} - \frac{1}{128}(1.2) = .2406$.

13. Let $f(x) = x^{3/4}$.

$f'(x) = \frac{3}{4}x^{-1/4}$.

Take $(83.2)^{3/4} \approx f(81) + f'(81)(83.2 - 81) = 27 + \frac{1}{4}(2.2) = 27.55$.

15. $R(p) = 800p - 3p^2$.

a. $R(100) = 800(100) - 3(100)^2 = 50,000$.

b. $R(105) - R(100) \approx R'(100)(105 - 100)$.

$R'(p) = 800 - 6p \implies R'(100) = 800 - 6(100) = 200$.

$R(105) - R(100) \approx 200(5) = 1,000$.

c. $R(120) - R(100) \approx R'(100)(120 - 100) = 200(20) = 4,000$.

17. $R(p) = x \cdot p = 1000p - 2p^{7/4}$.

The percent change is approximately

$$\frac{R'(81) \cdot [85 - 81]}{R(81)} \cdot 100.$$

$R(81) = 1000(81) - 2(81)^{7/4} = 81,000 - 2(2187) = 76,626$.

$R'(p) = 1000 - \dfrac{7}{2}p^{3/4}$.

$R'(81) = 1000 - \dfrac{7}{2}(81)^{3/4} = 905.5$.

Therefore the percent change is approximately

$$\frac{905.5(4)}{76,626} \cdot 1000 = 4.73\% .$$

19. $f(K) = 500\sqrt{K}$.

$f'(K) = 500 \cdot \frac{1}{2\sqrt{K}} = \frac{250}{\sqrt{K}}$.

$f(640,000) = 400,000$.

$f'(640,000) = .3125$.

The approximate percentage increases in the daily output is

$$\frac{f'(640,000) \cdot (700,000 - 640,000)}{f(640,000)} \cdot 100$$

$$= \frac{.3125(60,000)}{400,000} \cdot 100 = 4.6875\% .$$

21. $x = 2,000 - 40p \implies 40p = 2,000 - x \implies p = 50 - \frac{1}{40}x = 50 - .025x.$

a. $R(x) = x \cdot p = 50x - .025x^2.$

b. $MR(x) = R'(x) = 50 - .05x.$

$MR(100) = 50 - .05(100) = 45.$

c. $R(80) \approx R(100) + R'(100) \cdot (80 - 100) = 4,750 + 45(-20) = 3,850.$

d. $R(80) = 50 \cdot 80 - .025(80)^2 = 3,840.$

e. The approximation to $R(80)$ obtained in part c. comes from the linear approximation

$$L(x) = R(100) + R'(100) \cdot (x - 100)$$

to $R(x)$ for x near 100.

23. The volume V of a sphere of radius r is given by the formula

$$V = f(r) = \frac{4}{3}\pi r^3.$$

When the radius of the sphere is increased by 5% from a given value to r_0, the corresponding percentage increase in the volume is given by

$$\frac{f'(r_0) \cdot [1.05r_0 - r_0]}{f(r_0)} \cdot 100 = \frac{4\pi r_0^2 \cdot (.05r_0)}{\frac{4}{3}\pi r_0^3} \cdot 100 = 15\% \ .$$

25. a. $V(r) = 3 \cdot \pi r^2$ liters/min.

b. $V(4) = 3\pi(4)^2 = 48\pi$ liters/min.

c. $V(4.5) - V(4) \approx V'(4) \cdot (4.5 - 4) = 6\pi(4) \cdot (.5) = 12\pi$ liters/min.

27. The area A of the plot is given by the formula

$$A = s^2 = f(s).$$

Our approximation to the maximum absolute variation in the area of the plot is given by

$$f'(200) \cdot (2) = 800 \text{ ft}^2.$$

Selected Solutions to Exercise Set 3.11

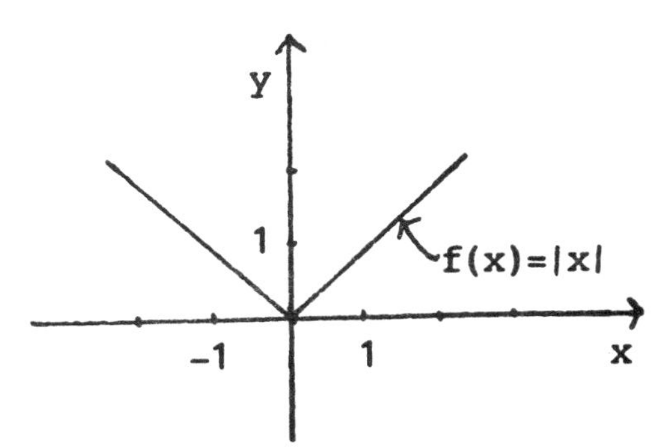

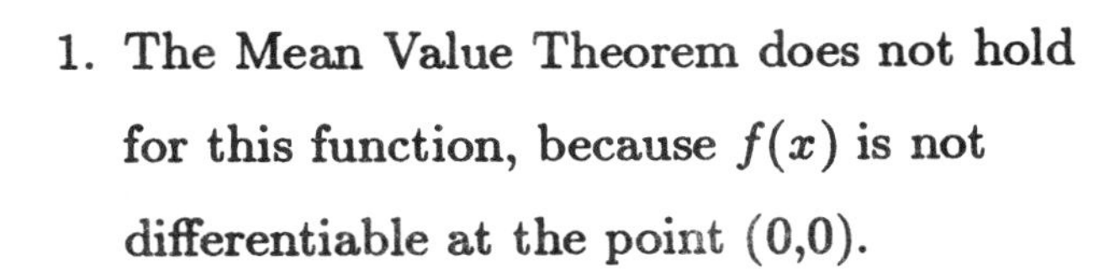

1. The Mean Value Theorem does not hold for this function, because $f(x)$ is not differentiable at the point (0,0).

3. $f'(x) = -2x$.

   $f'(c) = -2c = \frac{f(2)-f(0)}{2-0} = \frac{4-2^2-(4-0^2)}{2} = -2$.

   $c = 1$.

5. $f(x) = \sqrt{x+5}, \qquad x$ in $[-5, 4]$.

   $f'(x) = \frac{1}{2\sqrt{x+5}}$.

   $f'(c) = \frac{f(4)-f(-5)}{4-(-5)} \implies \frac{1}{2\sqrt{c+5}} = \frac{3-0}{9} \implies 2\sqrt{c+5} = 3 \implies \sqrt{c+5} = \frac{3}{2}$

   $\implies c+5 = \frac{9}{4} \implies c = -\frac{11}{4}$.

7. $f'(x) = 2x + 2$.

   $f'(c) = 2c + 2 = \frac{f(0)-f(-3)}{-3-0} = \frac{-3-(9-6-3)}{-3} = 1 \implies 2c + 2 = 1,\ c = -\frac{1}{2}$.

9. By the Mean Value Theorem there is at least one time t_0 during the 2 hours that the speed $v(t_0)$ equals the average speed during the 2 hours.

$$v(t_0) = \frac{120}{2} = 60 > 55 \text{ mph.}$$

   Therefore the driver violates the speed limit at least once.

11. Let x_1, x_2 be in (a, b) such that $x_1 < x_2$.

   Applying the Mean Value Theorem on the closed interval $[x_1, x_2]$ we have for at least one c in (x_1, x_2) that

$$f'(c) = \frac{f(x_2) - f(x_1)}{x_2 - x_1}.$$

   Since $f'(x) < 0$ for all x in (a, b), $f'(c) < 0$.

   So $\frac{f(x_2)-f(x_1)}{x_2-x_1} < 0 \implies f(x_2) - f(x_1) < 0$ since $x_2 - x_1 > 0$

   $\implies f(x_2) < f(x_1)$. So, $f(x)$ is decreasing on (a, b).

13. Suppose f is an even function on the interval $[-a, a]$ where $a > 0$. Suppose f is differentiable on the open interval $(-a, a)$, and suppose that the derivative function f' is continuous at $x = 0$.

    If h is any positive real number such that $h < a$, then there is at least one real number c satisfying $-h < c < h$ such that

$$f'(c) = \frac{f(h) - f(-h)}{h - (-h)} = 0.$$

    Therefore there are real numbers x arbitrarily close to 0 such that $f'(x) = 0$. In view of the supposed continuity of $f'(x)$ at $x = 0$, then $f'(0) = 0$. Therefore $f'(0) = \frac{f(a)-f(-a)}{a-(-a)}$.

15. Let x_1 and x_2 be any two real numbers in the interval (a, b) such that $x_1 < x_2$. Then there is a real number c satisfying $x_1 < c < x_2$ such that

$$f'(c) = \frac{f(x_2) - f(x_1)}{x_2 - x_1}.$$

    Since $f'(c) = 0$, it follows that $[f(x_2) - f(x_1)] = 0 \implies f(x_2) = f(x_1)$. Therefore f is a constant function on (a, b).

Selected Solutions to Review Exercises - Chapter 3

1. $f'(x) = 4 - 2x = 0$ for $x = 2$.

    For $x < 2$, $f'(x) > 0$, so $f(x)$ is increasing. For $x > 2$, $f'(x) < 0$, so $f(x)$ is decreasing. $f(2)$ is a relative maximum.

3. $f'(x) = 2x - 2 = 0$ for $x = 1$.

    For $x < 1$, $f'(x) < 0$, so $f(x)$ is decreasing. For $x > 1$, $f'(x) > 0$, so $f(x)$ is increasing. $f(1)$ is a relative minimum.

5. $f'(x) = \frac{1(x+3)-1(x-3)}{(x+3)^2} = \frac{6}{(x+3)^2}$.

    $f'(x) > 0$ for all $x \neq -3$, so $f'(x)$ is increasing on $(-\infty, -3)$ and on $(-3, \infty)$.

7. $f'(x) = 1 \cdot (16 - x^2)^{1/2} + x \cdot \left\{\frac{1}{2}(16 - x^2)^{-1/2} \cdot (-2x)\right\}$
$= (16 - x^2)^{-1/2} \cdot \{(16 - x^2) - x^2\}$
$= (16 - x^2)^{-1/2} \cdot (16 - 2x^2).$

$f'(x) = 0$ for $x = \pm\sqrt{8}$.

$f'(x)$ does not exist for $x = \pm 4$.

Interval	Test Number	Sign of $f'(x)$	Conclusion
$(-4, -\sqrt{8})$	-3	$f'(-3) < 0$	decreasing
$(-\sqrt{8}, \sqrt{8})$	0	$f'(0) > 0$	increasing
$(\sqrt{8}, 4)$	3	$f'(3) < 0$	decreasing

$f(-\sqrt{8})$ is a relative minimum. $f(\sqrt{8})$ is a relative maximum.

9. $f'(x) = 3x^2 - 6x = 3x(x - 2) = 0$ for $x = 0$ and $x = 2$.

Interval	Test Number	Sign of $f'(x)$	Conclusion
$(-\infty, 0)$	-1	$f'(-1) > 0$	increasing
$(0, 2)$	1	$f'(1) < 0$	decreasing
$(2, \infty)$	3	$f'(3) > 0$	increasing

$f(0)$ is a relative maximum. $f(2)$ is a relative minimum.

11. $f(x) = 4 - 2x^{2/3}$.

$f'(x) = -2 \cdot \frac{2}{3}x^{-1/3} = -\frac{4}{3}x^{-1/3}$.

$f'(x)$ is undefined for $x = 0$.

For $x < 0$, $f'(x) > 0$, so $f(x)$ is increasing.

For $x > 0$, $f'(x) < 0$, so $f(x)$ is decreasing.

$f(0) = 4$ is a relative maximum.

13. $f'(x) = -2x(x^2 - 4)^{-2} = 0$ for $x = 0$.

$f(-1) = -\frac{1}{3}$, $f(1) = -\frac{1}{3}$, $f(0) = -\frac{1}{4}$.

The maximum is $f(0) = -\frac{1}{4}$. The minimum is $f(-1) = f(1) = -\frac{1}{3}$.

15. $f(x) = x\sqrt{1 - x^2}$, x in $[-1, 1]$.

$f'(x) = 1 \cdot (1 - x^2)^{1/2} + x\left(\frac{1}{2}(1 - x^2)^{-1/2}(-2x)\right)$
$= (1 - x^2)^{-1/2}\left((1 - x^2) - x^2\right) = (1 - x^2)^{-1/2} \cdot (1 - 2x^2).$

$f'(x) = 0$ for $1 - 2x^2 = 0 \implies x^2 = \frac{1}{2} \implies x = \frac{-1}{\sqrt{2}}$ or $x = \frac{1}{\sqrt{2}}$.

$f'(-1) = 0$, $f(1) = 0$, $f\left(-\frac{1}{\sqrt{2}}\right) = -\frac{1}{2}$, $f\left(\frac{1}{\sqrt{2}}\right) = \frac{1}{2}$.

The maximum value is $f\left(\frac{1}{\sqrt{2}}\right) = \frac{1}{2}$, and the minimum value is $f\left(-\frac{1}{\sqrt{2}}\right) = -\frac{1}{2}$.

17. $f'(x) = 1 - \left\{\frac{1}{2}(1-x^2)^{-1/2} \cdot (-2x)\right\} = 1 + x \cdot (1-x^2)^{-1/2}$.

$f'(x) = 0$ for $\frac{x}{\sqrt{1-x^2}} = -1 \implies \frac{x^2}{1-x^2} = 1 \implies x^2 = 1 - x^2 \implies x^2 = \frac{1}{2} \implies$ $x = \pm\sqrt{\frac{1}{2}}$.

$f(-1) = -1$, $f(1) = 1$, $f\left(-\sqrt{\frac{1}{2}}\right) = -\sqrt{2}$, $f\left(\sqrt{\frac{1}{2}}\right) = 0$.

The maximum is $f(1) = 1$. The minimum is $f\left(-\sqrt{\frac{1}{2}}\right) = -\sqrt{2}$.

19. $f'(x) = 1 + 8x^{-3}$.

$f'(x) = 0$ for $x^{-3} = -\frac{1}{8} \implies x = -2$.

$f(-3) = -\frac{31}{9}$, $f(-1) = -5$, $f(-2) = -3$.

The maximum is $f(-2) = -3$. The minimum is $f(-1) = -5$.

21. $f'(x) = 4x^3 - 4x = 4x(x^2 - 1) = 0$ for $x = 0$ and $x = \pm 1$.

$f(-2) = 8$, $f(2) = 8$, $f(0) = 0$, $f(-1) = -1$, $f(1) = -1$.

The maximum is $f(-2) = f(2) = 8$. The minimum is $f(-1) = f(1) = -1$.

23. Vertical asymptote: $x = 7$.

$\lim_{x\to\pm\infty} f(x) = 0 \implies$ horizontal asymptote is $y = 0$.

25. $y = \frac{x^2}{4-x^2}$.

$\lim_{x\to\pm\infty} \frac{x^2}{4-x^2} = \lim_{x\to\pm\infty} \frac{x^2}{4-x^2} \cdot \frac{\frac{1}{x^2}}{\frac{1}{x^2}} = \lim_{x\to\pm\infty} \frac{1}{\frac{4}{x^2}-1} = \frac{1}{0-1} = -1$.

Therefore the line $y = -1$ is a horizontal asymptote.

The lines $x = 2$ and $x = -2$ are vertical asymptotes.

27. Vertical asymptotes: $x = \pm 3$.

$\lim_{x\to\pm\infty} f(x) = \lim_{x\to\pm\infty} \frac{1+\frac{9}{x^2}}{1-\frac{9}{x^2}} = \frac{1+0}{1-0} = 1 \implies$ horizontal asymptote is $y = 1$.

29. For $x < 0$, $f(x) = \frac{x}{-x} = -1$; for $x > 0$, $f(x) = \frac{x}{x} = 1$. So there is no vertical asymptote.

Horizontal asymptotes: $y = -1$ and $y = 1$.

31. Vertical asymptote: $x = 0$.

$\lim_{x\to\pm\infty} f(x) = 4 \implies$ horizontal asymptote is $y = 4$.

33. $\lim_{x\to+\infty} f(x) = \lim_{x\to+\infty} \frac{2+\frac{6}{x}}{\frac{9}{x}-1} = \frac{2+0}{0-1} = -2.$

35. $\lim_{x\to\infty} \frac{7}{3x-x^{-3/2}} = 0$ because $(3x) \to \infty$ and $x^{-3/2} \to 0$ as $x \to \infty$.

37. $\lim_{x\to4^-} \frac{x^2+5\sqrt{x}}{x^2-16} = -\infty$ because as $x \to 4^-$, $(x^2+5\sqrt{x}) \to 26 > 0$ and $(x^2-16) \to 0^-$.

39. As $x \to 2^-$, (x^2-4) approaches 0 through negative values. Hence,

$\lim_{x\to2^-} \frac{6}{x^2-4} = -\infty.$

41. $\lim_{x\to1^-} \frac{5x^2}{1-x} = +\infty$ because as $x \to 1^-$, $5x^2 \to 5 > 0$ and $(1-x) \to 0^+$.

43. $y = 4x - x^2$.

(1) The domain is all x.

(2) $y = 0$ for $x = 0$ and $x = 4$.

(3),(4) There are no asymptotes.

(5) $y' = 4 - 2x = 0$ for $x = 2$.

Interval	Test Number	Sign of $f'(x)$	Conclusion
$(-\infty, 2)$	-1	$f'(-1) > 0$	increasing
$(2, \infty)$	3	$f'(3) < 0$	decreasing

$f(2)$ is the relative maximum.

(6) $f''(x) = -2$

$f''(x) < 0$ for all x, so the curve is concave down on $(-\infty, \infty)$.

(7)

x	-1	0	1	2	3	4
y	-5	0	3	4	3	0

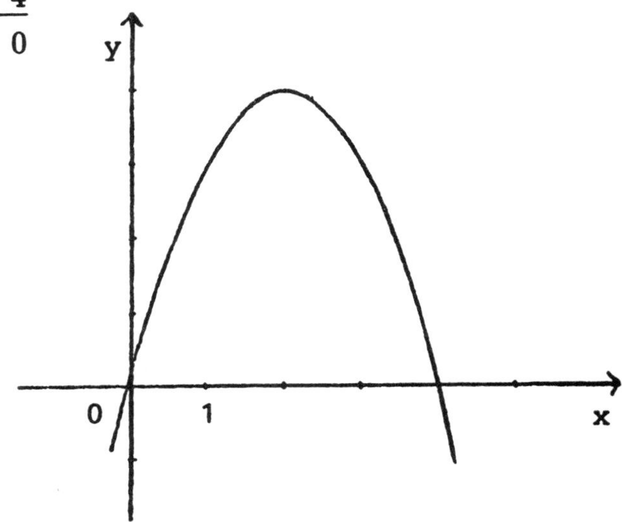

45. $f(x) = x^2 - 2x - 3 = (x+1)(x-3)$.

(1) The domain is all x.

(2) $f(x) = 0$ for $x = -1$, $x = 3$.

(3),(4) There are no asymptotes.

(5) $f'(x) = 2x - 2 = 0$ for $x = 1$.

For $x < 1$, $f'(x) < 0$, $f(x)$ is decreasing.

For $x > 1$, $f'(x) > 0$, $f(x)$ is increasing.

$f(1) = -4$ is the relative minimum.

(6) $f''(x) = 2 > 0$ for all x. So the graph is concave down on $(-\infty, \infty)$.

(7)

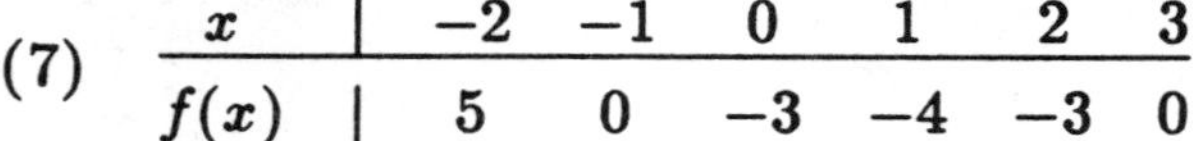

x	-2	-1	0	1	2	3
$f(x)$	5	0	-3	-4	-3	0

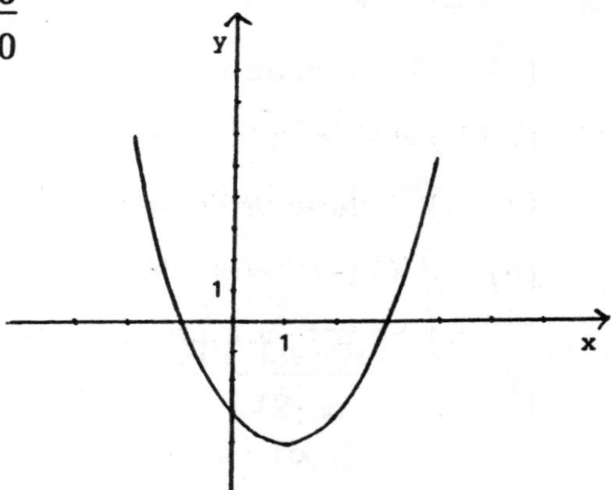

47. $y = \frac{x-3}{x+3}$.

(1) The domain is all $x \neq -3$.

(2) $y = 0$ for $x = 3$.

(3) The vertical asymptote is $x = -3$.

(4) $\lim_{x \to \pm\infty} \frac{x-3}{x+3} = 1$. So the horizontal asymptote is $y = 1$.

(5) $y' = \frac{6}{(x+3)^2} > 0$ for all $x \neq -3$.

So the graph of y is increasing on $(-\infty, -3)$ and $(-3, \infty)$,

and there is no relative extremum.

(6) $y'' = \frac{-12}{(x+3)^3}$.

For $x < -3$, $y'' > 0$, the graph is concave up.

For $x > -3$, $y'' < 0$, the graph is concave down.

There is not inflection point.

(7)

x	-5	-4	-3.5	-2.5	-2	-1	0	1	3
y	4	7	13	-11	-5	-2	-1	$-\frac{1}{2}$	0

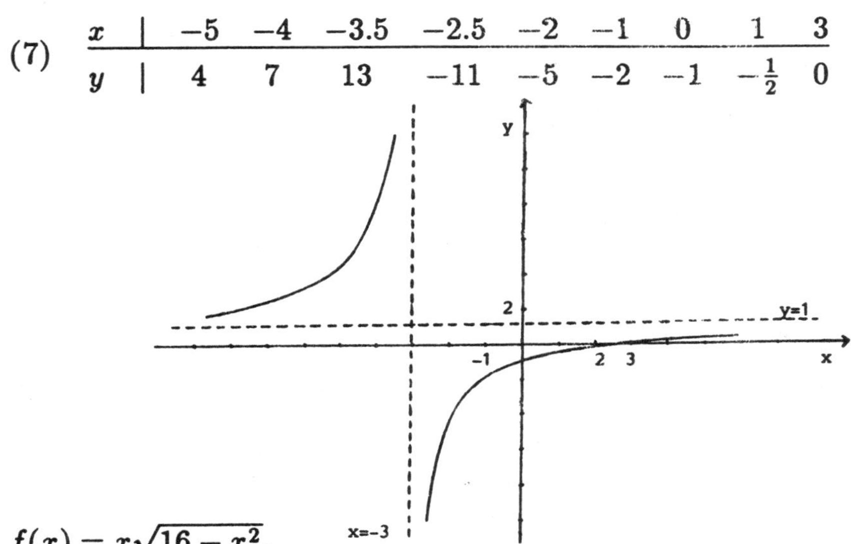

49. $f(x) = x\sqrt{16 - x^2}$.

(1) The domain is all x such that $-4 \le x \le 4$.

(2) $f(x) = 0$ for $x = 0$ and $x = \pm 4$.

(3),(4) There are no asymptotes.

(5) $f'(x) = 1 \cdot (16 - x^2)^{1/2} + x\left[\frac{1}{2}(16 - x^2)^{-1/2} \cdot (-2x)\right]$
$= (16 - x^2)^{-1/2} \cdot (16 - 2x^2)$

$f'(x) = 0$ for $x = \pm 2\sqrt{2}$. $f'(x)$ does not exist for $x = \pm 4$.

Interval	Sign of $f'(x)$	Conclusion
$(-4, -2\sqrt{2})$	$-$	decreasing
$(-2\sqrt{2}, 2\sqrt{2})$	$+$	increasing
$(2\sqrt{2}, 4)$	$-$	decreasing

$f\left(-2\sqrt{2}\right) = -8$ is the relative minimum.

$f\left(2\sqrt{2}\right) = 8$ is the relative maximum.

(6) $f''(x) = -\frac{1}{2}(16 - x^2)^{-3/2}(-2x)(16 - 2x^2) + (16 - x^2)^{-1/2}(-4x)$
$= x(16 - x^2)^{-3/2}\left[(16 - 2x^2) - 4(16 - x^2)\right]$
$= x(16 - x^2)^{-3/2}(-48 + 2x^2).$

$f''(x) = 0$ for $x = 0$. $f''(x)$ does not exist for $x = \pm 4$.

Interval	Sign of $f''(x)$	Conclusion
$(-4, 0)$	$+$	concave up
$(0, 4)$	$-$	concave down

The inflection point occurs when $x = 0$.

(7)

x	-4	-3	$-2\sqrt{2}$
$f(x)$	0	-7.94	-8

x	-2	-1	0
$f(x)$	-6.93	-3.87	0

x	1	2	$-2\sqrt{2}$
$f(x)$	$+3.87$		

$+6.93$ $+8$

x	3	4
$f(x)$	$+7.94$	0

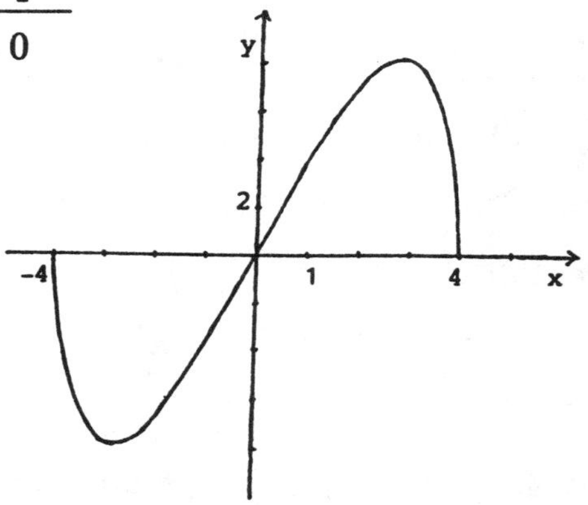

51. $y = \frac{x^2}{x^2+9}$.

(1) The domain is all x.

(2) $y = 0$ for $x = 0$.

(3),(4) There is no vertical asymptote.

Since $\lim_{x \to \pm\infty} \frac{x^2}{x^2+9} = 1$, the horizontal asymptote is $y = 1$.

(5) $y' = \frac{18x}{(x^2+9)^2} = 0$ for $x = 0$.

For $x < 0$, $y' < 0$, the graph is decreasing.

For $x > 0$, $y' > 0$, the graph is increasing.

$y = 0$ is the relative minimum when $x = 0$.

(6) $y'' = 18 \cdot \frac{1\cdot(x^2+9)^2 - x\cdot 2(x^2+9)\cdot 2x}{(x^2+9)^4} = \frac{18\cdot(x^2+9-4x^2)}{(x^2+9)^3} = \frac{54(3-x^2)}{(x^2+9)^3}$.

$y'' = 0$ for $x = \pm\sqrt{3}$.

Interval	Sign of $f''(x)$	Conclusion
$(-\infty, -\sqrt{3})$	$-$	concave down
$(-\sqrt{3}, \sqrt{3})$	$+$	concave up
$(\sqrt{3}, \infty)$	$-$	concave down

$(-\sqrt{3}, 0.25)$ and $(\sqrt{3}, 0.25)$ are the inflection points.

(7)

x	-3	-2	-1	0
y	0.5	0.31	0.1	0

x	1	2	3
y	0.1	0.31	0.5

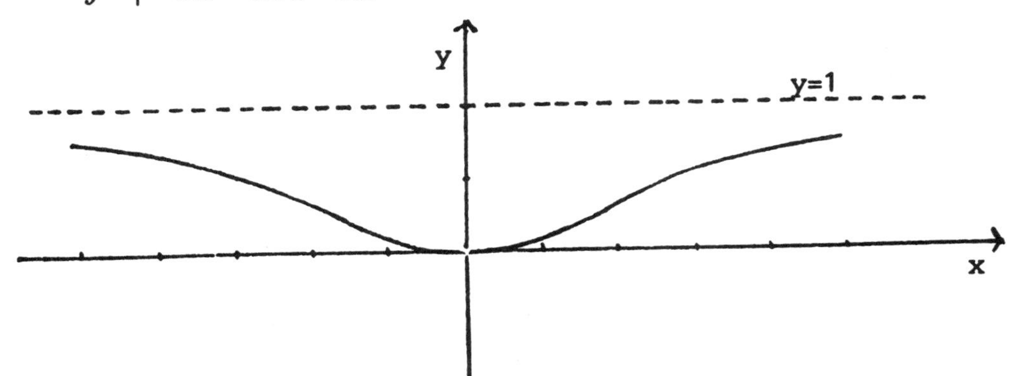

53. Let $\ell =$ the length of a side of the square.

$A =$ the area of the square.

Then $A = \ell^2$. $\frac{dA}{dt} = 2\ell \cdot \frac{d\ell}{dt}$.

When $\ell = 10$ and $\frac{d\ell}{dt} = 2$, $\frac{dA}{dt} = 2 \cdot 10 \cdot 2 = 40\text{cm}^2/\text{sec}$.

55. Let $r =$ the radius of the disc of oil,

$\ell =$ the thickness of the layer of oil.

Then $V = \pi r^2 \ell$. $\ell = \frac{V}{\pi r^2}$, $\frac{d\ell}{dt} = -\frac{2V}{\pi r^3} \cdot \frac{dr}{dt}$.

When $t = 9$, $r(9) = 6a$ and $\frac{dr}{dt} = \frac{a}{\sqrt{9}} = \frac{a}{3}$,

$\frac{d\ell}{dt} = -\frac{2V}{\pi(6a)^3} \cdot \frac{a}{3} = -\frac{V}{324\pi a^2}$ m/sec.

57. The weekly revenue $R(x) = x \cdot p(x) = x\left(5 + \frac{200}{x}\right) = 5x + 200$.

The weekly profit $P(x) = R(x) - C(x) = (5x + 200) - (5000 + 2x) = 3x - 4800$.

Since $P(x)$ is increasing with x for $0 \le x \le 1000$, the sales level $x = 1000$ maximizes profit.

59. Since the time the car and the truck travel is the same, then

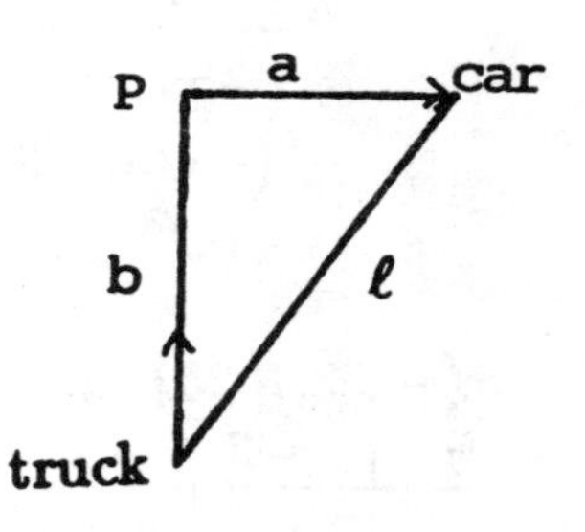

$\frac{20-b}{80} = \frac{a}{60} \implies a = \frac{3(20-b)}{4}$.

$\ell = \sqrt{a^2 + b^2}$.

$\frac{d\ell}{dt} = \frac{1}{2}(a^2 + b^2)^{-1/2} \cdot \left(2a\frac{da}{dt} + 2b\frac{db}{dt}\right)$

$= (a^2 + b^2)^{-1/2}(a \cdot 60 - b \cdot 80)$.

When $b = 0$, $a = \frac{3 \cdot 20}{4} = 15$, $\frac{d\ell}{dt} = (15^2 + 0)^{-1/2}(15 \cdot 60 - 0) = 60$ km/hr.

61. Let x = the number of times per year the store orders dishwashers

$C(x)$ = the yearly costs.

Then $C(x) = 150 \cdot \frac{1}{2x} \cdot 15 + 25x + 2 \cdot 150 = \frac{1125}{x} + 25x + 300$.

$C'(x) = -\frac{1125}{x^2} + 25 = \frac{25(x^2-45)}{x^2} = 0$ for $x = \sqrt{45} = 6.7$.

$C(6) = 637.5$, $C(7) = 635.7$.

So the store should order 7 times/yr to minimize cost.

63. $P(x) = R(x) - C(x) - 30x = -3x^2 + 794x - 100 - 30x$

$= -3x^2 + 764x - 100$.

$P'(x) = -6x + 764 = 0$ for $x = 127.3$.

For $0 < x < 127.3$, $P'(x) > 0$. For $127.3 < x < 200$, $P'(x) < 0$.

$P(127) = 48541$, $P(128) = 48540$.

So the optimal level would be changed to $x = 127$ items/week.

65. The revenue $R(x) = x \cdot p = x \cdot \sqrt{100 - x^2}$.

$R'(x) = \sqrt{100 - x^2} \cdot 1 + x \cdot \frac{1}{2}(100 - x^2)^{-1/2}(-2x) = \frac{(100-x^2)-x^2}{\sqrt{100-x^2}} = \frac{100-2x^2}{\sqrt{100-x^2}}$.

$R'(x) = 0$ for $x = \sqrt{50} \approx 7.07$, $p = \sqrt{100 - \left(\sqrt{50}\right)^2} = \sqrt{50} \approx 7.07$.

For $0 < x < 7.07$, $R'(x) > 0$; and for $7.07 < x < 10$, $R'(x) < 0$.

So the maximum revenue occurs when the price $p = 7.07$.

67. a. $v = s'(t) = 96 - 32t.$

b. $s'(t) = 0$ for $t = \frac{96}{32} = 3$ sec. $s(3) = 144.$

For $0 < t < 3$, $s'(t) > 0$; and for $3 < t < 6$, $s'(t) < 0$.

So the maximum height $s = 144$ ft. occurs when $t = 3$ sec after the ball is thrown.

c. $v' = -32 < 0$ for all $t \geq 0$.

So v is decreasing on (0,6) and has a maximum when $t = 0$ at height $s = 0$.

d. $s(t) = 0$ for $t = 0$ and $t = 6$.

So when $t = 6$ sec., $v = 96 - 32 \cdot 6 = -96$ ft/sec.

So the speed is 96 ft/sec. when it strikes the ground.

69. a. $c(x) = \frac{C(x)}{x} = \frac{30x+200+0.5x^2}{x} = 30 + \frac{200}{x} + 0.5x.$

b. $c'(x) = -\frac{200}{x^2} + 0.5 = \frac{0.5x^2-200}{x^2} = 0$ for $x = \sqrt{\frac{200}{0.5}} = 20.$

For $0 < x < 20$, $c'(x) < 0$; and for $x > 20$, $c'(x) > 0$.

So average cost will be a minimum for $x = 20$ tires/day.

71. $x^2y + xy^2 = -4. \qquad 2xy + x^2\frac{dy}{dx} + y^2 + x \cdot 2y\frac{dy}{dx} = 0.$

$(x^2 + 2xy)\frac{dy}{dx} = -y^2 - 2xy. \qquad \frac{dy}{dx} = \frac{-y(y+2x)}{x(x+2y)}.$

73. $\sqrt[3]{xy} - y = -6. \qquad \frac{1}{3}(xy)^{-2/3}\left(y + x\frac{dy}{dx}\right) - \frac{dy}{dx} = 0.$

$\left(\frac{1}{3}(xy)^{-2/3}x - 1\right)\frac{dy}{dx} = -\frac{1}{3}(xy)^{-2/3}y.$

$\frac{dy}{dx} = \frac{-\frac{1}{3}(xy)^{-2/3}y}{\frac{1}{3}(xy)^{-2/3}x-1} = \frac{-y}{x-3(xy)^{2/3}}.$

At (3,9), $\frac{dy}{dx} = \frac{-9}{3-3(3\cdot 9)^{2/3}} = \frac{3}{8}.$

75. $u(x) = \sqrt{x(x+24)} = \sqrt{x^2 + 24x}.$

$u'(x) = \frac{1}{2}(x^2 + 24x)^{-1/2} \cdot (2x + 24) = \frac{x+12}{\sqrt{x^2+24x}}.$

a. $u(x)$ is increasing for all $x > 0$.

b. There are no intervals on which $u(x)$ is decreasing.

c. $u(x)$ does not have any relative extrema.

d. $u''(x) = \frac{(x^2+24x)^{1/2}\cdot 1-(x+12)\cdot\frac{1}{2}(x^2+24x)^{-1/2}\cdot(2x+24)}{x^2+24x}$

$= \frac{(x^2+24x)^{-1/2}\cdot 1\left[(x^2+24x)-(x+12)^2\right]}{x^2+24x}$

$= \frac{\left[(x^2+24x)-(x^2+24x+144)\right]}{(x^2+24x)^{3/2}}$

$= \frac{-144}{(x^2+24x)^{3/2}}.$

$u''(x) < 0$ for all $x > 0$, so the graph of u is concave down on the interval $(0, \infty)$.

e. The employee satisfaction continually increases as the salary increases, but the rate of increase is continuously decreasing.

77. $D(p) = \frac{10\sqrt{p}}{16+p}, \quad p > 0.$

$D'(p) = \frac{(16+p)\cdot 10\cdot\frac{1}{2}\cdot p^{-1/2}-10\sqrt{p}\cdot 1}{(16+p)^2}$

$= \frac{5(16+p)p^{-1/2}-10p^{1/2}}{(16+p)^2}$

$= \frac{5p^{-1/2}[(16+p)-2p]}{(16+p)^2} = \frac{5p^{-1/2}[16-p]}{(16+p)^2}.$

$D'(p) = 0$ for $p = 16$.

For $0 < p < 16$, $D'(p) > 0$, so the demand $D(p)$ increases as the price p increases.

For $p > 16$, $D'(p) < 0$, so the demand $D(p)$ decreases as the price p increases.

The demand $D(p)$ in units of thousands of ounces of cologne per month is a maximum when the price p is \$16.

PRACTICE PROBLEMS FOR CHAPTER 3

1. Graph each of the following functions.

   a. $f(x) = x^3 + 6x^2 + 9x + 6$

   b. $f(x) = 2x^3 - 3x^2 - 6x + 1$

   c. $f(x) = 2x^4 - 8x^3 + 45$

   d. $f(x) = \frac{3x-12}{2x+3}$

   e. $f(x) = x^3 - 3x^2 + 3x - 9$

   f. $f(x) = \frac{x^2+1}{2x-1}$

   g. $f(x) = 4x^3 - 15x^2 + 12x - 5$

   h. $f(x) = \frac{4x-5}{x^2-1}$

   i. $f(x) = \frac{-20}{\sqrt{9+4x^2}}$

   j. $f(x) = \frac{2x-1}{x^2+1}$

2. Find where each of the following functions has an absolute maximum value and an absolute minimum value on the indicated interval.

   a. $f(x) = x^3 + 6x^2 + 9x + 6$ on [-4,1].

   b. $f(x) = \frac{2x-1}{x^2+1}$ on [-3,.25].

   c. $f(x) = -3x^4 + 8x^3 - 16$ on [1,3].

   d. $f(x) = \frac{1}{4}x^4 - x^3 - 3x^2 + 8x + 5$ on [-4,5].

   e. $f(x) = \frac{1}{5}x^5 + \frac{1}{4}x^4 - 6x^3 - 8x^2 + 32x + 10$ on [-3,6].

3. Assuming that each of the following equations defines y as a differentiable function of x, find a formula for the derivative $\frac{dy}{dx}$, and determine $\frac{dy}{dx}$ at the indicated point.

   a. $x^2y^2 + 4x = 3y^4$; the indicated point is (3,-2).

   b. $2x^2 - 4xy + xy^2 = 14$; the indicated point is (-2,1).

   c. $x^4 + (y+3)^4 + x^2y = 0$; the indicated point is (-1,-2).

4. For each of the following equations, suppose that x and y are functions of t. Use implicit differentiation to determine $\frac{dy}{dt}$ in terms of x, y, and $\frac{dx}{dt}$.

a. $x^2 + 2xy = y^3$.

b. $x^2y^2 = 2y^3 + 1$.

c. $x^3 + 2xy = x^2y^2$.

5. A manufacturer of computers estimates that t months from now it will be selling $x = 0.05t^2 + 2t + 5$ thousand computers per month. The profit P from manufacturing and selling x thousand computers is estimated to be given by $P = 0.001x^2 + 0.1x - 0.25$ million dollars. Determine the rate at which the profit will be increasing 5 months from now.

6. The volume V of a sphere of radius r is $V = \frac{4}{3}\pi r^3$. When helium is pumped into a spherical balloon, both the radius and volume change with respect to time. Suppose that the volume of the balloon is increasing at the rate of 500 cm^3 per minute. At what rate is the radius of the balloon changing when the radius is 10 cm?

7. A storage shed is to be built in the shape of a box with a square base and with a volume of 150 cubic feet. The concrete for the base costs $4 per square foot, the material for the roof costs $2 per square foot, and the material for the sides costs $2.50 per square foot. Find the dimensions of the most economical shed.

8. A certain toll road averages 36,000 cars per day when charging $1 per car. Increasing the toll will result in 300 fewer cars per day for each cent of increase. What toll should be charged in order to maximize the revenue?

9. Let x be the number of thousands of kilowatt-hours of electricity generated by an electric utility company, and let p be the corresponding price in dollars for a thousand kilowatt-hours. The relationship between x and p is given by the formula $p = 60 - (10^{-5})x$. The variable costs are $30 per thousand kilowatt-hours of electricity generated, and the fixed costs are $7,000,000 per month.

a. Find the value of x and the corresponding value of p for which the company's monthly profit is maximized.

b. Suppose that rising fuel costs cause the utility company's variable costs to increase by \$10 per thousand kilowatt-hours. How much of this increase should the utility company pass on to the consumers?

10. Let x be the number of units produced and sold, and let p be the corresponding price in dollars per unit. Suppose the demand function is $p = 200 - 3x$ and the cost function is $C(x) = 75 + 80x - x^2$, $0 \leq x \leq 40$.

a. Determine the value of x and the corresponding value of p for which the company's profit is maximized.

b. Suppose the government imposes a tax on the manufacturer of T dollars per unit produced. Determine the new value of x that maximizes the company's profit as a function of T. Assuming that the company cuts back production to this level, find the tax revenue received by the government as a function of T. Determine the value of T that will maximize the tax revenue received by the government.

11. Consider the graph of the equation

$$ax^2 - y^2 + bx + cy + d = 0$$

where a, b, c, d are real constants such that $a > 0$.

a. By writing the given equation as a quadratic equation in the unknown y and using the quadratic formula show that the graph of the given equation consists of the graphs of the two functions

$$y = f_1(x) = \frac{c + \sqrt{(4a)x^2 + (4b)x + (c^2 + 4d)}}{2}$$

and

$$y = f_2(x) = \frac{c - \sqrt{(4a)x^2 + (4b)x + (c^2 + 4d)}}{2} .$$

b. Show that

$$\lim_{x\to\infty}\left[\sqrt{(4a)x^2+(4b)x+(c^2+4d)}-2\sqrt{a}\,x\right]=\frac{b}{\sqrt{a}}\,.$$

c. Explain why

$$\lim_{x\to\infty}\left[\frac{\sqrt{(4a)x^2+(4b)x+(c^2+4d)}}{2}-\sqrt{a}\,x-\frac{b}{2\sqrt{a}}\right]=0.$$

d. Rewrite

$$f_1(x)=\left[\frac{c}{2}+\sqrt{a}\,x+\frac{b}{2\sqrt{a}}\right]+\left[\frac{\sqrt{(4a)x^2+(4b)x+(c^2+4d)}}{2}\right.$$
$$\left.-\sqrt{a}\,x-\frac{b}{2\sqrt{a}}\right]\,.$$

Explain why the straight line defined by the equation

$$y=\sqrt{a}\,x+\left(\frac{c}{2}+\frac{b}{2\sqrt{a}}\right)$$

is an oblique asymptote for the graph of the original equation.

e. By examining the behavior of $f_2(x)$ as $x\to\infty$, show that the straight line defined by the equation

$$y=-\sqrt{a}\,x+\left(\frac{c}{2}-\frac{b}{2\sqrt{a}}\right)$$

is also an oblique asymptote for the graph of the original equation.

f. Show that

$$\lim_{x\to\infty}f_1'(x)=\sqrt{a}\qquad\text{and}\qquad\lim_{x\to\infty}f_2'(x)=-\sqrt{a}\,.$$

g. Sketch the graph of the equation

$$2x^2 - y^2 + 9x - 4y + 5 = 0$$

and the two oblique asymptotes for this graph.

Discuss the symmetry of the graph of this equation.

h. Sketch the graph of the equation

$$2x^2 - y^2 - 9x + 7y + 5 = 0$$

and the two oblique asymptotes for this graph.

Discuss the symmetry of the graph of this equation.

i. Sketch the graph of the equation

$$4x^2 - y^2 - 10x + 3y + 4 = 0$$

and the two oblique asymptotes for this graph.

12. Consider the graph of the equation

$$ax^2 + bxy + cy^2 + d = 0$$

where $c \neq 0$ and $(b^2 - 4ac) > 0$.

a. By writing the given equation as a quadratic equation in the unknown y and using the quadratic formula, show that the graph of the given equation consists of the graphs of the two functions

$$y = \frac{1}{2c} \cdot \left\{-(bx) + \sqrt{(b^2 - 4ac)x^2 - 4cd}\right\} = f_1(x)$$

and

$$y = \frac{1}{2c} \cdot \left\{-(bx) - \sqrt{(b^2 - 4ac)x^2 - 4cd}\right\} = f_2(x).$$

b. Show that

$$\lim_{x\to\infty}\left[\sqrt{(b^2-4ac)x^2-4cd}-\sqrt{b^2-4ac}\;x\right]=0.$$

c. Rewrite

$$f_1(x)=\frac{1}{2c}\cdot\left\{-(bx)+\sqrt{b^2-4ac}\;x\right\}$$
$$+\frac{1}{2c}\cdot\left\{\sqrt{(b^2-4ac)x^2-4cd}-\sqrt{b^2-4ac}\;x\right\}.$$

Explain why the straight line defined by the equation

$$y=\frac{-b+\sqrt{b^2-4ac}}{2c}\,x$$

is an asymptote for the graph of the original equation.

d. By examining the behavior of $f_2(x)$ as $x\to\infty$, show that the straight line defined by the equation

$$y=\frac{-b-\sqrt{b^2-4ac}}{2c}\,x$$

is an asymptote for the graph of the original equation.

e. Sketch the graph of the equation

$$6x^2+15xy+4y^2+5=0$$

and the two asymptotes for this graph.

Discuss the symmetry of the graph of this equation.

13. a. Use a linear approximation to estimate $\sqrt[3]{1005}$.

b. Use a linear approximation to estimate $(26)^{3/2}$.

c. The radius of a hemispherical dome is measured as 100 meters with a maximum possible error of 1 centimeter. Use the differential to approximate the maximum resulting error in the calculated surface area of the dome.

14. Consider the demand function

$$D(p) = \frac{500}{10 + \sqrt{p(p+24)}}$$

of Problem 76 in the Review Exercises for Chapter 3.

a. Find the corresponding revenue function $R(p)$.

b. Find the corresponding elasticity of demand $E(p)$.

c. Explain why $E(p) < 1$ for all $p > 0$.

d. Explain why $R'(p) > 0$ for all $p > 0$.

e. Sketch the graph of the equation $y = R(p)$.

15. Consider the total cost function

$$C(x) = 400 + 10x + \frac{1}{2}x^2$$

discussed in Practice Problems 14 and 15 for Chapter 1.

Here use the derivative to show that the average cost $c(x) = \frac{C(x)}{x}$ has an absolute minimum value of $(10 + 20\sqrt{2})$ dollars per jacket when $x = 20\sqrt{2}$ jackets are produced per month.

SOLUTIONS TO PRACTICE PROBLEMS FOR CHAPTER 3

1.a. $f(x) = x^3 + 6x^2 + 9x + 6$.

(1) Since $f(x)$ is a polynomial, the domain consists of all real numbers x.

(2) The zeros of $f(x)$ are irrational.

(3) and (4) Since $f(x)$ is a polynomial, there are no asymptotes.

(5) $f'(x) = 3x^2 + 12x + 9 = 3(x^2 + 4x + 3) = 3(x + 3)(x + 1)$.
$f'(x) = 0$ for $x = -3$, $x = -1$.

Interval	Test number t	$f'(t)$	Conclusion
$(-\infty, -3)$	$t = -4$	$f'(-4) = 9 > 0$	f increasing
$(-3, -1)$	$t = -2$	$f'(-2) = -3 < 0$	f decreasing
$(-1, \infty)$	$t = 0$	$f'(0) = 9 > 0$	f increasing

$\therefore$ $f(x)$ has a relative maximum value for $x = -3$ and has a relative minimum value for $x = -1$.

(6) $f''(x) = 6x + 12 = 6(x + 2)$.
$f''(x) = 0$ for $x = -2$.

Interval	Test number t	$f''(t)$	Conclusion
$(-\infty, -2)$	$t = -3$	$f''(-3) = -6 < 0$	f concave down
$(-2, \infty)$	$t = 0$	$f''(0) = 12 > 0$	f concave up

$\therefore$ An inflection point is (-2,4).

(7)

x	-4	-3	-2	-1	0	1
$f(x)$	2	6	4	2	6	22

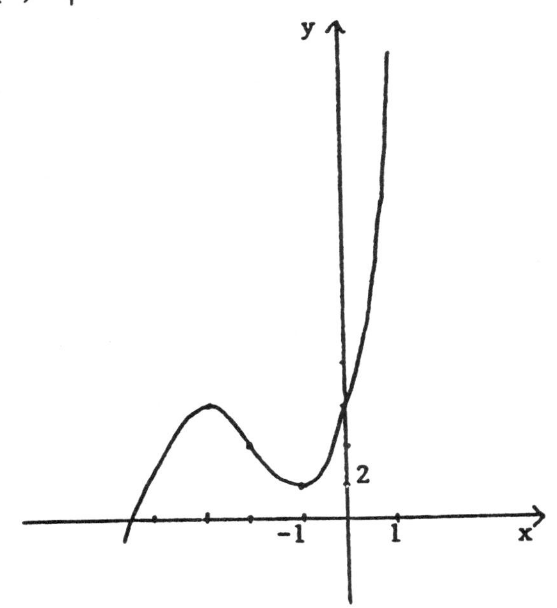

b. $f(x) = 2x^3 - 3x^2 - 6x + 1.$

(1) Since $f(x)$ is a polynomial, the domain consists of all real numbers x.

(2) The zeros of $f(x)$ are irrational.

(3) and (4) Since $f(x)$ is a polynomial, there are no asymptotes.

(5) $f'(x) = 6x^2 - 6x - 6 = 6(x^2 - x - 1)$

$f'(x) = 0$ for $x = \frac{1 \pm \sqrt{1+4}}{2} = \frac{1 \pm \sqrt{5}}{2}$

$x_1 = \frac{1-\sqrt{5}}{2} = -.618.$

$x_2 = \frac{1+\sqrt{5}}{2} = 1.618.$

Interval	Test number t	$f'(t)$	Conclusion
$\left(-\infty, \frac{1-\sqrt{5}}{2}\right)$	$t = -1$	$f'(-1) = 6 > 0$	f increasing
$\left(\frac{1-\sqrt{5}}{2}, \frac{1+\sqrt{5}}{2}\right)$	$t = 0$	$f'(0) = -6 < 0$	f decreasing
$\left(\frac{1+\sqrt{5}}{2}, \infty\right)$	$t = 2$	$f'(2) = 6 > 0$	f increasing

$\therefore$ $f(x)$ has a relative maximum value for $x = \frac{1-\sqrt{5}}{2}$ and has a relative minimum value for $x = \frac{1+\sqrt{5}}{2}$.

(6) $f''(x) = 12x - 6 = 6(2x - 1)$.

$f''(x) = 0$ for $x = \frac{1}{2}$.

Interval	Test number t	$f''(t)$	Conclusion
$(-\infty, \frac{1}{2})$	$t = 0$	$f''(0) = -6 < 0$	f concave down
$(\frac{1}{2}, \infty)$	$t = 1$	$f''(1) = 6 > 0$	f concave up

$\therefore$ An inflection point is $(\frac{1}{2}, -2.5)$.

(7)

x	-1	$-.618$	0	$.5$	1.618	2
$f(x)$	2	3.1	1	-2.5	-8.1	-7

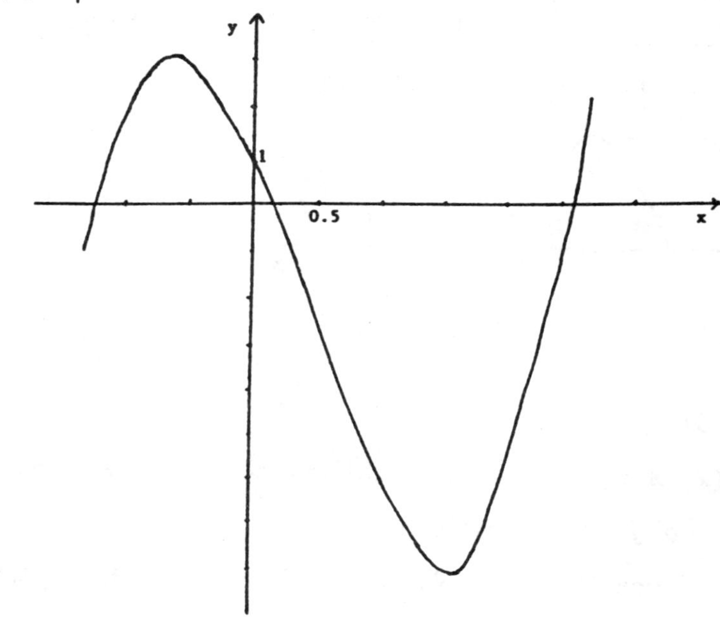

c. $f(x) = 2x^4 - 8x^3 + 45$.

(1) Since $f(x)$ is a polynomial, the domain consists of all real numbers x.

(2) The zeros of $f(x)$ are irrational.

(3) and (4) Since $f(x)$ is a polynomial, there are no asymptotes.

(5) $f'(x) = 8x^3 - 24x^2 = 8x^2(x - 3)$.

$f'(x) = 0$ for $x = 0$ and $x = 3$.

Interval	Test number t	$f'(t)$	Conclusion
$(-\infty, 0)$	$t = -1$	$f'(-1) = -32 < 0$	f decreasing
$(0, 3)$	$t = 1$	$f'(1) = -16 < 0$	f decreasing
$(3, \infty)$	$t = 4$	$f'(4) = 128 > 0$	f increasing

$\therefore f(x)$ has a relative minimum value for $x = 3$, and has neither a relative maximum value nor a relative minimum value for $x = 0$.

(6) $f''(x) = 24x^2 - 48x = 24x(x-2)$.

$f''(x) = 0$ for $x = 0$ and $x = 2$.

Interval	Test number t	$f''(t)$	Conclusion
$(-\infty, 0)$	$t = -1$	$f''(-1) = 72 > 0$	f concave up
$(0, 2)$	$t = 1$	$f''(1) = -24 < 0$	f concave down
$(2, \infty)$	$t = 3$	$f''(3) = 72 > 0$	f concave up

$\therefore$ The inflection points are (0,45) and (2,13).

(7)

x	-2	-1	0	1	2	3	4
$f(x)$	141	55	45	39	13	-9	45

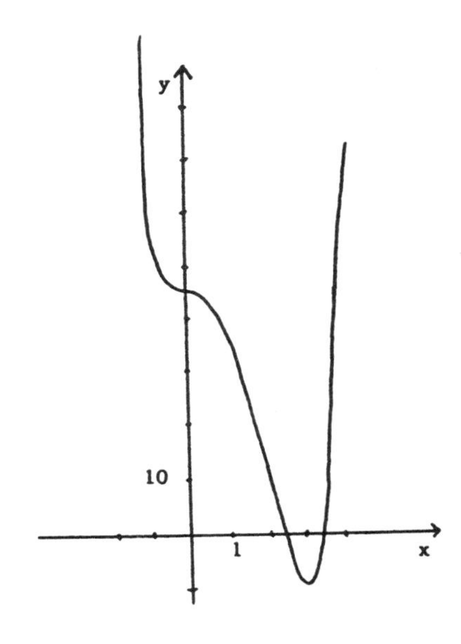

d. $f(x) = \frac{3x-12}{2x+3}$.

(1) The domain consists of all real numbers x such that $(2x+3) \neq 0 \implies x \neq -\frac{3}{2}$.

(2) The zero of $f(x)$ is the real number x such that $(3x - 12) = 0 \implies x = 4$.

(3) The vertical asymptote is the line $x = -\frac{3}{2}$.

(4) The horizontal asymptote is the line $y = \frac{3}{2}$.

(5) $f'(x) = \frac{(2x+3)\cdot 3-(3x-12)\cdot 2}{(2x+3)^2} = \frac{33}{(2x+3)^2}$.

$f'(x) > 0$ for all $x \neq -\frac{3}{2}$.

$f(x)$ is increasing on each of the intervals $(-\infty, -\frac{3}{2})$ and $(-\frac{3}{2}, \infty)$.

(6) $f''(x) = \frac{(2x+3)^2(0)-33[2(2x+3)2]}{(2x+3)^4} = \frac{-132}{(2x+3)^3}$.

Interval	Test number t	$f''(t)$	Conclusion
$(-\infty, -\frac{3}{2})$	$t = -2$	$f''(-2) = 132 > 0$	f concave up
$(-\frac{3}{2}, \infty)$	$t = 0$	$f''(0) = -\frac{132}{27} < 0$	f concave down

There is no inflection point.

(7)

x	-10	-5	-2	$-\frac{3}{2}$	-1	0	4	10
$f(x)$	$\frac{42}{17}$	$\frac{27}{7}$	18	Und.	-15	-4	0	$\frac{18}{23}$

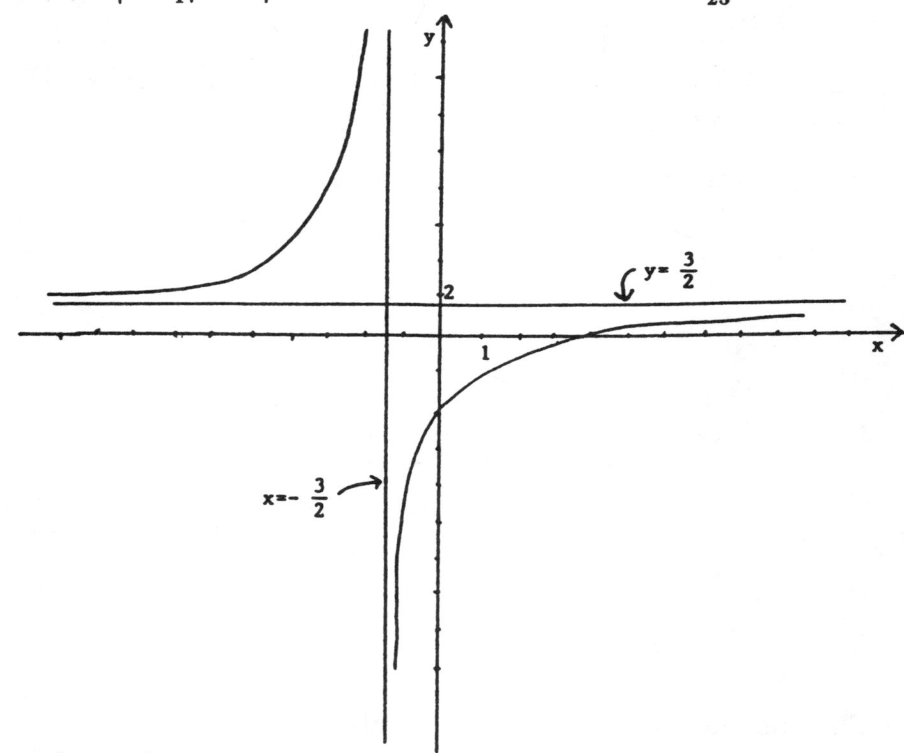

e. $f(x) = x^3 - 3x^2 + 3x - 9$.

(1) Since $f(x)$ is a polynomial, the domain consists of all real numbers x.

(2) $f(3) = 0$ by trial and error. $f(x) = (x-3)(x^2+3)$.

$x = 3$ is the only zero of $f(x)$.

(3) and (4) Since $f(x)$ is a polynomial, there are no asymptotes.

(5) $f'(x) = 3x^2 - 6x + 3 = 3(x^2 - 2x + 1) = 3(x-1)^2$.

$f'(x) = 0$ for $x = 1$.

Interval	Test number t	$f'(t)$	Conclusion
$(-\infty, 1)$	$t = 0$	$f'(0) = 3 > 0$	f increasing
$(1, \infty)$	$t = 2$	$f'(2) = 3 > 0$	f increasing

$\therefore$ $f(x)$ has neither a relative maximum value nor a relative minimum value for $x = 1$.

(6) $f''(x) = 6x - 6 = 6(x-1)$.

$f''(x) = 0$ for $x = 1$.

Interval	Test number t	$f''(t)$	Conclusion
$(-\infty, 1)$	$t = 0$	$f''(0) = -6 < 0$	f concave down
$(1, \infty)$	$t = 2$	$f''(2) = 6 > 0$	f concave up

$\therefore$ An inflection point is (1,-8).

(7)

x	-2	-1	0	1	2	3	4
$f(x)$	-35	-16	-9	-8	-7	0	19

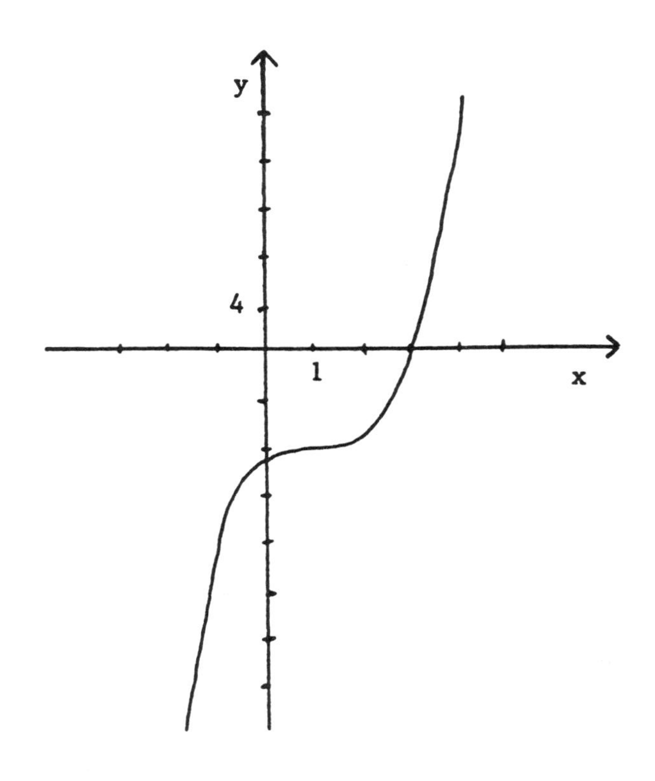

f. $f(x) = \frac{x^2+1}{2x-1}$.

(1) The domain consists of all real numbers x such that

$(2x-1) \neq 0 \implies x \neq \frac{1}{2}$.

(2) There are no zeros.

(3) The vertical asymptote is $x = \frac{1}{2}$.

(4) There are no horizontal asymptotes.

(5) $f'(x) = \frac{(2x-1)(2x)-(x^2+1)\cdot 2}{(2x-1)^2} = \frac{2(x^2-x-1)}{(2x-1)^2}$.

$f'(x) = 0$ for $(x^2 - x - 1) = 0 \implies x = \frac{1 \pm \sqrt{5}}{2}$

$x_1 = \frac{1-\sqrt{5}}{2} = -.618$.

$x_2 = \frac{1+\sqrt{5}}{2} = 1.618$.

Interval	Test number t	$f'(t)$	Conclusion
$\left(-\infty, \frac{1-\sqrt{5}}{2}\right)$	$t=-1$	$f'(-1) = \frac{2}{9} > 0$	f increasing
$\left(\frac{1-\sqrt{5}}{2}, \frac{1}{2}\right)$	$t=0$	$f'(0) = -2 < 0$	f decreasing
$\left(\frac{1}{2}, \frac{1+\sqrt{5}}{2}\right)$	$t=1$	$f'(1) = -2 < 0$	f decreasing
$\left(\frac{1+\sqrt{5}}{2}, \infty\right)$	$t=2$	$f'(2) = \frac{2}{9} > 0$	f increasing

$\therefore$ $f(x)$ has a relative maximum value for $x = \frac{1-\sqrt{5}}{2}$ and has a relative minimum value for $x = \frac{1+\sqrt{5}}{2}$.

(6) $f''(x) = \frac{(2x-1)^2(4x-2)-(2x^2-2x-2)\{2(2x-1)2\}}{(2x-1)^4}$.

$= \frac{(2x-1)(4x-2)-8(x^2-x-1)}{(2x-1)^3} = \frac{10}{(2x-1)^3}$.

$f''(x)$ is undefined at $x = \frac{1}{2}$.

Interval	Test number t	$f''(t)$	Conclusion
$(-\infty, \frac{1}{2})$	$t=0$	$f''(0) = -10 < 0$	f concave down
$(\frac{1}{2}, \infty)$	$t=1$	$f''(1) = 10 > 0$	f concave up

There is no inflection point.

(7)

x	-10	-5	$\frac{1-\sqrt{5}}{2}$	0	1	$\frac{1+\sqrt{5}}{2}$	5	10
$f(x)$	$-\frac{101}{21}$	$-\frac{26}{11}$	$-.618$	-1	2	1.618	$\frac{26}{9}$	$\frac{101}{19}$

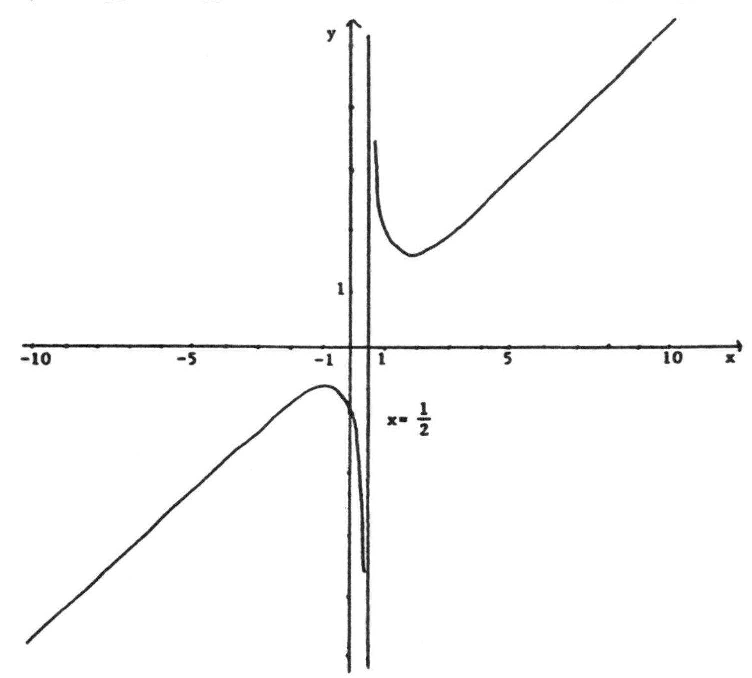

g. $f(x) = 4x^3 - 15x^2 + 12x - 5$.

(1) Since $f(x)$ is a polynomial, the domain consists of all real numbers x.

(2) The zeros of $f(x)$ are irrational.

(3) and (4) Since $f(x)$ is a polynomial, there are no asymptotes.

(5) $f'(x) = 12x^2 - 30x + 12 = 6(2x^2 - 5x + 2) = 6(2x - 1)(x - 2)$

$f'(x) = 0$ for $x = \frac{1}{2}$ and $x = 2$.

Interval	Test number x	$f'(x)$	Conclusion
$(-\infty, \frac{1}{2})$	$x = 0$	$f'(0) = 12 > 0$	f increasing
$(\frac{1}{2}, 2)$	$x = 1$	$f'(1) = -6 < 0$	f decreasing
$(2, \infty)$	$x = 3$	$f'(3) = 30 > 0$	f increasing

$\therefore$ $f(x)$ has a relative maximum value for $x = \frac{1}{2}$ and has a relative minimum value for $x = 2$.

(6) $f''(x) = 24x - 30 = 6(4x - 5)$.

$f''(x) = 0$ for $x = \frac{5}{4}$.

Interval	Test number x	$f''(x)$	Conclusion
$\left(-\infty, \frac{5}{4}\right)$	$x = 1$	$f''(1) = -6 < 0$	f concave down
$\left(\frac{5}{4}, \infty\right)$	$x = 2$	$f''(2) = 18 > 0$	f concave up

$\therefore$ The point $\left(\frac{5}{4}, -\frac{45}{8}\right)$ is an inflection point.

(7)

x	-1	0	$\frac{1}{2}$	1	$\frac{5}{4}$	2	3
$f(x)$	-36	-5	$-\frac{9}{4}$	-4	$-\frac{45}{8}$	-9	4

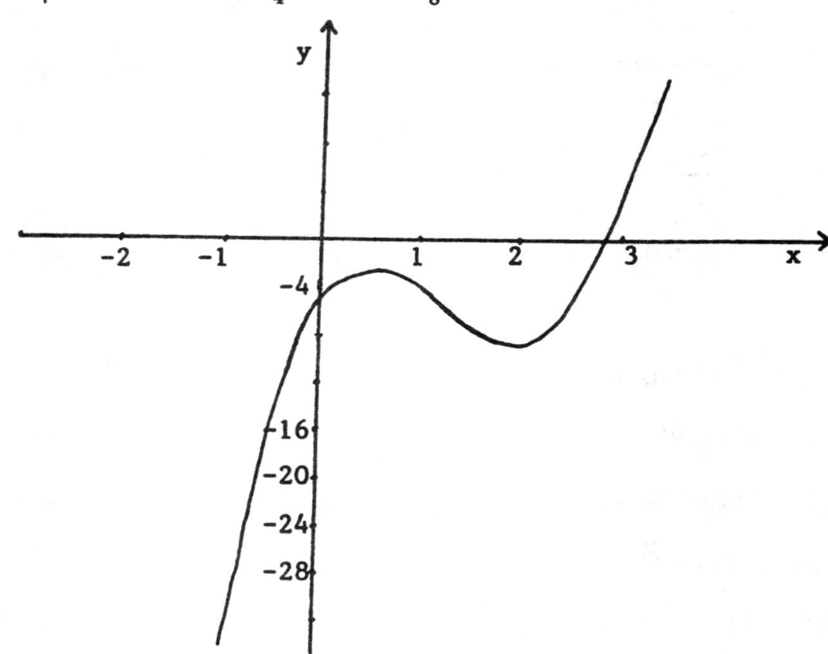

$f(x)$ has exactly one zero which is between $x = 2$ and $x = 3$. Since $f(2.85) < 0$, $f(2.86) > 0$, and $f(2.855) > 0$, this unique zero of $f(x)$ is approximately 2.85.

h. $f(x) = \frac{4x-5}{x^2-1}$.

(1) The domain consists of all real numbers x except $x = 1$ and $x = -1$.

(2) The zero of $f(x)$ is the real number x such that

$4x - 5 = 0 \implies x = \frac{5}{4}$.

(3) The vertical asymptotes are the lines $x = 1$ and $x = -1$.

(4) The horizontal asymptote is the line $y = 0$.

(5) $f'(x) = \frac{(x^2-1)4-(4x-5)(2x)}{(x^2-1)^2} = \frac{-4x^2+10x-4}{(x^2-1)^2}$.

$f'(x) = 0$ for $-4x^2 + 10x - 4 = 0$

$\Longrightarrow x = \frac{-10\pm\sqrt{10^2-4(-4)(-4)}}{2(-4)} = \frac{-10\pm6}{-8}$

$\Longrightarrow x_1 = \frac{-10-6}{-8} = 2$ and

$x_2 = \frac{-10+6}{-8} = \frac{1}{2}$.

Interval	Test number t	$f'(t)$	Conclusion
$(-\infty,-1)$	$t=-2$	$f'(-2) = \frac{-40}{9} < 0$	f decreasing
$(-1,\frac{1}{2})$	$t=0$	$f'(0) = -4 < 0$	f decreasing
$(\frac{1}{2},1)$	$t=\frac{3}{4}$	$f'\left(\frac{3}{4}\right) = \frac{320}{49} > 0$	f increasing
$(1,2)$	$t=\frac{3}{2}$	$f'\left(\frac{3}{2}\right) = \frac{32}{25} > 0$	f increasing
$(2,\infty)$	$t=3$	$f'(3) = -\frac{10}{64} < 0$	f decreasing

$\therefore$ $f(x)$ has a relative minimum value for $x = \frac{1}{2}$ and has a relative maximum value for $x = 2$.

(6) $f''(x) = \frac{(x^2-1)^2(-8x+10)-(-4x^2+10x-4)2(x^2-1)2x}{(x^2-1)^4}$

$= \frac{2(4x^3-15x^2+12x-5)}{(x^2-1)^3}$.

$f''(x) = 0$ for $4x^3 - 15x^2 + 12x - 5 = 0$.

We know from Problem 1(g) that there is exactly one x such that $4x^3 - 15x^2 + 12x - 5 = 0$ and this is $x \approx 2.85$.

Interval	Test number t	$f''(t)$	Conclusion
$(-\infty,-1)$	$t=-2$	$f''(-2) = -\frac{242}{27} < 0$	f concave down
$(-1,1)$	$t=0$	$f''(0) = -10 < 0$	f concave down
$(1,2.85)$	$t=2$	$f''(2) = -\frac{2}{3} < 0$	f concave down
$(2.85,\infty)$	$t=3$	$f''(3) = \frac{1}{64} > 0$	f concave up

An inflection point is (2.85, 0.8986).

(7)

x	-10	-5	-2	$-\frac{1}{2}$	0	$\frac{1}{2}$	$\frac{3}{4}$	$\frac{5}{4}$	2	2.85	5	10
$f(x)$	$-\frac{45}{99}$	$-\frac{25}{24}$	$-\frac{13}{3}$	$\frac{28}{3}$	5	4	$\frac{32}{7}$	0	1	0.8986	$\frac{15}{24}$	$\frac{35}{99}$

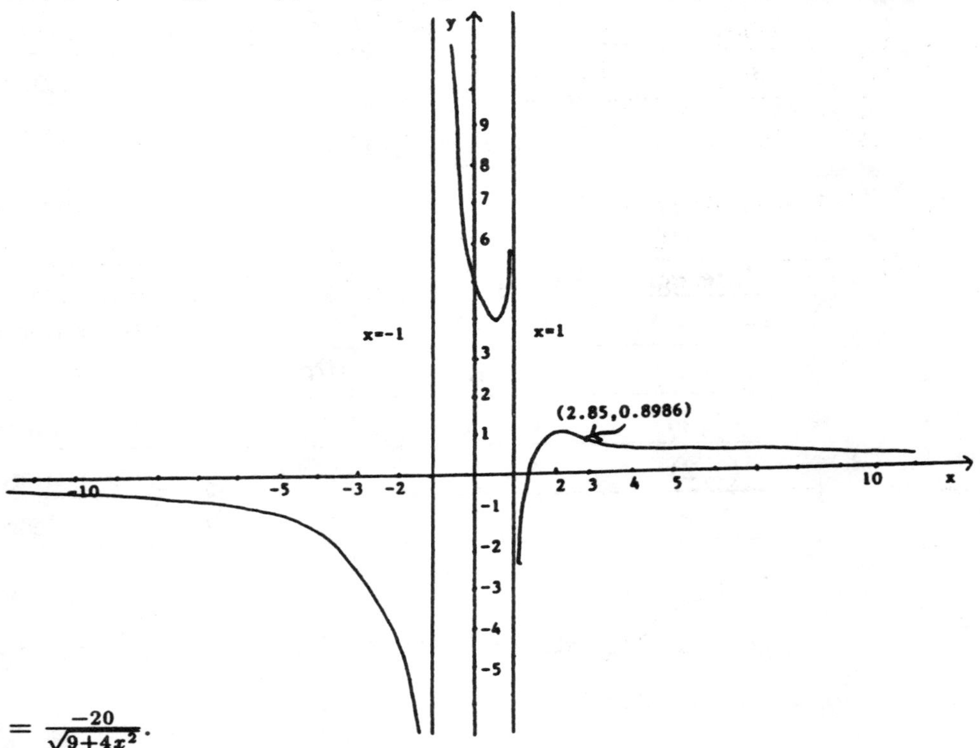

i. $f(x) = \frac{-20}{\sqrt{9+4x^2}}$.

(1) The domain consists of all real numbers x.

(2) There is no zero of $f(x)$.

(3) There is no vertical asymptote.

(4) The horizontal asymptote is the line $y = 0$.

(5) $f'(x) = \frac{-(-20)\frac{1}{2}(9+4x^2)^{-1/2}8x}{\left(\sqrt{9+4x^2}\right)^2} = \frac{80x}{(9+4x^2)^{3/2}}$.
$f'(x) = 0$ for $x = 0$.

Interval	Test number t	$f'(t)$	Conclusion
$(-\infty, 0)$	$t = -1$	$f'(-1) = -1.7 < 0$	f decreasing
$(0, \infty)$	$t = 1$	$f'(1) = 1.7 > 0$	f increasing

$\therefore f(x)$ has a relative minimum value for $x = 0$.

(6) $f''(x) = \frac{(9+4x^2)^{3/2} \cdot 80 - 80x \cdot \frac{3}{2}(9+4x^2)^{1/2} \cdot 8x}{(9+4x^2)^3} = \frac{80(9-8x^2)}{(9+4x^2)^{5/2}}$.
$f''(x) = 0$ for $9 - 8x^2 = 0 \implies x^2 = \frac{9}{8} \implies x = \pm 1.06$.

Interval	Test number t	$f''(t)$	Conclusion
$(-\infty, -1.06)$	$t = -2$	$f''(-2) = -\frac{1840}{5^5} < 0$	f concave down
$(-1.06, 1.06)$	$t = 0$	$f''(0) = \frac{720}{3^5} > 0$	f concave up
$(1.06, \infty)$	$t = 2$	$f''(2) = -\frac{1840}{5^5} < 0$	f concave down

$\therefore$ the points (-1.06,-5.44) and (1.06,-5.44) are the inflection points.

(7)

x	-10	-5	-1.06	-1	0	1	1.06	5	10
$f(x)$	$-.989$	-1.92	-5.44	-5.55	-6.67	-5.55	-5.44	-1.92	$-.989$

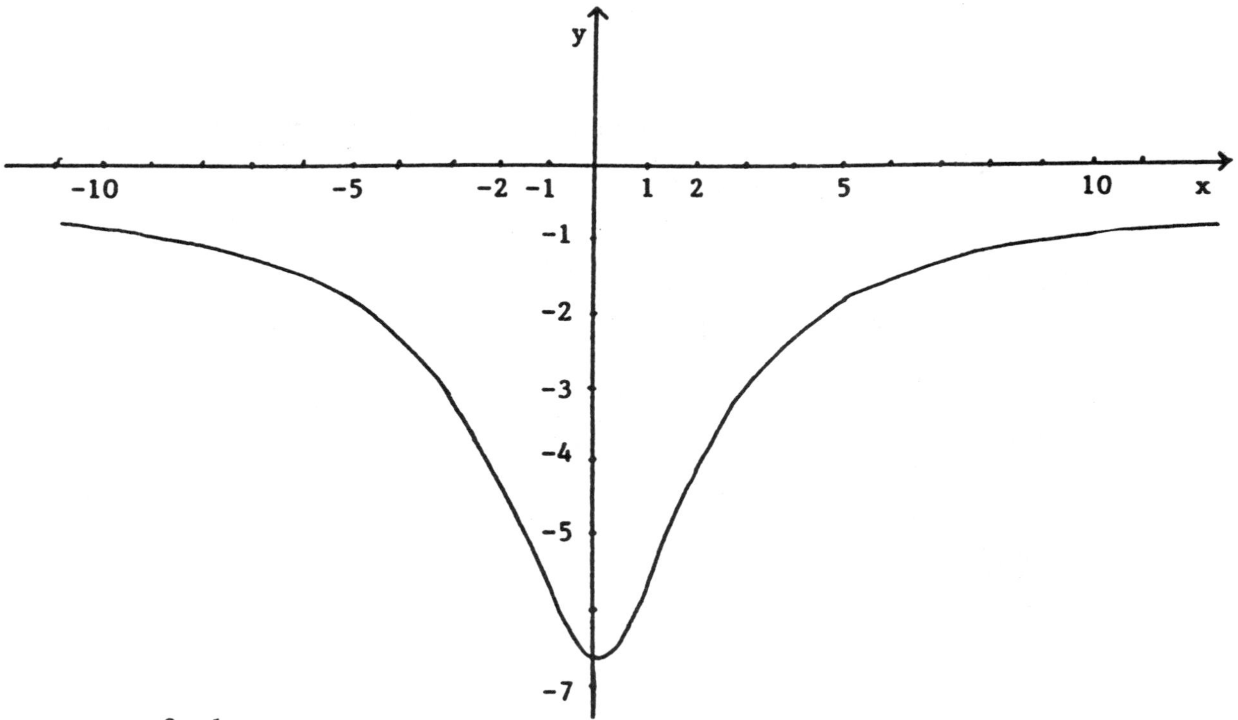

j. $f(x) = \frac{2x-1}{x^2+1}$.

(1) The domain consists of all real numbers x.

(2) The zero of $f(x)$ is the real number x such that $2x - 1 = 0 \implies x = \frac{1}{2}$.

(3) There is no vertical asymptote.

(4) The horizontal asymptote is the line $y = 0$.

(5) $f'(x) = \frac{2(x^2+1)-(2x-1)2x}{(x^2+1)^2} = \frac{2(1+x-x^2)}{(x^2+1)^2}$.

$f'(x) = 0$ for $1 + x - x^2 = 0 \implies x = \frac{-1\pm\sqrt{1^2-4(-1)(1)}}{2(-1)} = \frac{-1\pm\sqrt{5}}{-2}$

$\Longrightarrow x_1 = \frac{-1+\sqrt{5}}{-2} = -.618$ and $x_2 = \frac{-1-\sqrt{5}}{-2} = 1.618$.

Interval	Test number t	$f'(t)$	Conclusion
$(-\infty, -.618)$	$t = -1$	$f'(-1) = -\frac{1}{2} < 0$	f decreasing
$(-.618, 1.618)$	$t = 0$	$f'(0) = 2 > 0$	f increasing
$(1.618, \infty)$	$t = 2$	$f'(2) = -\frac{2}{25} < 0$	f decreasing

$\therefore$ $f(x)$ has a relative minimum value for $x = -.618$ and has a relative maximum value for $x = 1.618$.

(6) $f''(x) = 2 \cdot \frac{(1-2x)(x^2+1)^2-(1+x-x^2)2(x^2+1)2x}{(x^2+1)^4}$.

$= 2\frac{(1-2x)(x^2+1)-4x(1+x-x^2)}{(x^2+1)^3} = \frac{2(2x^3-3x^2-6x+1)}{(x^2+1)^3}$.

$f''(x) = 0$ for $2x^3 - 3x^2 - 6x + 1 = 0$.

We know from the graph of Problem 1.b. that the function $g(x) = 2x^3 - 3x^2 - 6x + 1$ has a zero t_1 between -1.5 and -1, a zero t_2 between 0 and 0.5, and a zero t_3 between 2.5 and 3.

Interval	Test number t	$f''(t)$	Conclusion
$(-\infty, t_1)$	$t = -2$	$f''(-2) = -\frac{30}{5^3} < 0$	f concave down
(t_1, t_2)	$t = 0$	$f''(0) = 2 > 0$	f concave up
(t_2, t_3)	$t = 1$	$f''(1) = -\frac{3}{2} < 0$	f concave down
(t_3, ∞)	$t = 3$	$f''(3) = \frac{1}{50} > 0$	f concave up

$\therefore$ $f(x)$ has inflection points for $x = t_1$, $x = t_2$ and $x = t_3$.

(7)

x	-10	-5	-2	-1	$-.618$	0	$.5$	1	1.618	2	3	5	10
$f(x)$	$-.21$	$-.42$	-1	-1.5	-1.62	-1	0	$.5$	$.618$	$.6$	$.5$	$.35$	$.19$

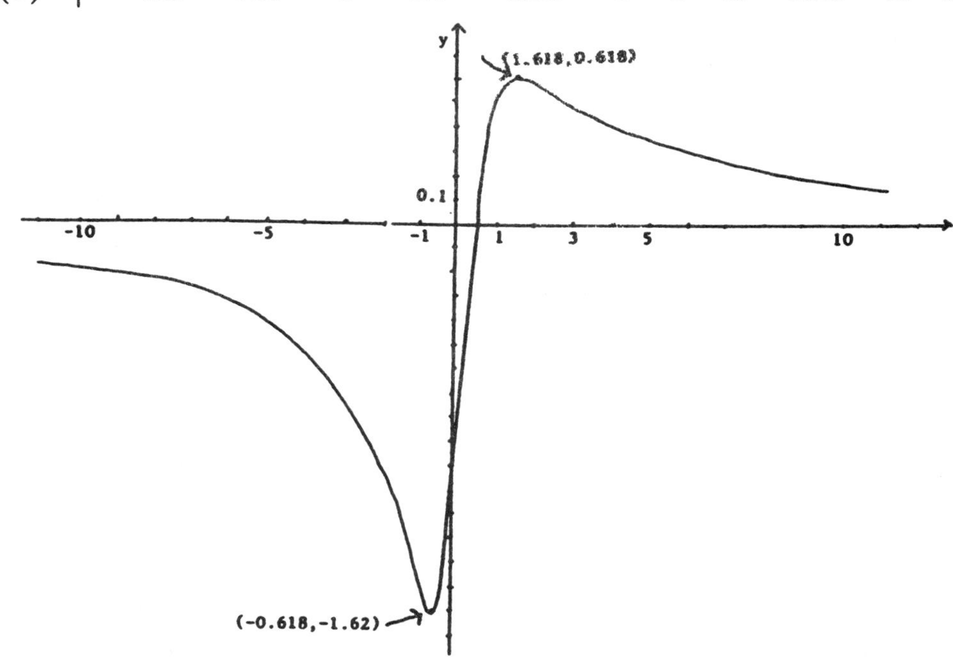

2.(a) $f(x) = x^3 + 6x^2 + 9x + 6$ on [-4,1].

$f'(x) = 3x^2 + 12x + 9 = 3(x^2 + 4x + 3) = 3(x+1)(x+3)$.

$f'(x) = 0$ for $x = -3$ and $x = -1$.

So the critical points of $f(x)$ are $x = -3$ and $x = -1$.

$f(-4) = (-4)^3 + 6(-4)^2 + 9(-4) + 6 = 2$.

$f(-3) = (-3)^3 + 6(-3)^2 + 9(-3) + 6 = 6$

$f(-1) = (-1)^3 + 6(-1)^2 + 9(-1) + 6 = 2$

$f(1) = 22$.

So $f(x)$ has an absolute maximum value for $x = 1$ and has an absolute minimum value for $x = -4$ and $x = -1$.

b. $f(x) = \frac{2x-1}{x^2+1}$ on [- 3, .25].

$f'(x) = \frac{2(x^2+1)-(2x-1)2x}{(x^2+1)^2} = \frac{-2(x^2-x-1)}{(x^2+1)^2}$.

$f'(x) = 0$ for $x^2 - x - 1 = 0 \implies x = \frac{1\pm\sqrt{(-1)^2-4(1)(-1)}}{2(1)} = \frac{1\pm\sqrt{5}}{2}$.

The critical points of $f(x)$ are $x = \frac{1+\sqrt{5}}{2}$ and $x = \frac{1-\sqrt{5}}{2}$.

$f(-3) = \frac{2(-3)-1}{(-3)^2+1} = \frac{-7}{10}$.

$f\left(\frac{1-\sqrt{5}}{2}\right) = -1.62$.

$f(.25) = -\frac{16}{34}$.

So $f(x)$ has an absolute maximum value for $x = .25$ and has an absolute minimum value for $x = \frac{1-\sqrt{5}}{2}$.

c. $f(x) = -3x^4 + 9x^3 - 16$ on [1,3].

$f'(x) = -12x^3 + 24x^2$.

$f'(x) = 0$ for $-12x^2(x-2) = 0 \implies x = 0$ and $x = 2$.

The critical points of $f(x)$ are $x = 0$ and $x = 2$.

$f(1) = -3(1)^4 + 8(1)^3 - 16 = -11$.

$f(2) = -3(2)^4 + 8(2)^3 - 16 = 0$.

$f(3) = -3(3)^4 + 8(3)^3 - 16 = -43$.

So $f(x)$ has an absolute maximum value for $x = 2$ and has an absolute minimum value for $x = 3$.

d. $f(x) = \frac{1}{4}x^4 - x^3 - 3x^2 + 8x + 5$ on [-4,5].

$f'(x) = x^3 - 3x^2 - 6x + 8 = (x-1)(x^2 - 2x - 8) = (x-1)(x-4)(x+2)$.

$f'(x) = 0$ for $x = 1$, $x = 4$, and $x = -2$.

The critical points of $f(x)$ are $x = 1$, $x = 4$ and $x = -2$.

$f(-4) = \frac{1}{4}(-4)^4 - (-4)^3 - 3(-4)^2 + 8(-4) + 5 = 53$.

$f(-2) = \frac{1}{4}(-2)^4 - (-2)^3 - 3(-2)^2 + 8(-2) + 5 = -11$.

$f(1) = \frac{1}{4}(1)^4 - (1)^3 - 3(1)^2 + 8(1) + 5 = 9.25$.

$f(4) = \frac{1}{4}(4)^4 - (4)^3 - 3(4)^2 + 8(4) + 5 = -11$.

$f(5) = \frac{1}{4}(5)^4 - (5)^3 - 3(5)^2 + 8(5) + 5 = 1.25$.

So $f(x)$ has an absolute maximum value for $x = -4$ and has an absolute minimum value for $x = -2$ and $x = 4$.

e. $f(x) = \frac{1}{5}x^5 + \frac{1}{4}x^4 - 6x^3 - 8x^2 + 32x + 10$ on [-3,6].

$f'(x) = x^4 + x^3 - 18x^2 - 16x + 32 = (x-1)(x+2)(x^2 - 16)$.

$f'(x) = 0$ for $x = 1$, $x = -2$, $x = 4$ and $x = -4$.

The critical points of $f(x)$ are $x = 1$, $x = -2$, $x = 4$, and $x = -4$.

$f(-3) = \frac{1}{5}(-3)^5 + \frac{1}{4}(-3)^4 - 6(-3)^3 - 8(-3)^2 + 32(-3) + 10 = -24.35.$

$f(-2) = \frac{1}{5}(-2)^5 + \frac{1}{4}(-2)^4 - 6(-2)^3 - 8(-2)^2 + 32(-2) + 10 = -40.4.$

$f(1) = \frac{1}{5}(1)^5 + \frac{1}{4}(1)^4 - 6(1)^3 - 8(1)^2 + 32(1) + 10 = 28.45.$

$f(4) = \frac{1}{5}(4)^5 + \frac{1}{4}(4)^4 - 6(4)^3 - 8(4)^2 + 32(4) + 10 = -105.2.$

$f(6) = \frac{1}{5}(6)^5 + \frac{1}{4}(6)^4 - 6(6)^3 - 8(6)^2 + 32(6) + 10 = 497.2.$

So $f(x)$ has an absolute maximum value for $x = 6$ and has an absolute minimum value for $x = 4$.

3.(a) $x^2y^2 + 4x = 3y^4$.

$$2xy^2 + x^2\left(2y\frac{dy}{dx}\right) + 4 = 12y^3\left(\frac{dy}{dx}\right)$$

$$(2x^2y - 12y^3)\frac{dy}{dx} = -2xy^2 - 4.$$

$$\frac{dy}{dx} = \frac{-2xy^2 - 4}{2x^2y - 12y^3}.$$

$\frac{dy}{dx}$ when $x = 3$ and $y = -2$ equals

$$\frac{-2(3)(-2)^2 - 4}{2(3)^2(-2) - 12(-2)^3} = -\frac{28}{60} = -\frac{7}{15}.$$

b. $2x^2 - 4xy + xy^2 = 14$.

$$4x - 4y - 4x\left(\frac{dy}{dx}\right) + y^2 + x\left(2y\frac{dy}{dx}\right) = 0$$

$$(2xy - 4x)\frac{dy}{dx} = 4y - 4x - y^2$$

$$\frac{dy}{dx} = \frac{4y - 4x - y^2}{2xy - 4x}.$$

$\frac{dy}{dx}$ when $x = -2$ and $y = 1$ equals

$$\frac{4(1) - 4(-2) - (1)^2}{2(-2)(1) - 4(-2)} = \frac{11}{4}.$$

c. $x^4 + (y+3)^4 + x^2y = 0.$

$$4x^3 + 4(y+3)^3\frac{dy}{dx} + 2xy + x^2\frac{dy}{dx} = 0.$$

$$\left(4(y+3)^3 + x^2\right)\frac{dy}{dx} = -4x^3 - 2xy.$$

$$\frac{dy}{dx} = \frac{-4x^3 - 2xy}{4(y+3)^3 + x^2}.$$

$\frac{dy}{dx}$ when $x = -1$ and $y = -2$ equals

$$\frac{-4(-1)^3 - 2(-1)(-2)}{4(-2+3)^3 + (-1)^2} = \frac{0}{5} = 0.$$

4.(a) $x^2 + 2xy = y^3.$

$$2x\frac{dx}{dt} + 2\left(\frac{dx}{dt}y + x\frac{dy}{dt}\right) = 3y^2\left(\frac{dy}{dt}\right).$$

$$2x\frac{dy}{dt} - 3y^2\frac{dy}{dt} = -2x\frac{dx}{dt} - 2y\frac{dx}{dt}.$$

$$(2x - 3y^2)\frac{dy}{dt} = (-2x - 2y)\frac{dx}{dt}.$$

$$\frac{dy}{dt} = \frac{(-2x - 2y)\frac{dx}{dt}}{(2x - 3y^2)}.$$

b. $x^2y^2 = 2y^3 + 1.$

$$2x\frac{dx}{dt}y^2 + 2y\frac{dy}{dt}x^2 = 6y^2\frac{dy}{dt}.$$

$$(2yx^2 - 6y^2)\frac{dy}{dt} = -(2xy^2)\frac{dx}{dt}.$$

$$\frac{dy}{dt} = \frac{-(2xy^2)\frac{dx}{dt}}{(2yx^2 - 6y^2)}.$$

c. $x^3 + 2xy = x^2y^2.$

$$3x^2\frac{dx}{dt} + 2\left(\frac{dx}{dt}y + x\frac{dy}{dt}\right) = 2x\frac{dx}{dt}y^2 + 2y\frac{dy}{dt}x^2.$$

$$2x\frac{dy}{dt} - 2yx^2\frac{dy}{dt} = 2xy^2\frac{dx}{dt} - 3x^2\frac{dx}{dt} - 2y\frac{dx}{dt}.$$

$$(2x - 2yx^2)\frac{dy}{dt} = (2xy^2 - 3x^2 - 2y)\frac{dx}{dt}.$$

$$\frac{dy}{dt} = \frac{(2xy^2 - 3x^2 - 2y)\frac{dx}{dt}}{2x - 2yx^2}.$$

5. $x = 0.05t^2 + 2t + 5.$

$P = 0.001x^2 + 0.1x - 0.25.$

$\frac{dP}{dt} = \frac{dP}{dx} \cdot \frac{dx}{dt}.$

$\frac{dP}{dx} = 0.002x + 0.1.$

$\frac{dx}{dt} = 0.1t + 2.$

$\frac{dP}{dt} = (0.002x + 0.1)(0.1t + 2).$

When $t = 5$, $x = 0.05(5)^2 + 2(5) + 5 = 16.25$, so

$\frac{dP}{dt} = (0.002(16.25) + 0.1) \cdot 2.5 = 0.33125$ million dollars per month.

6. $\frac{dV}{dt} = \frac{4}{3}\pi 3r^2 \frac{dr}{dt} = 4\pi r^2 \frac{dr}{dt} \implies \frac{dr}{dt} = \frac{1}{4\pi r^2} \frac{dV}{dt}.$

$\frac{dV}{dt} = 500.$

So when $r = 10$, $\frac{dr}{dt} = \frac{1}{4\pi(10)^2} \cdot 500 = \frac{1}{400\pi} \cdot 500 = \frac{5}{4\pi}$ cm/sec.

7. Let $x =$ the length of a side of the base in feet.

Let $y =$ the height of the shed in feet.

Let $C =$ the total cost of material for the shed in dollars.

$C = 4x^2 + 2x^2 + 2.5(4)(xy).$

The volume of the shed $= 150$ feet$^3 = x^2 y$.

So $y = \frac{150}{x^2}$.

$C = 6x^2 + 10x\left(\frac{150}{x^2}\right) = 6x^2 + \frac{1500}{x}.$

$\frac{dC}{dx} = 12x - \frac{1500}{x^2}.$

$\frac{dC}{dx} = 0 \implies 12x = \frac{1500}{x^2} \implies 12x^3 = 1500 \implies x^3 = 125 \implies x = 5.$

For $0 < x < 5$, $\frac{dC}{dx} < 0$; for $5 < x$, $\frac{dC}{dx} > 0$.

Therefore the minimum cost occurs when $x = 5$ feet and $y = 6$ feet.

8. $x =$ the toll in dollars per car.

$y =$ the number of cars per day using the toll road.

$y = mx + b.$

$m = -\frac{300}{0.01} \implies y = -30{,}000x + b.$

When $x = 1$, $y = 36{,}000 \implies 36{,}000 = -30{,}000(1) + b$

$\implies b = 66{,}000$. So $y = -30{,}000x + 66{,}000.$

$R =$ the revenue in dollars per day.

$R = x \cdot y = -30,000x^2 + 66,000x.$

$\frac{dR}{dx} = -60,000x + 66,000.$

$\frac{dR}{dx} = 0$ when $x = 1.1.$

For $0 < x < 1.1$, $\frac{dR}{dx} > 0$; for $x > 1.1$, $\frac{dR}{dx} < 0$.

Therefore the maximum revenue occurs when the toll is $1.10 per car.

9.(a) $R(x) =$ the revenue.

$P(x) =$ the profit.

$R(x) = p \cdot x = 60x - (10^{-5})x^2.$

$P(x) = R(x) - C(x)$

$= 60x - (10^{-5})x^2 - 7 \cdot 10^6 - 30x = -(10^{-5})x^2 + 30x - 7 \cdot 10^6.$

$P'(x) = -2(10^{-5})x + 30.$

$P'(x) = 0$ for $x = \frac{30}{2(10^{-5})} = 15 \cdot 10^5.$

The maximum profit occurs when $x = 15 \cdot 10^5$ thousand kilowatt-hours, and $p = 60 - (10^{-5}) \cdot 15 \cdot 10^5 = \$45$ per (thousand kilowatt-hours).

b. The new profit function $= P_1(x) = 60x - (10^{-5})x^2 - 7 \cdot 10^6 - 40x$

$= -(10^{-5}) \cdot x^2 + 20x - 7 \cdot 10^6.$

$P_1'(x) = -2 \cdot 10^{-5}x + 20.$

$P_1'(x) = 0$ for $x = \frac{20}{2(10^{-5})} = 10 \cdot 10^5.$

Now, the maximum profit occurs when $x = 10 \cdot 10^5$ thousant kilowatt-hours.

The corresponding price $p_1 = 60 - (10^{-5}) \cdot 10 \cdot 10^5 = \$50.$

So, the company should pass $5 of this increase on to the consumers.

10.(a) The profit $P(x) = (200 - 3x) \cdot x - (75 + 80x - x^2)$

$= 200x - 3x^2 - 75 - 80x + x^2$

$= -2x^2 + 120x - 75.$

$P'(x) = -4x + 120 = 0$ for $x = 30.$

Therefore, the company's maximum profit occurs when $x = 30$, and the corresponding price $p = 200 - 3 \cdot (30) = \$110.$

b. Now, the new profit function equals

$$\begin{aligned} P_1(x) &= (200-3x)\cdot x - (75+80x-x^2+Tx) \\ &= 200x - 3x^2 - 75 - 80x + x^2 - Tx \\ &= -2x^2 + (120-T)x - 75. \end{aligned}$$

$P_1'(x) = -4x + (120-T) = 0$ for $x = \frac{120-T}{4} = 30 - \frac{T}{4}$.

Now the company's profit is maximized when $x = 30 - \frac{T}{4}$. The tax revenue received by the government for this value of x is $T\left(30 - \frac{T}{4}\right) = -\frac{T^2}{4} + 30T$.

$\frac{d}{dT}\left(-\frac{T^2}{4} + 30T\right) = -\frac{T}{2} + 30 = 0$ for $T = 60$.

The value of T which maximizes the revenue received by the government is 60 dollars per unit produced.

11.a. $y^2 - cy - (ax^2 + bx + d) = 0$.

$y = \frac{c \pm \sqrt{(-c)^2 - 4\cdot 1\cdot(-(ax^2+bx+d))}}{2\cdot 1}$.

$y = \frac{c + \sqrt{(4a)x^2+(4b)x+(c^2+4d)}}{2} = f_1(x)$ or $y = \frac{c - \sqrt{(4a)x^2+(4b)x+(c^2+4d)}}{2} = f_2(x)$.

b. $\left[\sqrt{(4a)x^2+(4b)x+(c^2+4d)} - 2\sqrt{a}\,x\right]$

$= \left[\sqrt{(4a)x^2+(4b)x+(c^2+4d)} - 2\sqrt{a}\,x\right] \cdot \frac{\sqrt{(4a)x^2+(4b)x+(c^2+4d)} + 2\sqrt{a}\,x}{\sqrt{(4a)x^2+(4b)x+(c^2+4d)} + 2\sqrt{a}\,x}$

$= \frac{\left[(4a)x^2+(4b)x+(c^2+4d)\right] - \left[(4a)x^2\right]}{\sqrt{(4a)x^2+(4b)x+(c^2+4d)} + 2\sqrt{a}\,x}$

$= \frac{(4b)x+(c^2+4d)}{\sqrt{(4a)x^2+(4b)x+(c^2+4d)} + 2\sqrt{a}\,x}$

$= \frac{(4b)+(c^2+4d)\cdot\frac{1}{x}}{\sqrt{(4a)+(4b)\cdot\frac{1}{x}+(c^2+4d)\cdot\frac{1}{x^2}} + 2\sqrt{a}}$

for large x. Therefore

$\lim_{x\to\infty}\left[\sqrt{(4a)x^2+(4b)x+(c^2+4d)} - 2\sqrt{a}\,x\right]$

$= \frac{4b}{\sqrt{4a}+2\sqrt{a}} = \frac{4b}{2\sqrt{a}+2\sqrt{a}} = \frac{4b}{4\sqrt{a}} = \frac{b}{\sqrt{a}}$.

c. $\lim_{x\to\infty}\left[\frac{\sqrt{(4a)x^2+(4b)x+(c^2+4d)}}{2} - \sqrt{a}\,x - \frac{b}{2\sqrt{a}}\right]$

$= \frac{1}{2}\cdot\lim_{x\to\infty}\left[\sqrt{(4a)x^2+(4b)x+(c^2+4d)} - 2\sqrt{a}\,x\right] - \frac{b}{2\sqrt{a}}$

$= \frac{1}{2}\cdot\frac{b}{\sqrt{a}} - \frac{b}{2\sqrt{a}} = 0$.

d. $f_1(x) = \frac{c}{2} + \frac{\sqrt{(4a)x^2+(4b)x+(c^2+4d)}}{2}$

$= \left[\frac{c}{2} + \sqrt{a}\,x + \frac{b}{2\sqrt{a}}\right] + \left[\frac{\sqrt{(4a)x^2+(4b)x+(c^2+4d)}}{2} - \sqrt{a}\,x - \frac{b}{2\sqrt{a}}\right].$

The function

$$y = L_1(x) = \frac{c}{2} + \sqrt{a}\,x + \frac{b}{2\sqrt{a}} = \sqrt{a}\,x + \left(\frac{c}{2} + \frac{b}{2\sqrt{a}}\right)$$

is linear. As shown in part c,

$$\lim_{x\to\infty}\left[\frac{\sqrt{(4a)x^2 + (4b)x + (c^2 + 4d)}}{2} - \sqrt{a}\,x - \frac{b}{2\sqrt{a}}\right] = 0.$$

Therefore the straight line defined by

$$y = L_1(x) = \sqrt{a}\,x + \left(\frac{c}{2} + \frac{b}{2\sqrt{a}}\right)$$

is an oblique asymptote for the graph of $y = f_1(x)$. Hence this straight line is an oblique asymptote for the graph of the original equation

$$ax^2 - y^2 + bx + cy + d = 0.$$

e. $f_2(x) = \left[\frac{c}{2} - \sqrt{a}\,x - \frac{b}{2\sqrt{a}}\right] + \left[-\frac{\sqrt{(4a)x^2+(4b)x+(c^2+4d)}}{2} + \sqrt{a}\,x + \frac{b}{2\sqrt{a}}\right].$

The function

$$y = L_2(x) = \frac{c}{2} - \sqrt{a}\,x - \frac{b}{2\sqrt{a}} = -\sqrt{a}\,x + \left(\frac{c}{2} - \frac{b}{2\sqrt{a}}\right)$$

is linear. Note that

$$\lim_{x\to\infty}\left[-\frac{\sqrt{(4a)x^2 + (4b)x + (c^2 + 4d)}}{2} + \sqrt{a}\,x + \frac{b}{2\sqrt{a}}\right]$$
$$= -\lim_{x\to\infty}\left[\frac{\sqrt{(4a)x^2 + (4b)x + (c^2 + 4d)}}{2} - \sqrt{a}\,x - \frac{b}{2\sqrt{a}}\right] = 0$$

as shown in part c. Therefore the straight line defined by

$$y = L_2(x) = -\sqrt{a}\,x + \left(\frac{c}{2} - \frac{b}{2\sqrt{a}}\right)$$

is an oblique asymptote for the graph of $y = f_2(x)$. Hence this straight line is an oblique asymptote for the graph of the original equation

$$ax^2 - y^2 + bx + cy + d = 0.$$

f. $f_1'(x) = \frac{1}{2} \cdot \frac{1}{2} \cdot \left[(4a)x^2 + (4b)x + (c^2 + 4d)\right]^{-1/2} \cdot (8ax + 4b)$
$= \frac{2a + \frac{b}{x}}{\sqrt{(4a) + (4b) \cdot \frac{1}{x} + (c^2 + 4d) \cdot \frac{1}{x^2}}}$
for large x. Therefore

$$\lim_{x \to \infty} f_1'(x) = \frac{2a}{\sqrt{4a}} = \sqrt{a}\ .$$

Similarly

$$\lim_{x \to \infty} f_2'(x) = -\sqrt{a}\ .$$

g. $2x^2 - y^2 + 9x - 4y + 5 = 0.$
$2\left(x^2 + \frac{9}{2}x + \left(\frac{9}{4}\right)^2\right) - \left(y^2 + 4y + (2)^2\right) = -5 + \frac{81}{8} - 4.$
$2 \cdot \left(x + \frac{9}{4}\right)^2 - (y + 2)^2 = \frac{9}{8}.$
The graph is symmetric about the vertical straight line $x = -\frac{9}{4}$ and the horizontal straight line $y = -2$. The point $\left(-\frac{9}{4}, -2\right)$ is the center of symmetry.

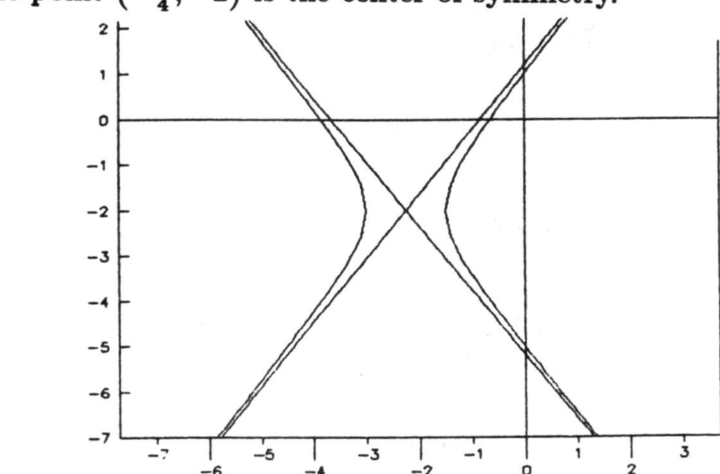

h. $2x^2 - y^2 - 9x + 7y + 5 = 0.$
$2\left(x^2 - \frac{9}{2}x + \left(-\frac{9}{4}\right)^2\right) - \left(y^2 - 7y + \left(-\frac{7}{2}\right)^2\right) = -5 + \frac{81}{8} - \frac{49}{4}.$
$2\left(x - \frac{9}{4}\right)^2 - \left(y - \frac{7}{2}\right)^2 = -\frac{57}{8}.$
The graph is symmetric about the vertical straight line $x = \frac{9}{4}$ and the horizontal

straight line $y = \frac{7}{2}$.

The point $\left(\frac{9}{4}, \frac{7}{2}\right)$ is the center of symmetry.

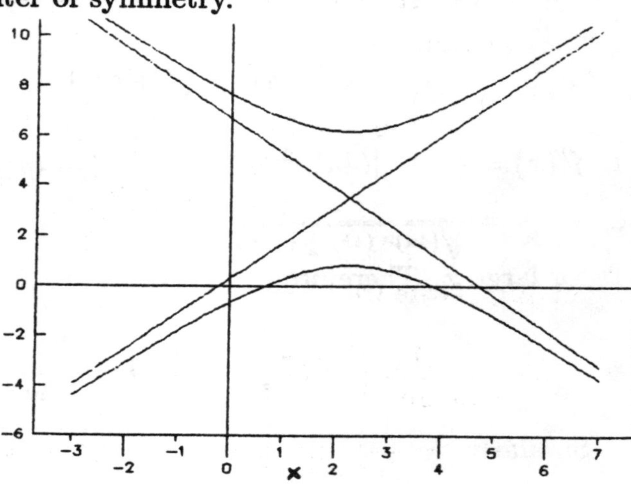

i.

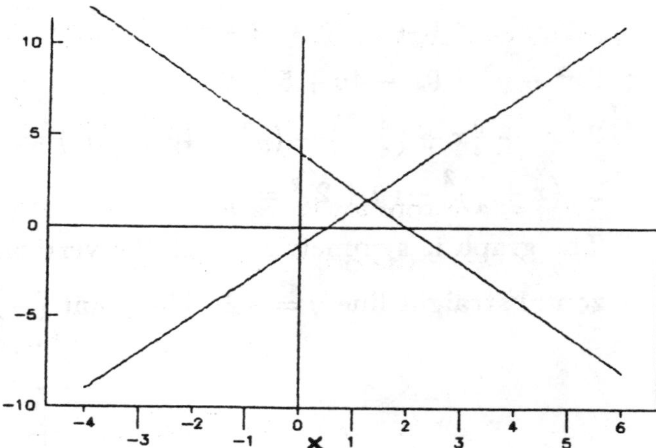

12.a. $cy^2 + (bx)y + (ax^2 + d) = 0$.

$y = \frac{-(bx) \pm \sqrt{(bx)^2 - 4 \cdot c \cdot (ax^2 + d)}}{2c}$.

$y = \frac{1}{2c} \cdot \left\{-(bx) + \sqrt{(b^2 - 4ac)x^2 - 4cd}\right\} = f_1(x)$ or

$y = \frac{1}{2c} \cdot \left\{-(bx) - \sqrt{(b^2 - 4ac)x^2 - 4cd}\right\} = f_2(x)$.

b. $\left[\sqrt{(b^2 - 4ac)x^2 - 4cd} - \sqrt{b^2 - 4ac}\ x\right]$

$= \left[\sqrt{(b^2 - 4ac)x^2 - 4cd} - \sqrt{b^2 - 4ac}\ x\right] \cdot \frac{\sqrt{(b^2-4ac)x^2-4cd} + \sqrt{b^2-4ac}\ x}{\sqrt{(b^2-4ac)x^2-4cd} + \sqrt{b^2-4ac}\ x}$

$= \frac{\left[(b^2-4ac)x^2-4cd\right] - \left[(b^2-4ac)x^2\right]}{\sqrt{(b^2-4ac)x^2-4cd} + \sqrt{b^2-4ac}\ x}$

$= \frac{-4cd}{\sqrt{(b^2-4ac)x^2-4cd} + \sqrt{b^2-4ac}\ x}$

$\to 0$ as $x \to \infty$.

c. $f_1(x) = \frac{1}{2c} \cdot \{-(bx) + \sqrt{b^2 - 4ac}\, x\}$
$+\frac{1}{2c} \cdot \left\{\sqrt{(b^2 - 4ac)x^2 - 4cd} - \sqrt{b^2 - 4ac}\, x\right\}.$

The function

$$y = L_1(x) = \frac{1}{2c} \cdot \left\{-(bx) + \sqrt{b^2 - 4ac}\, x\right\}$$
$$= \frac{-b + \sqrt{b^2 - 4ac}}{2c}\, x$$

is linear. Note that

$$\lim_{x \to \infty} \frac{1}{2c} \cdot \left\{\sqrt{(b^2 - 4ac)x^2 - 4cd} - \sqrt{b^2 - 4ac}\, x\right\}$$
$$= \frac{1}{2c} \cdot \lim_{x \to \infty} \left[\sqrt{(b^2 - 4ac)x^2 - 4cd} - \sqrt{b^2 - 4ac}\, x\right] = 0$$

in view of part b. Therefore the straight line defined by

$$y = L_1(x) = \frac{-b + \sqrt{b^2 - 4ac}}{2c}\, x$$

is an asymptote for the graph of $y = f_1(x)$. Hence this line is an asymptote for the graph of the original equation

$$ax^2 + bxy + cy^2 + d = 0.$$

d. $f_2(x) = \frac{1}{2c} \cdot \{-(bx) - \sqrt{b^2 - 4ac}\, x\}$
$+ \frac{1}{2c} \cdot \left\{-\left[\sqrt{(b^2 - 4ac)x^2 - 4cd} - \sqrt{b^2 - 4ac}\, x\right]\right\}.$

In view of part b, the straight line defined by

$$y = L_2(x) = \frac{1}{2c} \cdot \left\{-(bx) - \sqrt{b^2 - 4ac}\, x\right\}$$
$$= \frac{-b - \sqrt{b^2 - 4ac}}{2c}\, x$$

is an asymptote for the graph of $y = f_2(x)$. Hence this line is an asymptote for the graph of the original equation

$$ax^2 + bxy + cy^2 + d = 0.$$

e. $6x^2 + 15xy + 4y^2 + 5 = 0$.

The axes of symmetry can be seen in the graph. Each of these axes of symmetry is halfway between the asymptotes. In our later study of trigonometry we will show that these axes of symmetry are the straight lines defined by

$$y = -1.142183063x$$

and

$$y = 0.875516396x.$$

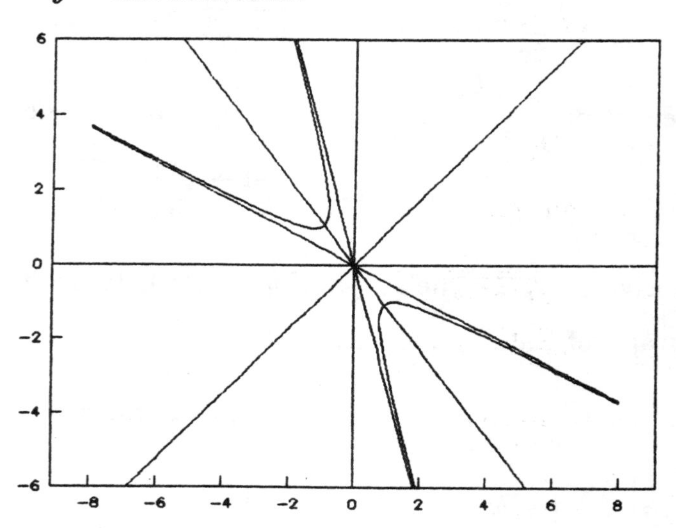

13.a. Let $f(x) = x^{1/3}$.

$f'(x) = \frac{1}{3}x^{-2/3}$.

$f(1000) = 10$ and $f'(1000) = \frac{(1000)^{-2/3}}{3} = \frac{\frac{1}{100}}{3} = \frac{1}{300}$.

$\sqrt[3]{1005} \approx f(1000) + f'(1000) \cdot (1005 - 1000)$

$= 10 + \frac{1}{300} \cdot (5) = \frac{601}{60} \approx 10.0167.$

b. Let $f(x) = x^{3/2}$.

$f'(x) = \frac{3}{2}x^{1/2}$.

$f(25) = 125$ and $f'(25) = \frac{3}{2} \cdot (25)^{1/2} = \frac{15}{2}$.

$(26)^{3/2} \approx f(25) + f'(25) \cdot (26 - 25)$

$= 125 + \frac{15}{2} \cdot (1) = 132.5.$

c. Let r be the radius of the hemispherical dome in meters.

Let S be the corresponding surface area of the dome in square meters. We have that

$$S = 2\pi r^2 = f(r).$$

The maximum resulting error in the calculated surface area of the dome is approximately

$$f'(100) \cdot (.01) = 4\pi \cdot 1000 \cdot .01 \approx 125.6637 \text{ meters}^2.$$

14.a. $R(p) = p \cdot D(p) = \frac{500p}{10+\sqrt{p(p+24)}}$

b. $D'(p) = 500 \cdot (-1)\left[10 + \sqrt{p^2+24p}\right]^{-2} \cdot \frac{1}{2}\left(p^2+24p\right)^{-1/2} \cdot (2p+24)$

$= -\frac{500\cdot(p+12)}{\sqrt{p^2+24p}\cdot\left[10+\sqrt{p^2+24p}\right]^2}.$

$E(p) = \frac{-p\cdot D'(p)}{D(p)} = \frac{500p\cdot(p+12)}{\sqrt{p^2+24p}\cdot\left[10+\sqrt{p^2+24p}\right]^2} \cdot \frac{10+\sqrt{p^2+24p}}{500}$

$= \frac{p^2+12p}{\sqrt{p^2+24p}\cdot\left[10+\sqrt{p^2+24p}\right]} = \frac{p^2+12p}{10\cdot\sqrt{p^2+24p}+(p^2+24p)}.$

c. Note that

$$10 \cdot \sqrt{p^2+24p} + (p^2+24p) > (p^2+12p)$$

for all $p > 0$. Therefore $E(p) < 1$ for all $p > 0$.

d. Observe that

$$\begin{aligned} R'(p) &= 1 \cdot D(p) + p \cdot D'(p) = D(p) \cdot \left[1 + \frac{p \cdot D'(p)}{D(p)}\right] \\ &= D(p) \cdot \left[1 - \left(-\frac{p \cdot D'(p)}{D(p)}\right)\right] \\ &= D(p) \cdot [1 - E(p)] > 0 \quad \text{for all } p > 0. \end{aligned}$$

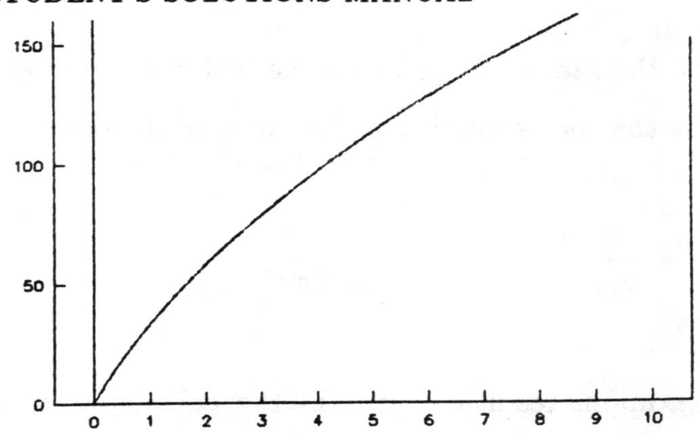

15. The average cost

$$c(x) = \frac{C(x)}{x} = \frac{400}{x} + 10 + \frac{1}{2}x .$$

Then

$$c'(x) = -\frac{400}{x^2} + \frac{1}{2} .$$

Note that

$$c'(x) = 0$$

for

$$\frac{400}{x^2} = \frac{1}{2}$$

$$x^2 = 800$$

$$x = \sqrt{800} = 20\sqrt{2}.$$

For $0 < x < 20\sqrt{2}$, $c'(x) < 0$; for $x > 20\sqrt{2}$, $c'(x) > 0$.

Thus the minimum average cost occurs when $x = 20\sqrt{2}$, and the minimum average cost is

$$\begin{aligned} c\left(20\sqrt{2}\right) &= \frac{400}{20\sqrt{2}} + 10 + \frac{1}{2} \cdot 20\sqrt{2} \\ &= \frac{20}{\sqrt{2}} + 10 + 10\sqrt{2} = \frac{20\sqrt{2}}{2} + 10 + 10\sqrt{2} \\ &= 10\sqrt{2} + 10 + 10\sqrt{2} = 10 + 20\sqrt{2}. \end{aligned}$$

CHAPTER 4

SUMMARY OF CHAPTER 4

1. The definitions of exponential functions on integers, rational numbers, and irrational numbers.

2. Properties of exponential functions.

3. Graphs of exponential functions.

4. Compound interest as an application of exponential functions.

5. The number $e = \lim_{x \to \infty} \left(1 + \frac{1}{x}\right)^x$ and continuous compounding of interest.

6. The definition of logarithm functions.
   Logarithm and exponential functions are inverses of each other.

7. Properties of logarithms.

8. Common logarithms and natural logarithms.

9. Fitting exponential models to data.

10. Effective annual yields on investments.

11. The derivative of the natural logarithm function, and then use of the Chain Rule to differentiate the composite function $f(x) = \ln[u(x)]$.

12. Use of the technique of logarithmic differentiation to determine the derivatives of two general types of functions.

13. The growth function as a logarithmic derivative.

14. The derivative of the natural exponential function, and then use of the Chain Rule to differentiate a composite function $f(x) = e^{u(x)}$.

15. Optimal harvesting as an application of the rate of growth of a function.

16. Derivatives of other exponential and logarithm functions.

17. The Law of Natural Growth (and Decay).

18. The idea of mathematical modelling.

19. Logistic curves, and the logistic model for bounded growth.

20. The Normal Distribution.

Selected Solutions to Exercise Set 4.1

1. a. $9^{3/2} = 27.$ b. $16^{1/4} = 2.$

   c. $49^{3/2} = 7^3 = 343.$ (d) $4^{-3/2} = 2^{-3} = \frac{1}{8}.$

3. a. $\left(\frac{1}{4}\right)^{3/2} = \frac{1}{2^3} = \frac{1}{8}.$ b. $\left(\frac{27}{8}\right)^{2/3} = \frac{27^{2/3}}{8^{2/3}} = \frac{3^2}{2^2} = \frac{9}{4}.$

   c. $\left(\frac{1}{16}\right)^{-3/4} = 16^{3/4} = 2^3 = 8.$ d. $\left(\frac{81}{36}\right)^{-3/2} = \frac{36^{3/2}}{81^{3/2}} = \frac{6^3}{9^3} = \frac{8}{27}.$

5. $f(x) = 3^{-x}.$

x	-1	$-\frac{1}{2}$	0	$\frac{1}{2}$	1
$f(x)$	3	1.73	1	0.58	$\frac{1}{3}$

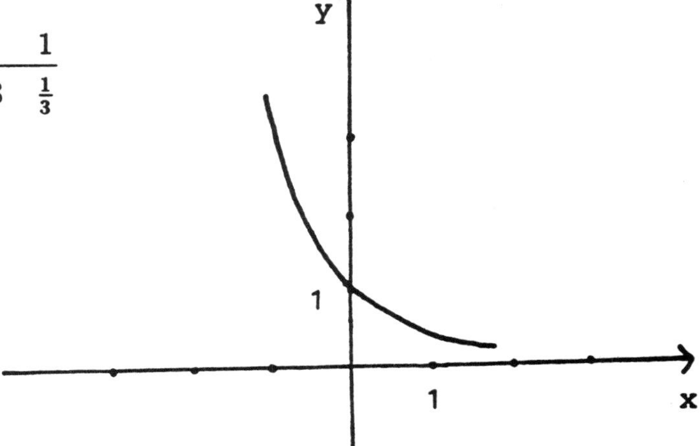

7. $f(x) = 10^{-x}.$

x	-1	0	1	2
$f(x)$	10	1	$\frac{1}{10}$	$\frac{1}{100}$

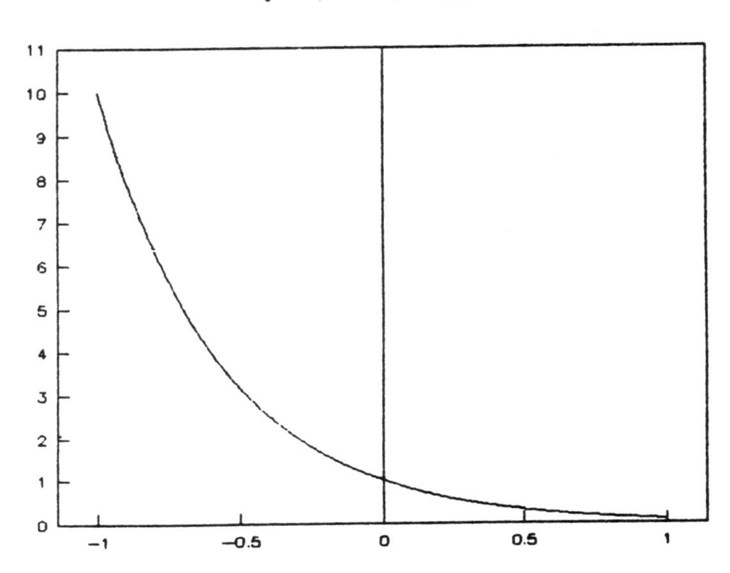

9. $f(x) = e^{-x}$.

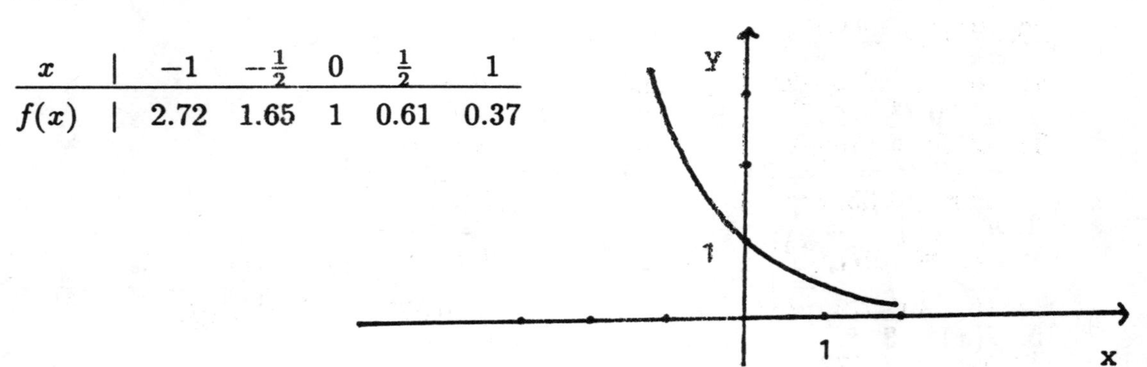

x	-1	$-\frac{1}{2}$	0	$\frac{1}{2}$	1
$f(x)$	2.72	1.65	1	0.61	0.37

11. $f(x) = -3e^{-x}$.

x	-1	$-\frac{1}{2}$	0	$\frac{1}{2}$	1	2
$f(x)$	-8.15	-4.95	-3	-1.82	-1.10	-0.41

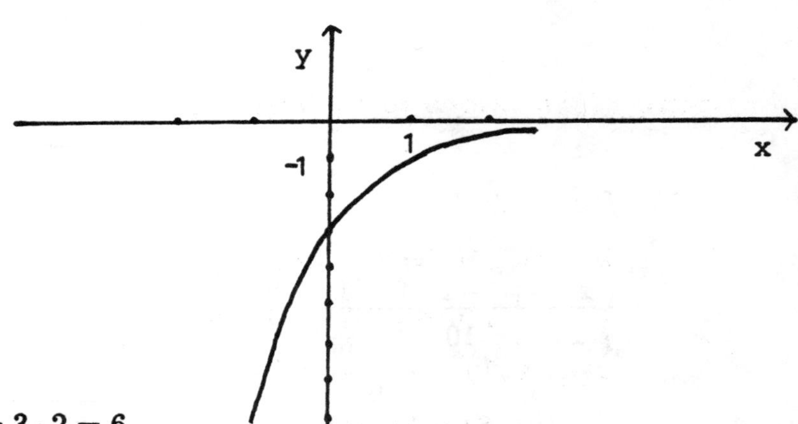

13. $\frac{27^{4/3} \cdot 4^3}{9^{3/2} \cdot 32} = \frac{3^4 \cdot (2^2)^3}{3^3 \cdot 2^5} = \frac{3^4 \cdot 2^6}{3^3 \cdot 2^5} = 3 \cdot 2 = 6.$

15. $\frac{[2^3 \cdot 3^{-2}]^2}{\frac{1}{3}(8)} = \frac{2^6 \cdot 3^{-4}}{3^{-1} 2^3} = \frac{2^{6-3}}{3^{-1+4}} = \frac{2^3}{3^3} = \frac{8}{27}.$

17. $e^3 \cdot e^x = e^{3+x}.$

19. $\sqrt{16e^{4x}} = (16)^{1/2}(e^{4x})^{1/2} = 4 \cdot e^{2x}.$

21. $\frac{(4e^{3x})^{2/3}}{\sqrt{8e^4}} = \frac{2^{4/3} e^{2x}}{2^{3/2} e^2} = 2^{\frac{4}{3} - \frac{3}{2}} \cdot e^{2x-2} = 2^{-1/6} e^{2x-2} = \frac{e^{2x-2}}{\sqrt[6]{2}}.$

23. a. $P(1) = \left(1 + \frac{0.08}{2}\right)^{1 \cdot 2} \cdot 500 = 1.0816 \cdot 500 = \$540.80.$

b. $P(2) = \left(1 + \frac{0.08}{2}\right)^{2 \cdot 2} \cdot 500 = 1.16986 \cdot 500 = \$584.93.$

c. $P(1.5) = \left(1 + \frac{0.08}{2}\right)^{1.5 \cdot 2} \cdot 500 = 1.124864 \cdot 500 = \$562.43.$

25. $P_0 = \frac{P(t)}{e^{rt}} = \frac{1000}{e^{0.10 \cdot 4}} = \frac{1000}{1.492} = \$670.32.$

27. $P_0 = \frac{P(t)}{e^{rt}} = \frac{10{,}000}{e^{0.08 \cdot 5}} = \frac{10{,}000}{1.492} = \$6703.20.$

29. $P(7) = e^{0.07 \cdot 7} \cdot 10{,}000 = 1.6323 \cdot 10{,}000 = 16{,}323.16.$

31. $P_0 = \frac{P(t)}{e^{rt}} = \frac{10{,}000}{e^{0.05 \cdot 6}} = \frac{10{,}000}{1.34986} = \$7408.18.$

33. $e^{rt}P_0 = (1 + 0.08)^t P_0$. For $t = 1$, $e^r = 1.08$.

    If $r = 0.05$, $e^{0.05} \approx 1.05$; if $r = 0.06$, $e^{0.06} \approx 1.06$;

    if $r = 0.07$, $e^{0.07} \approx 1.07$; if $r = 0.08$, $e^{0.08} \approx 1.083$;

    if $r = 0.073$, $e^{0.073} \approx 1.076$; if $r = 0.077$, $e^{0.077} \approx 1.08$.

    So $r \approx 0.077$.

35. a. $(1 + r) = \left(1 + \frac{0.10}{2}\right)^2$.

    $1 + r = 1.1025$, $r = 0.1025$ or 10.25%.

    b. $(1 + r) = e^{.1}$.

    $1 + r = 1.1052$, $r = 0.1052$ or 10.52%.

Selected Solutions to Exercise Set 4.2

1. a. $\log_{10} 100 = 2$ since $10^2 = 100$.

   b. $\log_{10} 10 = 1$ since $10^1 = 10$.

   c. $\log_2 16 = 4$ since $2^4 = 16$.

   d. $\log_4 64 = 3$ since $4^3 = 64$.

   e. $\log_3 81 = 4$ since $3^4 = 81$.

   f. $\ln e^2 = 2$ since $e^2 = e^2$.

3. a. $\ln e = 1$. b. $\ln 1 = 0$.

   c. $\ln(2.2) = 0.78846$. d. $\ln(0.3) = -1.20397$.

   e. $\ln(e^3) = 3$. f. $\ln\left(\frac{1}{e^2}\right) = -2$.

5. Set $a = 10$ in the formulas in Ex. 4; then

   $\log_{10} x = \frac{\ln x}{\ln 10}$ and $\ln x = (\ln 10) \log_{10} x$.

7. $e^x = 1$. $x = \ln 1 = 0$.

9. $e^{x^2-3} = 5 \implies x^2 - 3 = \ln 5 \implies x^2 = \ln 5 + 3 \implies x^2 = 4.6094 \implies x = \pm 2.147.$

11. $e^{x+\ln x} = 2x \implies x + \ln x = \ln 2x \implies x + \ln x = \ln 2 + \ln x \implies x = \ln 2.$

13. $\ln x^2 = 8. \quad x^2 = e^8, \quad x = \pm(e^8)^{1/2} = \pm e^4.$

15. $e^{\ln(2x+3)} = 7.$

By identity (4), $e^{\ln(2x+3)} = 2x + 3.$

Then $2x + 3 = 7$, $x = \frac{7-3}{2} = 2.$

17. $\log_{10} P = \frac{-A}{t+C} + B$. Since $\log_{10} P = \frac{\ln P}{\ln 10}$,

then $\frac{\ln P}{\ln 10} = \frac{-A}{t+C} + B. \quad \ln P = (\ln 10)\left(\frac{-A}{t+C} + B\right).$

19. $p = 100e^{-x}.$

$e^{-x} = \frac{p}{100}, \quad e^x = \frac{100}{p}, \quad x = \ln \frac{100}{p} = \ln 100 - \ln p.$

21. $T = \frac{1}{0.1} \ln\left(\frac{1500}{1000}\right) = 10(0.40547) \approx 4.05$ yrs.

23. $T = \frac{1}{0.1} \ln\left(\frac{2P_0}{P_0}\right) = 10 \ln 2 \approx 6.93$ yrs.

25. $p(x) = 100e^{-0.02x}$. When $p = 50, \quad 50 = 100e^{-.02x},$

$e^{-0.02x} = \frac{1}{2}, \quad e^{0.02x} = 2, \quad 0.02x = \ln 2, \quad x = \frac{\ln 2}{0.02} = 34.7 \approx 35$ items.

27. $f(t) = a + b \ln t.$

Since $f(1) = 15, \quad 15 = a + b \ln 1, \quad 15 = a + b \cdot 0, \quad a = 15.$

Since $f(e) = 20, \quad 20 = a + b \ln e, \quad 20 = 15 + b(1), \quad b = 5.$

29. Let $P(t)$ denote the value of the store after t years.

Let P_0 denote the original value of the store.

Let r denote the annual return.

$P(t) = e^{rt} P_0.$

$P_0 = \$1$ million.

$P(5) = \$2$ million $\implies 2 = e^{r(5)}1 \implies e^{5r} = 2 \implies 5r = \ln 2 \implies r = \frac{\ln 2}{5} = .1386$ or 13.86%.

31. $P(5) = 100{,}000e^{.15(5)} = \$211{,}700.$

33. Let $P(t)$ denote the value of the gift after t years.

Let P_0 denote the initial value of the gift.

Let r denote the nominal interest rate.

$P(t) = e^{rt}P_0$.

$P(6) = 2P_0 \implies 2P_0 = e^{r(6)}P_0 \implies e^{6r} = 2 \implies 6r = \ln 2 \implies r = \frac{\ln 2}{6} =$ $.1155$ or 11.55%.

35. $P(t) = 200,000e^{rt}$.

$P(4) = 140,000 \implies 140,000 = 200,000e^{r(4)} \implies e^{4r} = .7 \implies 4r = \ln .7 \implies r = \frac{\ln .7}{4} = -.089$ or -8.9%.

Therefore the house depreciated at the rate of 8.9% per year.

37. a. Let $u = \log_b x$ and $v = \log_b y$.

Then $x = b^u$ and $y = b^v$ by the definition of a logarithm.

b. Use the fact that $xy = b^u \cdot b^v = b^{u+v}$.

Then $\log_b(xy) = \log_b(b^u \cdot b^v) = \log_b(b^{u+v}) = u + v$.

c. From (a), $u = \log_b x$ and $v = \log_b y$.

Then by (b), $\log_b(xy) = \log_b x + \log_b y$.

39. a. Let $u = \log_b x$. Then $x = b^u$ by the definition of a logarithm.

b. Then $x^r = (b^u)^r = b^{ru}$. $\log_b x^r = \log_b b^{ru} = ru$.

c. Since $u = \log_b x$, then by b. $\log_b x^r = r \log_b x$.

Selected Solutions to Exercise Set 4.3

1. $y' = \frac{1}{2x} \cdot 2 = \frac{1}{x}$.

3. $f(x) = 10\ln(x^2 + 5)$.

$f'(x) = 10 \cdot \frac{1}{x^2+5} \cdot \frac{d}{dx}(x^2 + 5) = 10 \cdot \frac{1}{x^2+5} \cdot 2x = \frac{20x}{x^2+5}$.

5. $y' = x \cdot \frac{1}{x} + \ln x$. $y' = 1 + \ln x$.

7. $f'(x) = \frac{1}{\sqrt{x^3-x}} \cdot \frac{1}{2}(x^3 - x)^{-1/2}(3x^2 - 1) = \frac{3x^2-1}{2(x^3-x)}$.

9. $f'(x) = 3\left[\ln(x^2 - 2x)\right]^2 \cdot \frac{(2x-2)}{x^2-2x}$.
$f'(x) = \frac{3\ln^2(x^2-2x)(2x-2)}{x^2-2x}$.

11. $y' = \frac{1}{\ln t} \cdot \frac{1}{t} = \frac{1}{t \ln t}$.

13. $f(x) = \frac{x^3}{1+\ln x}$.
$f'(x) = \frac{(1+\ln x)\cdot(3x^2) - x^3 \cdot \left(\frac{1}{x}\right)}{(1+\ln x)^2} = \frac{x^2(3(1+\ln x)-1)}{(1+\ln x)^2} = \frac{x^2(2+3\ln x)}{(1+\ln x)^2}$.

15. $y' = 4\left(3\ln\sqrt{x}\right)^3 \cdot \left(3 \cdot \frac{1}{\sqrt{x}} \cdot \frac{1}{2}x^{-1/2}\right)$. $y' = \frac{162\ln^3\sqrt{x}}{x}$.

17. $f'(x) = x^2 \cdot \frac{1}{3x-6} \cdot 3 + \ln(3x-6) \cdot 2x = \frac{3x^2}{3x-6} + 2x\ln(3x-6)$.

19. $f'(x) = \ln x \frac{1}{\sqrt{x}} \cdot \frac{1}{2}x^{-1/2} + \ln\sqrt{x} \cdot \frac{1}{x}$
$= \frac{\ln x}{2x} + \frac{\ln\sqrt{x}}{x} = \frac{\ln x + 2\ln\sqrt{x}}{2x} = \frac{\ln x + \ln x}{2x} = \frac{\ln x}{x}$.

21. $f(x) = \frac{x+\ln^2(3x)}{\sqrt{1+\ln x}}$.
$f'(x) = \frac{(1+\ln x)^{1/2}\cdot\left(1+2\ln(3x)\cdot\frac{1}{x}\right) - \left[x^2+\ln^2(3x)\right]\cdot\left[\frac{1}{2}(1+\ln x)^{-1/2}\cdot\frac{1}{x}\right]}{(1+\ln x)}$.

23. $f(x) = \ln(x^3+3)^{4/3}$. $f(x) = \frac{4}{3}\ln(x^3+3)$.
$f'(x) = \frac{4}{3} \cdot \frac{1}{x^3+3} \cdot 3x^2 = \frac{4x^2}{x^3+3}$.

25. $f(x) = \ln(x-6)^{2/3} - \ln\sqrt{1+x}$.
$f'(x) = \frac{1}{(x-6)^{2/3}} \cdot \frac{2}{3}(x-6)^{-1/3} - \frac{1}{\sqrt{1+x}} \cdot \frac{1}{2}(1+x)^{-1/2} = \frac{2}{3(x-6)} - \frac{1}{2(1+x)}$.

27. $\ln y = \ln(x+3)^x$. $\ln y = x\ln(x+3)$.
$\frac{1}{y} \cdot \frac{dy}{dx} = x \cdot \frac{1}{x+3} + \ln(x+3)$. $\frac{1}{y} \cdot \frac{dy}{dx} = \frac{x}{x+3} + \ln(x+3)$.
$\frac{dy}{dx} = (x+3)^x\left[\frac{x}{x+3} + \ln(x+3)\right]$.

29. $\ln y = \sqrt{x+1}\ln x$.
$\frac{1}{y} \cdot \frac{dy}{dx} = \sqrt{x+1} \cdot \frac{1}{x} + \ln x \cdot \frac{1}{2}(x+1)^{-1/2}$.
$\frac{dy}{dx} = x^{\sqrt{x+1}}\left(\frac{\sqrt{x+1}}{x} + \frac{\ln x}{2\sqrt{x+1}}\right)$.

31. $y = \frac{x(x+1)(x+2)}{(x+3)(x+4)}$.
$\ln y = \ln x(x+1)(x+2) - \ln(x+3)(x+4)$.
$\ln y = \ln x + \ln(x+1) + \ln(x+2) - \ln(x+3) - \ln(x+4)$.
$\frac{1}{y} \cdot \frac{dy}{dx} = \frac{1}{x} + \frac{1}{x+1} + \frac{1}{x+2} - \frac{1}{x+3} - \frac{1}{x+4}$.
$\frac{dy}{dx} = \frac{x(x+1)(x+2)}{(x+3)(x+4)}\left[\frac{1}{x} + \frac{1}{x+1} + \frac{1}{x+2} - \frac{1}{x+3} - \frac{1}{x+4}\right]$.

33. $y = \frac{(x^2+3)(1-x)^{2/3}}{x\sqrt{1+x}}$.

$\ln y = \ln(x^2+3) + \frac{2}{3}\ln(1-x) - \ln x - \frac{1}{2}\ln(1+x)$.

$\frac{1}{y}\frac{dy}{dx} = \frac{2x}{x^2+3} + \frac{2}{3}\cdot\left(\frac{1}{1-x}\cdot(-1)\right) - \frac{1}{x} - \frac{1}{2}\cdot\frac{1}{1+x}$.

$\frac{dy}{dx} = \frac{(x^2+3)(1-x)^{2/3}}{x\sqrt{1+x}}\cdot\left(\frac{2x}{x^2+3} - \frac{2}{3}\cdot\frac{1}{1-x} - \frac{1}{x} - \frac{1}{2}\cdot\frac{1}{1+x}\right)$.

35. $\ln x + \ln y = x + y$. $\quad \frac{1}{x} + \frac{1}{y}\cdot\frac{dy}{dx} = 1 + \frac{dy}{dx}$.

$\left(\frac{1}{y} - 1\right)\frac{dy}{dx} = 1 - \frac{1}{x}$. $\quad \frac{dy}{dx} = \frac{1-\frac{1}{x}}{\frac{1}{y}-1}$. $\quad \frac{dy}{dx} = \frac{y(x-1)}{x(1-y)}$.

37. $\frac{dy}{dx} = x\cdot 2\ln x\cdot\frac{1}{x} + (\ln x)^2 + \frac{\ln x - x\cdot\frac{1}{x}}{(\ln x)^2}$.

$\frac{dy}{dx} = 2\ln x + \ln^2 x + \frac{\ln x - 1}{(\ln x)^2}$. At $x = e$, $\frac{dy}{dx} = 2 + 1 = 3$.

The equation is $(y - 2e) = 3(x - e)$. $\quad y = 3x - e$.

39. $y = x - \ln x$. $\quad y' = 1 - \frac{1}{x} = 0$ for $x = 1$.

For $0 < x < 1$, $y' < 0$; and for $x > 1$, $y' > 0$.

So $y = 1 - \ln 1 = 1$ is the relative minimum when $x = 1$.

41. $y = \ln(x^2 - x)$. $\quad y' = \frac{1}{x^2-x}\cdot(2x-1)$.

y' does not exist for $x = 0$ and $x = 1$.

Since the domain of y is $x < 0$ and $x > 1$, y has no relative extrema.

43. $y = \ln(4 - x)$.

x	-1	0	1	2	3	$3\frac{1}{3}$	$3\frac{2}{3}$
y	1.61	1.39	1.10	0.69	0	-0.41	-1.10

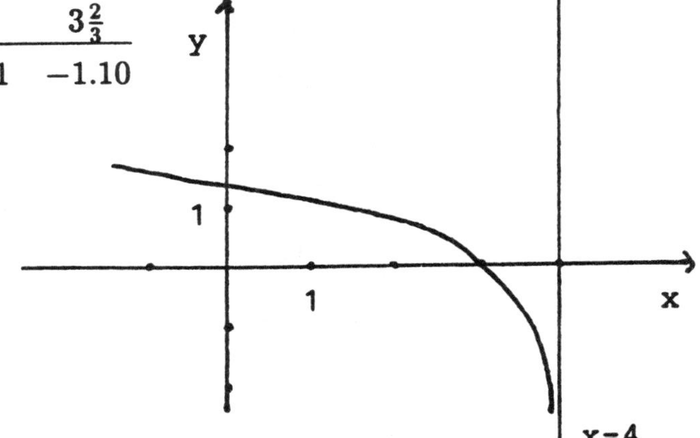

45. a. $C(x) = 20x + 300$.

$C'(x) = 20$.

$\frac{C'(20)}{C(20)} = \frac{20}{700} = .0286$ or 2.86%.

b. $R(x) = 60x - x^2.$

$R'(x) = 60 - 2x.$

$\frac{R'(20)}{R(20)} = \frac{20}{800} = .025$ or $2.5\%.$

c. $\frac{P'(20)}{P(20)} = \frac{R'(20)-C'(20)}{R(20)-C(20)} = \frac{20-20}{800-700} = 0.$

47. $P(t) = 750,000 + 30,000t + 10,000e^{-t}.$

$P'(t) = 30,000 + 10,000e^{-t}(-1).$

a. $\frac{P'(2)}{P(2)} \cdot 100 = \frac{28646.64717}{811353.3528} \cdot 100 = 3.53\%.$

b. $\frac{P'(5)}{P(5)} \cdot 100 = \frac{29932.62053}{900067.3795} \cdot 100 = 3.33\%.$

49. $V(t) = 10,000te^{-0.5t}.$

$V'(t) = 10,000\left((1)(e^{-0.5t} + t\left(e^{-0.5t}(-0.5)\right)\right)$

$= 10,000e^{-0.5t}(1 - 0.5t).$

Relative rate of growth $= \frac{V'(4)}{V(4)} = \frac{-1353.352832}{5413.41133} = -.25$

The relative rate of decline is .25.

The relative percent rate of decline is 25% .

Selected Solutions to Exercise Set 4.4

1. $f'(x) = 6e^{6x}.$

3. $f'(x) = 2x \cdot e^{x^2-4}.$

5. $f(x) = \frac{e^x}{x^2+2}.$

$f'(x) = \frac{(x^2+2)(e^x)-(e^x)(2x)}{(x^2+2)^2} = \frac{e^x(x^2-2x+2)}{(x^2+2)^2}.$

7. $f(x) = \sqrt{x + e^{-x}}.$

$f'(x) = \frac{1}{2}(x + e^{-x})^{-1/2}(1 - e^{-x}).$

9. $f'(x) = e^{x^2-\sqrt{x}} \cdot (2x - \frac{1}{2}x^{-1/2}).$

11. $f(x) = \ln \frac{e^x+1}{x^3+1} = \ln(e^x + 1) - \ln(x^3 + 1).$

$f'(x) = \frac{1}{e^x+1} \cdot e^x - \frac{1}{x^3+1} \cdot 3x^2.$

13. $y' = \frac{1}{2}(e^x - e^{-x}).$

15. $f'(x) = 4(x - e^{-2x})^3 \left(1 - e^{-2x}(-2)\right) = 4(x - e^{-2x})^3(1 + 2e^{-2x})$.

17. $f'(x) = 1 \cdot (e^{x^{-2}}) + (x)(e^{x^{-2}})(-2x^{-3}) = e^{x^{-2}}(1 - 2x^{-2})$.

19. $f(x) = \frac{xe^{-x}}{1+x^2}$.
$f'(x) = \frac{(1+x^2)\left((1)(e^{-x})+(x)\left(e^{-x}(-1)\right)\right)-(xe^{-x})(2x)}{(1+x^2)^2}$
$= \frac{e^{-x}(1+x^2-x-x^3-2x^2)}{(1+x^2)^2} = \frac{e^{-x}(1-x-x^2-x^3)}{(1+x^2)^2}$.

21. $e^{xy}\left(x \cdot \frac{dy}{dx} + y \cdot 1\right) = 1 \implies e^{xy}x\frac{dy}{dx} = 1 - e^{xy}y$.
$\implies \frac{dy}{dx} = \frac{1-e^{xy}y}{e^{xy}x}$.

23. $f'(x) = 1 \cdot (2^x) + x \cdot 2^x \cdot 1 \cdot \ln 2 = 2^x + x \cdot 2^x \cdot \ln 2$.

25. $y' = \frac{1}{2}(1 + 4^x)^{-1/2}(4^x)(1)\ln 4 = \frac{4^x \ln 4}{2\sqrt{1+4^x}}$.

27. $f'(x) = 2(\log_2 x^{1/2})\left\{\frac{1}{x^{1/2}\cdot\ln 2}\left(\frac{1}{2}x^{-1/2}\right)\right\}$
$= 2\left(\log_2 x^{1/2}\right)\frac{\frac{1}{2}\cdot x^{-1/2}}{x^{1/2}\cdot\ln 2} = \frac{\log_2 x^{1/2}}{x \ln 2}$.

29. $f(x) = (3 - x^2)e^x$.

$f'(x) = (3 - x^2)e^x + (-2x)e^x = e^x(3 - x^2 - 2x) = e^x(-x + 1)(x + 3)$.

$f'(x) = 0$ for $x = 1$ and $x = -3$.

Interval I	Test Number	$f'(x)$	Conclusion
$(-\infty, -3)$	-4	$f'(-4) < 0$	decreasing
$(-3, 1)$	-2	$f'(-2) > 0$	increasing
$(1, \infty)$	2	$f'(2) < 0$	decreasing

$\lim_{x\to-\infty} f(x) = 0$ and $f(1) = 2e > 0$, so the maximum is $f(1) = 2e$.

31. $y' = (1)(e^{1-x^3}) + (x)(e^{1-x^3})(-3x^2) = e^{1-x^3}(1 - 3x^3)$.

$y' = 0$ for $x = \sqrt[3]{\frac{1}{3}}$.

For $x < \sqrt[3]{\frac{1}{3}}$, $y' > 0$; for $x > \sqrt[3]{\frac{1}{3}}$, $y' < 0$.

So y has a relative maximum at $x = \sqrt[3]{\frac{1}{3}}$.

33. $p(x) = 100e^{-.5x}$.

$R(x) = x \cdot p(x) = 100xe^{-.5x}$.

$R'(x) = (100)(e^{-.5x}) + (100x)\left(e^{-.5x}(-.5)\right) = 100e^{-.5x}(1 - .5x)$.

$R'(x) = 0$ for $(1 - .5x) = 0 \implies x = 2$.

For $0 < x < 2$, $R'(x) > 0$, so $R(x)$ increases as x increases.

For $x > 2$, $R'(x) < 0$, so $R(x)$ decreases as x increases.

Therefore the maximum revenue occurs when $x = 2$.

The corresponding price is $100e^{-.5(2)} = 36.79$.

35. Revenue $R(x) = 10000x$. Cost $C(x) = 500 + 40x + e^{0.5x}$.

Profit $P(x) = 10000x - 500 - 40x - e^{0.5x} = 9960x - 500 - e^{0.5x}$.

$P'(x) = 9960 - e^{0.5x}(0.5) = 9960 - 0.5e^{0.5x}$.

The weekly production level x that maximizes profits occurs when

$P'(x) = 0 \implies 0.5e^{0.5x} = 9960 \implies \ln e^{0.5x} = \ln 19920$

$\implies 0.5x = \ln 19920 \implies x = 19.8$.

37. Rate of growth $= \frac{V'(4)}{V(4)} = \frac{2500 \ln 2\left(2^{\sqrt{4}}\right)\left(\frac{1}{\sqrt{4}}\right)}{5000\left(2^{\sqrt{4}}\right)} = \frac{\ln 2\left(\frac{1}{2}\right)}{2} = \frac{\ln 2}{4}$.

39. The present value of the revenue is

$R(t) = e^{-0.1t} \cdot 10,000(1.2)^{\sqrt{t}} = 10,000e^{-0.1t}(1.2)^{\sqrt{t}}$.

$$R'(t) = 10,000\{e^{-0.1t} \cdot (-0.1)(1.2)^{\sqrt{t}} + e^{-0.1t} \cdot (1.2)^{\sqrt{t}} \cdot \left(\tfrac{1}{2}t^{-1/2}\right) \cdot \ln 1.2\}$$
$$= 10,000\left[e^{-0.1t}(1.2)^{\sqrt{t}}\{-0.1 + \tfrac{1}{2}t^{-1/2} \ln 1.2\}\right].$$

$R'(t) = 0$ for $\frac{1}{2}t^{-1/2} \ln 1.2 = 0.1 \implies t^{-1/2} = \frac{0.1}{\frac{1}{2}\ln 1.2} \implies t^{1/2} = \frac{\frac{1}{2}\ln 1.2}{0.1} \implies$ $t = \frac{\frac{1}{4}\ln^2 1.2}{(0.1)^2} = 0.83$ year.

41. a. $f = \lim_{t\to\infty} y(t) = \frac{a}{1+0} + \frac{f-a}{1+0}$, since b_1 and b_2 are positive.

b. If $b_1 = b_2$ and $c_1 = c_2 = c$, then $y(t) = \frac{f}{1+e^{-b_1(t-c)}}$.

$$y'(t) = \frac{-f\cdot e^{-b_1(t-c)}\cdot(-b_1)}{\left(1+e^{-b_1(t-c)}\right)^2} = \frac{b_1\cdot f\cdot e^{-b_1(t-c)}}{\left(1+e^{-b_1(t-c)}\right)^2}.$$

$$y''(t) = b_1 f\left\{\left(1 + e^{-b_1(t-c)}\right)^2 \cdot e^{-b_1(t-c)} \cdot (-b_1) - e^{-b_1(t-c)} \cdot 2\left(1 + e^{-b_1(t-c)}\right) \cdot e^{-b_1(t-c)} \cdot (-b_1)\right\}$$
$$\div \left(1 + e^{-b_1(t-c)}\right)^4$$
$$= \frac{b_1 f\left\{\left(1+e^{-b_1(t-c)}\right)\cdot e^{-b_1(t-c)}\cdot(-b_1)+2b_1 e^{-2b_1(t-c)}\right\}}{\left(1+e^{-b_1(t-c)}\right)^3}$$
$$= \frac{b_1^2 f\{-e^{-b_1(t-c)} - e^{-2b_1(t-c)} + 2e^{-2b_1(t-c)}\}}{\left(1+e^{-b_1(t-c)}\right)^3}.$$
$$= \frac{b_1^2 f\cdot e^{-2b_1(t-c)}\{1-e^{b_1(t-c)}\}}{\left(1+e^{-b_1(t-c)}\right)^3}.$$

$y''(t) = 0$ for $\{1 - e^{b_1(t-c)}\} = 0 \implies e^{b_1(t-c)} = 1 \implies t = c$.

For $t < c$, $y''(t) > 0$; for $t > c$, $y''(t) < 0$.

So the maximum value of $y'(t)$ occurs when $t = c$.

43. $f(x) = \frac{1}{\sigma\sqrt{2\pi}} e^{-x^2/2\sigma^2}$.

$f'(x) = \frac{1}{\sigma\sqrt{2\pi}} e^{-x^2/2\sigma^2} \left(\frac{-2x}{2\sigma^2}\right) = \frac{-1}{\sigma^3\sqrt{2\pi}} x e^{-x^2/2\sigma^2}$.

$f'(x) = 0$ for $x = 0$.

For $x < 0$, $f'(x) > 0$, so $f(x)$ is increasing.

For $x > 0$, $f'(x) < 0$, so $f(x)$ is decreasing.

So $f(0)$ is a relative maximum.

$$f''(x) = -\frac{1}{\sigma^3\sqrt{2\pi}} \left\{1e^{-x^2/2\sigma^2} + x\left(e^{-x^2/2\sigma^2}\right)\left(-\frac{1}{2\sigma^2}(2x)\right)\right\}$$
$$= -\frac{1}{\sigma^3\sqrt{2\pi}} \left\{e^{-x^2/2\sigma^2}\left(1 - \frac{x^2}{\sigma^2}\right)\right\}.$$

$f''(x) = 0$ for $x = \pm\sigma$.

Interval I	Test Number	$f''(x)$	Conclusion
$(-\infty, -\sigma)$	-2σ	$f''(-2\sigma) > 0$	concave up
$(-\sigma, \sigma)$	0	$f''(0) < 0$	concave down
(σ, ∞)	2σ	$f''(2\sigma) > 0$	concave up

The inflection points occur at $x = \pm\sigma$.

Selected Solutions to Exercise Set 4.5

1. $N(t) = Ce^{4t}$.

3. $f(x) = Ce^{-2x}$.

5. $y = y(0)e^{-3t}$.

7. $N(t) = N(0)e^{4t} \implies N(t) = 4e^{4t}$.

9. $\frac{dN}{dt} + 10N = 0, \qquad N(0) = 6 \implies N(t) = 6e^{-10t}$.

11. A = amount of deposit today.

    $2500 = Ae^{0.1(8)}$. $\qquad A = 2500e^{-0.8}$. $\qquad A = \$1123.32$.

13. $e^r = 1 + i$. $\qquad i = e^r - 1$.

15. Let $P(t)$ denote the value of the deposit after t years.

$P(t) = P_0 e^{.05t}$.

$P(t) = 2P_0 \implies 2P_0 = P_0 e^{.05t} \implies e^{.05t} = 2 \implies .05t = \ln 2$

$\implies t = \frac{\ln 2}{.05} = 13.86$ years.

17. a. Let $Y(t)$ be the amount present after t days.

$Y(t) = Y(0)e^{Kt}$.

$Y(20) = \frac{1}{2}Y(0) \implies \frac{1}{2} = e^{20K} \implies 20K = \ln \frac{1}{2} = -\ln 2$

$\implies K = \frac{-\ln 2}{20}$.

$Y(10) = 50 \implies 50 = Y(0) \cdot e^{(-\ln 2/20)\cdot 10}$

$\implies Y(0) = 50e^{\ln 2/2} = 70.71$ mg.

b. $Y(30) = 70.71e^{(-\ln 2/20)\cdot 30} = 25$ mg.

19. Let $Y(t)$ be the number of bacteria in the colony after t hours.

$Y(0) = 50 \implies Y(t) = 50e^{Kt}$.

$Y(12) = 400 \implies 400 = 50e^{K(12)} \implies e^{12K} = 8 \implies 12K = \ln 8 \implies K = \ln 8/12$.

a. When $Y(t) = 100$, $100 = 50e^{(\ln 8/12)t}$

$\implies e^{(\ln 8/12)t} = 2 \implies (\ln 8/12)t = \ln 2$

$\implies t = 12 \ln 2 / \ln 2^3 = 12 \ln 2/(3 \ln 2) = 4$ hrs.

b. $Y(16) = 50e^{(\ln 8/12)\cdot 16} = 50e^{\ln 8(4/3)}$

$= 50e^{\ln (8)^{4/3}} = 50e^{\ln 16} = 50(16) = 800.$

21. Let $N(t) =$ the number of bacteria after t hours.

Then $N(t) = N(0)e^{Kt}$ and $N(0) = 100,000$.

Since $N(2) = 150,000$, $150,000 = 100,000e^{K\cdot 2}$, $1.5 = e^{2K}$, $2K = \ln 1.5$

$\implies K = \frac{1}{2} \ln 1.5$.

Then $N(t) = 100,000e^{\left(\frac{1}{2} \ln 1.5\right)t}$.

So $N(5) = 100,000e^{\left(\frac{1}{2} \ln 1.5\right)(5)} = 100,000e^{\ln (1.5)^{5/2}}$

$= 100,000(1.5)^{5/2} \approx 275,568.$

23. $P(t) = 1000e^{rt}$. $\quad P(3) = 1000 + 400 = 1400$.

Then $P(3) = 1400 = 1000e^{r3}$, $\quad 1.4 = e^{3r}$, $\quad r = \frac{1}{3}\ln 1.4 \approx 0.112 = 11.2\%$.

25. $A'(t) = -0.10A(t)$, $\quad A(0) = 10$ mg.

$A(t) = A(0)e^{-0.1t} = 10e^{-0.1t}$. Then $A(6) = 10e^{-0.1(6)} \approx 5.5$ mg.

27. The sales level t years after year one is $N(t) = 10,000e^{Kt}$.

$N(1) = 150,000$. $\quad 150,000 = 10,000e^{K\cdot 1}$, $\quad K = \ln 15$.

$N(t) = 10,000e^{(\ln 15)t} = 10,000 \cdot 15^t$.

$N(2) = 10,000 \cdot 15^2 = \$2,250,000$.

29. The amount of the substance $A(t) = Ce^{Kt}$.

Then $A(12) = 0.7C = Ce^{K(12)}$. $\quad 0.7 = e^{12K}$, $\quad K = \frac{1}{12}\ln 0.7$.

Thus $A(t) = Ce^{\left(\frac{1}{12}\ln 0.7\right)t}$.

Let t_0 be the half-life of the substance.

Then $A(t_0) = \frac{1}{2}C = Ce^{\left(\frac{1}{12}\ln 0.7\right)t_0}$, $\quad \frac{1}{2} = (0.7)^{t_0/12}$,

$t_0 = 12\log_{0.7}\left(\frac{1}{2}\right) = 12 \cdot \frac{\ln\frac{1}{2}}{\ln 0.7} \approx 23.3$ hours.

31. From Example 7 the amount of ^{14}C remaining t years after the fossil's death is $A(t) = A_0e^{\left(\frac{-\ln 2}{5760}t\right)}$. Let t_0 be the fossil's age.

Then $A(t_0) = 0.8A_0 = A_0e^{\left(\frac{-\ln 2}{5760}t_0\right)}$.

$\frac{-\ln 2}{5760}t_0 = \ln 0.8$, $\quad t_0 = -5760\frac{\ln 0.8}{\ln 2} \approx 1854.31$ years.

33. As in Exercise 32, let $N(t)$ be the number of micrograms of iodine in the bloodstream after t days. As shown in our discussion of Exercise 32,

$$N(t) = 50e^{\left(-\frac{1}{8}\ln 2\right)t}.$$

$$N(t) = 20 \implies 20 = 50e^{\left(-\frac{1}{8}\ln 2\right)t} \implies e^{\left(-\frac{1}{8}\ln 2\right)t} = .4 \implies$$

$$\left(-\frac{1}{8}\ln 2\right)t = \ln .4 \implies t = \frac{\ln .4}{-\frac{1}{8}\ln 2} = 10.58 \text{ days}.$$

Selected Solutions to Exercise Set 4.6

1. Let $P(t)$ denote the number of bats present after t years.

   The logistic model is defined by the equation

$$P(t) = \frac{M}{1 + Ce^{-kt}}.$$

   We are given that $M = 500$ bats.

$$P(0) = 10 \implies 10 = \frac{500}{1+C} \implies 10(1+C) = 500 \implies C = 49.$$

$$P\left(\frac{1}{2}\right) = 40 \implies 40 = \frac{500}{1+49e^{-k(.5)}} \implies 40\left(1+49e^{-.5k}\right) = 500 \implies$$

$$1960e^{-.5k} = 460 \implies e^{-.5k} = \frac{460}{1960} = .234693877 \implies$$

$$-.5k = \ln .234693877 \implies k = \frac{\ln .234693877}{-.5} \implies$$

$$k = 2.8989.$$

Therefore

$$P(t) = \frac{500}{1+49e^{-2.8989t}}.$$

   a. $P(1) = \frac{500}{1+49e^{-2.8989(1)}} = 135.17$ or 135.

   b. $P(3) = \frac{500}{1+49e^{-2.8989(3)}} = 495.94$ or 496.

3. Let $P(t)$ denote the number of bats persent after t years. Assuming the Law of Natural Growth, we have that

$$P(t) = P_0e^{kt}.$$

$$P(0) = 10 \implies P_0 = 10.$$

$$P(.5) = 40 \implies 40 = 10e^{k(.5)} \implies e^{.5k} = 4 \implies .5k = \ln 4 \implies$$

$$k = \frac{\ln 4}{.5} = 2.7726.$$

a. $P(1) = 10e^{2.7726(1)} = 160.$

b. $P(3) = 10e^{2.7726(3)} = 40,961.$

5. $f(t) = \frac{100}{1+5e^{-2t}}.$

$f'(t) = \frac{(1+5e^{-2t})\cdot 0 - 100\cdot(5e^{-2t}(-2))}{(1+5e^{-2t})^2} = 1000\frac{e^{-2t}}{(1+5e^{-2t})^2}.$

$f''(t) = 1000 \cdot \frac{(1+5e^{-2t})^2\cdot(e^{-2t}(-2)) - e^{-2t}\cdot(2(1+5e^{-2t})(5e^{-2t}(-2)))}{(1+5e^{-2t})^4}$

$= 1000 \cdot \frac{(1+5e^{-2t})(-2e^{-2t})+20e^{-4t}}{(1+5e^{-2t})^3}$

$= 1000 \cdot \frac{2e^{-2t}(-1+5e^{-2t})}{(1+5e^{-2t})^3}.$

$f''(t) = 0$ for $(-1+5e^{-2t}) = 0 \implies e^{-2t} = \frac{1}{5} \implies -2t = \ln\frac{1}{5}$

$\implies -2t = -\ln 5 \implies t = \frac{\ln 5}{2} \approx .8047.$

For $0 < t < \frac{\ln 5}{2}$, $f''(t) > 0$, so the graph is concave up.

For $t > \frac{\ln 5}{2}$, $f''(t) < 0$, so the graph is concave down.

$f\left(\frac{\ln 5}{2}\right) = \frac{100}{1+5e^{(-2)(\ln 5/2)}} = \frac{100}{1+5e^{-\ln 5}} = \frac{100}{1+5\cdot\frac{1}{5}} = 50.$

The inflection point is $\left(\frac{\ln 5}{2}, 50\right)$.

7. $f(x) = \frac{1}{\sigma\sqrt{2\pi}}e^{-\frac{1}{2}\left(\frac{x-\mu}{\sigma}\right)^2}$

$f(\mu - x) = \frac{1}{\sigma\sqrt{2\pi}}e^{-\frac{1}{2}\left(\frac{-x}{\sigma}\right)^2} = \frac{1}{\sigma\sqrt{2\pi}}e^{-\frac{1}{2\sigma^2}x^2}.$

$f(\mu + x) = \frac{1}{\sigma\sqrt{2\pi}}e^{-\frac{1}{2}\left(\frac{x}{\sigma}\right)^2} = \frac{1}{\sigma\sqrt{2\pi}}e^{-\frac{1}{2\sigma^2}x^2}.$

$f(\mu - x) = f(\mu + x).$

9. (Percent of the scores above 116) = 100 - (percent of the total area lying to the oeft of 116) = 100 - 84 = 16%.

11. (Percent of the scores below 132) = 100 - (percent of the scores above 132)= 100 - 2.5 = 97.5%.

13. (Percent of the scores between 68 and 100) = $\frac{1}{2}\cdot$ (percent of the total area between 68 and 132) = $\frac{1}{2} \cdot 95 = 45.5\%$.

15. (Percentage falling above $x = 0$) = (percent of the total area above $x = 0$) = 50%.

17. (Percentage falling between $x = -1$ and $x = 1$) = (percent of the total area between 4x=-1andx=1) = 68%.

19. (Percentage falling below $x = -2$) = 50 - (percent of the total area between $x = -2$ and $x = 0$) = 50 - 47.5 = 2.5%.

21.

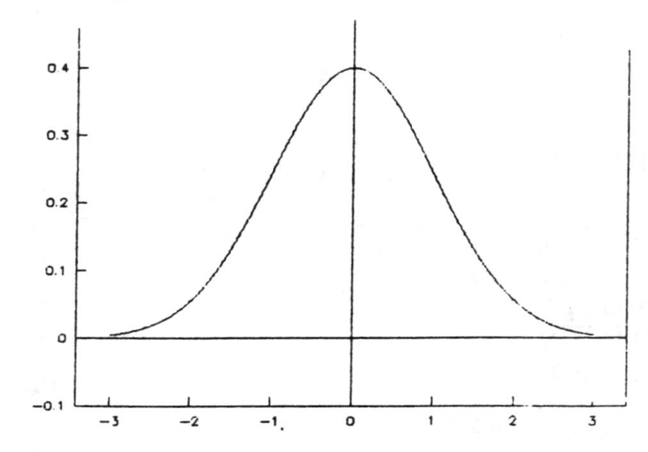

23. Suppose a particle is launched vertically upward from a height of 125 feet above ground elvel with an initial velocity of 144 feet/second. Let $s(t)$ denote the height in feet of the particle above ground level after t seconds. Then the position function for the particle is given by

$$s(t) = -16t^2 + 144t + 125.$$

The velocity function for the particle is given by

$$v(t) = -32t + 144.$$

This velocity function for predicting the velocity of the particle t seconds after it is launched provides an examples of a mathematical model from physics.

25. Suppose Linda tosses a pair of fair dice 25 times. Let n denote the number of the toss, and let $f(n)$ be the sum of the dots on the top faces of the two dice. This provides an example of a mathematical model discussed in probability. However, this example is not entirely precise in its predictions because Linda is probably not able to determine the outcome of a toss of the dice.

Selected Solutions to Review Exercises - Chapter 4

1. a. $\frac{1}{343}$.

   b. $8^{-4/3} = (8^{1/3})^{-4} = 2^{-4} = \frac{1}{2^4} = \frac{1}{16}$.

   c. $\frac{4}{9}$.

   d. $\frac{27}{8}$.

3. $P(3) = \left(1 + \frac{0.06}{2}\right)^6 \cdot 400$.

   $P(3) = (1.03)^6 \cdot 400 = \$477.62$.

5. $P(t) = P_0 e^{0.1t}$. $\quad P(5) = P_0 e^{0.1 \cdot 5}$. $\quad 10,000 = P_0 e^{0.5}$.

   $P_0 = 10,000e^{-0.5} \approx \$6,065.31$.

7. a. $\ln e^{3x} = \ln 1, \quad 3x \ln e = 0, \quad 3x = 0, \quad x = 0$.

   b. $\ln e^{x^2+x-2} = \ln 1, \quad (x^2 + x - 2) \ln e = 0,$

   $x^2 + x - 2 = 0, \quad (x+2)(x-1) = 0; \quad x = -2, 1$.

   c. $e^{3x} = e^{2x+5} \implies 3x = 2x + 5 \implies x = 5$.

   d. $(e^x)^2 - e^x - 2 = 0, \quad (e^x - 2)(e^x + 1) = 0,$

   $e^x = 2$ or $e^x = -1$.

   But e^x cannot be negative. So $e^x = 2, \qquad x = \ln 2$.

9. $R = 50e^{-2x} \cdot x. \qquad R' = 50e^{-2x} + x(50e^{-2x})(-2),$

   $R' = 50e^{-2x}(1 - 2x), \qquad x = \frac{1}{2}$ when $R' = 0$.

   $p = 50e^{-2\left(\frac{1}{2}\right)} = 50e^{-1} = \frac{50}{e} \approx \$18.39$.

11. $\frac{dy}{dx} = \frac{1}{6x} \cdot 6 = \frac{1}{x}$.

13. $\frac{dy}{dx} = -2xe^{6-x^2}$.

15. $f'(t) = (3t^2 - 3)e^{t^3-3t+2}$.

17. $f'(x) = \frac{1}{\sqrt{x+1}} \cdot \frac{1}{2}(x+1)^{-1/2} = \frac{1}{2(x+1)}$.

19. $y' = x^2 \cdot \frac{1}{2}x^{-1/2}e^{\sqrt{x}} + e^{\sqrt{x}} \cdot 2x = \frac{x^2 e^{\sqrt{x}}}{2\sqrt{x}} + e^{\sqrt{x}} \cdot 2x$

    $= \frac{x^2 e^{\sqrt{x}}}{2\sqrt{x}} + 2xe^{\sqrt{x}} = \left(2x + \frac{1}{2}x^{3/2}\right) e^{\sqrt{x}}$.

21. $y = \ln \frac{x}{x+2} = \ln x - \ln(x+2)$.

    $\frac{dy}{dx} = \frac{1}{x} - \frac{1}{x+2} = \frac{(x+2)-x}{x(x+2)} = \frac{2}{x(x+2)}$.

23. $f'(x) = \frac{1}{\ln\sqrt{x}} \cdot \frac{1}{\sqrt{x}} \cdot \frac{1}{2}x^{-1/2} = \frac{1}{2x\ln\sqrt{x}}.$

25. $y' = x \cdot e^{x-\sqrt{x}}\left(1 - \frac{1}{2}x^{-1/2}\right) + e^{x-\sqrt{x}} = xe^{x-\sqrt{x}}\left(1 - \frac{1}{2\sqrt{x}}\right) + e^{x-\sqrt{x}}$

$= xe^{x-\sqrt{x}} - \frac{xe^{x-\sqrt{x}}}{2\sqrt{x}} + e^{x-\sqrt{x}} = \left(x - \frac{1}{2}\sqrt{x} + 1\right)e^{x-\sqrt{x}}.$

27. $y = \ln^2(1 - e^{-2x}).$

$\frac{dy}{dx} = 2\ln(1 - e^{-2x}) \cdot \frac{1}{1-e^{-2x}} \cdot (2e^{-2x}) = \frac{4\ln(1-e^{-2x}) \cdot e^{-2x}}{1-e^{-2x}}.$

29. $f'(t) = e^{\sqrt{t}-\ln t} \cdot \left(\frac{1}{2}t^{-1/2} - \frac{1}{t}\right).$

31. $x\ln y + y\ln x = 4$

$x \cdot \frac{1}{y} \cdot \frac{dy}{dx} + \ln y \cdot 1 + y\frac{1}{x} + \ln x \cdot \frac{dy}{dx} = 0$

$\frac{x}{y} \cdot \frac{dy}{dx} + \ln x\frac{dy}{dx} = -\ln y - \frac{y}{x}$

$\left(\frac{x}{y} + \ln x\right)\frac{dy}{dx} = -\ln y - \frac{y}{x}$

$\frac{dy}{dx} = \frac{-\ln y - \frac{y}{x}}{\frac{x}{y} + \ln x} = \frac{-y^2 - xy\ln y}{x^2 + xy\ln x}.$

33. $e^{xy} = 3xy.$

$e^{xy}\left(1 \cdot y + x \cdot \frac{dy}{dx}\right) = 3 \cdot 1 \cdot y + 3x \cdot \frac{dy}{dx}.$

$\frac{dy}{dx}(xe^{xy} - 3x) = 3y - ye^{xy}.$

$\frac{dy}{dx} = \frac{3y - ye^{xy}}{xe^{xy} - 3x} = \frac{y(3 - e^{xy})}{x(e^{xy} - 3)} = -\frac{y}{x}.$

35. $y = x^2\ln x, \quad x > 0.$

$y' = 2x\ln x + x^2 \cdot \frac{1}{x} = x(2\ln x + 1).$

$y' = 0$ for $2\ln x + 1 = 0, \quad x = e^{-1/2}.$

For $0 < x < e^{-1/2}, \quad y' < 0$; and for $x > e^{-1/2}, \quad y' > 0.$

So $y = (e^{-1/2})^2\ln e^{-1/2} = -\frac{1}{2e}$ is the relative minimum for $x = e^{-1/2}.$

37. $y = \ln(1 + x^2). \qquad y' = \frac{2x}{1+x^2}. \qquad y'' = \frac{2(1+x^2) - 2x \cdot 2x}{(1+x^2)^2} = \frac{2-2x^2}{(1+x^2)^2}.$

$y'' = 0$ for $x = \pm 1.$

On $(-1, 1), \quad y'' > 0. \qquad$ On $(-\infty, -1)$ and $(1, \infty)$, $y'' < 0.$

So y is concave up on $(-1, 1).$

39. $2^x = e^{rx}, \qquad \ln 2^x = \ln e^{rx}, \qquad x\ln 2 = rx, \qquad r = \ln 2.$

41. The volume of the water inside the jar is

$V(t) = 6^2\pi \cdot H(t).$

$V'(t) = 36\pi \cdot H'(t).$

Now $V'(t) = -\ln(t^2).$

So $H'(t) = \frac{-1}{36\pi}V'(t) = \frac{-1}{36\pi}\ln(t^2).$

43. $\frac{dy}{dx} + y = 0. \qquad \frac{dy}{dx} = -y. \qquad y = Ce^{-x}.$

45. The population t years after 1970 is $N(t) = 203e^{kt}$ million.

$N(10) = 227 = 203e^{k\cdot 10}. \qquad k = \frac{1}{10}\ln\frac{227}{203}.$

So $N(t) = 203e^{\left(\frac{1}{10}\ln\frac{227}{203}\right)t}$

Then in 1990, $N(20) = 203e^{\left(\frac{1}{10}\ln\frac{227}{203}\right)(20)} = 203\left(\frac{227}{203}\right)^2 \approx 254$ million.

47. $N(t) =$ the number of fruit flies after t days.

$N(0) = 100. \qquad N(t) = 100e^{kt}.$

$N(10) = 500 \implies 500 = 100e^{k\cdot 10} \implies e^{10k} = 5 \implies 10k = \ln 5$

$\implies k = \frac{1}{10}\ln 5.$

$N(t) = 100e^{\left(\frac{1}{10}\ln 5\right)t}. \qquad N(4) = 100e^{\left(\frac{1}{10}\ln 5\right)4} = 190.36$ flies.

49. $f(x) = C \cdot e^{kx}. \qquad f'(x) = k \cdot C \cdot e^{kx}.$

$f'(x+a) = k \cdot C \cdot e^{k(x+a)} = k \cdot f(x+a).$

51. Let $P(t)$ denote the value of the bond after t years. Let P_0 denote the present value of the bond.

$$P(t) = P_0e^{.05t} \implies P_0 = e^{-.05t} \cdot P(t).$$

$$P(4) = 2000 \implies P_0 = e^{-.05(4)} \cdot 2000 = 1,637.46 \text{ dollars.}$$

PRACTICE PROBLEMS FOR CHAPTER 4

1. Simplify the following expressions.

   a. $4^3 \cdot 8^{-2}$ b. $\frac{[2^3 \cdot 3^{-2}]^2}{\frac{1}{3} \cdot 16^2}$

   c. $\frac{e^4 \cdot e^{-3x}}{e^{-x}}$ d. $\sqrt[3]{27 \cdot e^{7x}}$

2. Sketch the graphs of the following functions.

   a. $f(x) = \left(\frac{1}{3}\right)^x$ b. $f(x) = -e^{-2x}$

   c. $f(x) = 4 \cdot (3)^{2x}$ d. $f(x) = xe^{-2x}$

3. What is the value of $\lim_{x \to \infty} \left(1 + \frac{4}{x}\right)^x$?

4. Suppose you deposit \$500 in the First National Bank which pays an annual interest rate of 7.5 percent. Let $A(t)$ be the resulting amount in dollars you have in the bank after t years.

   a. Find $A(5)$ if the interest is compounded semiannually.

   b. Find $A(5)$ if the interest is compounded continuously.

   c. Find t such that $A(t) = 750$ if the interest is compounded semiannually.

   d. Find t such that $A(t) = 750$ if the interest is compounded continuously.

5. Use the definition of the logarithm to find the value of each of the following expressions.

   a. $\log_{27} 9$ b. $\log_4 \left(\frac{1}{32}\right)$

   c. $\ln \sqrt{e}$ d. $a^{x \log_a 3}$

6. Simplify the following expressions.

   a. $\log_b c \cdot \log_c d \cdot \log_d e$ b. $\log_2 \left(\frac{4}{8^x}\right)$

7. Find the derivative of each of the following functions.

   a. $f(x) = \ln(x^2 + 1)$ b. $f(x) = 6 \cdot \ln\left(\ln(x^2 + 1)\right)$

   c. $f(x) = x^3 \cdot e^{-x^{1/2}}$ d. $f(x) = x^2 \cdot \ln x$

   e. $f(x) = \frac{\ln(2x+1)}{e^{2x+1}}$ f. $f(x) = 2 \cdot e^{(4+5x^2)^{-1/2}}$

8. Find $\frac{dy}{dx}$ by logarithmic differentiation.

   a. $y = e^{-x^2/2} \cdot \sqrt{x^2+1}$ b. $y = \frac{(x+1)(x+2)}{(x+3)(x+4)}$

   c. $y = (2x^2+3)^{x^2}$ d. $y = (2x^2+3)^{x^2} \cdot e^{3x+2}$

9. Suppose a new product is being marketed. Let x be the number of months after this product has been introduced, and let $f(x)$ denote the number of units of this product sold during the first x months. Suppose $f(x) = \frac{7500}{1+750 \cdot e^{-x}}$.

   a. When is the rate of sales of this product the largest?

   b. Find the limiting value of $f(x)$ as $x \to \infty$.

   c. Sketch the graph of f.

10. Suppose you invest $1000 in a highly speculative venture. Would you rather receive 20% annual interest compounded semiannually or 19% annual interest compounded continuously?

11. Suppose the spread of an epidemic of flu in a city of 500,000 is being monitored. The number of cases of flu at the end of t weeks is given by the formula

$$f(t) = \frac{500,000}{1+Be^{-ct}}$$

where B and c are constants to be determined. At the beginning of the first week of monitoring, 200 cases had been reported. During the first week of monitoring, 300 new cases are reported. How many people will have been infected after 6 weeks?

12. If a company sells calculators at a price of p dollars per calculator, the number x of calculators it can sell is given by the formula

$$x = 1000p^2e^{-.02(p+5)} \quad \text{where } p \leq 500.$$

   a. For what interval of p values does the revenue increase as the price p increases?

   b. For what value of p is the revenue the largest?

c. For what value of p is the rate of increase of the revenue the largest as p increases?

13. Suppose a certan island is contaminated by strontium fallout from a nuclear explosion. Immediately after the explosion, the strontium level is 100 times the safe level. The half-life of strontium is 28 years. How many years will it take for the strontium level on the island to return to the safe level?

14. Suppose the value of a certain asset after t years is given by

$$V(t) = 1.8t + 8.$$

Assume that interest rates under continuous compounding remain constant at 12%. When should the asset be sold so as to maximize revenue to the owner?

15. Suppose the value of a certain historic treasure t centuries from now is given by

$$V(t) = 2000 + 96\sqrt{t} - 12t.$$

Assume that interest rates under continuous compounding remain constant at 10 percent. When should this treasure be sold so as to maximize revenue to the owner?

Solutions to Practice Problems for Chapter 4

1.a. $4^3 \cdot 8^{-2} = (2^2)^3(2^3)^{-2} = 2^6 \cdot 2^{-6} = 2^{6+(-6)} = 2^0 = 1.$

b. $\frac{\left[2^3 \cdot 3^{-2}\right]^2}{\frac{1}{3}\cdot 16^2} = \frac{3\cdot 2^6 \cdot 3^{-4}}{(2^4)^2} = \frac{3^{(1-4)}\cdot 2^6}{2^8} = \frac{3^{-3}}{2^{(8-6)}} = \frac{1}{3^3 \cdot 2^2} = \frac{1}{108}.$

c. $\frac{e^4 \cdot e^{-3x}}{e^{-x}} = e^{4+(-3x)-(-x)} = e^{4-2x}.$

d. $\sqrt[3]{27 \cdot e^{7x}} = \left(3^3 \cdot e^{7x}\right)^{1/3} = 3^{3\cdot\frac{1}{3}} \cdot e^{7x\cdot\frac{1}{3}} = 3^1 e^{7x/3} = 3e^{7x/3}.$

2.a. $f(x) = \left(\frac{1}{3}\right)^x.$

x	-3	-2	-1	0	1	2	3
$f(x)$	27	9	3	1	$\frac{1}{3}$	0.111	0.037

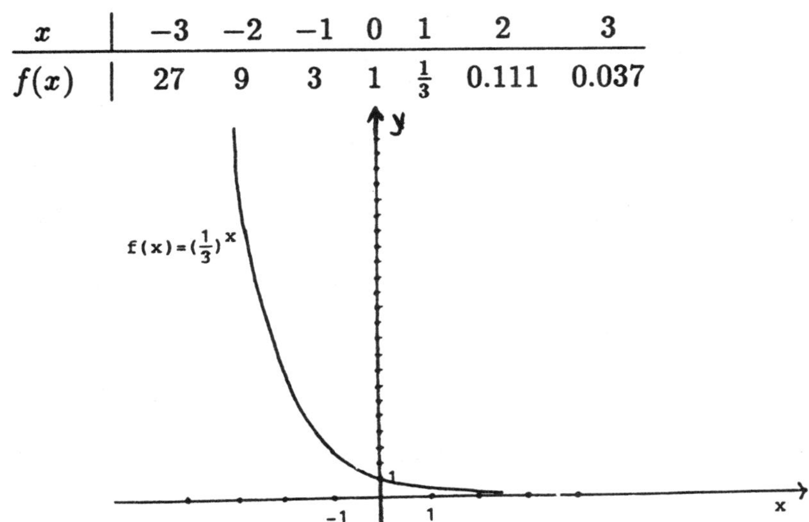

b. $f(x) = -e^{-2x}.$

x	-1	$-\frac{1}{2}$	0	$\frac{1}{2}$	1
$f(x)$	-7.39	-2.718	-1	-0.368	-0.135

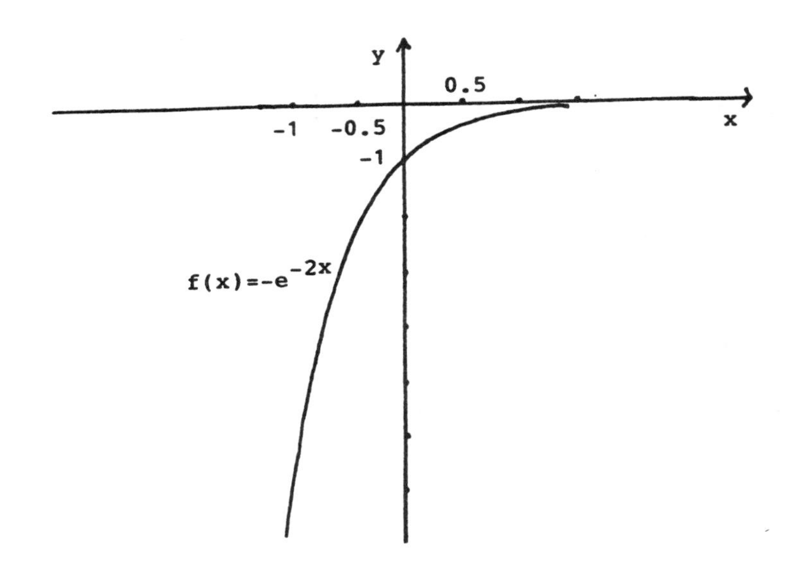

c. $f(x) = 4 \cdot (3)^{2x}$

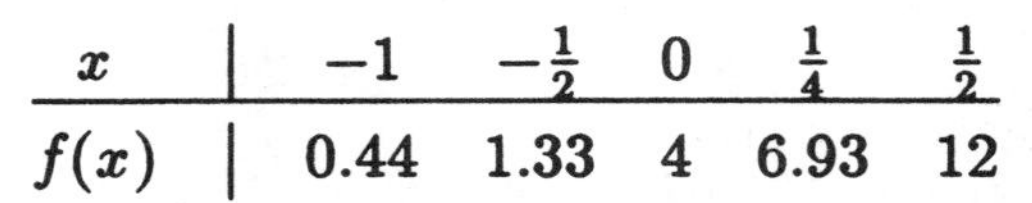

x	-1	$-\frac{1}{2}$	0	$\frac{1}{4}$	$\frac{1}{2}$
$f(x)$	0.44	1.33	4	6.93	12

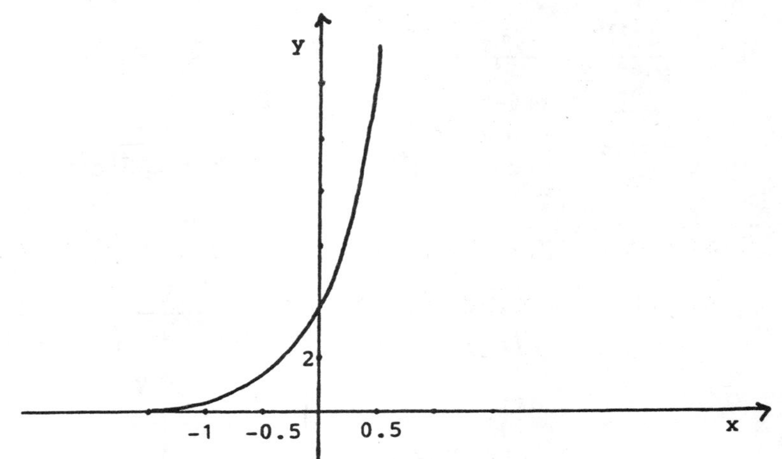

d. $f(x) = x \cdot e^{-2x}$

x	$-\frac{2}{3}$	$-\frac{1}{2}$	0	$\frac{1}{2}$	$\frac{2}{3}$	1
$f(x)$	-2.53	-1.36	0	0.184	0.176	0.135

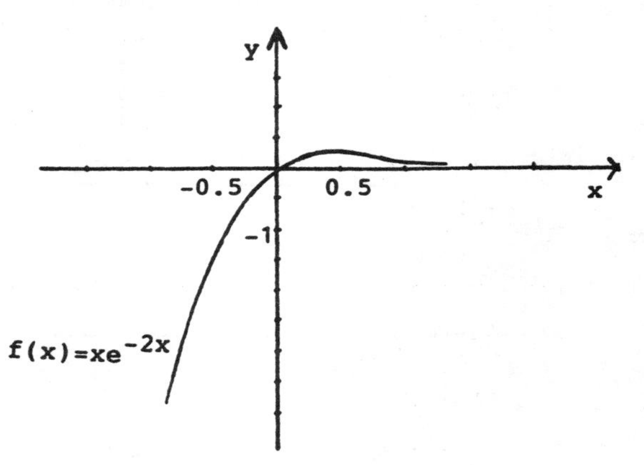

3. $\lim_{x\to\infty} \left(1+\frac{4}{x}\right)^x = \lim_{x\to\infty} \left[\left(1+\frac{1}{\frac{x}{4}}\right)^{x/4}\right]^4 = e^4.$

4.a. $A(5) = 500\left(1+\frac{0.075}{2}\right)^{10} = 500 \cdot (1.0375)^{10} = 500 \cdot (1.445) = \$722.52.$

b. $A(5) = 500e^{.075 \cdot 5} = 500 \cdot (1.455) = \$727.50.$

c. $A(t) = 500\left(1+\frac{0.075}{2}\right)^{2t} = 750$

$\Longrightarrow (1.0375)^{2t} = 1.5 \Longrightarrow \ln(1.0375)^{2t} = \ln 1.5$

$\Longrightarrow 2t \cdot \ln 1.0375 = \ln 1.5 \Longrightarrow t = \frac{\ln 1.5}{2 \ln 1.0375} = 5.507$ Yr.

d. $A(t) = 500e^{.075t} = 750$

$\Longrightarrow e^{.075t} = 1.5 \Longrightarrow .075t = \ln 1.5$

$\Longrightarrow t = \frac{\ln 1.5}{.075} = 5.406$ Yr.

5.a. $\log_{27} 9 = \frac{2}{3}$ since $(27)^{2/3} = 9$.

b. $\log_4\left(\frac{1}{32}\right) = -\frac{5}{2}$ since $(4)^{-5/2} = \frac{1}{32}$.

c. $\ln\sqrt{e} = \frac{1}{2}$ since $e^{1/2} = \sqrt{e}$.

d. $a^{x \log_a 3} = 3^x$.

6.(a) $\log_b c \cdot \log_c d \cdot \log_d e = \frac{\ln c}{\ln b} \cdot \frac{\ln d}{\ln c} \cdot \frac{\ln e}{\ln d} = \frac{\ln e}{\ln b} = \log_b e$.

b. $\log_2\left(\frac{4}{8^x}\right) = (\log_2 4) - (\log_2 8^x) = \log_2(2^2) - \log_2(2^{3x}) = 2 - 3x$.

7.a. $f(x) = \ln(x^2+1)$.

$f'(x) = \frac{2x}{x^2+1}$.

b. $f(x) = 6 \cdot \ln\left(\ln(x^2+1)\right)$.

$f'(x) = \frac{6}{\ln(x^2+1)} \cdot \frac{1}{x^2+1} \cdot 2x = \frac{12x}{(x^2+1)\ln(x^2+1)}$.

c. $f(x) = x^3 \cdot e^{-x^{1/2}}$.

$f'(x) = 3x^2 \cdot e^{-x^{1/2}} + x^3 \cdot e^{-x^{1/2}} \cdot \left(-\frac{1}{2}x^{-1/2}\right) = x^2 e^{-x^{1/2}}\left(3 - \frac{1}{2}x^{1/2}\right)$.

d. $f(x) = x^2 \cdot \ln x$.

$f'(x) = 2x \cdot \ln x + x^2 \cdot \frac{1}{x} = x(1 + 2\ln x)$.

e. $f(x) = \frac{\ln(2x+1)}{e^{2x+1}}$.

$f'(x) = \frac{\frac{2}{2x+1} \cdot e^{2x+1} - \ln(2x+1) \cdot e^{2x+1} \cdot 2}{(e^{2x+1})^2}$

$= \frac{2 - 2(2x+1)\ln(2x+1)}{(2x+1) \cdot e^{2x+1}}$.

f. $f(x) = 2 \cdot e^{(4+5x^2)^{-1/2}}$.

$f'(x) = 2e^{(4+5x^2)^{-1/2}} \cdot \left(-\frac{1}{2}\right)(4+5x^2)^{-3/2} \cdot (10x)$.

$= -10x(4+5x^2)^{-3/2} e^{(4+5x^2)^{-1/2}}$.

8.a. $y = e^{-x^2/2} \cdot \sqrt{x^2+1}$.

$\ln y = \ln e^{-x^2/2} + \ln\sqrt{x^2+1} = -\frac{x^2}{2} + \frac{1}{2}\ln(x^2+1)$.

$\frac{1}{y}\frac{dy}{dx} = -x + \frac{2x}{2(x^2+1)}$.

$\frac{dy}{dx} = xy\left(\frac{1}{x^2+1} - 1\right) = \frac{-x^3}{x^2+1} \cdot e^{-x^2/2} \cdot \sqrt{x^2+1} = \frac{-x^3 e^{-x^2/2}}{\sqrt{x^2+1}}$.

b. $y = \frac{(x+1)(x+2)}{(x+3)(x+4)}$.

$\ln|y| = \ln(|x+1|) + \ln(|x+2|) - \ln(|x+3|) - \ln(|x+4|)$.

$\frac{1}{y}\frac{dy}{dx} = \frac{1}{x+1} + \frac{1}{x+2} - \frac{1}{x+3} - \frac{1}{x+4}$.

$\frac{dy}{dx} = y\left[\frac{2(2x^2+10x+11)}{(x+1)(x+2)(x+3)(x+4)}\right]$

$= 2 \cdot \frac{(x+1)(x+2)}{(x+3)(x+4)} \cdot \frac{2x^2+10x+11}{(x+1)(x+2)(x+3)(x+4)}$

$= \frac{2(2x^2+10x+11)}{(x+3)^2(x+4)^2}$

c. $y = (2x^2+3)^{x^2}$.

$\ln y = \ln(2x^2+3)^{x^2} = x^2\ln(2x^2+3)$.

$\frac{1}{y} \cdot \frac{dy}{dx} = 2x\ln(2x^2+3) + x^2 \cdot \frac{4x}{2x^2+3}$.

$\frac{dy}{dx} = y\left[2x\ln(2x^2+3) + \frac{4x^3}{2x^2+3}\right]$

$= (2x^2+3)^{x^2}\left[2x\ln(2x^2+3) + \frac{4x^3}{2x^2+3}\right]$.

d. $y = (2x^2+3)^{x^2} \cdot e^{3x+2}$.

By the Multiplication Rule and Problem 8(c) we have

$\frac{dy}{dx} = \frac{d}{dx}\left[(2x^2+3)^{x^2}\right] \cdot e^{3x+2} + (2x^2+3)^{x^2} \cdot \frac{d}{dx}(e^{3x+2})$

$= (2x^2+3)^{x^2}\left[2x\ln(2x^2+3) + \frac{4x^3}{2x^2+3}\right] \cdot e^{3x+2}$

$+(2x^2+3)^{x^2} \cdot e^{3x+2} \cdot 3.$

$= (2x^2+3)^{x^2} \cdot e^{3x+2} \cdot \left[3 + 2x\ln(2x^2+3) + \frac{4x^3}{2x^2+3}\right]$.

9. $f(x) = \frac{7500}{1+750 \cdot e^{-x}}$.

a. $f'(x) = \frac{-7500(-750e^{-x})}{(1+750 \cdot e^{-x})^2} = \frac{750^2 \cdot 10 \cdot e^{-x}}{(1+750e^{-x})^2}$.

$f''(x) = \frac{-10 \cdot 750^2 \cdot e^{-x} \cdot (1+750 \cdot e^{-x})^2 - 10 \cdot 750^2 \cdot e^{-x} \cdot 2(1+750e^{-x}) \cdot (-750e^{-x})}{(1+750e^{-x})^4}$

$f''(x) = \frac{-10 \cdot 750^2 \cdot e^{-x} \cdot (1+750 \cdot e^{-x} - 1500 \cdot e^{-x})}{(1+750 \cdot e^{-x})^3}$

$= \frac{-10 \cdot 750^2 \cdot e^{-x}(1-750e^{-x})}{(1+750 \cdot e^{-x})^3}$.

$f''(x) = 0$ for $1 - 750e^{-x} = 0 \implies x = \ln 750 = 6.62$ mo.

So the rate of sales is the largest when $x = 6.62$ months.

b. $\lim_{x\to\infty} f(x) = \lim_{x\to\infty} \frac{7500}{1+750 \cdot e^{-x}} = \frac{7500}{1+7500 \cdot (0)} = 7500$.

c.

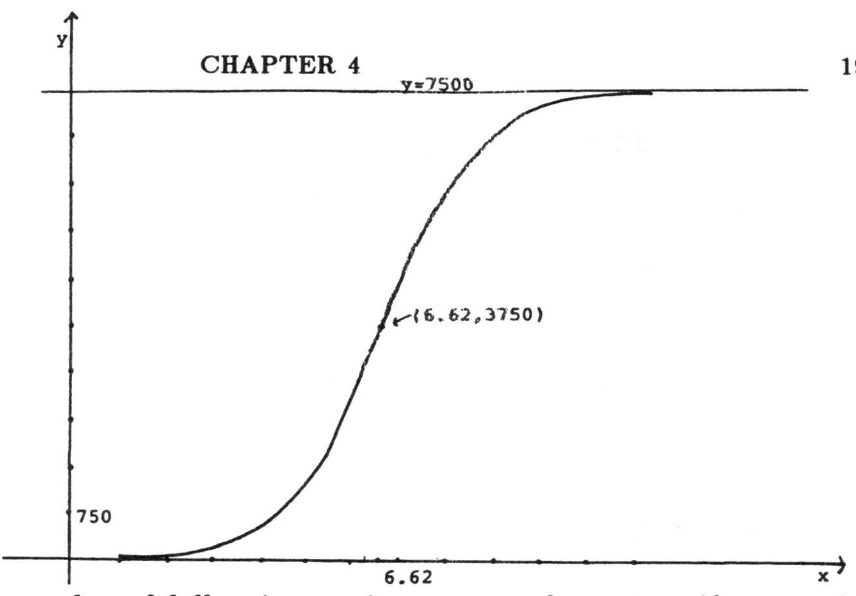

10. Let $A_1(t)$ be the number of dollars in your investment after t years if you receive 20% annual interest compounded semiannually, and let $A_2(t)$ be the number of dollars in your investment if you receive 19% annual interest compounded continuously. Then

$$A_1(t) = 1000 \cdot \left(1 + \frac{.20}{2}\right)^{2t} = 1000 \cdot (1.1)^{2t} = 1000 \cdot (1.21)^t$$
$$A_2(t) = 1000 \cdot e^{.19t} \doteq 1000 \cdot (1.20925)^t.$$

Note that

$$A_1(t) - A_2(t) = 1000\left[(1.21)^t - (1.20925)^t\right],$$

and that for any fixed value of t and a positive real number x

$$\frac{d}{dx}(x^t) = t \cdot x^{t-1} > 0.$$

Therefore $(1.21)^t > (1.20925)^t \implies A_1(t) > A_2(t)$.
So it will be better to get 20% annual interest compounded semiannually than 19% annual interest compounded continuously.

11. Since $f(0) = 200$, $200 = \frac{500{,}000}{1+B\cdot e^0} = \frac{500{,}000}{1+B}$.

$\Longrightarrow 200(1+B) = 500{,}000.$

$1 + B = 2{,}500 \Longrightarrow B = 2{,}499.$

So $f(t) = \frac{500{,}000}{1+2{,}499e^{-ct}}$.

Since $f(1) = 500$, $500 = \frac{500{,}000}{1+2{,}499e^{-c}}$.

$\Longrightarrow 500(1 + 2{,}499e^{-c}) = 500{,}000.$

$2{,}499e^{-c} = 999$

$e^{-c} = \frac{999}{2{,}499}$.

$e^c = \frac{2{,}499}{999} \Longrightarrow c = \ln\left(\frac{2{,}499}{999}\right) \doteq .92.$

Thus $f(t) = \frac{500{,}000}{1+2{,}499e^{-.92t}}$.

Therefore, $f(6) = \frac{500{,}000}{1+2{,}499e^{-.92\cdot(6)}} \doteq 45{,}411$ people.

12.a. The revenue R as a function of price p is given by the formula

$$\begin{aligned} R = px &= p\left[1000p^2e^{-.02(p+5)}\right] \\ &= 1000p^3e^{-.02(p+5)} = f(p). \end{aligned}$$

$$\begin{aligned} f'(p) &= 1000\left[3p^2e^{-.02(p+5)} + p^3e^{-.02(p+5)}(-.02)\right] \\ &= 1000e^{-.02(p+5)}(3p^2 - .02p^3) \\ &= 1000e^{-.02(p+5)}p^2(3 - .02p). \end{aligned}$$

$f'(p) > 0$ for $3 - .02p > 0 \Longrightarrow p < \frac{3}{.02} = \$150.$

So the revenue R increases as the price p increases when $p < 150$.

b. For $0 < p < 150$, $\frac{dR}{dp} > 0$;

for $p > 150$, $\frac{dR}{dp} < 0$.

Thus the maximum revenue occurs when the price p is \$150 per calculator.

c. $f''(p) = 1000\left[e^{-.02(p+5)} \cdot (-.02)(3p^2 - .02p^3) + e^{-.02(p+5)} \cdot (6p - .06p^2)\right]$

$= 1000e^{-.02(p+5)}\left[(-.06p^2 + .0004p^3) + (6p - .06p^2)\right]$

$= 1000e^{-.02(p+5)} \cdot .0004p \cdot (p^2 - 300p + 15000).$

$f''(p) = 0$ for $p^2 - 300p + 15000 = 0.$

$\Longrightarrow p = \frac{300 \pm \sqrt{300^2 - 4(1)(15000)}}{2}.$

$p = \frac{300 \pm \sqrt{30{,}000}}{2} = \frac{300 \pm 173.20}{2}$

$p_1 \approx \frac{300 + 173.20}{2} = 236.6$

$p_2 \approx \frac{300 - 173.20}{2} = 63.4.$

$f'(0) = 0, \quad f'(63.4) > 0, \quad f'(236.6) < 0, \quad f'(500) < 0.$

Therefore, the maximum value of the derivative $R'(p)$ occurs when $p \approx \$63.40$.

13. Let t be the number of years after the nuclear explosion; let $S(t)$ be the corresponding strontium level on the island; and let S_0 be the strontium level on the island immediately after the nuclear explosion.

$$S = S_0 e^{kt}.$$

$$\frac{1}{2}S_0 = S_0 e^{k(28)} \Longrightarrow e^{28k} = \frac{1}{2} \Longrightarrow 28k = \ln\frac{1}{2}$$

$$= \ln 1 - \ln 2 = -\ln 2 \Longrightarrow k = -\frac{\ln 2}{28}.$$

$$S(t) = S_0 e^{\left(-\frac{\ln 2}{28}\right)\cdot t}.$$

When $S = .01S_0$:

$$.01S_0 = S_0 e^{\left(-\frac{\ln 2}{28}\right)t}.$$

$$\left(-\frac{\ln 2}{28}\right) \cdot t = \ln .01.$$

$$t = -28 \cdot \frac{\ln .01}{\ln 2} = 186 years.$$

14. Let $R(t)$ be the present value of the revenue received by selling the asset after t years.

$$R(t) = V(t)e^{-0.12t} = (1.8t + 8)e^{-0.12t}.$$

$$\begin{aligned} R'(t) &= (1.8)e^{-0.12t} + (1.8t + 8)e^{-0.12t} \cdot (-0.12) \\ &= e^{-0.12t}\left[1.8 - 0.216t - 0.96\right] \\ &= e^{-0.12t}\left[0.84 - 0.216t\right]. \end{aligned}$$

$$R'(t) = 0 \implies t = \frac{0.84}{0.216} \implies t = 3.89 \text{ years.}$$

Therefore the asset should be sold after 3.89 years.

15. Let $R(t)$ be the present value of the revenue received by selling the treasure after t cenuries.

$$\begin{aligned} R(t) &= V(t)e^{-0.10t} = \left(2000 + 96\sqrt{t} - 12t\right) \cdot e^{-0.10t}. \\ R'(t) &= \left(48t^{-1/2} - 12\right)e^{-0.10t} + \left(2000 + 96\sqrt{t} - 12t\right)e^{-0.10t}(-0.10) \\ &= e^{-0.10t}\left(48t^{-1/2} - 12 - 200 - 9.6t^{1/2} + 1.2t\right) \\ &= e^{-0.10t}\left(48t^{-1/2} - 212 - 9.6t^{1/2} + 1.2t\right) \\ &= 4e^{-0.10t} \cdot t^{-1/2} \cdot \left(12 - 53t^{1/2} - 2.4t + 0.3t^{3/2}\right). \end{aligned}$$

Let

$$f(t) = \left(12 - 53t^{1/2} - 2.4t + 0.3t^{3/2}\right).$$

Note that

$$f(0.05) = 0.032$$

and

$$f(0.06) = -1.12.$$

Therefore $R'(t) = 0$ for $t \approx 0.05$. Thus, the treasure should be sold after about 5 years.

CHAPTER 5

SUMMARY OF CHAPTER 5

1. The definition of an antiderivative F for a function f, the result that any other antiderivative for f has the form $F + C$ for some constant C, notation for antiderivatives of f.

2. The power rule for antiderivatives, the sum rule for antiderivatives, and the constant multiple rule for antiderivatives.

3. Finding particular antiderivatives for a function.

4. Use of antiderivatives to determine total cost and total revenue from their marginal rates.

5. The method of substitution for finding antiderivatives.

6. Approximating the area of the region R bounded by the graph of the nonnegative continuous function $y = f(x)$, the x-axis, and the vertical straight lines $x = a$ and $x = b$ by the sum of areas of rectangular regions as described in Section 5.4; the actual area of R as the limit of such approximating sums.

7. The definite integral $\int_a^b f(x)dx$ of the continuous function f on the interval $[a, b]$ as the limit of sums, and the observation that when $f(x) \geq 0$ for all x in $[a, b]$ this definite integral is the area of the region R described in (6).

8. The Fundamental Theorem of Calculus for computing definite integrals.

9. The properties of the definite integral.

10. Finding the area of a region bounded by two graphs.

11. The total value of the utility of consumption.

12. Consumers' surplus and suppliers' surplus.

13. The nominal value of a revenue stream, and the present value of a revenue stream.

14. Use of the definite integral in determining the volume of the solid of revolution obtained by rotating about the x-axis the region R bounded by the graph of a continuous nonnegative function $y = f(x)$, the x-axis, and the vertical straight lines $x = a$ and $x = b$.

15. The average value of a continuous function on the closed interval $[a, b]$.

Selected Solutions to Exercise Set 5.1

1. $\int 5dx = 5x + C$.
3. $\int 2x^3 dx + \int 5x dx = \frac{1}{2}x^4 + \frac{5}{2}x^2 + C$.
5. $\int x^3 dx - \int 6x^2 dx + \int 2x dx - \int 1 dx = \frac{1}{4}x^4 - 2x^3 + x^2 - x + C$.
7. $\int x^{2/3} dx - \int 3x^{1/2} dx = \frac{3}{5}x^{5/3} - 2x^{3/2} + C$.
9. $\int \frac{4}{x} dx = 4\ln x + C, \qquad x > 0$.
11. $\int e^{5x} dx = \frac{1}{5}e^{5x} + C$.
13. $\int x^{1/3} dx - \int 2x^{-2/3} dx + \int x^{-5/3} dx = \frac{3}{4}x^{4/3} - 6x^{1/3} - \frac{3}{2}x^{-2/3} + C$.
15. $\int 4e^{-x} dx - \int \frac{1}{x} dx + \int \sqrt{x} dx = -4e^{-x} - \ln x + \frac{2}{3}x^{3/2} + C, \quad x > 0$.
17. $\int 4x^{-6} dx - \int 6x^{-4} dx = -\frac{4}{5}x^{-5} + 2x^{-3} + C$.
19. $\int (x - 2x^{5/6} + x^{2/3}) dx = \frac{1}{2}x^2 - \frac{12}{11}x^{11/6} + \frac{3}{5}x^{5/3} + C$.
21. $\int (\sqrt{x} - e^{-x})\, dx = \frac{1}{\frac{3}{2}}x^{3/2} - \frac{1}{-1}e^{-x} + C = \frac{2}{3}x^{3/2} + e^{-x} + C$.
23. $\int (6x + 3x^2 + 5) dx = 6 \cdot \left(\frac{1}{2}x^2\right) + 3 \cdot \left(\frac{1}{3}x^3\right) + 5x + C$

    $= 3x^2 + x^3 + 5x + C$.

    $f(x) = 3x^2 + x^3 + 5x$ and $g(x) = 3x^2 + x^3 + 5x + 1$ are functions such that

    $f'(x) = g'(x) = 6x + 3x^2 + 5$ for all x.

    $[f(x) - g(x)] = -1$ for all x.
25. $f(x) = Ae^{2x} + \sqrt{x} + 5$ and $g(x) = 6e^{2x} + 3Bx^{1/2} + C$.

    $f'(x) = 2Ae^{2x} + \frac{1}{2\sqrt{x}}$ and $g'(x) = 12e^{2x} + \frac{3B}{2\sqrt{x}}$.

    $f'(x) = g'(x)$ for all $x > 0 \implies 2A = 12$ and $1 = 3B$.

    a. $2A = 12 \implies A = 6$.

    b. $3B = 1 \implies B = \frac{1}{3}$.

    c. C is any constant.
27. $P'(t) = 4 + 2t$ for $t > 0$.

    a. $P(t) = 4t + t^2 + C$ where C is the population of the deer at time $t = 0$.

    b. The population is increasing because $P'(t) = 4 + 2t > 0$ for all $t > 0$.

Selected Solutions to Exercise Set 5.2

1. ii 3. iv 5. vi

7. $F(x) = \int 4dx = 4x + C.$

   $F(0) = 3 = 4(0) + C \implies C = 3.$

   $F(x) = 4x + 3.$

9. $F(x) = \int (x^2 - 2x)dx = \frac{1}{3}x^3 - x^2 + C.$

   $F(0) = -3 = \frac{1}{3}(0)^3 - (0)^2 + C \implies C = -3.$

   $F(x) = \frac{1}{3}x^3 - x^2 - 3.$

11. $F(x) = \int \frac{1}{x}dx = \ln x + C.$

    $F(e) = 5 = \ln e + C \implies C = 4.$

    $F(x) = \ln x + 4.$

13. $F(x) = \int (x^{2/3} - x^{-1/3})dx = \frac{3}{5}x^{5/3} - \frac{3}{2}x^{2/3} + C.$

    $F(0) = 5 = \frac{3}{5} \cdot (0)^{5/3} - \frac{3}{2} \cdot (0)^{2/3} + C \implies C = 5.$

    $F(x) = \frac{3}{5}x^{5/3} - \frac{3}{2}x^{2/3} + 5.$

15. $F(x) = \int (5e^{2x} + 4)dx = \frac{5}{2}e^{2x} + 4x + C.$

    $F(0) = 10 = \frac{5}{2}e^{2(0)} + 4(0) + C \implies C = \frac{15}{2}.$

    $F(x) = \frac{5}{2}e^{2x} + 4x + \frac{15}{2}.$

17. $F(x) = \int \left(\frac{3}{x+5} + 2e^{-3x}\right) dx = 3\ln(x+5) - \frac{2}{3}e^{-3x} + C.$

    $F(0) = 0 \implies 0 = 3\ln(0+5) - \frac{2}{3}e^{-3(0)} + C = 3\ln 5 - \frac{2}{3} + C$

    $\implies C = \frac{2}{3} - 3\ln 5.$

    $F(x) = 3\ln(x+5) - \frac{2}{3}e^{-3x} + \frac{2}{3} - 3\ln 5.$

19. $C(x) = \int MC(x)dx = \int 250dx = 250x + K.$

    $C(0) = 700 = 250(0) + K \implies K = 700.$

    $C(x) = 250x + 700$ dollars.

21. $C(x) = \int MC(x)dx = \int \left(400 + \frac{1}{4}x\right) dx = 400x + \frac{1}{8}x^2 + K.$

(a) The manufacturer's cost for 10 dishwashers $= C(10) = 5200$

$= 400(10) + \frac{1}{8}(10)^2 + K \implies K = \$1187.50$, fixed monthly cost.

(b) $C(x) = 400x + \frac{1}{8}x^2 + 1187.5.$

23. $MC(x) = 100 + 4x \implies C(x) = 100x + 2x^2 + K.$

$C(100) = 35,000 \implies 35,000 = 100(100) + 2(100)^2 + K \implies K = 5000.$

Therefore $C(x) = 100x + 2x^2 + 5000.$

25. $R(x) = \int MR(x)dx = \int \left(250 - \frac{1}{2}x\right) dx = 250x - \frac{1}{4}x^2 + K.$

$R(0) = 0 = 250(0) - \frac{1}{4}(0)^2 + K \implies K = 0.$

The total revenue function $R(x) = 250x - \frac{1}{4}x^2.$

27. $P'(t) = 4 + 2t, \quad t > 0.$

$P(t) = 4t + t^2 + K.$

$P(0) = 12 \implies K = 12.$

Therefore $P(t) = 4t + t^2 + 12.$

29. I matches up with c.

II matches up with a.

III matches up with b.

31. $s(t) = \int v(t)dt = \int (3t^2 + 6t + 2)dt = t^3 + 3t^2 + 2t + C.$

$s(0) = 4 = (0)^3 + 3(0)^2 + 2(0) + C \implies C = 4.$

$s(t) = t^3 + 3t^2 + 2t + 4.$

(a) Its location after t seconds $s(t) = t^3 + 3t^2 + 2t + 4.$

(b) Its location after 4 seconds $s(4) = (4)^3 + 3(4)^2 + 2(4) + 4 \implies s(4) = 124.$

33. $N(t) = \int \left(20 + 24\sqrt{t}\right) dt = 20t + 16t^{3/2} + C.$

$N(0) = 2 = 20(0) + 16(0)^{3/2} + C \implies C = 2.$

(a) $N(t) = 20t + 16t^{3/2} + 2.$

(b) $N(16) = 20(16) + (16)(16)^{3/2} + 2 \implies N(16) = 1346.$

Selected Solutions to Exercise Set 5.3

1. $\int x(3+x^2)^3 dx$.

   Let $u = 3 + x^2$. $du = 2x\,dx \implies x\,dx = \frac{1}{2}du$.

   $\int x(3+x^2)^3 dx = \int u^3 \left(\frac{1}{2}du\right) = \frac{1}{2}\cdot\frac{1}{4}u^4 + C = \frac{1}{8}(3+x^2)^4 + C$.

3. $\int 5x\sqrt{9+x^2}\,dx$.

   Let $u = 9 + x^2$. $du = 2x\,dx \implies x\,dx = \frac{1}{2}du$.

   $\int 5x\sqrt{9+x^2}\,dx = \int 5u^{1/2}\left(\frac{1}{2}du\right) = \frac{5}{2}\cdot\frac{2}{3}u^{3/2} + C = \frac{5}{3}\left(9+x^2\right)^{3/2} + C$.

5. $\int xe^{x^2}\,dx$.

   Let $u(x) = x^2$. $du = 2x\,dx$, $x\,dx = \frac{1}{2}du$.

   $\int xe^{x^2}\,dx = \frac{1}{2}\int e^u\,du = \frac{1}{2}e^u + C = \frac{1}{2}e^{x^2} + C$.

7. $\int \frac{\left(\sqrt{x}+5\right)^6}{\sqrt{x}}\,dx$.

   Let $u = \sqrt{x} + 5$. $du = \frac{1}{2\sqrt{x}}\,dx \implies \frac{1}{\sqrt{x}}\,dx = 2\,du$.

   $\int \frac{\left(\sqrt{x}+5\right)^6}{\sqrt{x}}\,dx = \int u^6(2\,du) = 2\cdot\frac{1}{7}u^7 + C = \frac{2}{7}\left(\sqrt{x}+5\right)^7 + C$.

9. $\int \frac{x}{1+3x^2}\,dx$.

   Let $u = 1 + 3x^2$. $du = 6x\,dx \implies x\,dx = \frac{1}{6}\,du$.

   $\int \frac{x}{1+3x^2}\,dx = \int \frac{1}{u}\left(\frac{1}{6}\,du\right) = \frac{1}{6}\ln|u| + C = \frac{1}{6}\ln(1+3x^2) + C$.

11. $\int e^{2x}(1+e^{2x})^3\,dx$.

    Let $u(x) = 1 + e^{2x}$. $du = 2e^{2x}\,dx$.

    $\int e^{2x}(1+e^{2x})^3\,dx = \int u^3\cdot\frac{1}{2}\,du = \frac{1}{2}\cdot\frac{1}{4}u^4 + C = \frac{1}{8}(1+e^{2x})^4 + C$.

13. $\int \frac{e^x}{\sqrt{e^x+1}}\,dx$.

    Let $u(x) = e^x + 1$. $du = e^x\,dx$.

    $\int \frac{e^x}{\sqrt{e^x+1}}\,dx = \int \frac{1}{\sqrt{u}}\,du = 2u^{1/2} + C = 2\sqrt{e^x+1} + C$.

15. $\int \frac{2x+3}{(x^2+3x+6)^3}\,dx$

    Let $u(x) = x^2 + 3x + 6$. $du = (2x+3)dx$.

    $\int \frac{2x+3}{(x^2+3x+6)^3}\,dx = \int \frac{1}{u^3}\,du = -\frac{1}{2}u^{-2} + C = -\frac{1}{2(x^2+3x+6)^2} + C$.

17. $\int \left(1-\frac{1}{x}\right)^3 \left(\frac{1}{x^2}\right) dx$

Let $u(x) = 1 - \frac{1}{x}$. $\quad du = \frac{1}{x^2}\, dx$.

$\int \left(1-\frac{1}{x}\right)^3 \left(\frac{1}{x^2}\right) dx = \int u^3\, du = \frac{1}{4}u^4 + C = \frac{1}{4}\left(1-\frac{1}{x}\right)^4 + C.$

19. $\int \frac{x^3}{\sqrt{5+x^4}}\, dx$

Let $u(x) = 5 + x^4$. $\quad du = 4x^3\, dx$.

$\int \frac{x^3}{\sqrt{5+x^4}}\, dx = \int \frac{1}{\sqrt{u}} \cdot \frac{1}{4}\, du = \frac{1}{4} \cdot 2u^{1/2} + C = \frac{1}{2}(5+x^4)^{1/2} + C.$

21. $\int \frac{4\ln x^2}{x}\, dx = \int \frac{8\ln x}{x}\, dx.$

Let $u = \ln x$. $\quad du = \frac{1}{x}\, dx$.

$\int \frac{4\ln x^2}{x}\, dx = 8 \cdot \int u\, du = 8 \cdot \frac{1}{2}u^2 + C = 4(\ln x)^2 + C.$

23. $\int \frac{(x^{2/3}-5)^{2/3}}{\sqrt[3]{x}}\, dx$

Let $u(x) = x^{2/3} - 5$. $\quad du = \frac{2}{3}x^{-1/3}\, dx = \frac{2}{3} \cdot \frac{1}{\sqrt[3]{x}}\, dx$.

$\int \frac{(x^{2/3}-5)^{2/3}}{\sqrt[3]{x}}\, dx = \int u^{2/3} \cdot \frac{3}{2}\, du = \frac{3}{2} \cdot \frac{3}{5}u^{5/3} + C = \frac{9}{10}(x^{2/3}-5)^{5/3} + C.$

25. $\int \frac{(1+e^{\sqrt{x}})}{\sqrt{x}}\, e^{\sqrt{x}}\, dx$

Let $u(x) = 1 + e^{\sqrt{x}}$. $\quad du = \frac{1}{2} \cdot \frac{e^{\sqrt{x}}}{\sqrt{x}}\, dx$.

$\int \frac{(1+e^{\sqrt{x}})e^{\sqrt{x}}}{\sqrt{x}}\, dx = \int u \cdot 2\, du = u^2 + C = \left(1+e^{\sqrt{x}}\right)^2 + C.$

27. $MC(x) = 40 + \frac{20x}{1+x^2}$.

(a) $\lim\limits_{x\to\infty} MC(x) = \lim\limits_{x\to\infty} \left(40 + \frac{20x}{1+x^2}\right) = 40 + \lim\limits_{x\to\infty} \frac{\frac{20x}{x^2}}{\frac{1}{x^2}+\frac{x^2}{x^2}}$

$= 40 + \lim\limits_{x\to\infty} \frac{\frac{20}{x}}{\frac{1}{x^2}+1} = 40 + 0 = 40.$

(b) $C(x) = \int C'(x)dx = \int \left(40 + \frac{20x}{1+x^2}\right) dx = \int 40dx + \int \frac{20x}{1+x^2}dx$

$= 40x + \int \frac{20x}{1+x^2}dx.$

Let $u(x) = 1 + x^2$. $\quad du = 2xdx$.

$\int \frac{20x}{1+x^2}dx = \int \frac{10}{u}du = 10\ln u + C = 10\ln(1+x^2) + C.$

Thus $C(x) = 40x + 10\ln(1+x^2) + C$.

Since $C(0) = 500 = 40 \cdot 0 + 10\ln(1+0^2) + C$, $\quad 500 = C$.

Thus, $C(x) = 40x + 10\ln(1+x^2) + 500$.

29. $MR(x) = R'(x) = \frac{10,000x}{100+0.2x^2}$.

(a) $\lim_{x\to\infty} MR(x) = \lim_{x\to\infty} \frac{10,000x}{100+0.2x^2} = \lim_{x\to\infty} \frac{\frac{10,000x}{x^2}}{\frac{100}{x^2}+\frac{0.2x^2}{x^2}} = \lim_{x\to\infty} \frac{\frac{10,000}{x}}{\frac{100}{x^2}+0.2} = 0.$

(b) $R(x) = \int MR(x)dx = \int \frac{10,000x}{100+0.2x^2}dx.$

Let $u(x) = 100 + 0.2x^2$. $du = 0.4x dx$.

$R(x) = \int \frac{10,000x}{100+0.2x^2}dx = \int \frac{10,000}{u} \cdot \frac{1}{0.4}du = 25,000 \ln u + C.$

$R(x) = 25,000 \ln(100 + 0.2x^2) + C$

Since $R(0) = 0$, $R(0) = 0 = 25,000 \ln(100 + 0.2 \cdot 0^2) + C.$

$0 = 25,000 \ln 100 + C.$ $C = -25,000 \ln 100 \approx -115,129.25.$

Thus $R(x) = 25,000 \ln(100 + 0.2x^2) - 115,129.25.$

31. $P'(t) = \frac{100e^{20t}}{1+e^{20t}}$, $P(0) = 40,000.$

$P(t) = \int P'(t)dt = \int \frac{100e^{20t}}{1+e^{20t}}dt.$

Let $u(t) = 1 + e^{20t}$, $du = 20e^{20t}dt.$

$P(t) = \int \frac{5}{u}du = 5 \ln u + C = 5 \ln(1 + e^{20t}) + C.$

Since $P(0) = 40,000$, $40,000 = 5 \ln(1 + e^{20\cdot 0}) + C.$

$C = -5 \ln 2 + 40,000.$

Thus $P(t) = 5 \ln(1 + e^{20t}) - 5 \ln 2 + 40,000$

$P(t) = 5 \ln \frac{1}{2}(1 + e^{20t}) + 40,000.$

33. $P'(t) = \sqrt{t}e^{t^{3/2}}$.

Let $u = t^{3/2}$. $du = \frac{3}{2}t^{1/2}\,dt \implies t^{1/2}\,dt = \frac{2}{3}du.$

$\int \sqrt{t}e^{t^{3/2}}\,dt = \int e^u \left(\frac{2}{3}\,du\right) = \frac{2}{3}e^u + C = \frac{2}{3}e^{t^{3/2}} + C.$

$P(0) = 20 \implies 20 = \frac{2}{3}e^{0^{3/2}} + C \implies C = \frac{58}{3}.$

Therefore

$$P(t) = \frac{2}{3}e^{t^{3/2}} + \frac{58}{3}.$$

Selected Solutions to Exercise Set 5.4

1. $f(x) = 2x + 5$.

   $\Delta x = \frac{4-0}{4} = 1$, and use the left endpoint of each subinterval as t_j. Then

$$\text{Approx. Sum} = [f(0) + f(1) + f(2) + f(3)]\,\Delta x = [5 + 7 + 9 + 11] \cdot 1 = 32.$$

3. $f(x) = 4 - x^2$.

   $\Delta x = \frac{2-(-2)}{8} = \frac{1}{2}$, $\quad t_j =$ left endpoint of each interval.

$$\begin{aligned}\text{Approx. Sum} &= \left[f(-2) + f\left(-\frac{3}{2}\right) + f(-1) + f\left(-\frac{1}{2}\right) + f(0)\right.\\ &\quad \left. + f\left(\frac{1}{2}\right) + f(1) + f\left(\frac{3}{2}\right)\right] \Delta x \\ &= \left[(4-(-2)^2) + \left(4 - \left(-\frac{3}{2}\right)^2\right) + (4-(-1)^2) + \left(4 - \left(-\frac{1}{2}\right)^2\right)\right. \\ &\quad \left. + (4-0^2) + \left(4 - \left(\frac{1}{2}\right)^2\right) + (4-1^2) + \left(4 - \left(\frac{3}{2}\right)^2\right)\right] \cdot \frac{1}{2} \\ &= \left(0 + \frac{7}{4} + 3 + \frac{15}{4} + 4 + \frac{15}{4} + 3 + \frac{7}{4}\right) \cdot \frac{1}{2} = \frac{21}{2}.\end{aligned}$$

5. $f(x) = \frac{1}{1+x^2}$.

   $\Delta x = \frac{1-(-1)}{4} = \frac{1}{2}$, $\quad t_j =$ left endpoint of each interval.

$$\begin{aligned}\text{Approx. Sum} &= \left[f(-1) + f\left(-\frac{1}{2}\right) + f(0) + f\left(\frac{1}{2}\right)\right] \Delta x \\ &= \left[\frac{1}{1+(-1)^2} + \frac{1}{1+\left(-\frac{1}{2}\right)^2} + \frac{1}{1+0^2} + \frac{1}{1+\left(\frac{1}{2}\right)^2}\right] \frac{1}{2} \\ &= \left(\frac{1}{2} + \frac{4}{5} + 1 + \frac{4}{5}\right) \frac{1}{2} = \frac{31}{20}.\end{aligned}$$

7. $f(x) = x^2 + 3.$

$\Delta x = \frac{2-0}{4} = \frac{1}{2}$, $\quad t_j$ = right endpoint of each interval.

$$\begin{aligned}
\text{Approx. Sum} &= \left[f\left(\frac{1}{2}\right) + f(1) + f\left(\frac{3}{2}\right) + f(2)\right]\Delta x \\
&= \left[\left(\left(\frac{1}{2}\right)^2 + 3\right) + (1^2+3) + \left(\left(\frac{3}{2}\right)^2 + 3\right) + (2^2+3)\right]\frac{1}{2} \\
&= \left(12 + \frac{1}{4} + 1 + \frac{9}{4} + 4\right)\frac{1}{2} = \frac{39}{4}.
\end{aligned}$$

9. $f(x) = \frac{1}{4-x}.$

$\Delta x = \frac{2-0}{4} = \frac{1}{2}$, $\quad t_j$ = right endpoint of each interval.

$$\begin{aligned}
\text{Approx. Sum} &= \left[f\left(\frac{1}{2}\right) + f(1) + f\left(\frac{3}{2}\right) + f(2)\right]\Delta x \\
&= \left[\frac{1}{4-\frac{1}{2}} + \frac{1}{4-1} + \frac{1}{4-\frac{3}{2}} + \frac{1}{4-2}\right]\frac{1}{2} \\
&= \left(\frac{2}{7} + \frac{1}{3} + \frac{2}{5} + \frac{1}{2}\right)\frac{1}{2} = \frac{319}{420}.
\end{aligned}$$

11. The approximating sums in Exercises 1 and 2 are lower sums.

13. None of the approximating sums in Exercises 6-10 are lower sums.

15. $\sum_{j=1}^{6}(3j+2) = (3(1)+2) + (3(2)+2) + (3(3)+2) + (3(4)+2) + (3(5)+2)$

$$+ (3(6)+2) = 5 + 8 + 11 + 14 + 17 + 20 = 75.$$

17. $\sum_{j=1}^{5}(2j-5)^2 = (2(1)-5)^2 + (2(2)-5)^2 + (2(3)-5)^2 + (2(4)-5)^2 + (2(5)-5)^2$

$$= (-3)^2 + (-1)^2 + (1)^2 + (3)^2 + (5)^2$$

$$= 9 + 1 + 1 + 9 + 25 = 45.$$

19. $\sum_{j=4}^{7}(j-5) = (4-5) + (5-5) + (6-5) + (7-5)$

$$= -1 + 0 + 1 + 2 = 2.$$

21. $\sum_{j=4}^{10}(j^3-10) = \left((4)^3-10\right)+\left((5)^3-10\right)+\left((6)^3-10\right)+\left((7)^3-10\right)+\left((8)^3-10\right)$

$$+\left((9)^3-10\right)+\left((10)^3-10\right)$$

$$= 54+115+206+333+502+719+990 = 2919.$$

23. $\int_0^2 x\,dx = \text{Area of } R = \frac{1}{2}\cdot 2\cdot 2 = 2.$

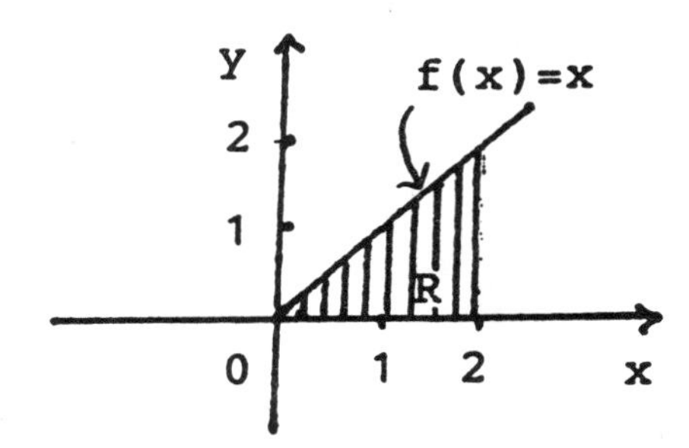

25. $\int_{-1}^{3} 4\,dx = 4\cdot 4 = 16.$

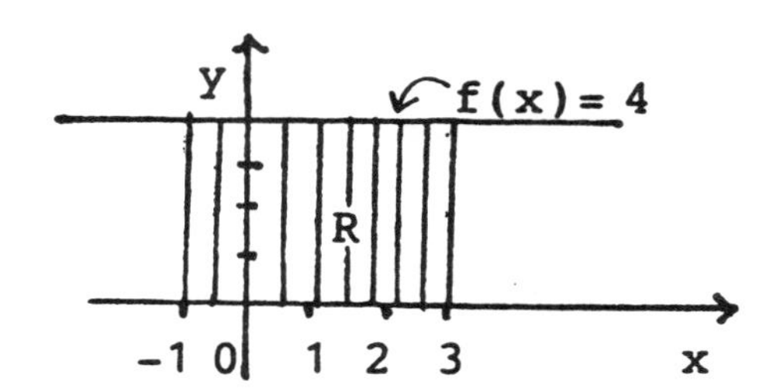

27. $\int_0^9 (9-x)\,dx = \text{Area of } R = \frac{1}{2}\cdot 9\cdot 9 = \frac{81}{2}.$

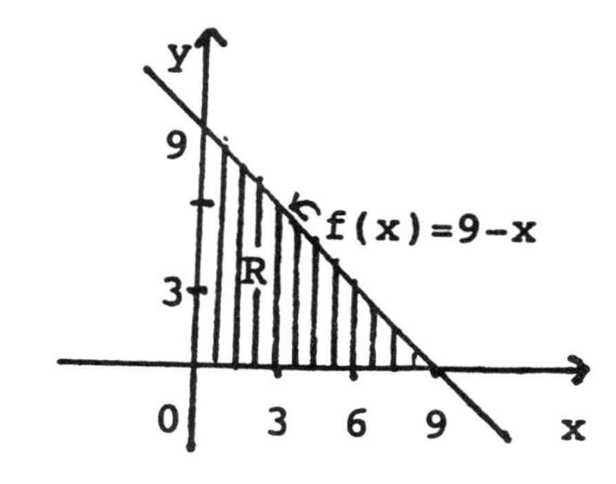

29. $\int_{-1}^{2} |x|dx = (\text{Area of } R_1) + (\text{Area of } R_2) = \frac{1}{2}1 \cdot 1 + \frac{1}{2}2 \cdot 2 = \frac{5}{2}$.

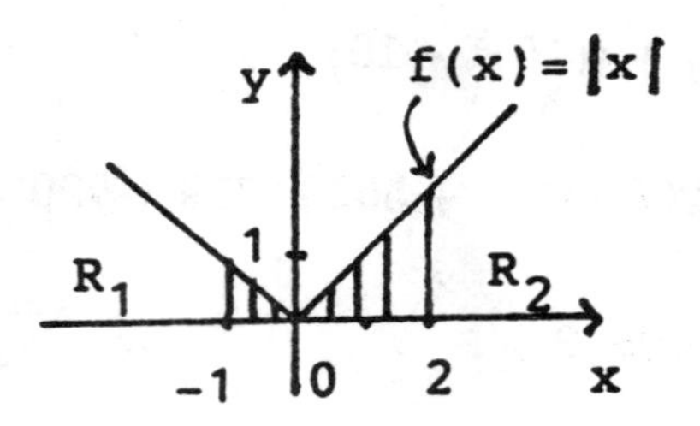

31. $\int_{0}^{4} |x - 2|dx = 2\left(\frac{1}{2} \cdot 2 \cdot 2\right) = 4$

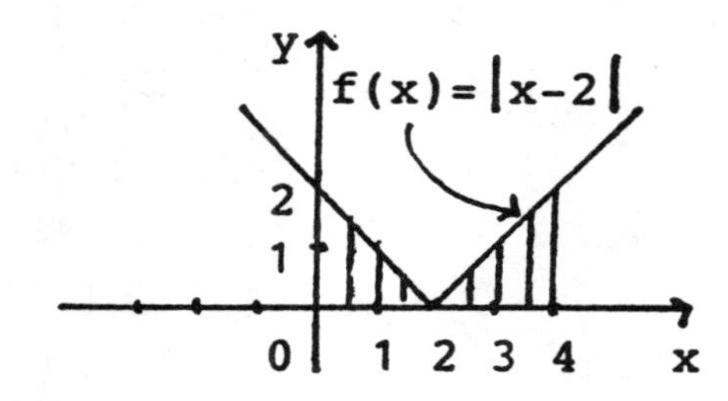

33. $\int_{-2}^{0} (-x)dx + \int_{0}^{2} (x)dx$.

35. $\int_{0}^{2} (-3x + 6)dx + \int_{2}^{4} (3x - 6)dx$.

37. $\int_{-2}^{-1} (-1 + x^2)dx + \int_{-1}^{1} (1 - x^2)dx + \int_{1}^{2} (-1 + x^2)dx$.

39. $\int_{0}^{2} \sqrt{x + 2}dx \approx 3.441853$ using $n = 100$ subintervals. Note that the exact answer is

$$\frac{2}{3}(x + 2)^{3/2}\Big|_0^2 = \frac{2}{3}\left(4^{3/2} - 2^{3/2}\right) = \frac{2}{3}\left(8 - 2\sqrt{2}\right) = 3.44771525.$$

41. $\int_0^4 \ln(x+3)dx \approx 6.308563$ using $n = 100$ subintervals.
Using integration by parts (see Section 7.2), we find that the exact answer is

$$\int_0^4 \ln(x+3)dx = \{(x+3)\ln(x+3) - x\}|_0^4 = 7\ln 7 - 3\ln 3 - 4$$
$$= 6.325534178.$$

43. $\int_0^4 \sqrt{5+x^2}dx \approx ?$

number of subintervals	approximation to integral
100	12.48087
1,000	12.52299
10,000	12.52723
100,000	12.52768

Using techniques available from Chapter 6 on trigonometric functions, we can show that

$$\int \sqrt{5+x^2}dx = \frac{1}{2}x\sqrt{5+x^2} + \frac{5}{2}\ln|\sqrt{5+x^2}+x| + C.$$

Hence the exact answer is

$$\frac{1}{2}\cdot 4\sqrt{21} + \frac{5}{2}\ln\left(\sqrt{21}+4\right) - \frac{5}{2}\ln\sqrt{5} = 2\sqrt{21} + \frac{5}{2}\ln\frac{\sqrt{21}+4}{\sqrt{5}}$$
$$= 12.52768916.$$

Selected Solutions to Exercise Set 5.5

1. $\int_0^2 (x+3)dx = \left(\frac{x^2}{2} + 3x\right)\Big]_0^2 = (\frac{4}{2}+6) - 0 = 8.$

3. $\int_1^5 (x^2-6)dx = \left(\frac{x^3}{3} - 6x\right)\Big]_1^5 = (\frac{125}{3} - 30) - (\frac{1}{3} - 6) = \frac{52}{3}.$

5. $\int_{-3}^2 (9+x^2)dx = \left(9x + \frac{x^3}{3}\right)\Big]_{-3}^2 = (18 + \frac{8}{3}) - (-27-9) = \frac{170}{3}.$

7. $\int_0^1 e^{4x}dx = \left.\frac{e^{4x}}{4}\right]_0^1 = \frac{1}{4}(e^4 - 1)$.

9. $\int_1^4 \left(\sqrt{x} - \frac{3}{\sqrt{x}}\right) dx = \left(\frac{2}{3}x^{3/2} - 6\sqrt{x}\right)\Big]_1^4 = \left(\frac{2}{3}\cdot 8 - 12\right) - \left(\frac{2}{3} - 6\right) = -\frac{4}{3}$.

11. $\int_0^1 (x^4 - 6x^3 + x)dx = \left(\frac{1}{5}x^5 - \frac{3}{2}x^4 + \frac{x^2}{2}\right)\Big]_0^1 = \left(\frac{1}{5} - 1\right) = -\frac{4}{5}$.

13. $\int_1^4 \left(\frac{x+3}{\sqrt{x}}\right) dx = \int_1^4 \left(\sqrt{x} + 3\cdot x^{-1/2}\right) dx$

$= \left(\frac{2}{3}x^{3/2} + 6\cdot x^{1/2}\right)\Big]_1^4 = \left(\frac{16}{3} + 12\right) - \left(\frac{2}{3} + 6\right) = \frac{32}{3}$.

15. $\int_0^2 (x^3 - 6x^2 + 3x + 3)dx$

$= \left(\frac{x^4}{4} - 2x^3 + \frac{3}{2}x^2 + 3x\right)\Big]_0^2 = (4 - 16 + 6 + 6) = 0$.

17. $\int_1^2 \frac{x+6}{x^2+12x}dx = \int_1^2 \frac{1}{2}\cdot\frac{2x+12}{x^2+12x}dx = \frac{1}{2}\ln(x^2+12x)\Big]_1^2$

$= \left(\frac{1}{2}\ln 28 - \frac{1}{2}\ln 13\right) = \frac{1}{2}\ln\frac{28}{13}$.

19. $\int_2^4 \frac{x-1}{\sqrt{x}-1}dx = \int_2^4 \frac{(\sqrt{x}+1)(\sqrt{x}-1)}{(\sqrt{x}-1)}dx = \int_2^4 (\sqrt{x}+1)\,dx$

$= \left(\frac{2}{3}x^{3/2} + x\right)\Big]_2^4 = \left(\frac{2}{3}\cdot 8 + 4\right) - \left(\frac{4}{3}\sqrt{2} + 2\right)$

$= \frac{22}{3} - \frac{4}{3}\sqrt{2} = \frac{2}{3}\left(11 - 2\sqrt{2}\right)$.

21. $\int_0^1 \left(x^{3/5} - x^{5/3}\right) dx = \left(\frac{5}{8}x^{8/5} - \frac{3}{8}x^{8/3}\right)\Big]_0^1 = \frac{1}{4}$.

23. $\int_0^4 \frac{dx}{\sqrt{2x+1}} = \left(\sqrt{2x+1}\right)\Big]_0^4 = 3 - 1 = 2$.

25. Then $du = 2x\,dx$.

$\int \frac{x}{\sqrt{16+x^2}}dx = \int \frac{1}{\sqrt{u}}\cdot\frac{1}{2}du = u^{1/2} + C = \sqrt{16+x^2} + C$.

Therefore, $\int_0^2 \frac{x}{\sqrt{16+x^2}}dx = \left(\sqrt{16+x^2}\right)\Big|_0^2 = \sqrt{20} - 4$.

27. Let $u = 9 - x^2$. Then $du = -2x dx \implies x dx = -\frac{1}{2} du$.

$$\int x\sqrt{9-x^2}dx = \int u^{1/2}\cdot\left(-\tfrac{1}{2}\right)du = -\tfrac{1}{2}\cdot\tfrac{2}{3}u^{3/2} + C$$
$$= -\tfrac{1}{3}(9-x^2)^{3/2} + C.$$
$$\int_0^3 x\sqrt{9-x^2}dx = \left(-\tfrac{1}{3}(9-x^2)^{3/2}\right)\Big]_0^3 = -\tfrac{1}{3}(0-27) = 9.$$

29. Let $u = x^2 - 1$. Then $du = 2x dx \implies x dx = \frac{1}{2} du$.

$$\int x(x^2-1)^{1/3}dx = \int u^{1/3}\cdot\tfrac{1}{2}du = \tfrac{1}{2}\cdot\tfrac{3}{4}\cdot u^{4/3} + C$$
$$= \tfrac{3}{8}(x^2-1)^{4/3} + C.$$
$$\int_1^2 x(x^2-1)^{1/3}dx = \tfrac{3}{8}(x^2-1)^{4/3}\Big]_1^2 = \tfrac{3}{8}(3^{4/3}-0) = \tfrac{9}{8}\sqrt[3]{3}.$$

31. $\int_0^8 \sqrt{x}\, dx = \frac{2}{3}x^{3/2}\Big]_0^8 = \frac{2}{3}\left(8^{3/2}\right) - \frac{2}{3}\left(0^{3/2}\right) = \frac{2}{3}\left(8^{3/2}\right) \approx 15.08.$

33. $\int_0^{\ln 5} e^{-x}\, dx = -e^{-x}\Big]_0^{\ln 5} = -e^{-\ln 5} + e^{-0} = -\frac{1}{5} + 1 = \frac{4}{5}.$

35. $\int_{-2}^2 (-x^2+4)dx = \left(-\frac{1}{3}x^3 + 4x\right)\Big]_{-2}^2 = \frac{16}{3} - \left(-\frac{16}{3}\right) = \frac{32}{3} \approx 10.67.$

37. $\int_0^1 (-x^3+x^2)dx = \left(-\frac{1}{4}x^4 + \frac{1}{3}x^3\right)\Big]_0^1 = \left(-\frac{1}{4}+\frac{1}{3}\right) - 0 = \frac{1}{12} \approx .0833.$

Selected Solutions to Exercise Set 5.6

1. Since $f(x) > 0$ for all x in $[0,2]$,

$$\text{area} = \int_0^2 (2x+5)dx = (x^2+5x)\Big]_0^2 = 14.$$

3. Since $f(x) > 0$ for all x in $[3,5]$,

$$\text{area} = \int_3^5 \tfrac{1}{x-2}dx = \ln(x-2)\Big]_3^5 = \ln 3 - \ln 1 = \ln 3.$$

5. Since $f(x) > 0$ on $[1,2]$,

$$\text{area} = \int_1^2 \tfrac{x+1}{x^2+2x}dx = \tfrac{1}{2}\ln(x^2+2x)\Big]_1^2$$
$$= \tfrac{1}{2}(\ln 8 - \ln 3) = \tfrac{1}{2}\ln\tfrac{8}{3}.$$

7. Since $f(x) > 0$ on $[-4, 4]$,

$$\text{area} = \int_{-4}^{4} \sqrt{5+x}\,dx = \tfrac{2}{3}(5+x)^{3/2}\Big]_{-4}^{4} = \tfrac{2}{3}(27-1) = \tfrac{52}{3}.$$

9. $$\int_0^{\ln 5} \frac{e^x}{1+e^x}\,x = \ln(1+e^x)]_0^{\ln 5} = \ln(1+e^{\ln 5}) - \ln(1+e^0)$$

$$= \ln(1+5) - \ln(1+1) = \ln 6 - \ln 2 = \ln \tfrac{6}{2} = \ln 3 \approx 1.0986.$$

11. $$\int_1^e \frac{\sqrt{1+\ln x}}{x}\,dx.$$

Let $u = (1 + \ln x)$. $du = \frac{1}{x}\,dx$.

$$\int \frac{\sqrt{1+\ln x}}{x}\,dx = \int u^{1/2}\,du = \tfrac{2}{3}u^{3/2} + C = \tfrac{2}{3}(1+\ln x)^{3/2} + C.$$

$$\int_1^e \frac{\sqrt{1+\ln x}}{x}\,dx = \tfrac{2}{3}(1+\ln x)^{3/2}\Big]_1^e = \tfrac{2}{3}(1+\ln e)^{3/2} - \tfrac{2}{3}(1+\ln 1)^{3/2}$$

$$= \tfrac{2}{3}(1+1)^{3/2} - \tfrac{2}{3}(1+0)^{3/2} = \tfrac{2}{3}\left(2^{3/2}-1\right) \approx 1.219.$$

13. Since $f(x) > g(x)$ on $[-2, 2]$,

$$\text{area} = \int_{-2}^{2} (f(x) - g(x))\,dx = \int_{-2}^{2} (9 - x^2 + 2)dx$$

$$= \left(9x - \tfrac{1}{3}x^3 + 2x\right)\Big]_{-2}^{2} = \left(22 - \tfrac{8}{3}\right) - \left(-22 + \tfrac{8}{3}\right)$$

$$= 44 - \tfrac{16}{3} = \tfrac{116}{3}.$$

15. Since $f(x) \geq g(x)$ on $[0, 2]$,

$$\text{area} = \int_0^2 (f(x) - g(x))\,dx = \int_0^2 (x + 1 + 2x - 1)dx$$

$$= \int_0^2 3x\,dx = \tfrac{3}{2}x^2\Big]_0^2 = 6.$$

17. Since $f(x) \geq g(x)$ on $[0, 4]$,

$$\text{area} = \int_0^4 (f(x) - g(x))\,dx = \int_0^4 \left(\sqrt{x} + x^2\right)dx$$

$$= \left(\tfrac{2}{3}x^{3/2} + \tfrac{1}{3}x^3\right)\Big]_0^4 = \left(\tfrac{2}{3}\cdot 8 + \tfrac{1}{3}\cdot 64\right) = \tfrac{80}{3}.$$

19. Since $f(x) \geq g(x)$ on $[0,3]$ and $g(x) \geq f(x)$ on $[-3,0]$,

$$\text{area} = \int_0^3 (f(x) - g(x))\,dx + \int_{-3}^0 (g(x) - f(x))\,dx$$
$$= \int_0^3 \left(x\sqrt{9-x^2} + x\right) dx + \int_{-3}^0 \left(-x - x\sqrt{9-x^2}\right) dx$$
$$= \left(\tfrac{x^2}{2} - \tfrac{1}{3}(9-x^2)^{3/2}\right)\Big]_0^3 + \left(-\tfrac{x^2}{2} + \tfrac{1}{3}(9-x^2)^{3/2}\right)\Big]_{-3}^0$$
$$= \left(\tfrac{9}{2} + \tfrac{1}{3}\cdot 27\right) + \left(\tfrac{1}{3}\cdot 27 + \tfrac{9}{2}\right) = (9 + 18) = 27.$$

21. The graphs intersect at two points. To find these points, we equate the two functions to obtain

$$4 - x^2 = x - 2 \implies x^2 + x - 6 = 0.$$

Thus $x^2 + x - 6 = 0$, so $x = -3$ or 2. Therefore, the points of intersection are $(-3,-5)$ and $(2,0)$.

Since $4 - x^2 \geq x - 2$ on the interval $(-3,2)$,

$$\text{area} = \int_{-3}^2 \left(4 - x^2 - (x-2)\right) dx$$
$$= \int_{-3}^2 (-x^2 - x + 6)dx = \left(-\tfrac{1}{3}x^3 - \tfrac{1}{2}x^2 + 6x\right)\Big]_{-3}^2$$
$$= \left(-\tfrac{8}{3} - 2 + 12\right) - \left(9 - \tfrac{9}{2} - 18\right)$$
$$= \left(-\tfrac{8}{3} + 10\right) - \left(-\tfrac{9}{2} - 9\right) = \tfrac{125}{6}.$$

23. First, let's find the intersection points.

$$x = x^3 \implies x(1 - x^2) = 0 \implies x(1+x)(1-x) = 0.$$

Thus the intersection points are (0,0), (1,1), and (-1,-1).

Since $x \geq x^3$ on [0,1] and $x^3 \geq x$ on [-1,0],

$$\text{area} = \int_0^1 (x - x^3)dx + \int_{-1}^0 (x^3 - x)dx$$
$$= \left(\tfrac{x^2}{2} - \tfrac{x^4}{4}\right)\Big]_0^1 + \left(\tfrac{x^4}{4} - \tfrac{x^2}{2}\right)\Big]_{-1}^0$$
$$= \left(\tfrac{1}{2} - \tfrac{1}{4}\right) + \left(0 - \left(\tfrac{1}{4} - \tfrac{1}{2}\right)\right) = \tfrac{1}{4} + \tfrac{1}{4} = \tfrac{1}{2}.$$

25. To find the intersection points, we have

$$x^2-4=2-x \implies x^2+x-6=0 \implies (x+3)(x-2)=0$$
$$\implies x=-3 \text{ or } x=2.$$

Since $2-x \geq x^2-4$ on $(-3,2)$,

$$\text{area} = \int_{-3}^{2}(2-x-x^2+4)dx = \int_{-3}^{2}(6-x-x^2)dx$$
$$= (6x-\tfrac{1}{2}x^2-\tfrac{1}{3}x^3)\Big]_{-3}^{2} = (10-\tfrac{8}{3})-(-9-\tfrac{9}{2})$$
$$= 19+\tfrac{11}{6} = \tfrac{125}{6}.$$

27. To find the intersection points, we have

$$x^{2/3}=x^2 \implies x^2=x^6 \implies x^2(x^4-1)=0$$
$$\implies x^2(x^2+1)(x+1)(x-1)=0$$
$$\implies x=0,-1,1.$$

Since $x^{2/3} \geq x^2$ on both [-1,0] and [0,1],

$$\text{area} = \int_{-1}^{0}(x^{2/3}-x^2)dx + \int_{0}^{1}(x^{2/3}-x^2)dx$$
$$= \left(\tfrac{3}{5}x^{5/3}-\tfrac{x^3}{3}\right)\Big]_{-1}^{0} + \left(\tfrac{3}{5}x^{5/3}-\tfrac{x^3}{3}\right)\Big]_{0}^{1}$$
$$= \{0-(-\tfrac{3}{5}+\tfrac{1}{3})\} + \{(\tfrac{3}{5}-\tfrac{1}{3})-0\}$$
$$= \tfrac{3}{5}-\tfrac{1}{3}+\tfrac{3}{5}-\tfrac{1}{3} = \tfrac{6}{5}-\tfrac{2}{3} = \tfrac{8}{15}.$$

29. Note that $f(x)=x^3e^x \leq 0$ for $-2 \leq x \leq 0$, and $f(x) \geq 0$ for $0 \leq x \leq 2$.

Therefore the area described is given by

$$\int_{-2}^{0}[-f(x)]\,dx + \int_{0}^{2}f(x)\,dx = \int_{-2}^{0}(-x^3e^x)dx + \int_{0}^{2}(x^3e^x)dx.$$

31. Note that $f(x) = e^{x^2} - e = 0$ for $e^{x^2} = e^1 \implies x^2 = 1 \implies x = \pm 1.$

We assume that the problem here is to find the area of the region bounded by the graph of $y = f(x)$ and the x-axis for $-1 \le x \le 1$.

Observe that $f(x) < 0$ for $-1 < x < 1$.

Therefore the desired area is given by

$$\int_{-1}^{1} [-f(x)]\, dx = \int_{-1}^{1} \left(-e^{x^2} + e\right) dx.$$

33.

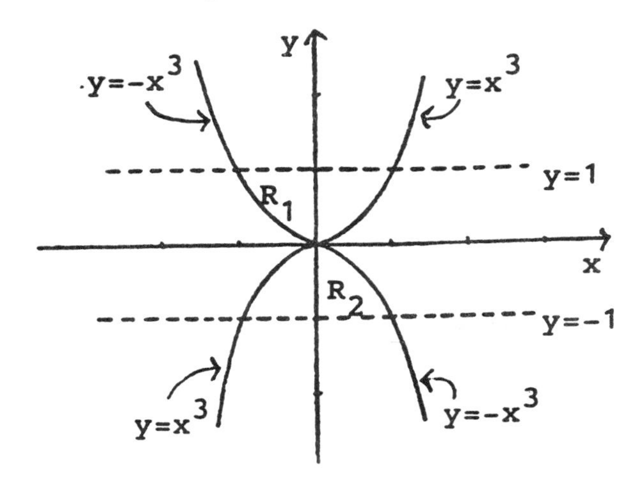

Area = (Area of R_1) + (Area of R_2)

$$= \int_{-1}^{0} (1-(-x^3))\,dx + \int_{0}^{1}(1-x^3)dx + \int_{-1}^{0} (x^3-(-1))\,dx + \int_{0}^{1} (-x^3-(-1))\,dx$$

$$= \int_{-1}^{0}(1+x^3)dx + \int_{0}^{1}(1-x^3)dx + \int_{-1}^{0}(x^3+1)dx + \int_{0}^{1}(-x^3+1)dx$$

$$= \left(x+\frac{x^4}{4}\right)\Big]_{-1}^{0} + \left(x-\frac{x^4}{4}\right)\Big]_{0}^{1} + \left(\frac{x^4}{4}+x\right)\Big]_{-1}^{0} + \left(-\frac{x^4}{4}+x\right)\Big]_{0}^{1}$$

$$= 0 - \left(-1+\tfrac{1}{4}\right) + \left(1-\tfrac{1}{4}\right) + 0 - \left(\tfrac{1}{4}-1\right) + \left(-\tfrac{1}{4}+1\right) - 0$$

$$= \tfrac{3}{4} + \tfrac{3}{4} + \tfrac{3}{4} + \tfrac{3}{4} = 3.$$

35.

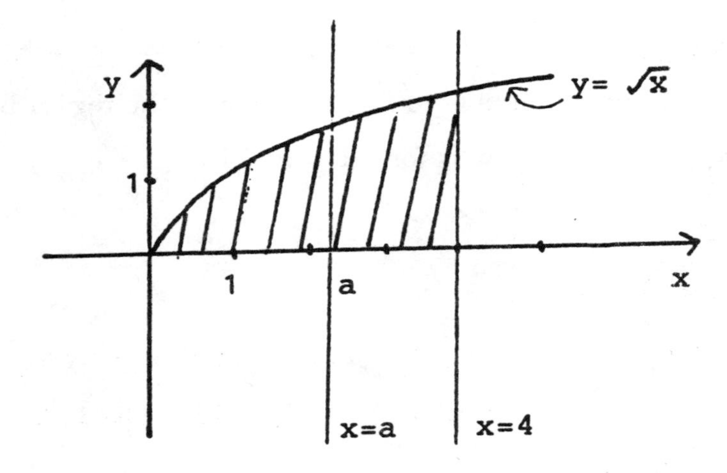

$$\int_0^a \sqrt{x}dx = \frac{1}{2}\int_0^4 \sqrt{x}dx \implies \frac{2}{3}x^{3/2}\Big]_0^a = \frac{1}{2}\left(\frac{2}{3}x^{3/2}\right)\Big]_0^4$$

$$\implies \frac{2}{3}a^{3/2} = \frac{1}{2}\left(\frac{2}{3}\cdot 8\right) \implies \frac{2}{3}a^{3/2} = \frac{8}{3} \implies a^{3/2} = 4 \implies a = 4^{2/3}.$$

Selected Solutions to Exercise Set 5.7

1. $\int_0^{10}(20 - 0.5x)dx - 10 \cdot D(10) = \left(20x - \frac{1}{4}x^2\right)\Big]_0^{10} - 10 \cdot 15$
   $= (200 - 25) - 150 = \$25.$

3. $\int_0^5 100(40 - x^2)dx - 5 \cdot D(5)$
   $= 100\left(40x - \frac{1}{3}x^3\right)\Big]_0^5 - 5 \cdot (100(40 - 25))$
   $= 100\left(200 - \frac{125}{3}\right) - 5 \cdot (1500)$
   $= \frac{47500}{3} - 7500 = \frac{25000}{3}$ dollars.

5. $\int_0^{12} \frac{40}{\sqrt{2x+1}}dx - 12 \cdot D(12) = 40 \cdot \sqrt{2x+1}\Big]_0^{12} - 12 \cdot \frac{40}{5}$
   $= 40 \cdot (5 - 1) - 96 = 160 - 96 = \$64.$

7. $6 \cdot S(6) - \int_0^6 \frac{1}{100}x^2dx = 6 \cdot \frac{36}{100} - \frac{1}{100}\left(\frac{x^3}{3}\right)\Big]_0^6$
   $= \frac{216}{100} - \frac{1}{100}(72) = \frac{144}{100} = \$1.44.$

9. $20 \cdot S(20) - \int_0^{20} (e^{0.2x} - 1)dx$

$= 20 \cdot (e^4 - 1) - (5 \cdot e^{0.2x} - x)\Big]_0^{20}$

$= 20 \cdot (e^4 - 1) - \{5 \cdot e^4 - 20 - 5\}$

$= 20 \cdot e^4 - 20 + 25 - 5e^4 = 15 \cdot e^4 + 5$ dollars.

11. a. $\int_{400}^{500} MC(x)dx = \int_{400}^{500} (200 + .4x)dx.$

b. $C(500) - C(400) = (200x + .2x^2)\Big]_{400}^{500} = (200(500) + .2(500)^2)$
$- (200(400) + .2(400)^2) = 150,000 - 112,000 = 38,000.$

13. $C(40) = C(30) + \int_{30}^{40} (200 + 0.4x)dx$

$= 7000 + (200x + 0.2x^2)\Big]_{30}^{40}$

$= 7000 + (8000 + 320) - (6000 + 180)$

$= 7000 + 2000 + 140 = \$9140.$

15. $R(50) - R(30) = \int_{30}^{50} (400 - 0.5x)dx$

$= (400x - \frac{1}{4}x^2)\Big]_{30}^{50}$

$= (20000 - \frac{2500}{4}) - (12000 - \frac{900}{4})$

$= 8000 - \frac{1600}{4} = \$7600.$

17. The total nominal value of the revenue flow is

$$\int_0^3 A(t)dt = \int_0^3 (3000t + 2000)dt = (1500t^2 + 2000t)\Big]_0^3$$
$$= (1500(3)^2 + 2000(3)) - 0 = \$19,500.$$

19. $\int_5^{10} 1000e^{-0.2t}dt = 1000 \cdot (-5 \cdot e^{-0.2t})\Big]_5^{10}$

$= 1000 \cdot (-5 \cdot e^{-2} + 5e^{-1}) = 5000(e^{-1} - e^{-2})$ dollars.

21. $\int_0^{10} 5000 \cdot e^{-rt}dt = \int_0^{10} 5000 \cdot e^{-0.1t}dt$

$= 5000 \cdot (-10)(e^{-0.1t})\Big]_0^{10} = 50000(1 - e^{-1})$ dollars.

23. a. The present value of this payment stream is

$$\int_0^{20} 2{,}000e^{-.08t}\,dt = 2{,}000 \cdot \frac{1}{-.08}e^{-.08t}\Big]_0^{20}$$

$$= -25{,}000\left(e^{-.08(20)} - e^{-.08(0)}\right) = \$19{,}952.59.$$

b. We proceed as indicated in Problem 52 in the Review Exercises for Chapter 4. Let n be some given positive integer. Let $\Delta t = \frac{20}{n}$. Let t_0, t_1, t_2, t_3, ..., t_n be the real numbers such that

$$0 = t_0 < t_1 < t_2 < t_3 < \cdots < t_n = 20$$

and $(t_j - t_{j-1}) = \Delta t$ for $j = 1, 2, 3, \ldots, n$. Note that $t_j = j \cdot \Delta t$ for $j = 0, 1, 2, \ldots, n$. Suppose that at each instant of time t_0, t_1, t_2, t_3, ..., t_{n-1} Robert deposits $2{,}000 \cdot \Delta t$ dollars into his IRA account which is paying 8% annual interest compounded continuously. As in Problem 52 in the Review Exercises for Chapter 4, the value of Robert's IRA account after 20 years would be given by

$$T(n) = 2{,}000 \cdot \Delta t \cdot \left(e^{(n \cdot .08)\cdot \Delta t} + e^{((n-1).08)\cdot\Delta t} + \cdots\right.$$
$$\left. + e^{(3\cdot .08)\cdot\Delta t} + e^{(2\cdot .08)\cdot\Delta t} + e^{(1\cdot .08)\cdot\Delta t}\right)$$
$$= 2{,}000 \cdot \left(e^{.08\cdot(1\cdot\Delta t)} + e^{.08\cdot(2\cdot\Delta t)} + e^{.08\cdot(3\cdot\Delta t)}\right.$$
$$\left. + \cdots + e^{.08\cdot((n-1)\cdot\Delta t)} + e^{.08\cdot(n\cdot\Delta t)}\right) \cdot \Delta t$$
$$= 2{,}000 \cdot \left(e^{.08t_1} + e^{.08t_2} + e^{.08t_3} + \cdots + e^{.08t_{n-1}}\right.$$
$$\left. + e^{.08t_n}\right) \cdot \Delta t.$$

As n approaches ∞, this value $T(n)$ approaches the definite integral

$$\int_0^{20} 2{,}000e^{.08t}\,dt.$$

We take this definite integral to be the nominal value of Robert's IRA account after 20 years. This nominal value is

$$2,000 \cdot \frac{1}{.08} \cdot \left(e^{.08(20)} - e^{.08(0)}\right) = \$98,825.81.$$

25. Let R denote the payment rate.

$$1,000,000 = \int_0^{20} Re^{-.08t}\, dt = R \cdot \frac{1}{-.08} \cdot e^{-.08t}\Big]_0^{20}$$

$$= R \cdot \frac{1}{.08} \cdot (1 - e^{-1.6}) = 9.9762935R.$$

Therefore $R = \$100,237.63$.

27. The present value of this payment stream is given by

$$\int_0^{60} 600e^{-.00833t}\, dt = -\frac{1}{.00833} \cdot 600 \cdot e^{-.00833t}\Big]_0^{60} = \$28,329.79.$$

29. Net present value for plan A is given by

$$PV_A = \int_0^5 (\$3 \text{ million})e^{(-.08t)}\, dt - \$10 \text{ million}$$

$$= \left\{3\left[\frac{-1}{.08}(e^{-.08t})\right]_0^5 - 10\right\} \text{ million dollars}$$

$$= \left\{3\left[12.5(1 - e^{-.4})\right] - 10\right\} \text{ million dollars}$$

$$= 2.362998274 \text{ million dollars}$$

$$= 2,362,998.27 \text{ dollars.}$$

Net present value for plan B is given by

$$PV_B = \int_0^5 (\$2\text{ million})e^{-.08t}dt - \$6\text{ million}$$

$$= \left\{2\left[\frac{-1}{.08}(e^{-.08t})\right]_0^5 - 6\right\} \text{ million dollars}$$

$$= \left\{2\left[12.5(1-e^{-.4})\right] - 6\right\} \text{ million dollars}$$

$$= 2.241998849 \text{ million dollars}$$

$$= 2,241,998.85 \text{ dollars.}$$

Plan A would be chosen.

31. The present value 30 years from now of the revenue stream which begins then is given by

$$PV_{30} = \int_0^{10} (30,000)e^{-.06t}\,dt = 30,000\left[\frac{-1}{.06}e^{-.06t}\right]_0^{10}$$

$$= 500,000(1-e^{-.6}) = 225,594.18 \text{ dollars.}$$

The present value now of the amount PV_{30} is

$$PV_0 = 225,584.18e^{-.06(30)} = \$37,290.47.$$

Selected Solutions to Exercise Set 5.8

1. $\int_1^4 \pi(2x+1)^2dx$

$$= \pi\int_1^4 (4x^2+4x+1)dx = \pi\left(\tfrac{4}{3}x^3+2x^2+x\right)\Big]_1^4$$

$$= \pi\left(\tfrac{256}{3}+32+4-\tfrac{4}{3}-3\right) = \pi\left(\tfrac{252}{3}+33\right) = \pi\cdot\tfrac{351}{3} = 117\pi.$$

3. $\int_0^3 \pi|x-1|^2dx = \int_0^3 (x-1)^2\cdot\pi\,dx$

$$= \pi\int_0^3 (x^2-2x+1)dx = \pi\left(\tfrac{x^3}{3}-x^2+x\right)\Big]_0^3$$

$$= \pi\cdot(9-9+3) = 3\pi.$$

5. Volume $= \pi \int_1^2 \left[\frac{\sqrt{x+1}}{x}\right]^2 dx = \pi \int_1^2 \frac{x+1}{x^2}\, dx = \pi \int_1^2 \left(\frac{1}{x} + \frac{1}{x^2}\right) dx$
$= \pi \left(\ln x - \frac{1}{x}\right)\Big]_1^2 = \pi\left(\left(\ln 2 - \frac{1}{2}\right) - (\ln 1 - 1)\right) = \pi\left(\ln 2 + \frac{1}{2}\right).$

7. First, find the intersections of the two graphs.

$f(x) = g(x) \implies \frac{1}{4}x^3 = x \implies x^3 - 4x = 0 \implies x(x^2-4) = 0 \implies$

$x = 0$ or $x = 2$ or $x = -2$.

For $-2 < x < 0$, $[g(x)]^2 > [f(x)]^2$.

For $0 < x < 2$, $[g(x)]^2 > [f(x)]^2$.

Therefore the volume is

$$\pi \int_{-2}^{2} \left\{[g(x)]^2 - [f(x)]^2\right\} dx = \pi \int_{-2}^{2} \left(x^2 - \frac{1}{16}x^6\right) dx$$
$$= \pi \left(\frac{1}{3}x^3 - \frac{1}{112}x^7\right)\Bigg]_{-2}^{2}$$
$$= \pi \left\{\left(\frac{8}{3} - \frac{128}{112}\right) - \left(-\frac{8}{3} + \frac{128}{112}\right)\right\}$$
$$= 16\pi\left(\frac{1}{3} - \frac{1}{7}\right) = 16\pi \cdot \frac{4}{21} = \frac{64\pi}{21}.$$

9. Volume $= \pi \int_0^4 (4-x)^2 dx = \frac{-\pi}{3}(4-x)^3\Big]_0^4 = \frac{64\pi}{3}.$

11. Average value $= \dfrac{\int_1^5 f(x)dx}{5-1} = \dfrac{\int_1^5 \sqrt{2x-1}\, dx}{4} = \dfrac{\frac{1}{3}(2x-1)^{3/2}\Big]_1^5}{4}$
$= \dfrac{\frac{1}{3}\cdot(27-1)}{4} = \dfrac{13}{6}.$

13. $\dfrac{\int_{-2}^{2} (x+2)^{1/2} dx}{2-(-2)} = \dfrac{\frac{2}{3}(x+2)^{3/2}\Big]_{-2}^{2}}{4} = \dfrac{\frac{16}{3}}{4} = \dfrac{4}{3}.$

15. $\dfrac{\int_0^2 (x^2-x+1)dx}{2-0} = \dfrac{\left(\frac{1}{3}x^3 - \frac{1}{2}x^2 + x\right)\Big]_0^2}{2} = \dfrac{\left(\frac{8}{3}-2+2\right)-0}{2} = \dfrac{4}{3}.$

17. The area of $R = \frac{1}{2}\pi(3)^2 = \frac{9}{2}\pi$.

Average value $= \frac{\frac{9\pi}{2}}{6} = \frac{3\pi}{4}$.

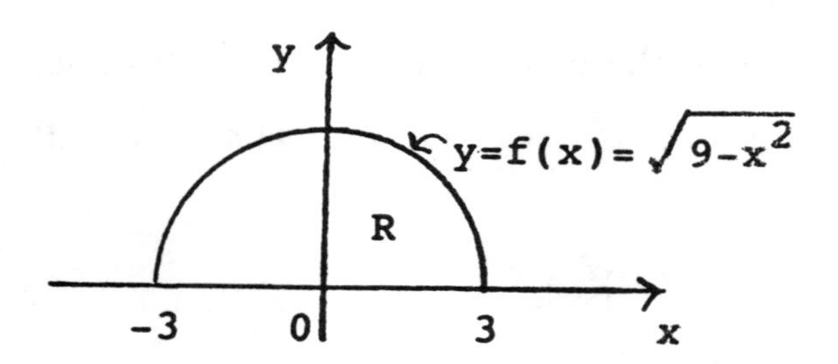

19. $\frac{\int_0^{10}(120x-4x^2)dx}{10-0} = \frac{\left(60x^2-\frac{4}{3}x^3\right)\Big]_0^{10}}{10} = \frac{6000-\frac{4000}{3}}{10} = \frac{1400}{3}$.

21. The average value of the marginal cost function $C'(x)$ on the interval $[0, x]$ is $\frac{\int_0^x C'(u)du}{x} = \frac{C(x)-C(0)}{x} = \frac{C(x)}{x} = c(x)$ in the case $C(0) = 0$.

23. Revolve R about the x-axis.

$$\text{Volume} = \pi \cdot \int_{-r}^{r} (f(x))^2\, dx = \pi \cdot \int_{-r}^{r} (r^2 - x^2)dx$$
$$= \pi \cdot \left(r^2x - \tfrac{1}{3}x^3\right)\Big]_{-r}^{r}$$
$$= \pi\left[\left(r^3 - \tfrac{1}{3}r^3\right) - \left(-r^3 + \tfrac{1}{3}r^3\right)\right]$$
$$= \pi\left(\tfrac{4}{3}r^3\right).$$

This is the formula for the volume of a sphere of radius r.

25. Let $P(t)$ denote the population of the earth in billions of people t years from now. We have that

$$P(t) = P_0e^{rt} \implies P(t) = 5e^{rt}.$$

$$P(50) = 10 \implies 10 = 5e^{r(50)} \implies e^{50r} = 2 \implies 50r = \ln 2$$
$$\implies r = \frac{\ln 2}{50}.$$

Therefore

$$P(t) = 5e^{\left(\frac{\ln 2}{50}\right)t}.$$

The average population of the earth during this 50-year period would be

$$\frac{\int_0^{50} P(t)dt}{50} = \frac{5 \cdot \frac{50}{\ln 2} \cdot e^{\left(\frac{\ln 2}{50}\right)t}\Big]_0^{50}}{50} = \frac{5}{\ln 2} \cdot \left(e^{\ln 2} - 1\right)$$
$$= \frac{5}{\ln 2} = 7.2135 \text{ billion.}$$

Selected Solutions to the Review Exercises - Chapter 5

1. $\int (6x^2 - 2x + 1)dx = 2x^3 - x^2 + x + C.$

3. $\int x\sqrt{3x^2+5}\,dx.$

   Let $u = 3x^2 + 5.$ $du = 6x\,dx \implies x\,dx = \frac{1}{6}\,du.$

   $\int x\sqrt{3x^2+5}\,dx = \int u^{1/2}\left(\frac{1}{6}\,du\right) = \frac{1}{6}\cdot\frac{2}{3}u^{3/2} + C = \frac{1}{9}(3x^2+5)^{3/2} + C.$

5. $\int \left(t + \sqrt[3]{t}\right)^2 dt = \int (t^2 + 2t^{4/3} + t^{2/3})dt = \frac{1}{3}t^3 + \frac{6}{7}t^{7/3} + \frac{3}{5}t^{5/3} + C.$

7. $\int \frac{x^3 - 7x^2 + 6}{x}dx = \int \left(x^2 - 7x + \frac{6}{x}\right)dx = \frac{1}{3}x^3 - \frac{7}{2}x^2 + 6\ln|x| + C.$

9. $\int (2x-1)(2x+3)dx = \int (4x^2 + 4x - 3)dx = \frac{4}{3}x^3 + 2x^2 - 3x + C.$

11. Let $u = 1 - x^2.$ $du = -2x\,dx \implies x\,dx = -\frac{1}{2}du.$

    $\int \frac{x}{1-x^2}dx = \int \frac{1}{u}\left(-\frac{1}{2}du\right) = -\frac{1}{2}\ln|u| + C = -\frac{1}{2}\ln|1-x^2| + C.$

13. $\int \sqrt{e^x}dx = \int e^{x/2}dx = 2e^{x/2} + C.$

15. Let $u = \ln x \implies du = \frac{1}{x}dx.$

    $\int \frac{1}{x\sqrt{\ln x}}dx = \int u^{-1/2}du = 2u^{1/2} + C = 2\sqrt{\ln x} + C.$

17. $\int \frac{x^3-1}{x+1}dx = \int \left(\frac{x^3+1}{x+1} - \frac{2}{x+1}\right)dx = \int \left(x^2 - x + 1 - \frac{2}{x+1}\right)dx$

    $= \frac{1}{3}x^3 - \frac{1}{2}x^2 + x - 2\ln|x+1| + C.$

19. $\int e^{2x}(1-e^{2x})^2\,dx.$

    Let $u = 1 - e^{2x}.$ $du = -2e^{2x}\,dx \implies e^{2x}\,dx = -\frac{1}{2}\,du.$

    $\int e^{2x}(1-e^{2x})^2\,dx = \int u^2\left(-\frac{1}{2}\,du\right) = -\frac{1}{6}u^3 + C = -\frac{1}{6}(1-e^{2x})^3 + C.$

21. $v(t) = 2t - (t+1)^{-2}$. $v(t) = s'(t)$.

(a) $s(t) = \int \left(2t - (t+1)^{-2}\right) dt = t^2 + \frac{1}{t+1} + C$.

$s(0) = 0 \implies 0 = 0^2 + \frac{1}{0+1} + C \implies C = -1$.

$s(t) = t^2 + \frac{1}{t+1} - 1$.

(b) $a(t) = v'(t) = 2 + \frac{2}{(t+1)^3}$.

23. (a) $C(x) = \int MC(x)dx = \int(120 + 6x)dx = 120x + 3x^2 + K$.

$C(0) = 500 \implies K = 500$.

$C(x) = 120x + 3x^2 + 500$.

(b) $C(20) = 120(20) + 3(20)^2 + 500 = 4100$.

25. $P(x) = R(x) - C(x) = 250x - (120x + 3x^2 + 500) = -3x^2 + 130x - 500$.

$P(20) = R(20) - C(20) = 5000 - 4100 = 900 > 0$.

The operation is profitable when $x = 20$.

27. $\frac{dV}{dt} = \frac{-80{,}000}{(t+1)^2}$. $V = \frac{80{,}000}{(t+1)} + C$.

$V(0) = 100{,}000 \implies 100{,}000 = \frac{80{,}000}{0+1} + C \implies C = 20{,}000$

(a) $V(t) = \frac{80{,}000}{t+1} + 20{,}000$ dollars.

(b) $V(3) = \frac{80{,}000}{3+1} + 20{,}000 = 40{,}000$ dollars.

(c) $\lim_{t\to\infty} V(t) = 20{,}000$.

29. (a) $C(x) = \int MC(x)dx = \int(70 + 2x)dx = 70x + x^2 + K$.

$C(10) = 1000 \implies 1000 = 70(10) + 10^2 + K \implies K = 200$.

$C(x) = 70x + x^2 + 200$ dollars.

(b) $C(0) = \$200$.

31. $\int_0^3 (2x - 3)dx = (x^2 - 3x)\Big]_0^3 = \left(3^2 - 3(3)\right) - \left(0^2 - 3(0)\right) = 0$.

33. $\int_0^1 (x^2 - 6)^2\, dx = \int_0^1 (x^4 - 12x^2 + 36)dx = \left(\frac{1}{5}x^5 - 4x^3 + 36x\right)\Big]_0^1$

$= \left(\frac{1}{5} - 4 + 36\right) = \frac{161}{5}$.

35. $\int_2^9 \sqrt{x+7}\, dx = \frac{2}{3}(x+7)^{3/2}\Big]_2^9 = \frac{2}{3}(9+7)^{3/2} - \frac{2}{3}(2+7)^{3/2} = \frac{2}{3}(64 - 27) = \frac{74}{3}$.

37. $\int_0^3 \frac{x}{x+3}dx = \int_0^3 \left(1 - \frac{3}{x+3}\right) dx = (x - 3\ln(x+3))]_0^3$

$= [3 - 3\ln(3+3)] - [0 - 3\ln(0+3)] = 3 - 3(\ln 6 - \ln 3)$

$= 3 - 3\ln\frac{6}{3} = 3(1 - \ln 2).$

39. $\int_3^4 \frac{x+2}{x-2}\,dx = \int_3^4 \left(\frac{(x-2)+4}{x-2}\right) dx = \int_3^4 \left(\frac{x-2}{x-2} + \frac{4}{x-2}\right) dx = \int_3^4 \left(1 + \frac{4}{x-2}\right) dx$

$= (x + 4\ln(x-2))]_3^4 = (4 + 4\ln 2) - (3 + 4\ln 1) = 1 + 4\ln 2.$

41. $\int_{-1}^3 (x-1)(x+5)dx = \int_{-1}^3 (x^2 + 4x - 5)dx = \left(\frac{1}{3}x^3 + 2x^2 - 5x\right)\Big]_{-1}^3$

$= \left(\frac{1}{3}\cdot 3^3 + 2\cdot 3^2 - 5\cdot 3\right) - \left(\frac{1}{3}\cdot(-1)^3 + 2\cdot(-1)^2 - 5(-1)\right)$

$= \frac{27}{3} + 3 + \frac{1}{3} - 7 = \frac{16}{3}.$

43. $\int_1^3 |x-4|dx = \int_1^3 (4-x)dx = \left(4x - \frac{1}{2}x^2\right)\Big]_1^3 = \left(4(3) - \frac{1}{2}(3)^2\right) - \left(4(1) - \frac{1}{2}(1)^2\right)$

$= \frac{15}{2} - \frac{7}{2} = 4.$

45. $\int_{-8}^{-1} (x^{1/3} - x^{5/3})dx = \left(\frac{3}{4}x^{4/3} - \frac{3}{8}x^{8/3}\right)\Big]_{-8}^{-1}$

$= \left(\frac{3}{4}\cdot(-1)^{4/3} - \frac{3}{8}\cdot(-1)^{8/3}\right) - \left(\frac{3}{4}\cdot(-8)^{4/3} - \frac{3}{8}\cdot(-8)^{8/3}\right)$

$= \frac{3}{4} - \frac{3}{8} - \frac{3}{4}\cdot 16 + \frac{3}{8}\cdot 2^8 = \frac{675}{8}.$

47. $\int_1^8 \frac{\sqrt[3]{x}}{5+x^{4/3}}dx.$

Let $u(x) = 5 + x^{4/3}$. $du = \frac{4}{3}x^{1/3}dx$. $\sqrt[3]{x}dx = \frac{3}{4}du$.

$\int \frac{\sqrt[3]{x}}{5+x^{4/3}}dx = \frac{3}{4}\int \frac{1}{u}du = \frac{3}{4}\ln u + C = \frac{3}{4}\ln(5 + x^{4/3}) + C.$

$\int_1^8 \frac{\sqrt[3]{x}}{5+x^{4/3}}dx = \frac{3}{4}\ln(5 + x^{4/3})\Big]_1^8$

$= \frac{3}{4}\ln(5 + 8^{4/3}) - \frac{3}{4}\ln(5 + 1^{4/3}) = \frac{3}{4}\ln(5+16) - \frac{3}{4}\ln 6$

$= \frac{3}{4}\ln\left(\frac{21}{6}\right) = \frac{3}{4}\ln\left(\frac{7}{2}\right).$

49. $\int_1^8 \frac{x^{2/3} + 3x^{5/2}}{x}dx = \int_1^8 (x^{-1/3} + 3x^{3/2})dx = \left(\frac{3}{2}x^{2/3} + \frac{6}{5}x^{5/2}\right)\Big]_1^8$

$= \left(\frac{3}{2}\cdot 8^{2/3} + \frac{6}{5}\cdot(8)^{5/2}\right) - \left(\frac{3}{2}\cdot 1 + \frac{6}{5}\cdot 1\right) = \frac{9}{2} + \frac{6}{5}(8^{5/2} - 1).$

51. $\int_{-4}^{4} \sqrt{16-x^2}\,dx = \text{Area of } R$

$= \frac{1}{2}\pi \cdot 4^2 = 8\pi.$

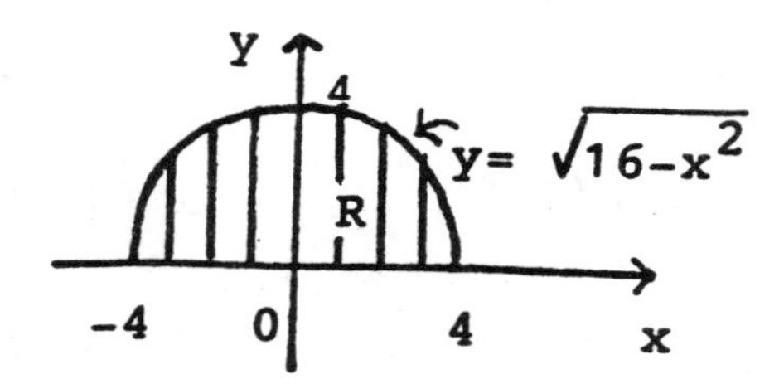

53. Area of $R = \int_0^9 (9-\sqrt{x})\,dx = \left(9x - \frac{2}{3}x^{3/2}\right)\Big]_0^9 = (9\cdot 9 - \frac{2}{3}\cdot 9^{3/2}) - 0 = 81 - \frac{2}{3}\cdot 27 =$ 63.

55. From a sketch of the graphs it can be seen that there are two points of intersection. These points are (1,4) and (-1,4).

Area of $R = 2\int_0^1 \left[(3+x^{2/3}) - (3x^2+1)\right]dx.$

Area of $R = 2\left(3x + \frac{3}{5}x^{5/3} - x^3 - x\right)\Big]_0^1$

$= 2\left(3\cdot 1 + \frac{3}{5}\cdot 1 - 1 - 1\right) - 0$

$= 2\left(\frac{8}{5}\right) = \frac{16}{5}.$

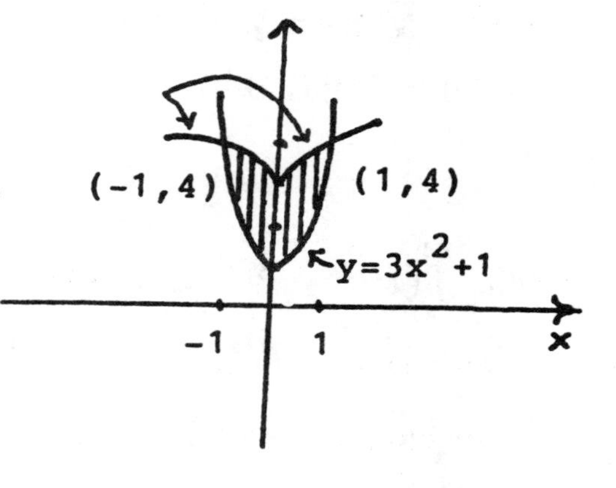

57. For intersection point P_1, solve

$$\begin{cases} y = -x \\ y = \frac{2}{3}x - \frac{10}{3} \end{cases}.$$

$x = 2$, $y = -2$, $P_1 = (2,-2)$.

For intersection point P_2, solve

$$\begin{cases} y = \sqrt[3]{x} & (1) \\ y = \frac{2}{3}x - \frac{10}{3} & (2) \end{cases}.$$

From (1), $x = y^3$. Substitute into (2).

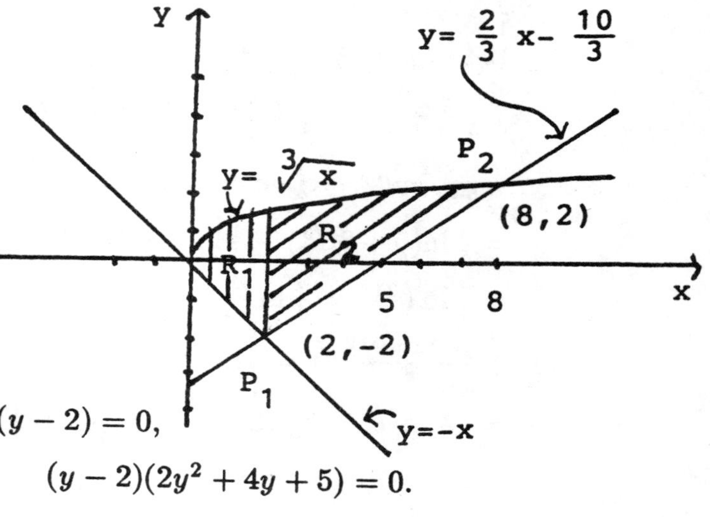

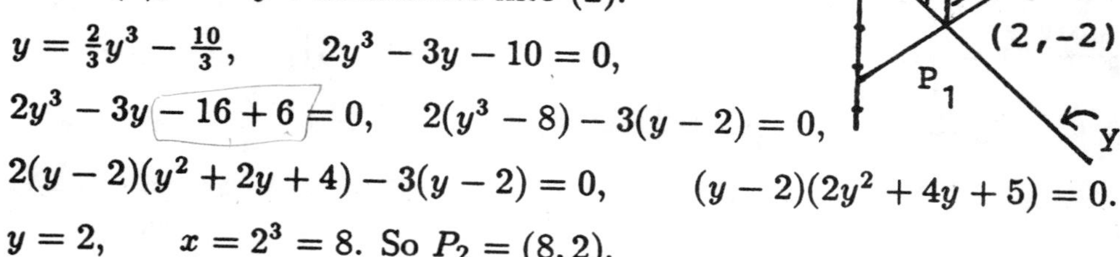

$y = \frac{2}{3}y^3 - \frac{10}{3}$, $\quad 2y^3 - 3y - 10 = 0$,

$2y^3 - 3y - 16 + 6 = 0$, $\quad 2(y^3-8) - 3(y-2) = 0$,

$2(y-2)(y^2+2y+4) - 3(y-2) = 0$, $\quad (y-2)(2y^2+4y+5) = 0$.

$y = 2$, $\quad x = 2^3 = 8$. So $P_2 = (8,2)$.

Area of $R =$ (Area of R_1) + (Area of R_2)

$$= \int_0^2 [\sqrt[3]{x} - (-x)]\,dx + \int_2^8 [\sqrt[3]{x} - (\tfrac{2}{3}x - \tfrac{10}{3})]\,dx$$

$$= (\tfrac{3}{4}x^{4/3} + \tfrac{1}{2}x^2)\Big]_0^2 + (\tfrac{3}{4}x^{4/3} - \tfrac{1}{3}x^2 + \tfrac{10}{3}x)\Big]_2^8$$

$$= (\tfrac{3}{4}\cdot 2^{4/3} + \tfrac{1}{2}\cdot 2^2) - 0 + (\tfrac{3}{4}\cdot 8^{4/3} - \tfrac{1}{3}\cdot 8^2 + \tfrac{10}{3}\cdot 8) - (\tfrac{3}{4}\cdot 2^{4/3} - \tfrac{1}{3}\cdot 2^2 + \tfrac{10}{3}\cdot 2)$$

$$= 2 + 12 - \tfrac{64}{3} + \tfrac{80}{3} + \tfrac{4}{3} - \tfrac{20}{3} = 14.$$

59. $C(50) - C(40) = \int_{40}^{50} MC(x)dx = \int_{40}^{50} \left(110 + \frac{50}{x+10}\right) dx$

$C(50) - C(40) = (110x + 50\ln(x+10))\Big]_{40}^{50}$

$$= 110\cdot 50 + 50\ln(50+10) - 110\cdot 40 - 50\ln(40+10)$$

$$= 1100 + 50\ln\tfrac{6}{5} = 50\left(22 + \ln\tfrac{6}{5}\right).$$

61. $D(x_0) = D(6) = 40(50 - 6^2) = 560.$

Consumers' surplus $= \int_0^6 40(50 - x^2)dx - 6\cdot(560)$

$$= 40\left(50x - \tfrac{1}{3}x^3\right)\Big]_0^6 - 3360$$

$$= 40\left(300 - \tfrac{1}{3}\cdot 216 - 0\right) - 3360 = 5760.$$

63. $S(x_0) = S(15) = \frac{1}{\sqrt{15+1}}e^{\sqrt{15+1}} = \frac{1}{4}e^4.$

Suppliers' surplus $= 15\cdot(\frac{1}{4}e^4) - \int_0^{15} \frac{e^{\sqrt{x+1}}}{\sqrt{x+1}}dx.$

Let $u(x) = \sqrt{x+1}$. $\quad du = \frac{1}{2\sqrt{x+1}}dx.$

$\int \frac{e^{\sqrt{x+1}}}{\sqrt{x+1}}dx = 2\int e^u du = 2e^u + C = 2e^{\sqrt{x+1}} + C.$

So the suppliers' surplus $= \frac{15}{4}e^4 - \left(2e^{\sqrt{x+1}}\right)\Big]_0^{15} = \frac{15}{4}e^4 - 2\left(e^{\sqrt{15+1}} - e^{\sqrt{0+1}}\right)$

$$= \tfrac{15}{4}e^4 - 2\left(e^4 - e\right) = \tfrac{7}{4}e^4 + 2e.$$

65. The value of revenue $= \int_0^{10} A(t)dt = \int_0^{10} 100e^{0.2t}dt$

$$= 100\cdot\tfrac{1}{0.2}e^{0.2t}\Big]_0^{10} = 500(e^2 - e^0) = 500(e^2 - 1) \text{ dollars.}$$

67. The total value of the annuity $A = \int_0^5 10,000t\sqrt{1+t^2}dt.$

Let $u(t) = 1 + t^2$. $\quad du = 2t\ dt.$

$\int 10,000t\sqrt{1+t^2}dt = 5,000\int\sqrt{u}du = \frac{10,000}{3}u^{3/2} + C = \frac{10,000}{3}(1+t^2)^{3/2} + C.$

$A = \int_0^5 10,000t\sqrt{1+t^2}dt = \frac{10,000}{3}(1+t^2)^{3/2}\Big]_0^5 = 10,000(26^{3/2} - 1)/3$ dollars.

69. Average of $f = \frac{1}{3-0}\int_0^3 \frac{x}{x^2+7}dx = \frac{1}{3}\cdot\frac{1}{2}\cdot \ln(x^2+7)\Big]_0^3$
$= \frac{1}{6}\left[\ln(3^2+7) - \ln(0+7)\right] = \frac{1}{6}(\ln 16 - \ln 7) = \frac{1}{6}\ln\left(\frac{16}{7}\right).$

71. $\ln(kx) = \ln x + C.$

$\ln x + \ln k = \ln x + C \implies C = \ln k.$

73. Suppose you manufacture and sell x thousand items per day.

Suppose your total daily cost in thousands of dollars for manufacturing these x thousand items is given by

$$C(x) = -x^3 + 12x^2 + 20x + 15.$$

Suppose your revenue in thousands of dollars from the sale of these x thousand items is given by

$$R(x) = 60x.$$

Then your profit in thousands of dollars from the manufacture and sale of these x thousand items is given by

$$P(x) = R(x) - C(x) = x^3 - 12x^2 + 40x - 15.$$

Let I denote the open interval consisting of all real numbers x such that $4 < x < 5.633$. Throughout the interval I, the marginal profit $MP(x)$ increases as x increases, but the profit $P(x)$ decreases as x increases. Throughout the interval I, the total cost $C(x)$ is positive and increases as x increases, and the profit $P(x)$ is positive. The graphs of the cost function $C(x)$ and the profit function $P(x)$ are shown.

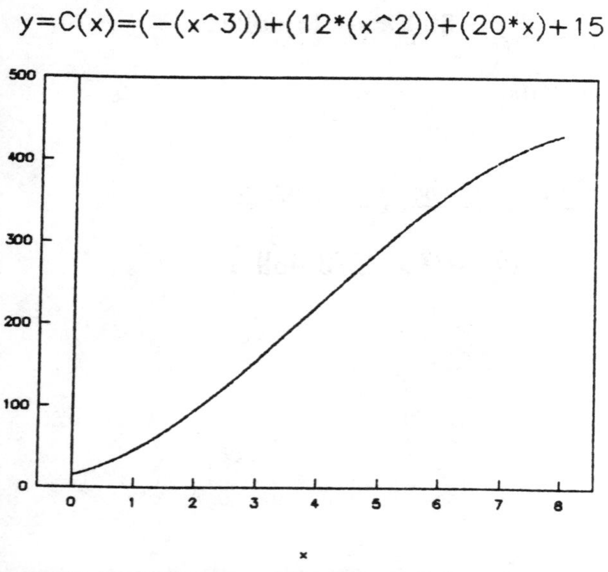

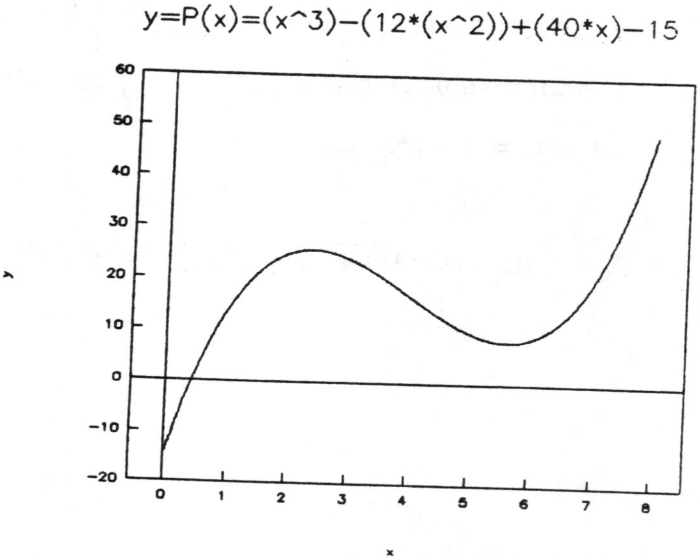

y=C(x)=(-(x^3))+(12*(x^2))+(20*x)+15

y=P(x)=(x^3)-(12*(x^2))+(40*x)-15

PRACTICE PROBLEMS FOR CHAPTER 5

1. Find the general form of the antiderivatives for each of the following integrals:

   a. $\int \left(12x^3 + 6x^2 - 8x + 5\right) dx$

   b. $\int \left(3x^{1/2} - 4x^{3/2}\right) dx$

   c. $\int \left(-\frac{6}{x} + \frac{3}{x^2} + \frac{2}{\sqrt{x}}\right) dx$

   d. $\int \left(3x^2 - \frac{2}{x^2}\right)^2 dx$

   e. $\int \left(-4 \cdot e^{2x}\right) dx$

   f. $\int \frac{x^2-x-2}{x+1} dx$

2. a. Find the function F such that $F'(x) = \sqrt{x} - \frac{2}{\sqrt{x}}$ and $F(9) = 10$.

   b. Find the function F such that $F'(x) = \frac{4}{x} - e^{-2x}$ and $F(2) = 12$.

   c. Suppose that Esmerelda's Hamburger Barn has a marginal profit function given by $P'(x) = -2x + 20$, where x is the number of sales in thousands of hamburgers, and the corresponding profit P is in dollars. Suppose that when $x = 10$, the corresponding profit is \$50. Find the profit function.

   d. Suppose an object is thrown downward from a height of 1100 feet with initial speed of 20 feet/second. Suppose the acceleration of the object is approximately 32 feet/second/second. Find the approximate number of seconds it takes the object to reach the ground, and find the approximate speed of the object when it reaches the ground.

3. Find each of the following antiderivatives by the method of substitution.

   a. $\int 4x^2(x^3 + 1)^{3/2} dx$

   b. $\int \frac{x^3-3x}{-x^4+6x^2-9} dx$

   c. $\int \frac{x}{e^{x^2}} dx$

   d. $\int e^{(x^2-3x+5)}(-4x + 6) dx$

   e. $\int \frac{e^{\sqrt[3]{x}}}{x^{2/3}} dx$

f. $\int \frac{x-1}{(2x^2-4x)^2}dx$

4. Suppose that a rumor spreads through a town of 10,000 people. Denote by $N(t)$ the number of people who have heard the rumor after t days. Suppose the rate of spread of the rumor after t days is equal to $\frac{178982100e^{-1.79t}}{(1+9999e^{-1.79t})^2}$ people per day. Suppose that originally only one person knows the rumor.

   a. Find the formula for $N(t)$.

   b. How long will it take for 5000 people to have heard the rumor?

5. For each of the following definite integrals, find an approximating sum using the indicated number of subintervals and indicated choice of the test number in each subinterval.

   a. $\int_1^4 \frac{1}{\sqrt{x^3+9}}dx$ for $n = 12$, and the left endpoint of each subinterval as the test number.

   b. $\int_2^4 \sqrt{\ln x}\, dx$ for $n = 16$, and the right endpoint of each subinterval as the test number.

   c. $\int_0^2 e^{e^{(2x-1)}}dx$ for $n = 10$, and the left endpoint of each subinterval as the test number.

   d. $\int_{.25}^{.75} \frac{x}{\ln x}dx$ for $n = 10$, and the midpoint of each subinterval as the test number.

6. a. Find the area of the region bounded by the curve $y = f(x) = \sqrt{9-4x}$ and the x-axis between $x = 0$ and $x = 2$.

   b. Find the area of the region bounded by the graph of $y = f(x) = e^{x^{2/3}} \big/ \sqrt[3]{x}$ and the x- axis between $x = 1$ and $x = 8$.

   c. Find the definite integral $\int_{-1}^{2} \frac{x-1}{x^2-2x+2}dx$ and give a geometric interpretation of the answer.

   d. Find the area of the region bounded by the graphs of $f(x) = x^2 - 20x + 108$ and $g(x) = \frac{1}{2}x^2 - 12x + 75$ between $x = 10$ and $x = 15$.

e. Find the area of the region bounded by the graphs of $f(x) = x^2 + x - 13$ and $g(x) = x + 3$.

7. Suppose it was predicted that the annual rate of oil consumption since 1974 would be given by the formula

$$R(t) = 16.1e^{.07t}, \qquad t \geq 0$$

where $t = 0$ corresponds to the beginning of 1974, and $R(t)$ is in billions of barrels of oil per year.

   a. Use this formula to estimate the total amount of oil that would have been consumed between 1974 and 1980.

   b. How many years after the beginning of 1974 would exactly 200 billion barrels of oil have been consumed?

8. Suppose the current price per unit of a certain product is \$45, and suppose it is predicted that this price will increase at the rate of \$1.25 per month. Suppose 25,000 units per month are to be produced and sold during the next 4 years. Find the resulting total revenue for the seller.

9. a. Find the consumers' surplus for the demand curve $p = D(x) = \sqrt{16 - .02x}$ at consumption level $x = 350$.

   b. Find the suppliers' surplus for the supply curve $p = \sqrt{.1x + 9} - 2$ at supply level $x = 200$.

   c. Find the point of intersection of the demand cuve in (a) and the supply curve in (b).

10. a. Find the present value of \$2000 payable at the end of 10 years if the money is invested at 15 percent annual interest compounded continuously. Find the average value of the investment during the next 20 years.

   b. Suppose that an investment produces a continuous stream of income at the rate of 1000 dollars per year at time t. Find the present value of the investment income over the next 5 years using a 10% annual interest rate.

11. A package of frozen strawberries is taken from a freezer which is at $-5°$ C. The package is put into a room which is at $20°$ C. Let t denote the number of hours after the strawberries are taken from the freezer. Suppose the average temperature of the strawberries is increasing at the rate of $10e^{-0.4t}$ degrees Celsius per hour. Find the temperature of the strawberries at time t.

12. Let n be a positive integer. Let

$$S_n = \frac{1}{n} \cdot \sum_{j=1}^{n} e^{\sqrt{j/n}}.$$

   a. Find S_{25}.

   b. Find S_{100}.

   c. Give the definite integral for which the S_n are approximating sums.

13. A business note for $50,000 carries an annual interest rate of 18 percent compounded continuously. For the first 5 years the business pays only the interest. For the next 5 years the business makes continuous payments at the rate of r dollars per year so that the note is paid off after 5 years.

   a. Find r.

   b. For an annual interest rate of 12% compounded continuously, find the present value of all the interest payments made during the first 5 years and the present value of all the payments made during the second 5 years under the original 18% interest rate.

14. On December 1, 1980, a philanthropist sets up a permanent trust fund. The fund will pay $6,000 per year continuously beginning on December 1, 1990. Suppose the money in the fund earns 6% annual interest compounded continuously. How much money must be set aside on December 1, 1980?

15. This problem is a continuation of Problem 72 in the Review Exercises for Chapter 5. Suppose you are operating a manufacturing company. Let x denote the number of items you manufacture and (presumably) sell per day. Let

$$M(x) = \frac{1}{12}x^2 - 10x + 325$$

be the cost of the material in dollars per item for your manufacturing x items per day. Let \$50 be your labor costs per item produced. Let \$400 be your fixed daily costs. Suppose we have the demand equation $p = D(x) = 1500 - 5x$ for the selling price p in dollars per item for your product as a function of x.

a. Let $r(x)$ denote the revenue in dollars to the supplier of your material for the material used in the production of x items per day. Sketch the graph of $y = r(x)$.

b. Find your daily profit function $P(x)$ in dollars for the manufacture and sale of x items per day.

c. Sketch the graph of $y = P(x)$.

d. Find the value of x for which $P(x)$ is a maximum, and find this maximum profit.

SOLUTIONS TO PRACTICE PROBLEMS FOR CHAPTER 5

1.a. $\int(12x^3+6x^2-8x+5)dx$

$= 12\cdot\frac{x^4}{4}+6\cdot\frac{x^3}{3}-8\cdot\frac{x^2}{2}+5x+C.$

$= 3x^4+2x^3-4x^2+5x+C.$

b. $\int(3x^{1/2}-4x^{3/2})dx$

$= 3\cdot\frac{2}{3}x^{3/2}-4\cdot\frac{2}{5}x^{5/2}+C$

$= 2x^{3/2}-\frac{8}{5}x^{5/2}+C.$

c. $\int\left(-\frac{6}{x}+\frac{3}{x^2}+\frac{2}{\sqrt{x}}\right)dx$

$= -6\ln|x|+3\cdot(-x^{-1})+2(2\cdot x^{1/2})+C$

$= -6\ln|x|-3x^{-1}+4x^{1/2}+C.$

d. $\int\left(3x^2-\frac{2}{x^2}\right)^2 dx$

$= \int\left(9x^4-12+\frac{4}{x^4}\right)dx$

$= 9\cdot\frac{x^5}{5}-12x+4\cdot\left(\frac{x^{-3}}{-3}\right)+C$

$= \frac{9}{5}x^5-12x-\frac{4}{3}x^{-3}+C.$

e. $\int(-4e^{2x})dx = -4\frac{e^{2x}}{2}+C = -2e^{2x}+C.$

f. $\int\frac{x^2-x-2}{x+1}dx = \int\frac{(x-2)(x+1)}{(x+1)}dx = \int(x-2)dx = \frac{x^2}{2}-2x+C.$

2.a. $F(x) = \int F'(x)dx = \int\left(\sqrt{x}-\frac{2}{\sqrt{x}}\right)dx$

$= \int\left(x^{1/2}-2\cdot x^{-1/2}\right)dx$

$= \frac{2}{3}x^{3/2}-4x^{1/2}+C.$

From $F(9)=10$, $F(9)=\frac{2}{3}\cdot 3^3-4\cdot 3+C=10$

$\Longrightarrow 18-12+C=10 \therefore C=4.$

Hence, $f(x)=\frac{2}{3}x^{3/2}-4x^{1/2}+4.$

b. $F(x)=\int F'(x)dx=\int\left(\frac{4}{x}-e^{-2x}\right)dx = 4\cdot\ln|x|+\frac{e^{-2x}}{2}+C.$

From $F(2)=12$, $F(2)=4\cdot\ln 2+\frac{e^{-4}}{2}+C=12$

$\Longrightarrow c=12-4\ln 2+\frac{1}{2}e^{-4}.$

Hence, $F(x)=4\ln|x|+\frac{e^{-2x}}{2}+12-4\ln 2+\frac{1}{2}e^{-4}.$

c. $P(x) = \int P'(x)dx = \int(-2x+20)dx$

$= -x^2 + 20x + C.$

From $P(10) = 50$, $P(10) = -100 + 200 + C = 50$

$\Longrightarrow C = -50.$

Hence, $P(x) = -x^2 + 20x - 50.$

d. Let t be the number of seconds since the object has been thrown, and let $f(t)$ be the corresponding number of feet the object has fallen. Then, $f'(0) = 20$ and $f''(t) = 32$. From $f''(t) = 32$, $f'(t) = 32t + C$. From $f'(0) = 20$, $f'(t) = 32t + 20.$

This implies that $f(t) = 16t^2 + 20t + C_1.$

Since $f(0) = 0$, $f(t) = 16t^2 + 20t.$

Since the distance the object is to fall is 1100 feet, we want to find t such that $f(t) = 1100$. Then

$$16t^2 + 20t = 1100$$

$$16t^2 + 20t - 1100 = 0$$

$$4t^2 + 5t - 275 = 0$$

$$t = \frac{-5 \pm \sqrt{25+4400}}{8} = \frac{-5 \pm \sqrt{4425}}{8}$$

$$t \doteq \frac{61.5}{8} \quad \text{or} \quad \frac{-71.5}{8}$$

$$\therefore t \doteq 7.69 \quad \text{or} \quad -8.94.$$

Hence, the object reaches the ground in about 7.69 seconds, and the speed of the object when it reaches the ground is $f'(7.69) = 32 \cdot (7.69) + 20 = 266.1$ feet/sec.

3.a. $\int (4x^2)(x^3+1)^{3/2}dx$

Let $u = x^3 + 1$. $du = 3x^2dx \Longrightarrow x^2dx = \frac{1}{3}du.$

$\int (4x^2)(x^3+1)^{3/2}dx = \int \frac{4}{3}u^{3/2}du = \frac{4}{3}\cdot\frac{2}{5}u^{5/2} + C$

$= \frac{8}{15}(x^3+1)^{5/2} + C.$

b. $\int \frac{x^3-3x}{-x^4+6x^2-9}dx$

Let $u = -x^4 + 6x^2 - 9$.

$du = (-4x^3 + 12x)dx \implies -\frac{1}{4}du = (x^3 - 3x)dx$.

$\int \frac{x^3-3x}{-x^4+6x^2-9}dx = \int \left(-\frac{1}{4}\right) \cdot \frac{1}{u}du = -\frac{1}{4}\ln|u| + C$

$= -\frac{1}{4}\ln|-x^4 + 6x^2 - 9| + C.$

c. $\int \frac{x}{e^{x^2}}dx.$

Let $u = x^2 \implies du = 2x\ dx.$

$\int \frac{x}{e^{x^2}}dx = \int \frac{1}{e^u} \cdot \frac{1}{2}du = \frac{1}{2}\int e^{-u}du$

$= -\frac{1}{2}e^{-u} + C = -\frac{1}{2}e^{-x^2} + C.$

d. Let $u = x^2 - 3x + 5 \implies du = (2x - 3)dx.$

$\int e^{(x^2-3x+5)}(-4x + 6)dx = \int e^u(-2)du = -2e^u + C$

$= -2 \cdot e^{(x^2-3x+5)} + C.$

e. $\int \frac{e^{\sqrt[3]{x}}}{x^{2/3}}dx.$

Let $u = x^{1/3} \implies du = \frac{1}{3}x^{-2/3}dx.$

$\int \frac{e^{\sqrt[3]{x}}}{x^{2/3}}dx = \int e^u \cdot 3du = 3 \cdot e^u + C$

$= 3 \cdot e^{x^{1/3}} + C.$

f. Let $u = 2x^2 - 4x \implies du = (4x - 4)dx = 4(x - 1)dx.$

$\int \frac{x-1}{(2x^2-4x)^2}dx = \int \frac{1}{4} \cdot \frac{1}{u^2}du = -\frac{1}{4u} + C = -\frac{1}{4(2x^2-4x)} + C.$

4. $N'(t) = \frac{178982100e^{-1.79t}}{(1+9999e^{-1.79t})^2}, \qquad N(0) = 1.$

a. $N(t) - N(0) = \int_0^t N'(x)dx \implies N(t) = \int_0^t N'(x)dx + N(0).$

$\int N'(x)dx = \int \frac{178982100e^{-1.79x}}{(1+9999e^{-1.79x})^2}dx$

$= \int(-10000) \cdot \frac{1}{u^2}du$

where $u = 1 + 9999e^{-1.79x}$

$= \frac{10000}{u} + C = \frac{10000}{1+9999e^{-1.79x}} + C.$

$\therefore N(t) = \left.\frac{10000}{1+9999e^{-1.79x}}\right]_0^t + N(0)$

$= \frac{10000}{1+9999e^{-1.79t}} - \frac{10000}{1+9999e^0} + 1$

$= \frac{10000}{1+9999e^{-1.79t}}.$

b. Suppose that after t days 5000 people have heard the rumor. That is, $N(t) = 5000$.

$$\Longrightarrow N(t) = 5000 = \frac{10000}{1 + 9999e^{-1.79t}}$$

$$1 + 9999e^{-1.79t} = \frac{10000}{5000}$$

$$e^{-1.79t} = \frac{1}{9999}$$

$$t = -\frac{\ln\left(\frac{1}{9999}\right)}{1.79}$$

$$t \doteq 5.145.$$

So it will take about 5.145 days for 5000 people to have heard the rumor.

5.a. $f(x) = \frac{1}{\sqrt{x^3+9}}$, $a = 1$, $b = 4$, $n = 12$.

$\Delta x = \frac{b-a}{n} = \frac{4-1}{12} = \frac{1}{4}$.

Test number t_j is the left endpoint of each subinterval:

$t_1 = 1$, $t_2 = \frac{5}{4}$, $t_3 = \frac{3}{2}$, ..., $t_{11} = \frac{7}{2}$, $t_{12} = \frac{15}{4}$.

Then the aproximating sum for this definite integral is

$$\int_1^4 f(x)dx \approx \left[f(1) + f\left(\frac{5}{4}\right) + f\left(\frac{3}{2}\right) + f\left(\frac{7}{4}\right) + f(2) + f\left(\frac{9}{4}\right)\right.$$
$$\left. + f\left(\frac{5}{2}\right) + f\left(\frac{11}{4}\right) + f(3) + f\left(\frac{13}{4}\right) + f\left(\frac{7}{2}\right) + f\left(\frac{15}{4}\right)\right]\Delta x$$

$$\int_1^4 f(x)dx \approx \left[\frac{1}{\sqrt{1^3+9}} + \frac{1}{\sqrt{\left(\frac{5}{4}\right)^3+9}} + \frac{1}{\sqrt{\left(\frac{3}{2}\right)^3+9}} + \frac{1}{\sqrt{\left(\frac{7}{4}\right)^3+9}} + \frac{1}{\sqrt{2^3+9}} + \frac{1}{\sqrt{\left(\frac{9}{4}\right)^3+9}} + \frac{1}{\sqrt{\left(\frac{5}{2}\right)^3+9}} + \frac{1}{\sqrt{\left(\frac{11}{4}\right)^3+9}} + \frac{1}{\sqrt{3^3+9}} + \frac{1}{\sqrt{\left(\frac{13}{4}\right)^3+9}} + \frac{1}{\sqrt{\left(\frac{7}{2}\right)^3+9}} + \frac{1}{\sqrt{\left(\frac{15}{4}\right)^3+9}}\right] \cdot \frac{1}{4}$$

$$\approx (.316 + .302 + .284 + .264 + .243 + .221 + .202 + .183 + .167 + .152 + .139 + .127) \cdot \frac{1}{4}$$

$$= 2.6 \cdot \frac{1}{4} = .65.$$

b. $f(x) = \sqrt{\ln x}, \quad a = 2, \quad b = 4, \quad n = 16.$

$\Delta x = \frac{b-a}{n} = \frac{4-2}{16} = \frac{1}{8}.$

Test number t_j is the right endpoint of each subinterval:

$t_1 = \frac{17}{8}, \quad t_2 = \frac{9}{4}, \quad t_3 = \frac{19}{8}, \quad t_4 = \frac{5}{2} \ldots, t_{15} = \frac{31}{8}, \quad t_{16} = 4.$

The approximating sum for this definite integral is then

$$\int_2^4 f(x)dx \approx \left[f\left(\frac{17}{8}\right) + f\left(\frac{9}{4}\right) + f\left(\frac{19}{8}\right) + f\left(\frac{5}{2}\right) + f\left(\frac{21}{8}\right) + f\left(\frac{11}{4}\right) + f\left(\frac{23}{8}\right) + f(3) + f\left(\frac{25}{8}\right) + f\left(\frac{13}{4}\right) + f\left(\frac{27}{8}\right) + f\left(\frac{7}{2}\right) + f\left(\frac{29}{8}\right) + f\left(\frac{15}{4}\right) + f\left(\frac{31}{8}\right) + f(4)\right] \cdot \Delta x$$

$$= \left(\sqrt{\ln\left(\frac{17}{8}\right)} + \sqrt{\ln\left(\frac{9}{4}\right)} + \sqrt{\ln\left(\frac{19}{8}\right)} + \sqrt{\ln\left(\frac{5}{2}\right)} + \sqrt{\ln\left(\frac{21}{8}\right)} + \sqrt{\ln\left(\frac{11}{4}\right)} + \sqrt{\ln\left(\frac{23}{8}\right)} + \sqrt{\ln 3} + \sqrt{\ln\left(\frac{25}{8}\right)} + \sqrt{\ln\left(\frac{13}{4}\right)} + \sqrt{\ln\left(\frac{27}{8}\right)} + \sqrt{\ln\left(\frac{7}{2}\right)} + \sqrt{\ln\left(\frac{29}{8}\right)} + \sqrt{\ln\left(\frac{15}{4}\right)} + \sqrt{\ln\left(\frac{31}{8}\right)} + \sqrt{\ln 4}\right) \cdot \frac{1}{8}$$

$$\int_2^4 f(x)dx \approx (.868 + .901 + .930 + .957 + .982 + 1.006 + 1.028 + 1.048$$
$$+ 1.067 + 1.086 + 1.103 + 1.119 + 1.135 + 1.150$$
$$+ 1.164 + 1.177) \cdot \frac{1}{8} = 16.721 \cdot \frac{1}{8} \doteq 2.09.$$

c. $f(x) = e^{e^{2x-1}}, \quad a = 0, \quad b = 2, \quad n = 10.$

$\Delta x = \frac{b-a}{n} = \frac{2-0}{10} = \frac{1}{5}.$

Test number t_j is the left endpoint of each subinterval:

$t_1 = 0,\ t_2 = \frac{1}{5},\ t_3 = \frac{2}{5},\ \ldots,\ t_9 = \frac{8}{5},\ t_{10} = \frac{9}{5}.$

The approximating sum for the definite integral is then

$$\int_0^2 f(x)dx \approx \left[f(0) + f\left(\frac{1}{5}\right) + f\left(\frac{2}{5}\right) + f\left(\frac{3}{5}\right) + f\left(\frac{4}{5}\right) + f(1)\right.$$
$$\left. + f\left(\frac{6}{5}\right) + f\left(\frac{7}{5}\right) + f\left(\frac{8}{5}\right) + f\left(\frac{9}{5}\right)\right] \cdot \Delta x$$
$$= \left[e^{e^{(2\cdot 0-1)}} + e^{e^{\left(2\cdot\frac{1}{5}-1\right)}} + e^{e^{\left(2\cdot\frac{2}{5}-1\right)}} + e^{e^{\left(2\cdot\frac{3}{5}-1\right)}} + e^{e^{\left(2\cdot\frac{4}{5}-1\right)}} + e^{e^{(2\cdot 1-1)}}\right.$$
$$\left. + e^{e^{\left(2\cdot\frac{6}{5}-1\right)}} + e^{e^{\left(2\cdot\frac{7}{5}-1\right)}} + e^{e^{\left(2\cdot\frac{8}{5}-1\right)}} + e^{e^{\left(2\cdot\frac{9}{5}-1\right)}}\right] \cdot \frac{1}{5}$$
$$\approx (1.445 + 1.731 + 2.268 + 3.392 + 6.185 + 15.154$$
$$+ 57.697 + 423.964 + 8308.327 + 703440.120) \cdot \frac{1}{5}$$
$$= 712260.283\left(\frac{1}{5}\right) \doteq 142452.057.$$

d. $f(x) = \frac{x}{\ln x}, \quad a = .25, \quad b = .75, \quad n = 10.$

$\Delta x = \frac{b-a}{n} = \frac{.75-.25}{10} = .05.$

The test number t_j is the midpoint of each subinterval:

$t_1 = \frac{.25+.3}{2} = .275,\ t_2 = \frac{.3+.35}{2} = .325,\ t_3 = .375,\ t_4 = .425,$

$\ldots,\ t_9 = .675,\ t_{10} = .725.$

The approximating sum for the definite integral is then

$$\int_{.25}^{.75} f(x)dx \approx [f(.275) + f(.325) + f(.375) + f(.425) + f(.475)$$

$$+ f(.525) + f(.575) + f(.625) + f(.675) + f(.725)] \cdot (.05)$$

$$= \left[\frac{.275}{\ln(.275)} + \frac{.325}{\ln(.325)} + \frac{.375}{\ln(.375)} + \frac{.425}{\ln(.425)} + \frac{.475}{\ln(.475)}\right.$$

$$\left. + \frac{.525}{\ln(.525)} + \frac{.575}{\ln(.575)} + \frac{.625}{\ln(.625)} + \frac{.675}{\ln(.675)} + \frac{.725}{\ln(.725)}\right] \cdot (.05)$$

$$\approx (-.213 - .289 - .382 - .497 - .638 - .815$$

$$-1.039 - 1.330 - 1.717 - 2.254) \cdot (.05)$$

$$= -9.174 \cdot (.05) \doteq -0.459.$$

6.a. Area of $R = \int_0^2 \sqrt{9-4x}\, dx = -\frac{1}{4} \cdot \frac{2}{3}(9-4x)^{3/2}\Big]_0^2$

$$= -\frac{1}{6}\left[(9-4(2))^{3/2} - (9-4(0))^{3/2}\right]$$

$$= -\frac{1}{6}\left(1 - 9^{3/2}\right) = \frac{26}{6} = \frac{13}{3}.$$

b. Area of $R = \int_1^8 \frac{e^{x^{2/3}}}{\sqrt[3]{x}}\, dx = \int_1^8 \frac{3}{2}e^u\, du$

where $u = x^{2/3}$

$$= \frac{3}{2}e^u\Big]_{x=1}^{x=8} = \frac{3}{2}e^{x^{2/3}}\Big]_1^8$$

$$= \frac{3}{2}\left(e^{8^{2/3}} - e^{1^{2/3}}\right) = \frac{3}{2}(e^4 - e) \approx 77.82.$$

c. $\int_{-1}^2 \frac{x-1}{x^2-2x+2}\, dx = \int_{-1}^2 \frac{1}{2} \cdot \frac{1}{u}\, du = \frac{1}{2}\ln|u|\Big]_{x=-1}^{x=2}$

where $u = x^2 - 2x + 2$

$$= \frac{1}{2}\ln|x^2 - 2x + 2|\Big]_{-1}^2$$

$$= \frac{1}{2}(\ln 2 - \ln 5) \approx -.458.$$

Note that the function $f(x) = \frac{x-1}{x^2-2x+2} \leq 0$ for $x \leq 1$, and $f(x) \geq 0$ for $x \geq 1$. Therefore $\int_{-1}^1 f(x)dx$ is the negative of the area of the region bounded by the graph of $y = f(x)$ and the x-axis between $x = -1$ and $x = 1$, and $\int_1^2 f(x)dx$ is the area of the region bounded by the graph of $y = f(x)$ and the x axis between $x = 1$ and $x = 2$.

Note that the definite integral $\int_{-1}^{2} \frac{x-1}{x^2-2x+2}\,dx$ is the sum of the two definite integrals $\int_{-1}^{1} f(x)dx$ and $\int_1^2 f(x)dx$.

d. For $10 \le x \le 15$,

$$\begin{aligned}[f(x) - g(x)] &= (x^2 - 20x + 108) - \left(\frac{1}{2}x^2 - 12x + 75\right) \\ &= \frac{1}{2}x^2 - 8x + 33 \ge 0.\end{aligned}$$

Therefore,

$$\begin{aligned}\text{Area of } R &= \int_{10}^{15} [f(x) - g(x)]\,dx \\ &= \int_{10}^{15} \left(\frac{1}{2}x^2 - 8x + 33\right) dx \\ &= \left(\frac{1}{6}x^3 - 4x^2 + 33x\right)\Big]_{10}^{15} \\ &= \left[\frac{1}{6}(15)^3 - 4(15)^2 + 33 \cdot 15\right] - \left[\frac{1}{6}(10)^3 - 4(10)^2 + 33 \cdot 10\right] \\ &= \left[\frac{3375}{6} - 900 + 495\right] - \left[\frac{1000}{6} - 400 + 330\right] \\ &= \frac{365}{6} = 60.83.\end{aligned}$$

e. First find where the graphs of f and g intersect.

$$\begin{aligned}f(x) = g(x) &\implies x^2 + x - 13 = x + 3 \\ &\implies x^2 = 16 \implies x = \pm 4.\end{aligned}$$

For $-4 \le x \le 4$,

$$[g(x) - f(x)] = [-x^2 + 16] \ge 0.$$

Therefore

$$\begin{aligned}
\text{Area of } R &= \int_{-4}^{4} (-x^2 + 16)dx \\
&= \left(-\frac{1}{3}x^3 + 16x\right)\Bigg]_{-4}^{4} \\
&= \left[-\frac{1}{3} \cdot 4^3 + 16 \cdot 4\right] - \left[-\frac{1}{3}(-4)^3 + 16(-4)\right] \\
&= \left[-\frac{64}{3} + 64\right] - \left[\frac{64}{3} - 64\right] \\
&= \frac{256}{3} = 85.33.
\end{aligned}$$

7.a. The total amount of oil that would have been consumed between 1974 and 1980 is

$$\begin{aligned}
\int_0^6 16.1 \cdot e^{.07t} dt &= \frac{16.1 \cdot e^{0.07t}}{0.07}\Bigg]_0^6 \\
&= 230 \cdot e^{0.07t}\Big]_0^6 \\
&= 230 \cdot e^{0.07(6)} - 230 \cdot e^{0.07(0)} \\
&= 230 \cdot e^{0.42} - 230 = 230(e^{0.42} - 1) \\
&= 120.05 \text{ billions of barrels of oil.}
\end{aligned}$$

b. Let T be the number of years such that

$$\int_0^T 16.1e^{0.07t}dt = 200.$$

$$230 \cdot e^{0.07t}\Big]_0^T = 200$$

$$230 \cdot e^{0.07T} = 200 + 230$$

$$e^{0.07T} = \frac{430}{230}$$

$$0.07T = \ln\frac{43}{23} = 0.6257$$

$$T = \frac{.6257}{0.07} = 8.94 \text{ years.}$$

8. 4 years = 48 months.

The price per unit after t months $= 45 + 1.25 \cdot t$.

Total revenue for the next 4 years equals

$$25000\int_0^{48}(45 + 1.25t)dt$$

$$= 25000(45t + .625t^2)\Big]_0^{48}$$

$$= 25000\left(45 \cdot 48 + .625(48)^2\right)$$

$$= 25000(2160 + 1440) = \$90,000,000.$$

9.a. When $x = 350$, $p = \sqrt{16 - .02(350)} = \sqrt{16 - 7} = 3$.

$$\text{Consumers' Surplus} = \int_0^{350} (16 - .02x)^{1/2}\,dx - 350 \cdot 3$$

$$= -50 \cdot \frac{2}{3} \cdot (16 - 0.02x)^{3/2}\Big]_0^{350} - 1050$$

$$= -\frac{100}{3}\left[(9)^{3/2} - (16)^{3/2}\right] - 1050$$

$$= -\frac{100}{3}[27 - 64] - 1050$$

$$= -\frac{100}{3} \cdot (-37) - 1050$$

$$= \frac{3700}{3} - \frac{3150}{3} = \frac{550}{3}$$

$\doteq$ 183.33 dollars.

b. When $x = 200$, $p = \sqrt{.1 \cdot 200 + 9} - 2 = \sqrt{29} - 2 = 3.385$.

$$\text{Suppliers' Surplus} = 200 \cdot 3.385 - \int_0^{200} \left[(.1x + 9)^{1/2} - 2\right] dx$$

$$= 677 - \left(10 \cdot \frac{2}{3}(.1x + 9)^{3/2} - 2x\right)\Big]_0^{200}$$

$$= 677 - \left(\frac{20}{3}(29)^{3/2} - 400 - \frac{20}{3} \cdot 9^{3/2}\right)$$

$$= 677 - \left(\frac{20}{3} \cdot 156.17 - 400 - 180\right)$$

$$= 677 - (461.133) \doteq 215.87 \text{ dollars.}$$

c. The x-coordinate of the point of intersection of the demand curve in (a) and the supply curve in (b) is

$$\sqrt{16 - .02x} = \sqrt{.1x + 9} - 2$$

$$16 - .02x = .1x + 9 - 4\sqrt{.1x + 9} + 4$$

$$3 - .12x = -4\sqrt{.1x + 9}$$

$$9 - .72x + .0144x^2 = 16(.1x + 9)$$

$$.0144x^2 - 2.32x - 135 = 0$$

$$\begin{aligned} x &= \frac{2.32 \pm \sqrt{(2.32)^2 + 4 \cdot (.0144) \cdot (135)}}{.0288} \\ &= \frac{2.32 \pm 3.63}{.0288} \end{aligned}$$

$\therefore \ x \doteqdot 206.51.$

The p-coordinate of this point is given by

$$p \doteqdot \sqrt{16 - .02 \cdot (206.51)} \doteqdot 3.445.$$

10.a. The present value equals

$$e^{-.15 \cdot (10)} \cdot 2000 = .22313 \cdot (2000) = 446.26 \text{ dollars.}$$

The average value equals

$$\begin{aligned} \frac{1}{20}\int_0^{20} 446.26 \cdot e^{.15t}\,dt &= \frac{1}{20}\left[\frac{446.26}{.15}e^{.15t}\right]_0^{20} \\ &= \frac{1}{20}\left[\frac{446.26}{.15}(e^3 - 1)\right] \\ &= \frac{1}{20}[2975.07 \cdot (19.086)] \\ &\doteqdot 2839.04 \text{ dollars.} \end{aligned}$$

b. The present value of the investment income is

$$\int_0^5 1000 \cdot e^{-.1t} dt = \frac{1000}{-.1} e^{-.1t} \Big]_0^5 = -10000(e^{-.5} - 1)$$

$$= -10000(.606 - 1) = 3934.69 \text{ dollars.}$$

11. Let $T(t)$ denote the average temperature of the strawberries at time t. Then $T'(t) = 10e^{-0.4t}$. It follows that

$$T(t) = 10 \cdot \frac{1}{-0.4} e^{-0.4t} + K$$
$$= -25e^{-0.4t} + K.$$

Note that $T(0) = -5$. Therefore

$$-5 = -25e^{-0.4 \cdot (0)} + K \implies K = 20.$$

Thus

$$T(t) = -25e^{-0.4t} + 20.$$

12.a. $S_{25} = \frac{1}{25} \cdot \sum_{j=1}^{25} e^{\sqrt{j/25}} = 2.03282.$

b. $S_{100} = \frac{1}{100} \cdot \sum_{j=1}^{100} e^{\sqrt{j/100}} = 2.00839.$

c. The sum S_n is an approximating sum for the definite integral

$$\int_0^1 e^{\sqrt{x}} \, dx.$$

13.a. $\int_0^5 re^{0.18t}\,dt = 50,000e^{0.18(5)}$.

$r \cdot \left.\frac{e^{0.18t}}{0.18}\right]_0^5 = 50,000 \cdot e^{0.9}$.

$r \cdot (e^{0.9} - 1) = 9,000 \cdot e^{0.9}$.

$r = \frac{9,000 \cdot e^{0.9}}{e^{0.9} - 1} = \$15,166.06$.

b. The present value of all the interest payments made during the first five years is given by

$$PV_1 = \int_0^5 50,000 \cdot (0.18)e^{-0.12t}\,dt$$

$$= 9,000 \cdot \frac{1}{-0.12} \cdot e^{-0.12t}\Big]_0^5 = -75,000 \cdot \left[e^{-0.6} - 1\right]$$

$$= \$33,839.13.$$

The present value of all the payments made during the second five years is given by

$$PV_2 = \int_5^{10} re^{-0.12t}\,dt$$

$$= \int_5^{10} (15,166.06)e^{-0.12t}\,dt$$

$$= \frac{15,166.06}{-0.12}e^{-0.12t}\Big]_5^{10}$$

$$= \frac{15,166.06}{-0.12}(e^{-1.2} - e^{-0.6}) = \$31,294.84.$$

14. Suppose the fund would run out T years after December 1, 1990. The present value of the payment stream on December 1, 1990 would be

$$PV_1 = \int_0^T (6,000)e^{-0.06t}\,dt$$

$$= 6,000 \cdot \frac{1}{-0.06} \cdot (e^{-0.06T} - 1)$$

$$= 100,000 \cdot (1 - e^{-0.06T}).$$

The present value of PV_1 on December 1, 1980 would be

$$PV_2 = 100,000 \cdot \left(1 - e^{-0.06T}\right) \cdot e^{-0.06(10)}.$$

As $T \to \infty$, $PV_2 \to 100,000(1-0) \cdot e^{-0.6} = \$54,881.16$.

15.a. $r(x) = x \cdot M(x) = \frac{1}{12}x^3 - 10x^2 + 325x$.

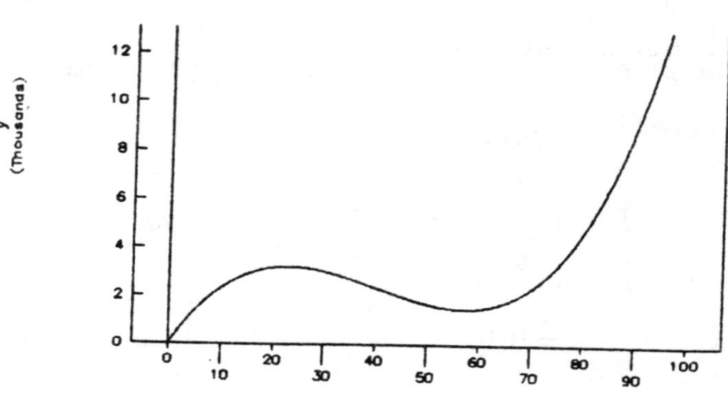

b. The total daily cost function is given by

$$\begin{aligned} C(x) &= x \cdot M(x) + 50x + 400 \\ &= \frac{1}{12}x^3 - 10x^2 + 375x + 400. \end{aligned}$$

The daily revenue function is given by

$$R(x) = x \cdot D(x) = 1500x - 5x^2.$$

The daily profit function is given by

$$\begin{aligned} P(x) &= R(x) - C(x) \\ &= -\frac{1}{12}x^3 + 5x^2 + 1125x - 400. \end{aligned}$$

c.

y=C(x)=((1/12)(x^3))-(10(x^2))+(375x) +400 (Total daily cost function)

y=P(x)=((-(1/12))*(x^3))+(5*(x^2)) +(1125*x)-400 (Daily profit function)

d. $P'(x) = -\frac{1}{4}x^2 + 10x + 1125.$

$P'(x) = 0 \implies x = \frac{-10 \pm \sqrt{10^2 - 4(-\frac{1}{4})(1125)}}{2(-\frac{1}{4})}$

$= 20 \pm 2\sqrt{1225} = 20 \pm 2(35).$

It follows that $P(x)$ is a maximum when $x = 90$ items per day.

This maximum profit is

$$P(90) = -\frac{1}{12}(90)^3 + 5(90)^2 + 1125(90) - 400$$
$$= \$80,600.$$

CHAPTER 6

SUMMARY OF CHAPTER 6

1. Radian and degree measurement for angles, the formulas for converting between degree and radian measure, extending radian measure to all real numbers, identifying the radian measure of an angle with its principal angle.

2. The definitions of $\sin\theta$ and $\cos\theta$ for $0 \leq \theta < 2\pi$ by using the unit circle, the values of $\sin\theta$ and $\cos\theta$ for some special angles given by Table 2.1 of Section 6.2, extending the definitions of $\sin\theta$ and $\cos\theta$ to the interval $(-\infty, \infty)$ by using principal angles and identities, right angle interpretations of $\sin\theta$ and $\cos\theta$.

3. Identities involving $\sin\theta$ and $\cos\theta$.

4. The limit $\lim_{\theta\to 0} \frac{\sin\theta}{\theta} = 1$, the limit $\lim_{\theta\to 0} \frac{\cos\theta - 1}{\theta} = 0$; use of these limits to find the derivative formula $\frac{d}{dx}\sin x = \cos x$; finding the derivative formula $\frac{d}{dx}\cos x = -\sin x$ by using the preceding derivative formula, the Chain Rule, and the identities $\sin\left(\frac{\pi}{2} - x\right) = \cos x$, $\cos\left(\frac{\pi}{2} - x\right) = \sin x$.

5. The antidifferentiation formulas $\int \sin x\, dx = -\cos x + C$ and $\int \cos x\, dx = \sin x + C$.

6. Definitions of the other four trigonometric functions $\tan\theta$, $\cot\theta$, $\csc\theta$, $\sec\theta$; the differentiation formulas for these functions; the antidifferentiation formulas for these functions.

7. Predator-prey models as an application of trigonometric functions.

8. Use of the trigonometric functions to study blood pressure and heart-beat rate.

Selected Solutions to Exercise Set 6.1

1.a. $\theta_r = \frac{\pi}{180} \cdot 90 = \frac{\pi}{2}$.

b. $\theta_r = \frac{\pi}{180} \cdot 45 = \frac{\pi}{4}$.

c. $\theta_r = \frac{\pi}{180} \cdot (-135) = -\frac{3}{4}\pi$.

d. $\theta_r = \frac{\pi}{180} \cdot 30 = \frac{\pi}{6}$.

e. $\theta_r = \frac{\pi}{180} \cdot 60 = \frac{\pi}{3}$.

f. $\theta_r = \frac{\pi}{180} \cdot (-150) = -\frac{5}{6}\pi$.

g. $\theta_r = \frac{\pi}{180} \cdot 180 = \pi$.

h. $\theta_r = \frac{\pi}{180} \cdot 210 = \frac{7}{6}\pi$.

3.a. $\theta_d = \frac{180}{\pi} \cdot \frac{\pi}{4} = 45°$.

b. $\theta_d = \frac{180}{\pi} \cdot \frac{3\pi}{2} = 270°$.

c. $\theta_d = \frac{180}{\pi} \cdot \left(-\frac{\pi}{12}\right) = -15°$.

d. $\theta_d = \frac{180}{\pi} \cdot \frac{7\pi}{6} = 210°$.

e. $\theta_d = \frac{180}{\pi} \cdot \frac{7\pi}{8} = \frac{315}{2}°$.

f. $\theta_d = \frac{180}{\pi} \cdot \left(-\frac{5\pi}{6}\right) = -150°$.

g. $\theta_d = \frac{180}{\pi} \cdot \frac{11\pi}{6} = 330°$.

h. $\theta_d = \frac{180}{\pi} \cdot \left(-\frac{3}{4}\pi\right) = -135°$.

5.a. Clockwise rotation through 45° corresponds to a radian measure $t = -45 \cdot \frac{\pi}{180} = -\frac{\pi}{4}$.

b. $t = -270 \cdot \frac{\pi}{180} = -\frac{3}{2}\pi$.

c. $t = -30 \cdot \frac{\pi}{180} = -\frac{\pi}{6}$.

d. $t = -150 \cdot \frac{\pi}{180} = -\frac{5}{6}\pi$.

e. $t = -390 \cdot \frac{\pi}{180} = -\frac{13}{6}\pi$.

f. $t = -135 \cdot \frac{\pi}{180} = -\frac{3}{4}\pi$.

g. $t = -330 \cdot \frac{\pi}{180} = -\frac{11}{6}\pi$.

h. $t = -540 \cdot \frac{\pi}{180} = -3\pi$.

7.a. $-6 \cdot \frac{\pi}{6} = -\pi$.

b. $-12 \cdot \frac{\pi}{6} = -2\pi$.

c. $-(12+3) \cdot \frac{\pi}{6} = -\frac{5}{2}\pi$.

d. $-(12+8) \cdot \frac{\pi}{6} = -\frac{10}{3}\pi$.

9. $-\left(2 \cdot 2\pi + \frac{\pi}{2}\right) = -\frac{9}{2}\pi$.

11. $(2\pi + \pi) = 3\pi$.

13. $-\left(2\pi + \frac{\pi}{4}\right) = -\frac{9}{4}\pi$.

15.

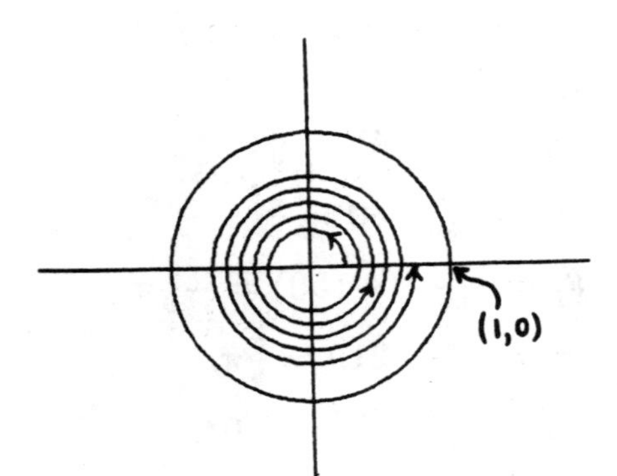

17.

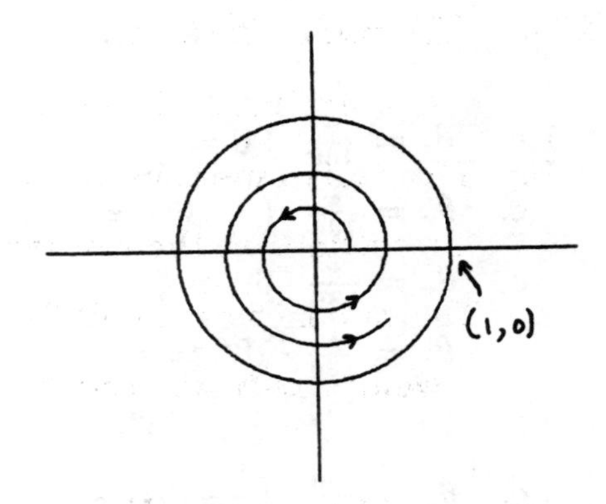

19.

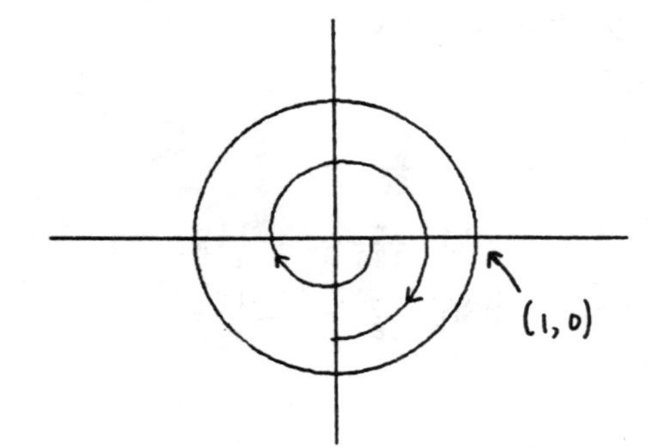

21.

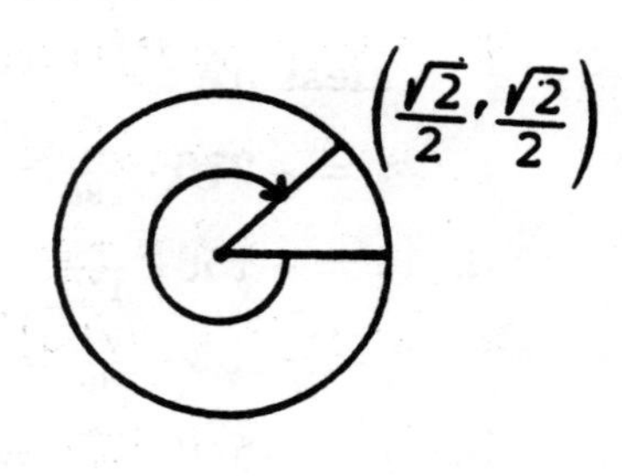

23.

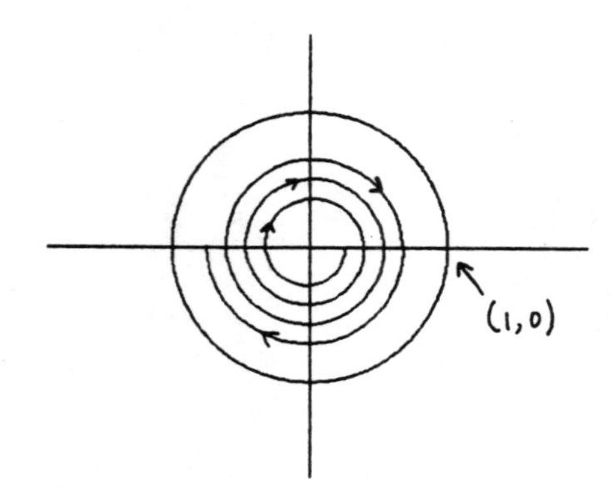

25. The radian measure of the angle is $(-3.25) \cdot 2\pi = -6.5\pi$.

27.a. The radian measure of -180 degrees is $-\pi$, corresponding to $\frac{1}{2}$ of a revolution in the clockwise direction.

b. The radian measure of -540 degrees is -3π, corresponding to $\frac{3}{2}$ revolutions in the clockwise direction.

c. The radian measure of 270 degrees in $\frac{3}{2}\pi$, corresponding to $\frac{3}{4}$ revolutions in the counterclockwise direction.

Selected Solutions to Exercise Set 6.2

1.a. $\sin\theta = \frac{3}{5}$, b. $\cos\theta = \frac{4}{5}$.

3. From the Pythagorean theorem, we have

$y^2 + 3^2 = \left(\sqrt{13}\right)^2 = 13 \implies y^2 = 4 \implies y = 2.$

$\therefore$ a. $\sin\theta = \frac{y}{\sqrt{13}} = \frac{2}{\sqrt{13}}$, b. $\cos\theta = \frac{3}{\sqrt{13}}$.

5. $x^2 + 7^2 = \left(\sqrt{65}\right)^2 = 65 \implies x^2 = 65 - 49 = 16 \implies x = 4.$

a. $\sin\theta = \frac{7}{\sqrt{65}}$, b. $\cos\theta = \frac{x}{\sqrt{65}} = \frac{4}{\sqrt{65}}$.

7. Since $t = \frac{3}{4}\pi$, $\cos t = -\frac{\sqrt{2}}{2}$ and $\sin t = \frac{\sqrt{2}}{2}$.

9. Since $t = -\frac{5}{3}\pi$, $\cos t = \cos\left(\frac{\pi}{3} - 2\pi\right) = \cos\frac{\pi}{3} = \frac{1}{2}$

and $\sin t = \sin\left(\frac{\pi}{3} - 2\pi\right) = \sin\frac{\pi}{3} = \frac{\sqrt{3}}{2}$.

11. $\sin\left(-\frac{\pi}{4}\right) = -\sin\left(\frac{\pi}{4}\right) = -\frac{\sqrt{2}}{2}$.

13. $\frac{9\pi}{2} = \left(\frac{\pi}{2} + 4\pi\right) = (\theta + 2n\pi)$ where $\theta = \frac{\pi}{2}$, $n = 2$.

$\cos\left(\frac{9\pi}{2}\right) = \cos\left(\frac{\pi}{2}\right) = 0.$

15. $\sin\left(-\frac{5}{3}\pi\right) = \sin\left(-\frac{5}{3}\pi + 2\pi\right) = \sin\frac{\pi}{3} = \frac{\sqrt{3}}{2}$.

17. $\cos\left(\frac{11}{4}\pi\right) = \cos\left(2\pi + \frac{3}{4}\pi\right) = \cos\frac{3}{4}\pi = -\frac{\sqrt{2}}{2}$.

19. $f(t)$ has amplitude 2 since $|f(t)| \leq 2$ for all t and $\left|f\left(\frac{\pi}{2}\right)\right| = \left|f\left(\frac{3\pi}{2}\right)\right| = 2$, and $f(t)$ has period 2π.

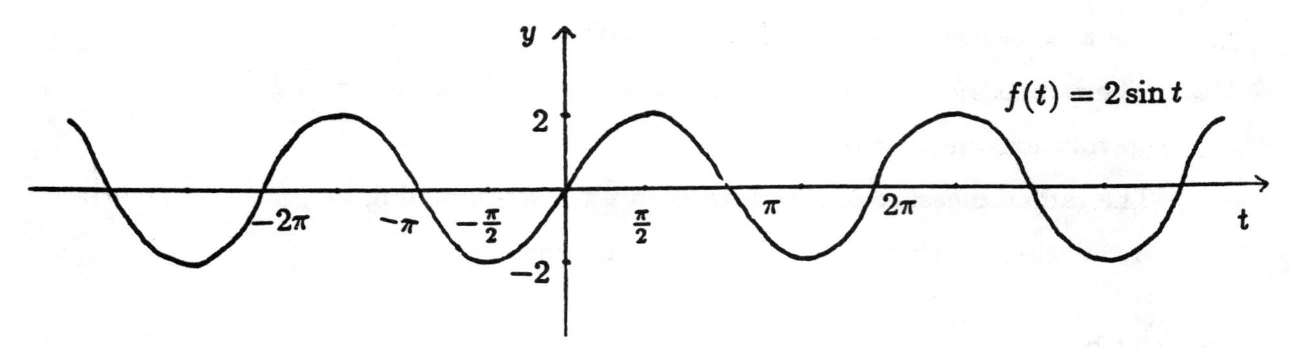

21. $f(t) = \sin 2t$ has amplitude 1 and period $T = \frac{2\pi}{2} = \pi$.

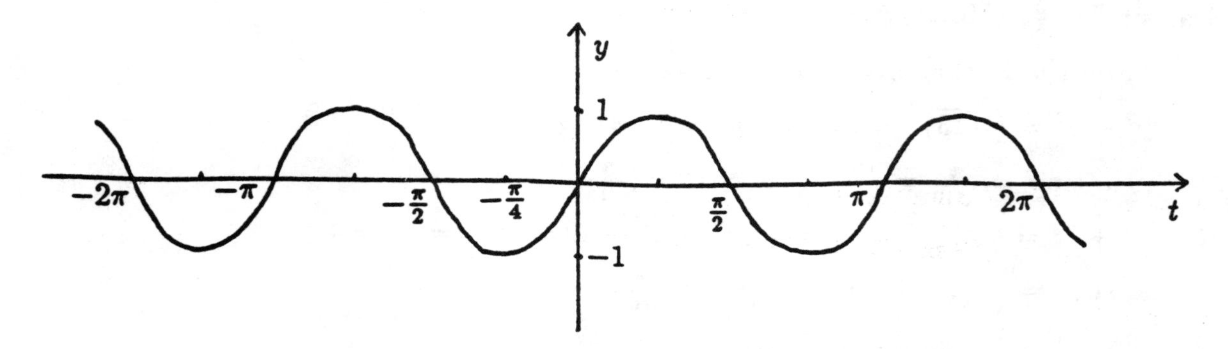

23. $f(t) = 4\cos(-t) = 4 \cdot \cos t$ has amplitude 4 and period $T = 2\pi$.

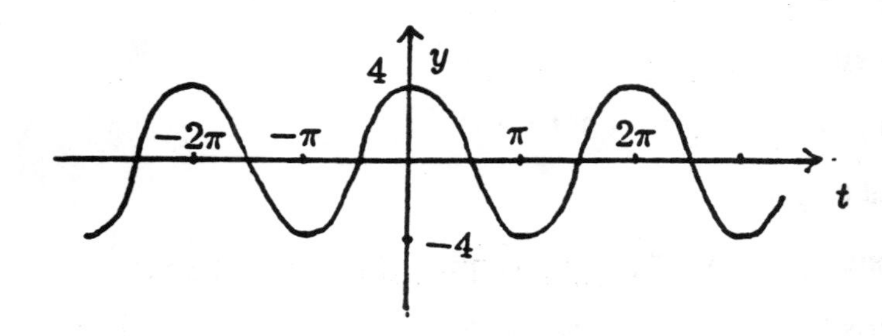

25. $\frac{y}{10} = \sin\frac{\pi}{4} = \frac{\sqrt{2}}{2} \implies y = \frac{\sqrt{2}}{2} \cdot 10 = 5\sqrt{2}.$

$\frac{x}{10} = \cos\frac{\pi}{4} = \frac{\sqrt{2}}{2} \implies x = \frac{\sqrt{2}}{2} \cdot 10 = 5\sqrt{2}.$

27. $\frac{y}{5} = \sin\frac{\pi}{6} = \frac{1}{2} \implies y = \frac{5}{2}$.

$\frac{x}{5} = \cos\frac{\pi}{6} = \frac{\sqrt{3}}{2} \implies x = \frac{5}{2}\sqrt{3}$.

29. $\sin 15° = \sin\left(15 \cdot \frac{\pi}{180}\right) = \sin\left(\frac{\pi}{12}\right) = 0.2588$.

31. $\cos\left(\frac{\pi}{7}\right) = .9010$.

33. $\sin\left(\frac{2\pi}{9}\right) = .6428$.

35. $x = \sin\frac{\pi}{12} = 0.2588$ miles.

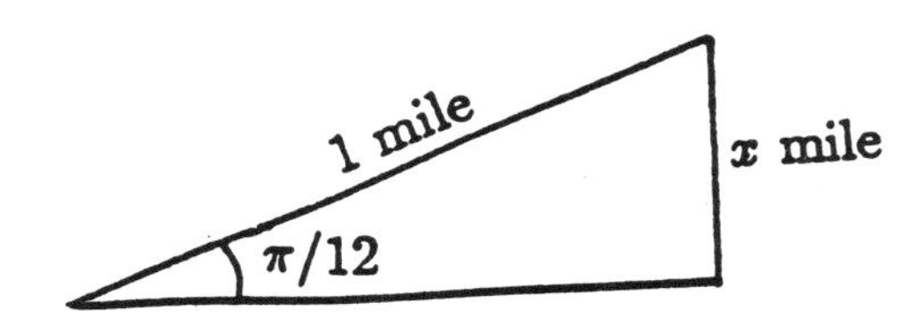

37.a. $D\left(\frac{5}{2}\right) = 12 + 3\sin\left(\frac{\pi}{6}\cdot\frac{5}{2} - \frac{5}{12}\pi\right)$

$= 12 + 3\sin 0 = 12$ hours.

b. $D\left(\frac{11}{2}\right) = 12 + 3\cdot\sin\left(\frac{\pi}{6}\cdot\frac{11}{2} - \frac{5}{12}\pi\right)$

$= 12 + 3\sin\left(\frac{\pi}{2}\right) = 15$ hours.

c. $D\left(\frac{23}{2}\right) = 12 + 3\sin\left(\frac{\pi}{6}\cdot\frac{23}{2} - \frac{5}{12}\pi\right)$

$= 12 + 3\cdot\sin\left(\frac{18}{12}\pi\right) = 12 + 3\sin\left(\frac{3}{2}\pi\right) = 9$ hours.

39.a. $h(0) = 0 + \sin\left(\frac{\pi\cdot 0}{4}\right) + B = 0 + B = B$.

b. $h(4) = 4 + \sin\frac{4\pi}{4} + B = 4 + B$.

c. $h(6) = 6 + \sin\frac{6\pi}{4} + B = 5 + B$.

41. $A = \frac{1}{2}xy$. $\frac{y}{h} = \sin\theta \implies y = h\sin\theta$. Therefore $A = \frac{1}{2}xh\sin\theta$.

43. The period T is $\frac{2\pi}{5}$ seconds. So, the frequency is $\frac{1}{T} = \frac{5}{2\pi}$ beats per second. Hence the heart rate in beats per minute is $\frac{5}{2\pi}\cdot 60 \approx 48$.

45.a. To find when high tide occurs, solve for t such that

$0 \leq t \leq 24$ and $\sin \frac{\pi(t-4)}{6} = 1$.

Case 1: $\frac{\pi(t-4)}{6} = \frac{\pi}{2} \Longrightarrow \frac{t-4}{6} = \frac{1}{2} \Longrightarrow t = 7$.

Case 2: $\frac{\pi(t-4)}{6} = \frac{5\pi}{2} \Longrightarrow \frac{t-4}{6} = \frac{5}{2} \Longrightarrow t = 19$.

The high tide occurs at 7am and 7pm.

b. To find when low tide occurs, solve for t such that

$0 \leq t \leq 24$ and $\sin \frac{\pi(t-4)}{6} = -1$.

Case 1: $\frac{\pi(t-4)}{6} = \frac{3\pi}{2} \Longrightarrow \frac{t-4}{6} = \frac{3}{2} \Longrightarrow t = 13$.

Case 2: $\frac{\pi(t-4)}{6} = -\frac{\pi}{2} \Longrightarrow \frac{t-4}{6} = -\frac{1}{2} \Longrightarrow t = 1$.

The low tide occurs at 1am and 1pm.

c. H(t)=15+(4*sin(Pi*(t−4)/6))

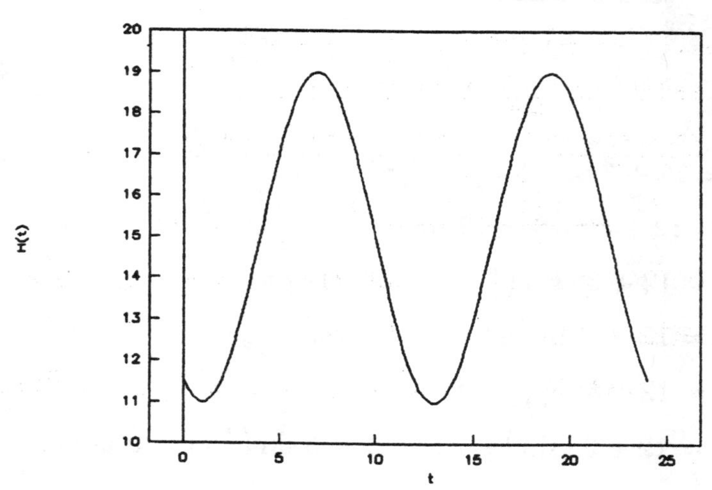

Selected Solutions to Exercise Set 6.3

1. $f'(x) = 3\cos 3x$.

3. $y = x\sin(x+\pi)$.

$y' = 1 \cdot \sin(x+\pi) + x \cdot \cos(x+\pi) = \sin(x+\pi) + x \cdot \cos(x+\pi)$.

5. $f'(x) = 3x^2(\cos 2x) + x^3 \cdot (-\sin 2x)(2) = 3x^2 \cos 2x - 2x^3 \sin 2x$.

7. $y = \sin^3 x \cos^2 x$.

$y' = (3\sin^2 x \cos x) \cdot \cos^2 x + (\sin^3 x) \cdot (2\cos x \cdot (-\sin x))$

$= 3\sin^2 x \cos^3 x - 2\sin^4 x \cos x$.

9. $f'(x) = \cos\sqrt{1+x^2}\{\frac{1}{2}(1+x^2)^{-1/2}\}\cdot\{2x\} = x(1+x^2)^{-1/2}\cdot\cos\sqrt{1+x^2}$.

11. $f(t) = \cos^2(\ln t)$.

    $f'(t) = 2\cos(\ln t)\cdot(-\sin(\ln t))\cdot\frac{1}{t} = -\frac{2\cos(\ln t)\cdot\sin(\ln t)}{t}$.

13. $f(x) = 1$ for all $x \implies f'(x) = 0$ for all x.

15. $f(t) = \ln(t^4 + \sin t)$.

    $f'(t) = \frac{1}{t^4+\sin t}\cdot(4t^3 + \cos t) = \frac{4t^3+\cos t}{t^4+\sin t}$.

17. $y' = \frac{\cos x(3+x^2)-(1+\sin x)(2x)}{(3+x^2)^2}$.

19. $f(x) = \cos x\cdot\sin\frac{1}{x}$.

    $f'(x) = (-\sin x)\cdot\sin\frac{1}{x} + \cos x\cdot\left(\cos\frac{1}{x}\right)\cdot\left(-\frac{1}{x^2}\right)$

    $= -\sin x\cdot\sin\frac{1}{x} - \frac{1}{x^2}\cdot\cos x\cdot\cos\frac{1}{x}$.

21. Slope $m = y' = 2\cos 2x \implies y' = 2\cos\frac{2\pi}{6} \implies y' = 1$.

    $y - \frac{\sqrt{3}}{2} = 1\cdot\left(x - \frac{\pi}{6}\right) \implies y = x - \frac{\pi}{6} + \frac{\sqrt{3}}{2}$.

23. $s'(t) = \frac{\cos 2t(2)\cdot(3+\cos^2 t)-\sin 2t\cdot(2\cos t)\cdot(-\sin t)}{(3+\cos^2 t)^2}$

    $s'\left(\frac{\pi}{4}\right) = \frac{2\cos\frac{2\pi}{4}\left(3+\cos^2\frac{\pi}{4}\right)-\sin\frac{2\pi}{4}\cdot\left(2\cos\frac{\pi}{4}\right)\left(-\sin\frac{\pi}{4}\right)}{\left(3+\cos^2\frac{\pi}{4}\right)^2}$

    $= \frac{2\cdot 0\cdot\left(3+\frac{1}{2}\right)-1\cdot\left(2\cdot\frac{1}{2}\sqrt{2}\right)\left(-\frac{1}{2}\sqrt{2}\right)}{\left(3+\frac{1}{2}\right)^2} = \frac{1}{\frac{49}{4}} = \frac{4}{49}$ m/s.

25. Because $\sin^2 x + \cos^2 x = 1$ for all x, $f(x) = g(x)$ for all x.

27. $f'(t) = \cos(\ln t)\cdot\left(\frac{1}{t}\right)$.

    $f''(t) = -\sin(\ln t)\cdot\left(\frac{1}{t}\right)\cdot\left(\frac{1}{t}\right) + \cos(\ln t)\cdot(-t^{-2}) = \frac{-\sin(\ln t)-\cos(\ln t)}{t^2}$

29. $\cos(xy) = x + y$.

    $-\sin(xy)\cdot\left(1\cdot y + x\cdot\frac{dy}{dx}\right) = 1 + \frac{dy}{dx}$.

    $\frac{dy}{dx}\cdot(1 + x\cdot\sin(xy)) = -y\cdot\sin(xy) - 1$.

    $\frac{dy}{dx} = -\frac{1+y\sin(xy)}{1+x\sin(xy)}$.

31. $\cos y\cdot\left(\frac{dy}{dx}\right) = 1\cdot\cos y + x\cdot(-\sin y)\left(\frac{dy}{dx}\right)$,

    $\frac{dy}{dx}(\cos y + x\sin y) = \cos y$. $\qquad \frac{dy}{dx} = \frac{\cos y}{\cos y+x\sin y}$.

33. $f'(x) = \cos x - \sin x$. $\quad f'(x) = 0$ when $x = \frac{\pi}{4}$ and $\frac{5\pi}{4}$.

$f(0) = 1, \quad f\left(\frac{\pi}{4}\right) = \sqrt{2}, \quad f\left(\frac{5\pi}{4}\right) = -\sqrt{2}, \quad f(2\pi) = 1.$

The maximum value is $f\left(\frac{\pi}{4}\right) = \sqrt{2}$.

The minimum value is $f\left(\frac{5\pi}{4}\right) = -\sqrt{2}$.

35. $f'(x) = 1 - \cos x$.

$f'(x) = 0$ when $\cos x = 1 \implies x = 0$ and 2π.

$f\left(\frac{\pi}{2}\right) = \frac{\pi}{2} - 1, \qquad f\left(\frac{3\pi}{2}\right) = \frac{3\pi}{2} + 1.$

The maximum value is $f\left(\frac{3\pi}{2}\right) = \frac{3\pi}{2} + 1$.

The minimum value is $f\left(\frac{\pi}{2}\right) = \frac{\pi}{2} - 1$.

37. $f'(x) = -\sin\left(x - \frac{\pi}{2}\right)$ for $0 \le x \le 2\pi$.

$f'(x) = 0$ for $x = \frac{\pi}{2}$ and $\frac{3\pi}{2}$.

Interval	Sign of $f'(x)$	Conclusion
$0 < x < \frac{\pi}{2}$	$f'(x) > 0$	increasing
$\frac{\pi}{2} < x < \frac{3\pi}{2}$	$f'(x) < 0$	decreasing
$\frac{3\pi}{2} < x < 2\pi$	$f'(x) > 0$	increasing

$f\left(\frac{\pi}{2}\right)$ is the relative maximum. $f\left(\frac{3\pi}{2}\right)$ is the relative minimum.

$f''(x) = -\cos\left(x - \frac{\pi}{2}\right)$.

$f''(x) = 0$ for $x = \pi$ and 2π.

Intervals	Sign of $f''(x)$	Conclusion
$0 < x < \pi$	$f''(x) < 0$	concave down
$\pi < x < 2\pi$	$f''(x) > 0$	concave up

The inflection point is when $x = \pi$.

x	0	$\frac{\pi}{2}$	π	$\frac{3\pi}{2}$	2π
y	0	1	0	−1	0

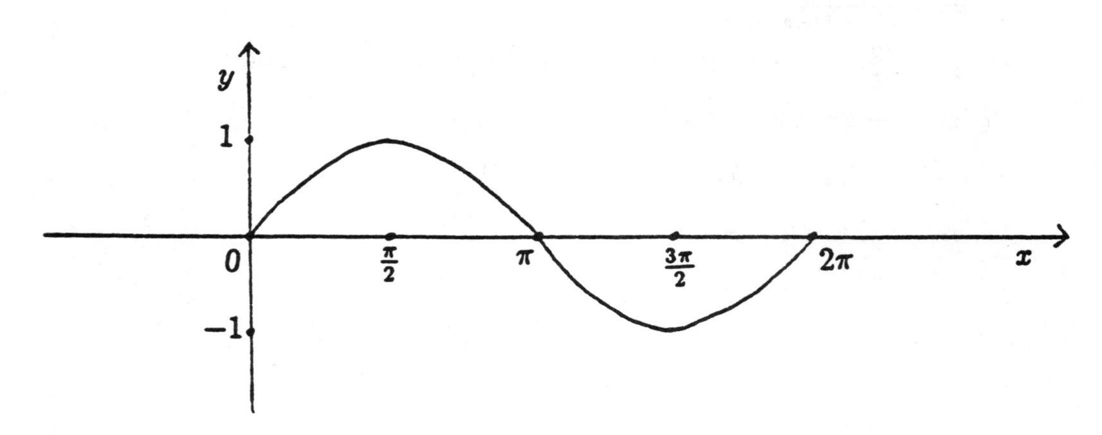

39. $f(x) = x + \sin x$ in $[0, 2\pi]$. $\quad f'(x) = 1 + \cos x$.

$f'(x) = 0$ for $x = \pi$, and $f'(x) > 0$ for all $x \neq \pi$.

So $f(x)$ has no relative extremum.

$f''(x) = -\sin x = 0$ for $x = \pi$.

Interval	Sign of $f''(x)$	Conclusion
$0 < x < \pi$	$-$	concave down
$\pi < x < 2\pi$	$+$	concave up

The inflection point occurs when $x = \pi$

x	0	$\frac{\pi}{4}$	$\frac{\pi}{2}$	π	$\frac{3\pi}{2}$	$\frac{7\pi}{4}$	2π
$f(x)$	0	1.49	2.57	π	3.71	4.79	2π

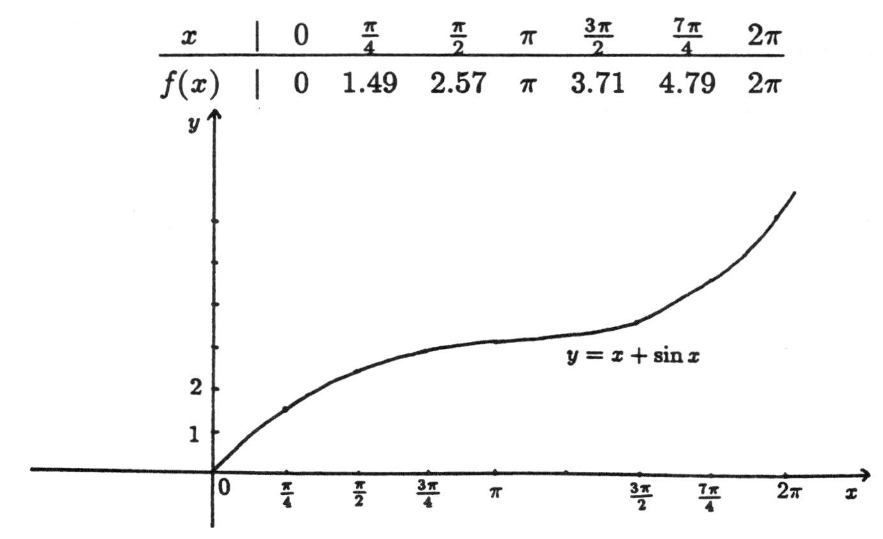

41. $f(x) = \frac{\sin x}{2+\cos x}$.

$f'(x) = \frac{\cos x(2+\cos x)-(-\sin x)\cdot\sin x}{(2+\cos x)^2} = \frac{2\cos x+\cos^2 x+\sin^2 x}{(2+\cos x)^2} = \frac{2\cos x+1}{(2+\cos x)^2}$.

$f'(x) = 0$ for $\cos x = -\frac{1}{2} \implies x = \frac{2\pi}{3}$ and $-\frac{2\pi}{3}$

Interval	Sign of $f'(x)$	Conclusion
$-\pi < x < -\frac{2\pi}{3}$	$-$	decreasing
$-\frac{2\pi}{3} < x < \frac{2\pi}{3}$	$+$	increasing
$\frac{2\pi}{3} < x < \pi$	$-$	decreasing

$f\left(-\frac{2\pi}{3}\right)$ is the relative minimum.

$f\left(\frac{2\pi}{3}\right)$ is the relative maximum.

$f''(x) = \frac{-2\sin x(2+\cos x)^2-2(2+\cos x)(-\sin x)\cdot(2\cos x+1)}{(2+\cos x)^4}$.

$f''(x) = \frac{-2\sin x(2+\cos x)+2\sin x(2\cos x+1)}{(2+\cos x)^3}$.

$f''(x) = \frac{-4\sin x-2\sin x\cos x+4\sin x\cos x+2\sin x}{(2+\cos x)^3}$.

$f''(x) = \frac{-2\sin x(1-\cos x)}{(2+\cos x)^3}$.

$f''(x) = 0$ for $x = -\pi$, $x = 0$, and $x = \pi$.

Interval	Sign of $f''(x)$	Conclusion
$-\pi < x < 0$	$+$	concave up
$0 < x < \pi$	$-$	concave down

The inflection point is when $x = 0$.

x	$-\pi$	$-\frac{2\pi}{3}$	$-\frac{\pi}{2}$	$-\frac{\pi}{4}$	0	$\frac{\pi}{4}$	$\frac{\pi}{2}$	$\frac{2\pi}{3}$	π
$f(x)$	0	-0.577	-0.5	-0.261	0	0.261	0.5	0.577	0

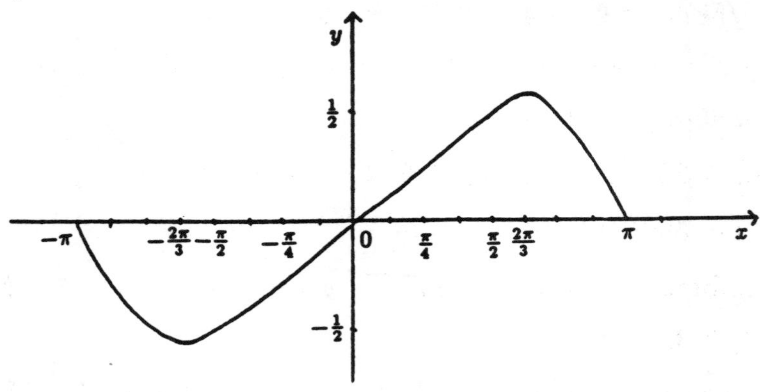

43. $y = f(t) = t + \sin\left(\frac{\pi t}{4}\right) + B$.

$y' = f'(t) = 1 + \cos\left(\frac{\pi t}{4}\right)\left(\frac{\pi}{4}\right)$.

The maximum value of $f'(t)$ occurs when $\cos\frac{\pi t}{4} = 1 \implies \frac{\pi t}{4} = 0$ and $2\pi \implies$ $t = 0$ and 8 months.

The minimum value of $f'(t)$ occurs when $\cos\frac{\pi t}{4} = -1 \implies \frac{\pi t}{4} = \pi$ and $3\pi \implies t = 4$ and 12 months.

45. $D(t) = 12 + 3\sin\left(\frac{\pi t}{6} - \frac{5\pi}{12}\right)$

(a) The days are longest when $\sin\left(\frac{\pi t}{6} - \frac{5\pi}{12}\right) = 1$

$\implies \left(\frac{\pi t}{6} - \frac{5\pi}{12}\right) = \frac{\pi}{2} \implies \frac{\pi t}{6} = \frac{11\pi}{12} \implies t = \frac{11}{2}$ months.

(b) The days are shortest when $\sin\left(\frac{\pi t}{6} - \frac{5\pi}{12}\right) = -1$

$\implies \left(\frac{\pi t}{6} - \frac{5\pi}{12}\right) = \frac{3\pi}{2} \implies \frac{\pi t}{6} = \frac{23\pi}{12} \implies t = \frac{23}{2}$ months.

47. $F(t) = 200 + 20\cos 2t + 20\sin 2t.$

$F'(t) = -40\sin 2t + 40\cos 2t.$

$F'(t) = 0 \implies \sin 2t = \cos 2t \implies 2t = \frac{\pi}{4}, \frac{5\pi}{4}, \frac{9\pi}{4}, \frac{13\pi}{4}, \ldots.$

$F''(t) = -80\cos 2t - 80\sin 2t.$

$F''\left(\frac{\pi}{8}\right) = -80 \cdot \cos\left(\frac{\pi}{4}\right) - 80\sin\left(\frac{\pi}{4}\right) < 0.$

$F''\left(\frac{5\pi}{8}\right) = -80 \cdot \cos\left(\frac{5\pi}{4}\right) - 80\sin\left(\frac{5\pi}{4}\right) > 0.$

Therefore, the function $F(t)$ has relative maxima at $t = \frac{\pi}{8} + n \cdot \pi$ and has relative minima at $t = \frac{5\pi}{8} + n \cdot \pi$ where $n = 0, 1, 2, 3, \ldots.$

49. $F(t) = 500 - 50\cos t + 50\sin t.$

a. $F'(t) = 50\sin t + 50\cos t.$

$F'(t) = 0 \implies \sin t = -\cos t \implies t = \frac{3\pi}{4}, \frac{7\pi}{4}, \frac{11\pi}{4}, \frac{15\pi}{4}, \ldots.$

$F''(t) = 50\cos t - 50\sin t.$

$F''\left(\frac{3\pi}{4}\right) = 50 \cdot \left(\frac{-\sqrt{2}}{2}\right) - 50 \cdot \left(\frac{\sqrt{2}}{2}\right) = -50\sqrt{2} < 0.$

$F''\left(\frac{7\pi}{4}\right) = 50 \cdot \left(\frac{\sqrt{2}}{2}\right) - 50 \cdot \left(\frac{-\sqrt{2}}{2}\right) = 50\sqrt{2} > 0.$

The numbers t for which $R(t)$ has relative maxima are $t = \frac{3\pi}{4} + 2n\pi$ where $n = 0, 1, 2, 3, \ldots.$

b. The numbers t for which $R(t)$ has relative minima are $t = \frac{7\pi}{4} + 2n\pi$ where $n = 0, 1, 2, 3, \ldots.$

c. These relative maxima are

$$500 - 50 \cdot \left(\frac{-\sqrt{2}}{2}\right) + 50 \cdot \left(\frac{\sqrt{2}}{2}\right) = 500 + 50\sqrt{2}.$$

These relative minima are

$$500 - 50 \cdot \left(\frac{\sqrt{2}}{2}\right) + 50 \cdot \left(\frac{-\sqrt{2}}{2}\right) = 500 - 50\sqrt{2}.$$

51. $f(t) = 55 + 35\sin(2\pi(t-90)/365)$.

a. $f'(t) = 35\cos(2\pi(t-90)/365)\cdot\frac{2\pi}{365}$.

$f'(t) = 0 \implies \frac{2\pi}{365}\cdot(t-90) = \frac{\pi}{2}, \frac{3\pi}{2}, \frac{5\pi}{2}, \frac{7\pi}{2}, \cdots \implies$

$t = 90 + \frac{365}{4},\ 90 + \frac{3\cdot 365}{4},\ 90 + \frac{5\cdot 365}{4},\ 90 + \frac{7\cdot 365}{4}, \cdots$

$\implies t = 181.25, \quad t = 363.75.$

$f''(t) = -35\sin(2\pi(t-90)/365)\cdot\left(\frac{2\pi}{365}\right)^2$.

$f''(181.25) = -35\cdot\sin\left(\frac{\pi}{2}\right)\cdot\left(\frac{2\pi}{365}\right)^2 < 0.$

$f''(363.75) = -35\cdot\sin\left(\frac{3\pi}{2}\right)\cdot\left(\frac{2\pi}{365}\right)^2 > 0.$

The maximum temperature occurs when $t = 181.25$.

b. The minimum temperature occurs when $t = 363.75$.

Selected Solutions to Exercise Set 6.4

1. $\int \cos 6x\,dx = \frac{1}{6}\sin 6x + C.$

3. $\int_0^{\pi} 3\sin(\pi - x)dx = 3\cos(\pi - x)\Big]_0^{\pi} = 3\cos(\pi-\pi) - 3\cos(\pi - 0)$

$$= 3\cos 0 - 3\cos\pi = 3 + 3 = 6.$$

5. Let $u = 1 + x^3, \quad du = 3x^2dx, \quad x^2dx = \frac{1}{3}du.$

$\int x^2\sin(1+x^3)dx = \int \sin u\cdot\frac{1}{3}du = -\frac{1}{3}\cos(1+x^3) + C.$

7. $\int_0^{\sqrt{\pi}} t\cos(\pi - t^2)dt.$

Let $u = \pi - t^2$.

$du = -2t\,dt \implies t\,dt = -\frac{1}{2}\,du.$

$\int t\cos(\pi - t^2)dt = \int \cos u\left(-\frac{1}{2}\,du\right) = -\frac{1}{2}\sin u + C = -\frac{1}{2}\sin(\pi - t^2) + C.$

$\int_0^{\sqrt{\pi}} t\cos(\pi - t^2)dt = -\frac{1}{2}\sin(\pi - t^2)\Big]_0^{\sqrt{\pi}} = -\frac{1}{2}\cdot\sin(0) + \frac{1}{2}\cdot\sin(\pi) = 0.$

9. Let $u = \cos t, \quad du = -\sin t\ dt.$

$$\int_0^{\pi} \cos^3 t \sin t\ dt = \int_0^{\pi} u^3(-du) = -\tfrac{1}{4}u^4\Big]_0^{\pi} = -\tfrac{1}{4}\cos^4 t\Big]_0^{\pi}.$$

$$= -\tfrac{1}{4}\cos^4 \pi + \tfrac{1}{4}\cos^4 0 = 0.$$

11. Let $u = 1 - \sin x, \quad du = -\cos x dx.$

$$\int \cos x\sqrt{1 - \sin x}\ dx = \int \sqrt{u}(-du) = -\tfrac{2}{3}u^{3/2} + C = -\tfrac{2}{3}(1 - \sin x)^{3/2} + C.$$

13. $\int \frac{\sin(3+\ln x)}{x}\ dx.$

Let $u = 3 + \ln x.$

$du = \frac{1}{x}\ dx.$

$$\int \frac{\sin(3+\ln x)}{x}\ dx = \int \sin(u)du = -\cos(u) + C = -\cos(3 + \ln x) + C.$$

15. $\int \frac{\cos(\ln^3 x)}{x}\ dx.$?

Suppose the author means $\int \frac{\cos(\ln(3x))}{x}\ dx.$

Let $u = \ln(3x).$

$du = \frac{1}{3x} \cdot (3)dx = \frac{1}{x}\ dx.$

$$\int \frac{\cos(\ln(3x))}{x}\ dx = \int \cos(u)du = \sin(u) + C = \sin(\ln(3x)) + C.$$

17. Let $u = \cos t, \qquad du = -\sin t\ dt.$

$$\int_0^{\pi/2} (\sin t)e^{\cos t}\ dt = \int_{t=0}^{t=\pi/2} e^u(-du) = -e^{\cos t}\Big]_0^{\pi/2}$$

$$= -e^{\cos \pi/2} + e^{\cos 0} = e - 1 = 1.718.$$

19. Identity: $\sin^2 x = \frac{1}{2} - \frac{1}{2}\cos 2x.$

Let $u = 2x, \qquad du = 2\ dx.$

$$\int \sin^2 x\ dx = \int \left(\tfrac{1}{2} - \tfrac{1}{2}\cos 2x\right) dx = \int \tfrac{1}{2}\ dx - \int \tfrac{1}{4}\cos u\ du = \tfrac{1}{2}x - \tfrac{1}{4}\sin 2x + C.$$

21. $\int_0^{\sqrt{\pi}/2} x \sin^2 x^2 \cos x^2\ dx.$

Let $u = \sin x^2.$

$du = \cos x^2(2x)dx \implies x\cos x^2\ dx = \frac{1}{2}\ du.$

$$\int x \sin^2 x^2 \cos x^2\ dx = \int u^2\left(\tfrac{1}{2}\ du\right) = \tfrac{1}{2} \cdot \tfrac{1}{3}u^3 + C = \tfrac{1}{6}\sin^3 x^2 + C.$$

$$\int_0^{\sqrt{\pi}/2} x\sin^2 x^2 \cos x^2\,dx = \tfrac{1}{6}\sin^3 x^2\Big]_0^{\sqrt{\pi}/2} = \tfrac{1}{6}\cdot\sin^3\left(\tfrac{\pi}{4}\right) - \tfrac{1}{6}\cdot\sin^3(0)$$

$$= \tfrac{1}{6}\cdot\left(\tfrac{\sqrt{2}}{2}\right)^3 - 0 = \tfrac{\sqrt{2}}{24}.$$

23. Let $u = \pi x$, $\quad du = \pi\,dx$.

$$\int_0^1 \sin\pi x\,dx = \int_{x=0}^{x=1} \sin u\left(\tfrac{1}{\pi}\,du\right) = -\tfrac{1}{\pi}\cos u\Big]_{x=0}^{x=1} = -\tfrac{1}{\pi}\cos\pi x\Big]_0^1$$

$$= -\tfrac{1}{\pi}\cos\pi + \tfrac{1}{\pi}\cos 0 = \tfrac{2}{\pi}.$$

25. Average value $= \int_0^1 \cos\pi t\,dt = \frac{\sin\pi t}{\pi}\Big]_0^1 = 0.$

27.

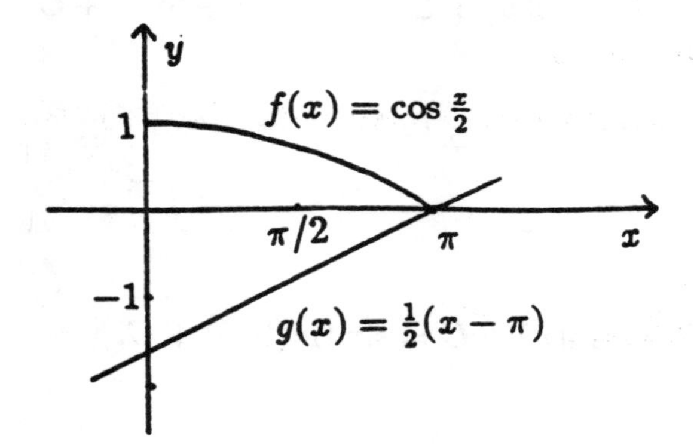

$$\text{Area} = \int_0^{\pi}\left(\cos\tfrac{x}{2} - \tfrac{1}{2}x + \tfrac{1}{2}\pi\right)dx$$
$$= \left(2\sin\tfrac{x}{2} - \tfrac{1}{4}x^2 + \tfrac{\pi}{2}x\right)\Big]_0^{\pi}$$
$$= \left(2 - \tfrac{1}{4}\pi^2 + \tfrac{\pi^2}{2}\right) = 2 + \tfrac{1}{4}\pi^2.$$

29. Volume $= \int_0^{\pi/2} \cos^2 x\cdot\pi\,dx = \pi\int_0^{\pi/2} \frac{1+\cos 2x}{2}\,dx.$

$$= \tfrac{\pi}{2}\cdot\left(x + \tfrac{\sin 2x}{2}\right)\Big]_0^{\pi/2} = \tfrac{\pi}{2}\cdot\left(\tfrac{\pi}{2} - 0\right) = \tfrac{\pi^2}{4}.$$

31. $P(t) = 40 + 60\sin\pi t$ for $0 \le t \le 10$.

a. Total profit $= \int_0^{10} P(t)dt = \int_0^{10}(40 + 60\sin\pi t)dt = \left(40t - \frac{60}{\pi}\cos\pi t\right)\Big]_0^{10}.$

$$= \left(40(10) - \tfrac{60}{\pi}\cos(10\pi)\right) - \left(40(0) - \tfrac{60}{\pi}\cos(0)\right) = 400.$$

b. Average profit $= \dfrac{\int_0^{10} P(t)dt}{10-0} = \frac{400}{10} = 40.$

33. $P(t) = 100 + 5t + 200\cos(\pi(t-5)/5)$ for $0 \le t \le 10$.

a. $P(0) = 100 + 5(0) + 200\cos\left(\frac{\pi}{5}(0-5)\right)$

$$= 100 + 0 + 200\cos(-\pi) = 100 + 200(-1) = -100.$$

At time $t = 0$ the company is losing money at the rate of 100,000 dollars per year.

b. Total earnings for the decade

$$= \int_0^{10} P(t)dt = \int_0^{10} \left(100 + 5t + 200\cos\left(\tfrac{\pi}{5}(t-5)\right)\right) dt$$

$$= \left(100t + \tfrac{5}{2}t^2 + 200\cdot\tfrac{5}{\pi}\cdot\sin\left(\tfrac{\pi}{5}(t-5)\right)\right)\Big]_0^{10}$$

$$= \left(100(10) + \tfrac{5}{2}(10)^2 + \tfrac{1000}{\pi}\sin(\pi)\right) - \left(0 + 0 + \tfrac{1000}{\pi}\sin(-\pi)\right)$$

$$= 1000 + 250 = 1250.$$

Therefore the total earnings are predicted to be 1,250,000 dollars.

Selected Solutions to Exercise Set 6.5

1. a. $\tan\theta = \frac{1}{\sqrt{3}}$. b. $\cot\theta = \frac{\sqrt{3}}{1} = \sqrt{3}$.

c. $\sec\theta = \frac{2}{\sqrt{3}}$. d. $\csc\theta = \frac{2}{1} = 2$.

3. a. $\tan\theta = \frac{3}{4}$. b. $\cot\theta = \frac{4}{3}$.

c. $\sec\theta = \frac{5}{4}$. d. $\csc\theta = \frac{5}{3}$.

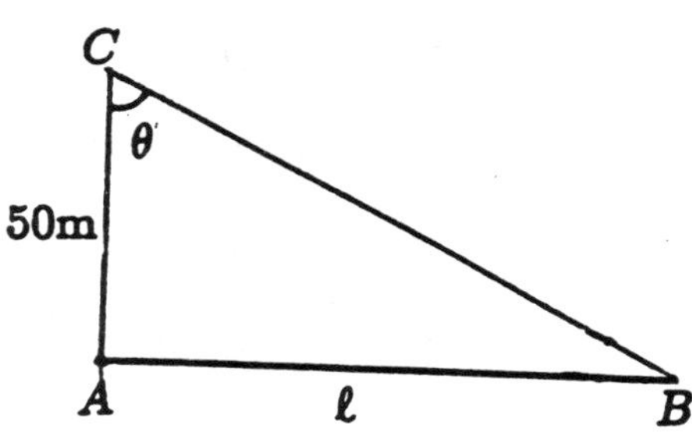

5. $\tan\theta = \frac{\ell}{50}$.

$\ell = 50\cdot\tan\theta = 50\tan\frac{\pi}{6} = 50\cdot\frac{\sqrt{3}}{3} \approx 28.87$ m.

The distance between A and B is 28.87 m.

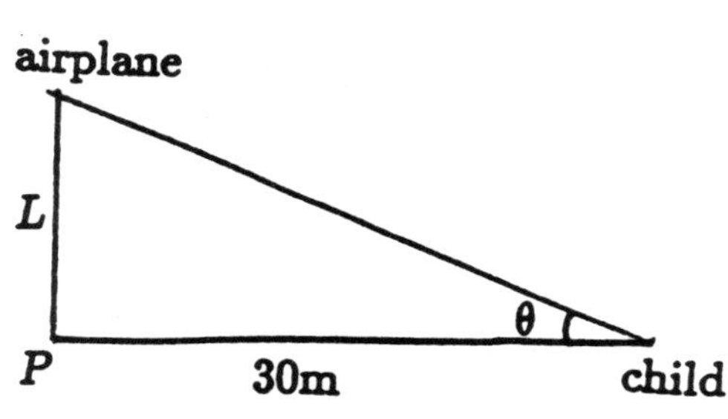

7. $\tan\theta = \frac{L}{30}$.

$L = 30\cdot\tan\theta = 30\cdot\tan 30° = 30\cdot\frac{\sqrt{3}}{3} \approx 17.32$ m.

The altitude of the airplane is 17.32 m.

9. $f'(x) = (2x)\cdot(\sec x) + x^2\cdot(\sec x\tan x)$.

11. $\frac{dy}{dx} = 3\cot^2(6x) \cdot \frac{d}{dx}(\cot(6x)) = 3\cot^2(6x) \cdot \left(-\csc^2(6x)\frac{d}{dx}(6)\right)$
$= -18\cot^2(6x) \cdot \csc^2(6x).$

13. $f(x) = \sec x \cdot \tan x.$

$f'(x) = \sec x \cdot \tan x \cdot \tan x + \sec x \cdot \sec^2 x = \sec x(\tan^2 x + \sec^2 x).$

15. $\frac{dy}{dx} = \frac{\sec x \cdot (1) - x \cdot (\sec x \tan x)}{(\sec x)^2} = \frac{1 - x\tan x}{\sec x} = \cos x - x\sin x.$

17. $f(x) = \cot^2 3x.$

$f'(x) = 2\cot 3x \cdot (-\csc^2 3x)(3) = -6\cot 3x \cdot \csc^2 3x.$

19. $y = \tan x^3.$

$\frac{dy}{dx} = \sec^2 x^3 \cdot 3x^2 = 3x^2 \sec^2 x^3.$

21. $f(x) = \sec\sqrt{1+x^2}.$

$f'(x) = \sec\sqrt{1+x^2} \cdot \tan\sqrt{1+x^2} \cdot \frac{1}{2}(1+x^2)^{-1/2} \cdot 2x.$
$= x(1+x^2)^{-1/2}\sec\sqrt{1+x^2} \cdot \tan\sqrt{1+x^2}.$

23. $f(x) = \cot\ln\sqrt{x}.$

$f'(x) = -\csc^2\ln\sqrt{x} \cdot \frac{1}{\sqrt{x}} \cdot \frac{1}{2\sqrt{x}} = -\frac{\csc^2\ln\sqrt{x}}{2x}.$

25. $\int_0^{\pi/16} \sec^2 4x\,dx = \frac{1}{4}\tan 4x\Big]_0^{\pi/16} = \frac{1}{4} \cdot \left(\tan\left(4 \cdot \frac{\pi}{16}\right) - \tan(0)\right)$

$= \frac{1}{4} \cdot \tan\frac{\pi}{4} = \frac{1}{4} \cdot 1 = \frac{1}{4}.$

27. $\int x\sec^2 x^2\,dx.$

Let $u(x) = x^2.$ $\qquad du = 2x\,dx.$

$\int x\sec^2 x^2\,dx = \frac{1}{2}\int \sec^2 u\,du = \frac{1}{2}\tan u + C = \frac{1}{2}\tan x^2 + C.$

29. $\int_0^{\pi/4} \sec^2(\pi + x)dx = \tan(\pi + x)\Big]_0^{\pi/4} = \tan\left(\frac{5\pi}{4}\right) - \tan(\pi) = 1 - 0 = 1.$

31. $\int(x - \csc^2 x)dx = \int x\,dx - \int \csc^2 x\,dx$

$= \frac{1}{2}x^2 + C_1 - (-\cot x + C_2) = \frac{1}{2}x^2 + \cot x + C.$

33. $\int \csc^3(\pi x)\cot(\pi x)dx.$

Let $u = \csc(\pi x).$

$du = -\csc(\pi x) \cdot \cot(\pi x) \cdot \pi\,dx \implies \csc(\pi x) \cdot \cot(\pi x)dx = -\frac{1}{\pi}\,du.$

$\int \csc^3(\pi x)\cot(\pi x)dx = \int u^2\left(-\frac{1}{\pi}\,du\right) = -\frac{1}{\pi} \cdot \frac{1}{3}u^3 + C = -\frac{1}{3\pi}\csc^3(\pi x) + C.$

35. $\int \left(\csc^2 \pi x - \cot \pi x\right) dx = \int \csc^2 \pi x dx - \int \cot \pi x \; dx$

$= \left(-\frac{1}{\pi} \cot \pi x + C_1\right) - \left(\frac{1}{\pi} \ln|\sin \pi x| + C_2\right).$

$= -\frac{1}{\pi}(\cot \pi x + \ln|\sin \pi x|) + C.$

37. Average value $= \dfrac{\int_{-\pi/4}^{\pi/4} \sec^2 x \, dx}{\frac{\pi}{4} - \left(-\frac{\pi}{4}\right)} = \dfrac{\tan x\big]_{-\pi/4}^{\pi/4}}{\frac{\pi}{2}} = \frac{2}{\pi} \cdot (1 - (-1)) = \frac{4}{\pi}.$

39. $y = f(x) = \tan(\sin x)$ for $-\frac{\pi}{2} < x < \frac{\pi}{2}$.

$f'(x) = \sec^2(\sin x) \cdot \cos x.$

$f''(x) = (2\sec(\sin x) \cdot \sec(\sin x) \cdot \tan(\sin x) \cdot \cos x) \cdot \cos x + \sec^2(\sin x) \cdot (-\sin x)$

$= 2\sec^2(\sin x) \cdot \tan(\sin x) \cdot \cos^2 x - \sec^2(\sin x) \cdot (\sin x).$

Using Lotus, we find that $f''(x) = 0$ for $x \approx -.89$, $x = 0$, $x \approx .89$.

For $-\frac{\pi}{2} < x < -.89$, $f''(x) > 0$, so the graph of $y = f(x)$ is concave up.

For $-.89 < x < 0$, $f''(x) < 0$, so the graph of $y = f(x)$ is concave down.

For $0 < x < .89$, $f''(x) > 0$, so the graph of $y = f(x)$ is concave up.

For $.89 < x < \frac{\pi}{2}$, $f''(x) < 0$, so the graph of $y = f(x)$ is concave down.

41. Volume of $S = 2 \int_0^{\pi/4} \pi \sec^2 x \; dx = 2\pi \tan x\big]_0^{\pi/4}$

$= 2\pi \left(\tan \frac{\pi}{4} - \tan 0\right) = 2\pi.$

43. Average Value $= \dfrac{\int_0^{\pi/4} \sec x \; dx}{\pi/4} = \left[\ln|\sec x + \tan x|\big]_0^{\pi/4}\right] \cdot \frac{4}{\pi}$

$= \left\{\ln\left|\sec \frac{\pi}{4} + \tan \frac{\pi}{4}\right| - \ln|\sec 0 + \tan 0|\right\} \cdot \frac{4}{\pi}$

$= \{\ln|\sqrt{2} + 1| - \ln|1 + 0|\} \cdot \frac{4}{\pi} = \frac{4}{\pi} \cdot \ln\left(\sqrt{2} + 1\right).$

45. (8) $\frac{d}{dx} \cot x = \frac{d}{dx}\left(\frac{\cos x}{\sin x}\right) = \dfrac{-\sin x \cdot \sin x - \cos x \cdot \cos x}{\sin^2 x}$

$= \dfrac{-\left(\sin^2 x + \cos^2 x\right)}{\sin^2 x} = -\frac{1}{\sin^2 x} = -\csc^2 x.$

(9) $\frac{d}{dx} \sec x = \frac{d}{dx}\left(\frac{1}{\cos x}\right) = \dfrac{-(-\sin x)}{\cos^2 x} = \frac{1}{\cos x} \cdot \frac{\sin x}{\cos x} = \sec x \cdot \tan x.$

(10) $\frac{d}{dx} \csc x = \frac{d}{dx}\left(\frac{1}{\sin x}\right) = \frac{d}{dx}(\sin x)^{-1} = -\sin^{-2} x \cdot \cos x = -\frac{1}{\sin x} \cdot \frac{\cos x}{\sin x}$

$= -\csc x \cdot \cot x.$

Selected Solutions to the Review Exercises - Chapter 6

1.

a. $\theta_r = 60 \cdot \frac{\pi}{180} = \frac{\pi}{3}$.

b. $\theta_r = 35 \cdot \frac{\pi}{180} = \frac{7}{36}\pi$.

c. $\theta_r = 120 \cdot \frac{\pi}{180} = \frac{2}{3}\pi$.

d. $\theta_r = 210 \cdot \frac{\pi}{180} = \frac{7}{6}\pi$.

e. $\theta_r = 10 \cdot \frac{\pi}{180} = \frac{1}{18}\pi$.

f. $\theta_r = 75 \cdot \frac{\pi}{180} = \frac{5}{12}\pi$.

3.

a. $\theta = -\frac{\pi}{2} + 2\pi = \frac{3}{2}\pi$.

b. $\theta = \frac{7\pi}{2} - 2\pi = \frac{3}{2}\pi$.

c. $\theta = \frac{11\pi}{4} - 2\pi = \frac{3}{4}\pi$.

d. $\theta = -\frac{5\pi}{3} + 2\pi = \frac{\pi}{3}$.

e. $\theta = -7\pi + 4 \cdot 2\pi = \pi$.

f. $\theta = 9\pi - 4 \cdot 2\pi = \pi$.

5. $y' = -\cos(\pi - x)$.

7. $f'(x) = 3 \cdot \sec^2 3x$.

9. $y = \cos(\pi \sin x)$.

$\frac{dy}{dx} = -\sin(\pi \sin x) \cdot (\pi \cos x)$.

11. $y = \ln(tan^2 x)$.

$\frac{dy}{dx} = \frac{1}{\tan^2 x} \cdot 2\tan x \cdot \sec^2 x = \frac{2\sec^2 x}{\tan x} = 2 \cdot \frac{1}{\cos^2 x} \cdot \frac{\cos x}{\sin x}$

$= \frac{2}{\cos x \cdot \sin x} = 2\sec x \cdot \csc x$.

13. $f'(x) = e^{\sec \pi x} \cdot \left(\frac{1}{\cos \pi x}\right)' = e^{\sec \pi x} \cdot \frac{\pi \cdot \sin \pi x}{\cos^2 \pi x} = e^{\sec \pi x} \cdot \pi \cdot (\sec \pi x)(\tan \pi x)$.

15. $y' = 3(x + \sec x)^2 \cdot \left(1 + \frac{\sin x}{\cos^2 x}\right) = 3(x + \sec x)^2 \cdot (1 + \sec x \tan x)$.

17. $f(x) = \frac{e^{\cos x}}{1+\sin^2 x}$.

$f'(x) = \frac{(1+\sin^2 x)\cdot(e^{\cos x}\cdot(-\sin x)) - (e^{\cos x})\cdot(2\sin x \cos x)}{(1+\sin^2 x)^2}$

$= \frac{-e^{\cos x} \cdot \sin x(1+\sin^2 x + 2\cos x)}{(1+\sin^2 x)^2}$

19. $y' = \frac{1}{\sqrt{4+\cos x}} \cdot \frac{-\sin x}{2\sqrt{4+\cos x}} = \frac{-sin x}{2(4+\cos x)}$.

21. $\int \sin 4x \; dx = -\frac{\cos 4x}{4} + C$.

23. $\int \sec \pi x \cdot \tan \pi x \, dx = \int \frac{\sin \pi x}{\cos^2 \pi x} dx.$

Let $u = \cos \pi x \implies du = -\pi \sin \pi x \, dx$. Thus we have

$\int \sec \pi x \cdot \tan \pi x \, dx = \int \frac{1}{u^2} \cdot \frac{-1}{\pi} \, du = -\frac{1}{\pi} \int u^{-2} \, du.$

$= -\frac{1}{\pi}(-u^{-1}) + C = \frac{1}{\pi}(\cos \pi x)^{-1} + C = \frac{1}{\pi} \sec \pi x + C.$

25. Let $u = \tan x \implies du = \sec^2 x \, dx$. Thus $\int \sqrt{\tan x} \cdot \sec^2 x \, dx = \int u^{1/2} \cdot du =$

$\frac{2}{3} u^{3/2} + C = \frac{2}{3}(\tan x)^{3/2} + C.$

27. Let $u = \tan 3x \implies du = 3 \sec^2 3x \, dx.$

$\int \sec^2(3x) \cdot e^{\tan 3x} \, dx = \int e^u \cdot \frac{1}{3} \, du = \frac{1}{3} e^u + C = \frac{1}{3} \cdot e^{\tan 3x} + C.$

29. $\int x \cos^{10}(x^2) \sin(x^2) dx.$

Let $u = \cos(x^2)$.

$du = -\sin(x^2) \cdot (2x \, dx) \implies x \sin(x^2) dx = -\frac{1}{2} du.$

$\int x \cos^{10}(x^2) \sin(x^2) dx = \int u^{10} \left(-\frac{1}{2} du\right) = -\frac{1}{2} \cdot \frac{1}{11} u^{11} + C$

$$= -\frac{1}{22} \cos^{11}(x^2) + C.$$

31. $\int_0^{\pi/4} \cos\left(x + \frac{\pi}{2}\right) dx = \sin\left(x + \frac{\pi}{2}\right)\Big]_0^{\pi/4} = \sin \frac{3}{4}\pi - \sin \frac{\pi}{2} = \frac{\sqrt{2}}{2} - 1.$

33. Let $u = \sec x \implies du = \sec x \tan x \, dx.$

$\int \sec x \tan x e^{\sec x} \, dx = \int e^u \, du = e^u + C = e^{\sec x} + C.$

Thus $\int_0^{\pi/4} \sec x \cdot \tan x \cdot e^{\sec x} \, dx = e^{\sec x}\Big]_0^{\pi/4} = e^{\sec \pi/4} - e^{\sec 0} = e^{\sqrt{2}} - e^1.$

35. Let $u = \cos \pi x \implies du = -\pi \sin \pi x \, dx.$

$\int \sin \pi x \cdot \cos \pi x \, dx = \int u \cdot \left(-\frac{1}{\pi}\right) du = -\frac{1}{\pi} \cdot \frac{u^2}{2} + C = -\frac{1}{\pi} \cdot \frac{\cos^2 \pi x}{2} + C.$

$\int_0^1 \sin \pi x \cdot \cos \pi x \, dx = -\frac{1}{\pi} \cdot \frac{\cos^2 \pi x}{2}\Big]_0^1 = -\frac{1}{2\pi}(1 - 1) = 0.$

37. a.

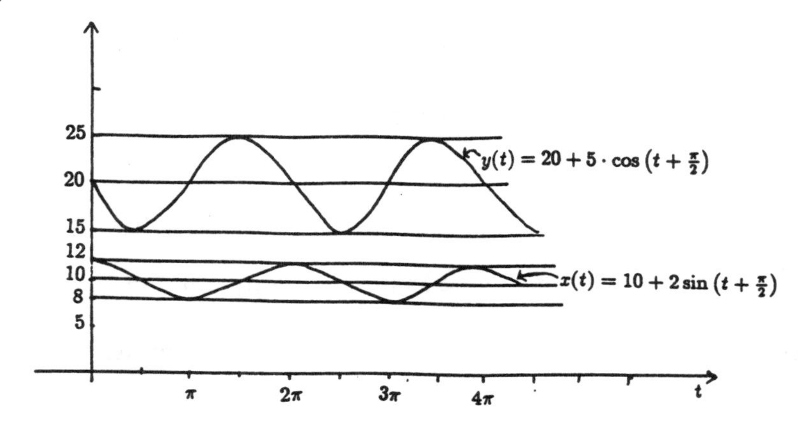

b. maximum of $x(t) = 12$.

minimum of $x(t) = 8$.

c. maximum of $y(t) = 25$.

minimum of $y(t) = 15$.

39. $y' = \sec^2 x \implies y'\left(\frac{\pi}{4}\right) = \sec^2\left(\frac{\pi}{4}\right) = 2$.

41. Area $= \int_0^{\sqrt{\pi/3}} x \cdot \sec x^2 \, dx$.

To find antiderivative $\int x \cdot \sec x^2 \, dx$, let $u = x^2 \implies du = 2x \, dx$.

$$\int x \cdot \sec x^2 \, dx = \int \sec u \cdot \tfrac{1}{2} \, du = \tfrac{1}{2} \ln|\sec u + \tan u| + C$$
$$= \tfrac{1}{2} \ln|\sec x^2 + \tan x^2| + C.$$

Thus
$$\text{Area} = \int_0^{\sqrt{\pi/3}} x \cdot \sec x^2 \, dx$$
$$= \tfrac{1}{2}\left(\ln|\sec x^2 + \tan x^2|\right)\Big]_0^{\sqrt{\pi/3}}$$
$$= \tfrac{1}{2}\left(\ln|\sec \tfrac{\pi}{3} + \tan \tfrac{\pi}{3}| - \ln 1\right)$$
$$= \tfrac{1}{2}\left(\ln(2+\sqrt{3}) - \ln 1\right)$$
$$= \tfrac{1}{2}\ln(2+\sqrt{3}).$$

43. Area $= \int_{\pi/6}^{\pi/3} \csc x \, dx = \left(\ln|\csc x - \cot x|\right)\Big]_{\pi/6}^{\pi/3}$

$= \ln|\csc \frac{\pi}{3} - \cot \frac{\pi}{3}| - \ln|\csc \frac{\pi}{6} - \cot \frac{\pi}{6}|$

$= \ln \frac{1}{\sqrt{3}} - \ln(2-\sqrt{3})$.

Thus the average value of $f(x) = \csc x$ is $(\text{Area})/\left(\frac{\pi}{3} - \frac{\pi}{6}\right)$.

$= \frac{6}{\pi}\left(\ln \frac{1}{\sqrt{3}} - \ln(2-\sqrt{3})\right)$.

45. $f(x) = \sin x + \cos x$.

a. $f'(x) = \cos x - \sin x$. $\quad f''(x) = -\sin x - \cos x$.

$f''(x) = 0 \implies \sin x = -\cos x \implies \tan x = -1 \implies x = \frac{3\pi}{4}$.

For $0 < x < \frac{3\pi}{4}$, $f''(x) < 0$. $\quad$ For $\frac{3\pi}{4} < x < \pi$, $f''(x) > 0$.

Therefore the graph of $y = f(x)$ is concave up on the interval $\left(\frac{3\pi}{4}, \pi\right)$.

b. The graph of $y = f(x)$ is concave down on the interval $\left(0, \frac{3\pi}{4}\right)$.

47. $P(t) = 300 + 120\sin\left(\frac{\pi}{5}\cdot(2+t)\right)$.

a. The population size oscillates because of the term $120\sin\left(\frac{\pi}{5}\cdot(2+t)\right)$.

b. The population maximum occurs when $\sin\left(\frac{\pi}{5}\cdot(2+t)\right) = 1 \implies$
$\frac{\pi}{5}\cdot(2+t) = \frac{\pi}{2}, \frac{5\pi}{2}, \frac{9\pi}{2}, \cdots \implies t = \frac{1}{2},\ t = \frac{21}{2},\ t = \frac{41}{2}, \ldots$.
Therefore the time between population maxima is 10 years.

c. The minimum population size occurs when $\sin\left(\frac{\pi}{5}\cdot(2+t)\right) = -1$
$\implies \frac{\pi}{5}\cdot(2+t) = \frac{3\pi}{2}, \frac{7\pi}{2}, \frac{11\pi}{2}, \cdot \implies t = \frac{11}{2},\ t = \frac{31}{2},\ t = \frac{51}{2}, \cdots$.
Therefore the minimum population size occurs after 5.5 years.

PRACTICE PROBLEMS FOR CHAPTER 6

1. For each of the following degree measures, find an equivalent radian measure.

(a) $135°$ (b) $-60°$ (c) $315°$ (d) $-240°$

2. Find each of the following values without the use of a calculator or a table.

(a) $\cos 225°$ (b) $\tan 150°$ (c) $\sin\left(-\frac{\pi}{6}\right)$
(d) $\sec\left(\frac{4\pi}{3}\right)$ (e) $\cot\left(\frac{7\pi}{4}\right)$ (f) $\csc(300°)$

3. Sketch the graphs of the following functions.

(a) $f(t) = -2\sin t$ (b) $f(t) = \cos 3t$
(c) $f(t) = 4\cos\left(\frac{t}{3}\right)$ (d) $f(t) = 2\sin(-2t)$
(e) $f(t) = \frac{1}{2} + \frac{1}{2}\cos 2t$

4. Find the derivative for each of the following functions.

(a) $f(x) = 2\sin^2 3x$ (b) $f(x) = \frac{1}{2}x - \frac{1}{4}\sin 2x$
(c) $f(x) = 4\cos^2\left(-\frac{3}{x} + x\right)$ (d) $y = 3\sin^3\left(\frac{t}{3}\right)\cdot\cos^2\left(\frac{t}{3}\right)$
(e) $y = \sin t + \cos(e^{\sin 2t})$ (f) $f(x) = \frac{1-x}{1-\cos x}$
(g) $y = \ln(x^3\sin x)$ (h) $f(x) = \sqrt{\frac{1}{x}\cos x^2}$

5. (a) Find the maximum height of the curve

$$y = 6\cos x - 8\sin x$$

above the x-axis.

(b) Find the maximum and minimum values of the function

$$y = 2\sin x + \cos 2x$$

for $0 \leq x \leq 2$.

(c) Show that the function

$$f(x) = x + \sin x$$

has no relative maxima or relative minima.

6. Find the derivative for each of the following functions.

(a) $f(x) = \tan(3x^2)$

(b) $f(x) = 3\sqrt{\sec(x^2+1)}$

(c) $f(x) = \sec^2(x^2) - \tan^2(x^2)$

(d) $f(x) = (\csc x + \cos x)$

(e) $\frac{dy}{dx}$ when $x + \tan(xy) = 0$

(f) $\frac{dy}{dx}$ when $x^2 = \sin y + \sin 2y$.

7. A revolving light 3 miles from a straight shoreline makes 8 revolutions per minute. Find the velocity of the beam of light along the shore at the instant it makes an angle of 45° with the shoreline.

8. Find the following antiderivatives.

(a) $\int \sin(3x)dx$

(b) $\int \tan^3(2x)\sec^2(2x)dx$

(c) $\int \frac{3}{\cos(2x)}dx$

(d) $\int x\tan\left(\frac{x^2}{2}\right)\sec\left(\frac{x^2}{2}\right)dx$

(e) $\int \frac{dx}{\sin^2\left(\frac{x}{3}\right)}dx$

(f) $\int \frac{\sec^2(x+1)}{2\tan(x+1)}dx$

(g) $\int \frac{\tan(e^{-2x})}{e^{2x}}dx$

(h) $\int \sin(2x)\tan(2x)dx$

9. Find the area of the region bounded by the graphs $y = \sec x$, $y = x$, $x = -\frac{\pi}{4}$, and $x = \frac{\pi}{4}$.

10. A man on a dock is pulling in a boat by means of a rope attached to the bow of the boat and passing through a pulley on the dock. The bow of the boat is 1 foot above the water level, and the pulley is 8 feet above the water level. Let θ be the angle that the rope makes with the horizontal. Suppose the man is pulling the rope at the rate of 2 feet/sec. Find the rate at which θ is changing at the instant when $\theta = \frac{\pi}{6}$.

11. Suppose a major league baseball pitcher releases a fastball pitch of 100 miles per hour from a height of 6 feet above ground level at a distance of 58 feet from homeplate. Suppose that if the ball was not acted on by gravity or any other force it would travel on a straight line to homeplate and reach there at a height of 4 feet above ground level. Suppose the only force acting on the ball is gravity which provides an acceleration of 32 feet per second per second.
    (a) How many seconds after the ball is released will it reach homeplate?
    (b) Find the height of the ball above ground level in feet when it reaches homeplate.

12. Suppose a person's blood pressure is given by the function

$$P(t) = 110 + 20\sin(ct),$$

where c is a constant to be determined, and where t is time in seconds.
    a. Find the person's systolic blood pressure.
    b. Find the person's diastolic blood pressure.
    c. Determine the constant c so that the heart-beat rate is 72 beats per minute.

13. Let the rabbit population as a function of time be given by

$$R(t) = 625 - 200sin(1.5t) - 50\cos(1.5t)$$

and the fox population as a function of time be given by

$$F(t) = 450 + 125\cos(1.5t) - 30\sin(1.5t),$$

where t is time measured in years.

a. Find the numbers t for which $R(t)$ has a maximum value, and find this maximum value.

b. Find the numbers t for which $R(t)$ has a minimum value, and find this minimum value.

c. Find the numbers t for which $F(t)$ has a maximum value, and find this maximum value.

d. Find the numbers t for which $F(t)$ has a minimum value, and find this minimum value.

e. Graph the two population functions on the same set of axes for $t \geq 0$.

f. Set up a table of values for t, $x = R(t)$, and $y = F(t)$. Sketch the locus of the points (x, y) so determined.

Solutions to Practice Problems for Chapter 6

1. a. $135° = 135 \cdot \frac{\pi}{180} = \frac{3}{4}\pi$ radians.

   b. $-60° = -60 \cdot \frac{\pi}{180} = -\frac{\pi}{3}$ radians.

   c. $315° = 315 \cdot \frac{\pi}{180} = \frac{7}{4}\pi$ radians.

   d. $-240° = -240 \cdot \frac{\pi}{180} = -\frac{4}{3}\pi$ radians.

2. a. $\cos 225° = -\frac{\sqrt{2}}{2}$. b. $\tan 150° = -\frac{1}{\sqrt{3}}$.

   c. $\sin\left(-\frac{\pi}{6}\right) = -\sin\frac{\pi}{6} = -\frac{1}{2}$.

   d. $\sec\left(\frac{4\pi}{3}\right) = \frac{1}{\cos\left(\frac{4}{3}\pi\right)} = \frac{1}{-.5} = -2$.

   e. $\cot\left(\frac{7\pi}{4}\right) = \frac{1}{\tan\left(\frac{7\pi}{4}\right)} = \frac{1}{-1} = -1$.

   f. $\csc(300°) = \frac{1}{\sin(300°)} = \frac{1}{-\frac{\sqrt{3}}{2}} = -\frac{2}{\sqrt{3}}$.

3. a. $f(t) = -2\sin t$

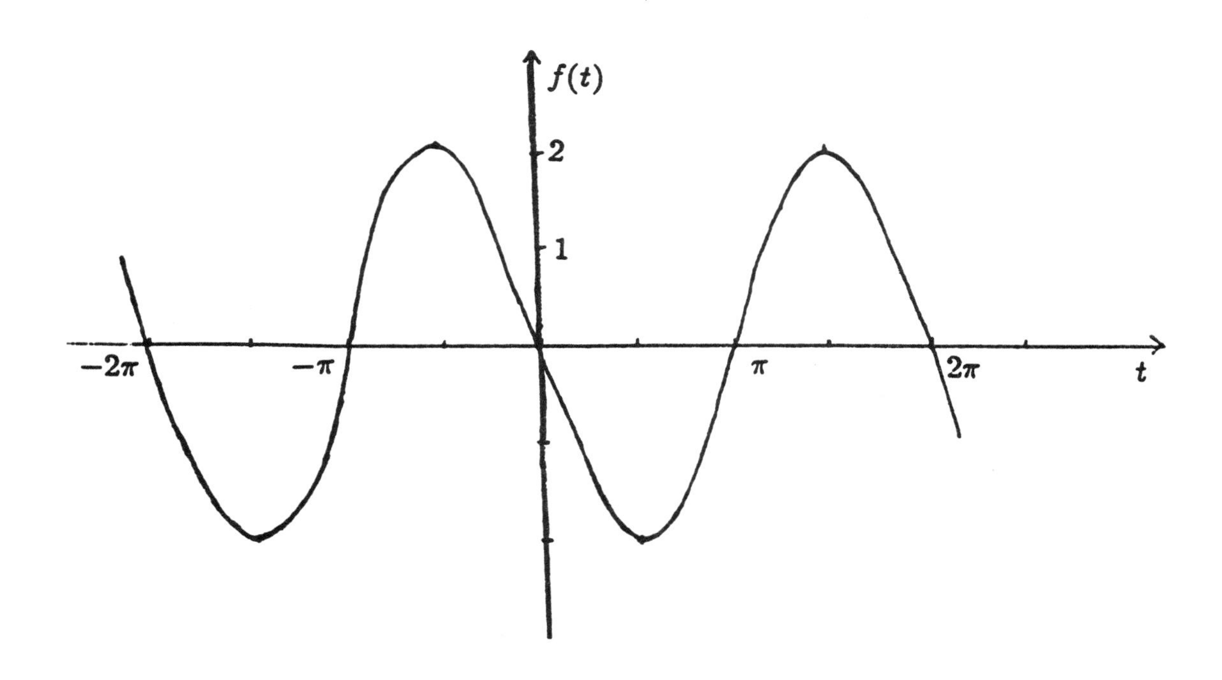

b. $f(t) = \cos 3t$

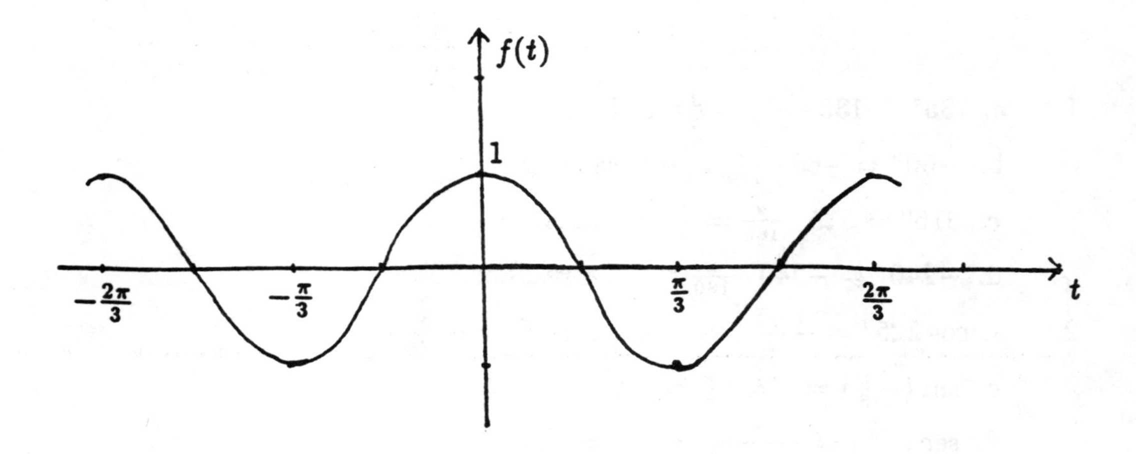

c. $f(t) = 4\cos\left(\frac{t}{3}\right)$

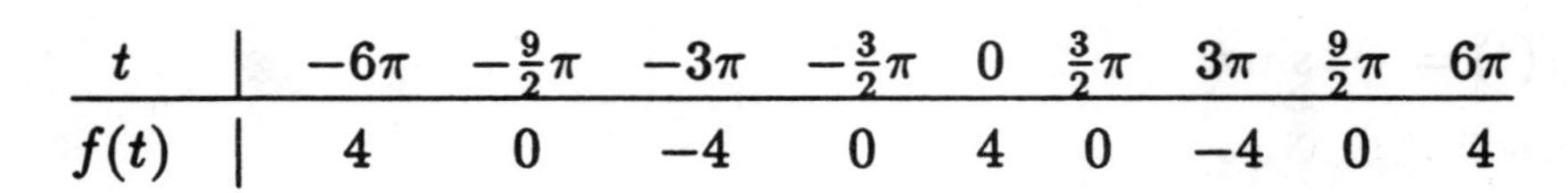

t	-6π	$-\frac{9}{2}\pi$	-3π	$-\frac{3}{2}\pi$	0	$\frac{3}{2}\pi$	3π	$\frac{9}{2}\pi$	6π
$f(t)$	4	0	−4	0	4	0	−4	0	4

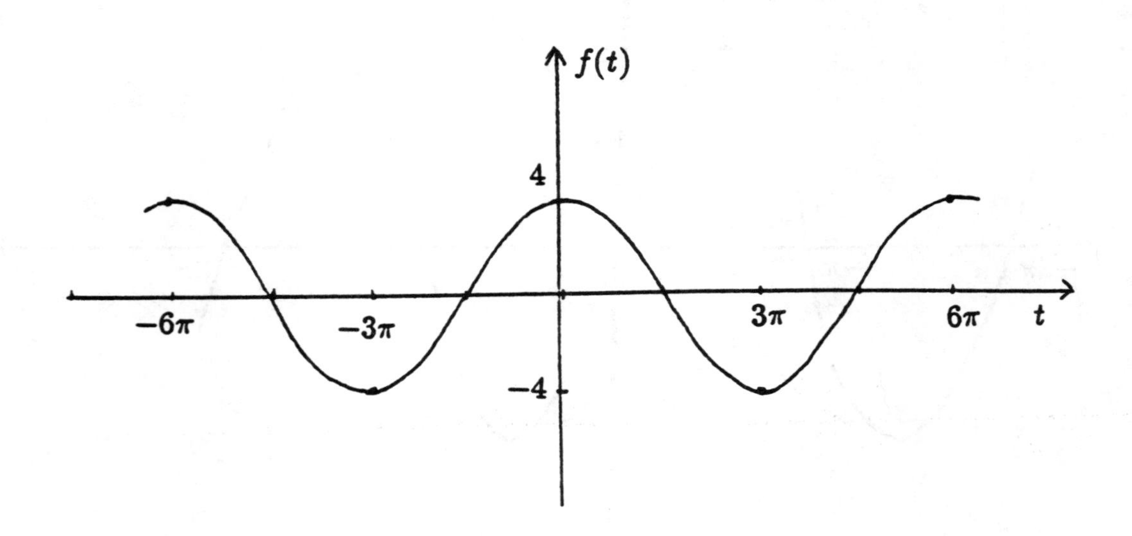

d. $f(t) = 2\sin(-2t) = -2\sin 2t$

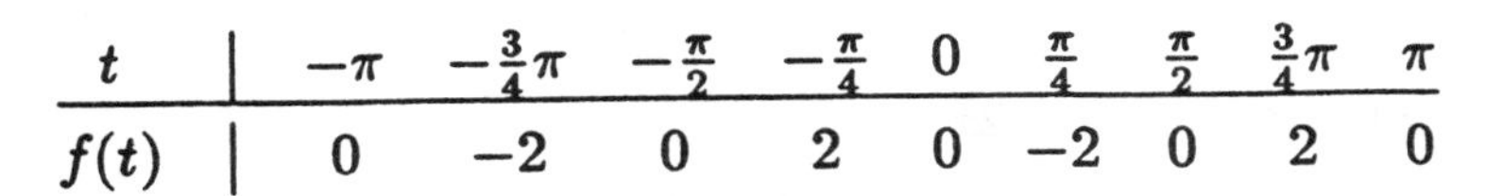

t	$-\pi$	$-\frac{3}{4}\pi$	$-\frac{\pi}{2}$	$-\frac{\pi}{4}$	0	$\frac{\pi}{4}$	$\frac{\pi}{2}$	$\frac{3}{4}\pi$	π
$f(t)$	0	-2	0	2	0	-2	0	2	0

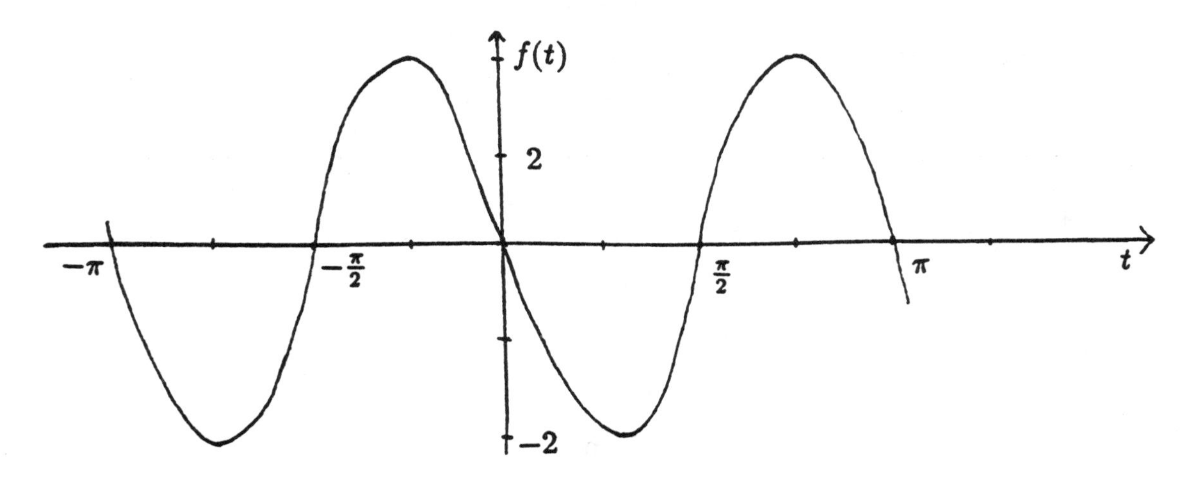

e. $f(t) = \frac{1}{2} + \frac{1}{2}\cos 2t.$

t	$-\pi$	$-\frac{3\pi}{4}$	$-\frac{\pi}{2}$	$-\frac{\pi}{4}$	0	$\frac{\pi}{4}$	$\frac{\pi}{2}$	$\frac{3}{4}\pi$	π
$f(t)$	1	$\frac{1}{2}$	0	$\frac{1}{2}$	1	$\frac{1}{2}$	0	$\frac{1}{2}$	1

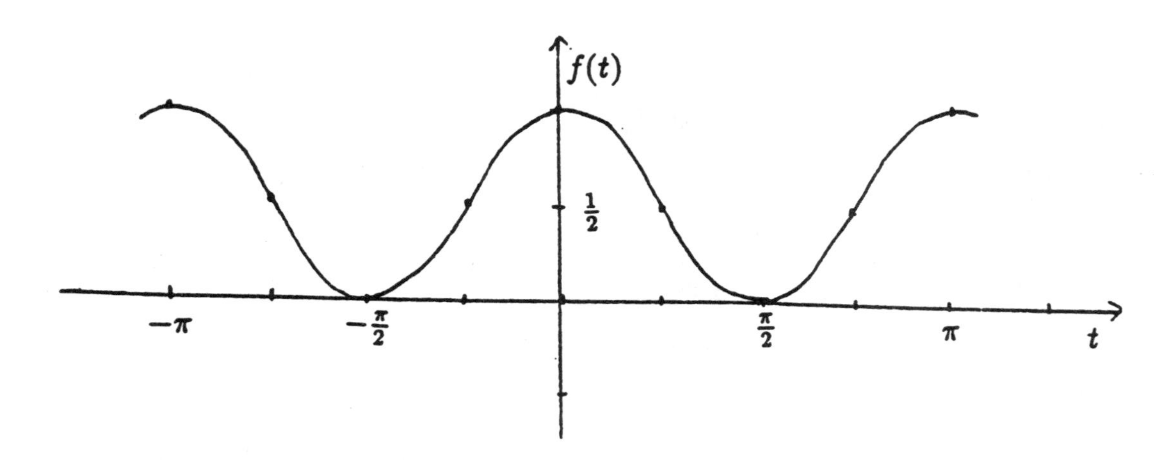

4. a. $f(x) = 2\sin^2 3x.$

$f'(x) = 4\sin 3x \cdot (\cos 3x \cdot 3) = 12\sin 3x \cdot \cos 3x = 6\sin 6x.$

b. $f(x) = \frac{1}{2}x - \frac{1}{4}\sin 2x.$

$f'(x) = \frac{1}{2} - \frac{1}{4}\cos 2x \cdot 2 = \frac{1}{2} - \frac{1}{2}\cos 2x.$

c. $f'(x) = \left[4\cos^2\left(-\frac{3}{x}+x\right)\right]' = 8\cdot\cos\left(-\frac{3}{x}+x\right)\left(-\sin\left(-\frac{3}{x}+x\right)\right)\left(\frac{3}{x^2}+1\right)$

$= -8\left(\frac{3}{x^2}+1\right)\cos\left(-\frac{3}{x}+x\right)\cdot\sin\left(-\frac{3}{x}+x\right)$

$= -4\left(\frac{3}{x^2}+1\right)\sin\left(-\frac{6}{x}+2x\right).$

d. $y = 3\sin^3\left(\frac{t}{3}\right)\cdot\cos^2\left(\frac{t}{3}\right).$

$y' = 3\cdot\left(\sin^3\left(\frac{t}{3}\right)\right)'\cdot\cos^2\left(\frac{t}{3}\right) + 3\sin^3\left(\frac{t}{3}\right)\cdot\left(\cos^2\left(\frac{t}{3}\right)\right)'$

$= 3\cdot\left[3\cdot\sin^2\left(\frac{t}{3}\right)\cdot\cos\left(\frac{t}{3}\right)\cdot\frac{1}{3}\right]\cdot\cos^2\left(\frac{t}{3}\right)$

$+3\cdot\sin^3\left(\frac{t}{3}\right)\left(\frac{2}{3}\cos\frac{t}{3}\cdot\left(-\sin\frac{t}{3}\right)\right)$

$= 3\cdot\sin^2\left(\frac{t}{3}\right)\cdot\cos^3\left(\frac{t}{3}\right) - 2\sin^4\left(\frac{t}{3}\right)\cos\frac{t}{3}.$

e. $y = \sin t + \cos(e^{\sin 2t}).$

$y' = \cos t - \sin(e^{\sin 2t})\cdot e^{\sin 2t}\cdot\cos 2t\cdot 2$

$= \cos t - 2\cos 2t\cdot\sin(e^{\sin 2t})\cdot e^{\sin 2t}.$

f. $f(x) = \frac{1-x}{1-\cos x}.$

$f'(x) = \frac{-(1-\cos x)-(1-x)\sin x}{(1-\cos x)^2} = \frac{(\cos x-1)+(x-1)\sin x}{(1-\cos x)^2}.$

g. $y = \ln(x^3\cdot\sin x).$

$y' = \frac{3x^2\cdot\sin x + x^3\cdot\cos x}{x^3\cdot\sin x} = \frac{3\sin x + x\cdot\cos x}{x\sin x}.$

h. $f(x) = \sqrt{\frac{1}{x}\cdot\cos x^2} = \left(\frac{1}{x}\cdot\cos x^2\right)^{1/2}.$

$f'(x) = \frac{1}{2}\left(\frac{1}{x}\cdot\cos x^2\right)^{-1/2}\cdot\left(-\frac{1}{x^2}\cos x^2 + \frac{1}{x}(-\sin x^2)\cdot 2x\right)$

$= \frac{1}{2}\left(\frac{1}{x}\cdot\cos x^2\right)^{-1/2}\cdot\left(-\frac{1}{x^2}\cos x^2 - 2\sin x^2\right).$

5. a. $y = 6\cos x - 8\sin x = 10\left(\frac{3}{5}\cos x - \frac{4}{5}\sin x\right).$

Since there exists a θ such that $\cos\theta = \frac{3}{5}$ and $\sin\theta = \frac{4}{5}$,

$y = 10\left(\frac{3}{5}\cos x - \frac{4}{5}\sin x\right) = 10(\cos\theta\cdot\cos x - \sin\theta\sin x)$

$= 10\cdot\cos(\theta + x).$

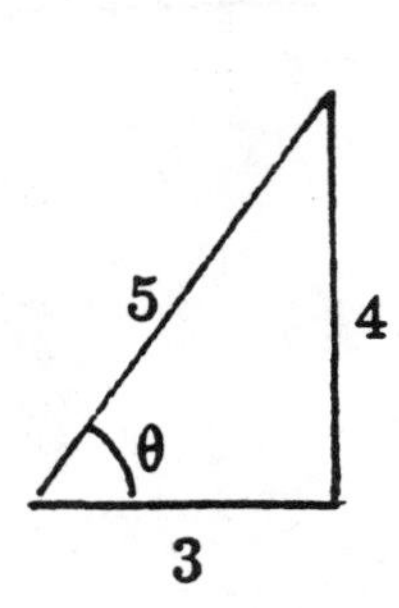

$\therefore$ The maximum height above the x-axis is 10.

b. $y = 2\sin x + \cos 2x,\ 0 \le x \le 2.$

$y' = 2\cos x - 2\sin 2x = 2\cos x - 4\sin x \cdot \cos x$

$= 2\cos x(1 - 2\sin x) = 0$ for $x = \frac{\pi}{2}, \frac{\pi}{6}$.

Checking the value of y at each critical number and endpoints gives

$$y_{x=0} = 1, \quad y_{x=2} = 2\sin 2 + \cos 4 = 1.1649,$$
$$y_{x=\frac{\pi}{2}} = 2 - 1 = 1, \quad y_{x=\frac{\pi}{6}} = 1 + \frac{1}{2} = \frac{3}{2}.$$

Hence the maximum of y is $\frac{3}{2}$ at $x = \frac{\pi}{6}$ and the minimum of y is 1 at $x = 0$ and $x = \frac{\pi}{2}$.

c. $f(x) = x + \sin x$.

Since $f'(x) = 1 + \cos x > 0$ for all $x \neq (2n+1)\pi$, $n = 0, \pm 1, \pm 2, \ldots$, $f(x)$ increases all the way along the x- axis without either a lower bound or an upper bound. Therefore, $f(x)$ has no relative maxima or relative minima.

6. a. $f(x) = \tan(3x^2)$.

$f'(x) = \left[\sec^2(3x^2)\right] \cdot 6x = 6x \cdot \sec^2(3x^2)$.

b. $f(x) = 3\sqrt{\sec(x^2+1)}$.

$$f'(x) = 3 \cdot \frac{1}{2\sqrt{\sec(x^2+1)}} \cdot \sec(x^2+1)\tan(x^2+1) \cdot 2x$$
$$= 3x \cdot \tan(x^2+1) \cdot \sqrt{\sec(x^2+1)}.$$

c. $f(x) = \sec^2(x^2) - \tan^2(x^2)$.

$$f'(x) = 2\sec(x^2) \cdot \left[\sec(x^2) \cdot \tan(x^2)\right] \cdot 2x$$
$$-2\left[\tan(x^2) \cdot \sec^2(x^2)\right] \cdot 2x = 0.$$

Another way:

Since $\sec^2 x = 1 + \tan^2 x$ for all x,

we have $f(x) = \sec^2(x^2) - \tan^2(x^2) = 1$.

So $f'(x) = 0$.

d. $f(x) = (\csc x + \cot x)$.

$f'(x) = -\csc x \cdot \cot x - \csc^2 x = -\csc x \cdot (\cot x + \csc x)$.

e. $x + \tan(xy) = 0$.

$1 + \sec^2(xy) \cdot \left(y + x\frac{dy}{dx}\right) = 0$

$\Longrightarrow\ y + x \cdot \frac{dy}{dx} = -\frac{1}{\sec^2(xy)} \Longrightarrow y + x \cdot \frac{dy}{dx} = -\cos^2(xy)$

$\Longrightarrow\ \frac{dy}{dx} = -\frac{1}{x}\left(y + \cos^2(xy)\right)$

f. $x^2 = \sin y + \sin 2y.$

$2x = (\cos y)\frac{dy}{dx} + (2\cos 2y)\frac{dy}{dx}.$

$2x = (\cos y + 2\cos 2y)\frac{dy}{dx} \Longrightarrow \frac{dy}{dx} = \frac{2x}{\cos y + 2\cos 2y}.$

7. Let x be the distance of the beam of light along the shore from the point O.

Then $x = 3\tan\theta$.

The velocity of the beam of light along the shore at any instant t is then

$$V = \frac{dx}{dt} = \frac{dx}{d\theta}\cdot\frac{d\theta}{dt} = 3\sec^2\theta\cdot\frac{d\theta}{dt}.$$

Note that $\frac{d\theta}{dt} = 8 \cdot 2\pi = 16\pi$ radians/min.

So $V = 48\pi\sec^2\theta$.

Also note that $\theta = \frac{\pi}{2} - \alpha$, so when $\alpha = 45° = \frac{\pi}{4}$, $\theta = \frac{\pi}{2} - \frac{\pi}{4} = \frac{\pi}{4}$, and

$$V = 48\pi\cdot\sec^2\left(\frac{\pi}{4}\right) = 48\pi\left(\sqrt{2}\right)^2$$

$$= 96\pi \text{ mi/min} = 1.6\pi \text{ mi/sec.}$$

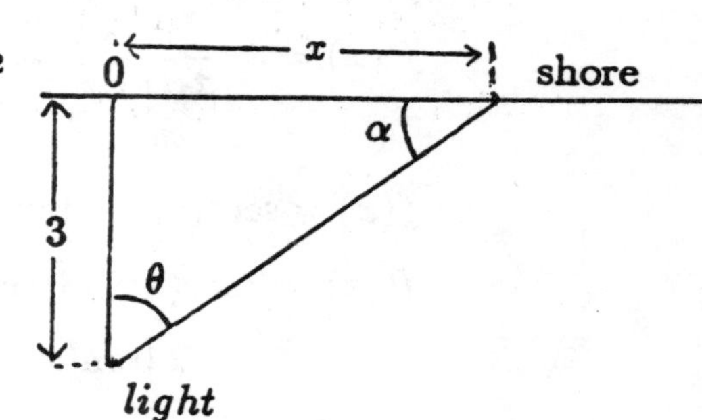

8. a. $\int \sin(3x)dx = -\frac{1}{3}\cos(3x) + C.$

b. $\int \tan^3(2x)\sec^2(2x)dx.$

Let $\tan(2x) = u$. $du = 2\cdot\sec^2(2x)dx.$

$\int \tan^3(2x)\sec^2(2x)dx = \frac{1}{2}\int u^3 du = \frac{1}{8}u^4 + C = \frac{1}{8}\tan^4(2x) + C.$

c. $\int \frac{3}{\cos(2x)}dx = \int 3\sec(2x)dx = \frac{3}{2}\ln|\sec(2x) + \tan(2x)| + C.$

d. $\int x\tan\left(\frac{x^2}{2}\right)\cdot\sec\left(\frac{x^2}{2}\right)dx.$

Let $u = \sec\left(\frac{x^2}{2}\right)$. $du = \left[\sec\left(\frac{x^2}{2}\right)\cdot\tan\left(\frac{x^2}{2}\right)\right]\cdot\frac{2x}{2}dx.$

So $\int x\tan\left(\frac{x^2}{2}\right)\cdot\sec\left(\frac{x^2}{2}\right)dx = \int du = u + C = \sec\left(\frac{x^2}{2}\right) + C.$

e. $\int \frac{dx}{\sin^2\left(\frac{x}{3}\right)} = \int \csc^2\left(\frac{x}{3}\right)dx = -3\cot\left(\frac{x}{3}\right) + C.$

f. $\int \frac{\sec^2(x+1)}{2\tan(x+1)}dx.$

Let $u = \tan(x+1)$. $\quad du = \sec^2(x+1)dx.$

$\int \frac{\sec^2(x+1)}{2\tan(x+1)}dx = \int \frac{du}{2u} = \frac{1}{2}\ln|u| + C = \frac{1}{2}\ln|\tan(x+1)| + C.$

g. $\int \frac{\tan(e^{-2x})}{e^{2x}}dx = \int \tan(e^{-2x}) \cdot e^{-2x}dx.$

Let $u = e^{-2x}$. $\quad du = -2e^{-2x}dx.$

$$\int \frac{\tan(e^{-2x})}{e^{2x}}dx = -\frac{1}{2}\int \tan u \; du = -\frac{1}{2}\ln|\sec u| + C$$
$$= -\frac{1}{2}\ln|\sec(e^{-2x})| + C.$$

h. $\int \sin(2x)\cdot\tan(2x)dx = \int \sin(2x)\cdot\frac{\sin(2x)}{\cos(2x)}dx$

$$= \int \frac{\sin^2(2x)}{\cos(2x)}dx = \int \frac{1-\cos^2(2x)}{\cos(2x)}dx$$
$$= \int \left[\frac{1}{\cos(2x)} - \frac{\cos^2(2x)}{\cos(2x)}\right]dx = \int [\sec(2x) - \cos(2x)]\,dx$$
$$= \frac{1}{2}\ln|\sec(2x) + \tan(2x)| - \frac{1}{2}\sin(2x) + C$$

9. Area of the region $= \int_{-\frac{\pi}{4}}^{\frac{\pi}{4}} (\sec x - x)dx$

$$= \left(\ln|\sec x + \tan x| - \frac{x^2}{2}\right)\Big|_{-\frac{\pi}{4}}^{\frac{\pi}{4}}$$
$$= \left[\ln\left|\sec\left(\frac{\pi}{4}\right) + \tan\left(\frac{\pi}{4}\right)\right| - \frac{\left(\frac{\pi}{4}\right)^2}{2}\right] - \left[\ln\left|\sec\left(-\frac{\pi}{4}\right) + \tan\left(-\frac{\pi}{4}\right)\right| - \frac{\left(-\frac{\pi}{4}\right)^2}{2}\right]$$
$$= \ln\left(\sqrt{2}+1\right) - \ln\left(\sqrt{2}-1\right) = \ln\left(\frac{\sqrt{2}+1}{\sqrt{2}-1}\right) = \ln\left(3 + 2\sqrt{2}\right).$$

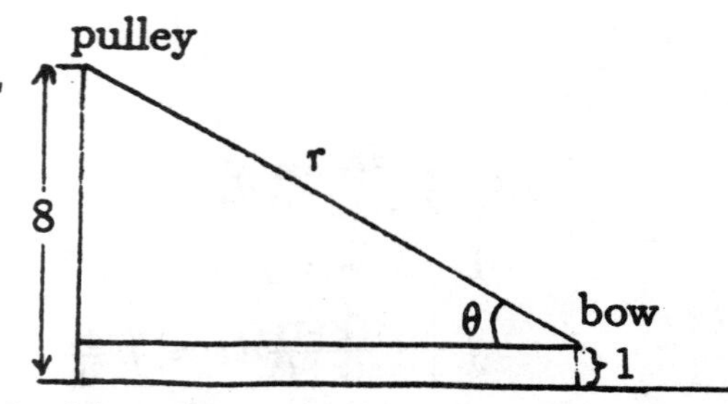

10. Let r be the length of the segment of the rope between the bow and the pulley. Then we have

$$\sin\theta = \frac{8-1}{r} = \frac{7}{r},$$

and to find the changing rate $\frac{d\theta}{dt}$, differentiating both sides gives

$$\cos\theta \cdot \frac{d\theta}{dt} = -\frac{7}{r^2} \cdot \frac{dr}{dt} \implies \frac{d\theta}{dt} = -\frac{7}{r^2\cos\theta} \cdot \frac{dr}{dt}.$$

When $\theta = \frac{\pi}{6}$, $r = 7/\sin\left(\frac{\pi}{6}\right) = 7/\frac{1}{2} = 14$, and note that $\frac{dr}{dt} = 2$. We have

$$\frac{d\theta}{dt} = -\frac{7}{14^2\cos\left(\frac{\pi}{6}\right)} \cdot 2 = -\frac{1}{14 \cdot \frac{\sqrt{3}}{2}} = -.082 \text{ radians/sec} \approx -4.7°/\text{sec}.$$

So the rate at which θ is changing when $\theta = \frac{\pi}{6}$ is 4.7° per second.

11. Let $f(t)$ be the horizontal distance in feet the ball has traveled t seconds after it was released until it reaches home plate, and let $g(t)$ be the height in feet of the ball above ground level t seconds after it was released until it reaches home plate.

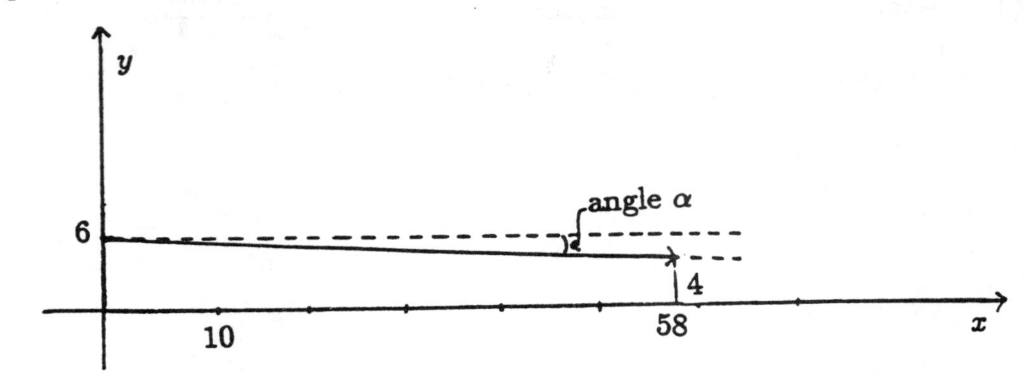

$\tan\alpha = -\frac{2}{58}$.

$\sin\alpha = \frac{-2}{\sqrt{2^2+58^2}} = -.03446, \quad \cos\alpha = \frac{58}{\sqrt{2^2+58^2}} = .9994.$

$f''(t) = 0, \quad g''(t) = -32.$

The initial speed is 100 miles per hour which is 146.66667 feet per second which we denote by v_0.

$$f'(0) = v_0 \cos\alpha, \qquad g'(0) = v_0 \sin\alpha.$$

$$f'(t) = v_0 \cos\alpha, \qquad g'(t) = v_0 \sin\alpha - 32t.$$

$$f(0) = 0, \qquad g(0) = 6.$$

$$f(t) = (v_0 \cos\alpha)t, \qquad g(t) = (v_0 \sin\alpha)t - 16t^2 + 6.$$

a. Let t_1 be the number of seconds it takes the ball to reach home plate.

$$t_1 = \frac{58}{v_0 \cos\alpha} = 0.3957 \text{ seconds}.$$

b. The height of the ball in feet above ground level when it reaches home plate is

$$(v_0 \sin\alpha)t_1 - 16t_1^2 + 6 = 1.4948 \text{ feet}.$$

12. $P(t) = 110 + 2sin(ct)$.

a. The maximum (systolic) blood pressure occurs when $\sin(ct) = 1$. Thus the person's systolic blood pressure is 130.

b. The minimum (diastolic) blood pressure occurs when $\sin(ct) = -1$. Thus the person's diastolic blood pressure is 90.

c. The period $T = \frac{2\pi}{c}$ seconds. Thus the heart-beat rate is $\frac{c}{2\pi}$ beats per second. Therefore the heart-beat is $\frac{c}{2\pi} \cdot 60$ beats per minute.

$$\frac{c}{2\pi} \cdot 60 = 72 \implies c = \frac{72 \cdot 2\pi}{60} = 7.54.$$

13. a. $R(t) = 625 - 200\sin(1.5t) - 50\cos(1.5t)$.

$$R'(t) = -200 \cdot \cos(1.5t) \cdot (1.5) + 50 \cdot \sin(1.5t) \cdot (1.5)$$
$$= 75 \cdot (-4\cos(1.5t) + \sin(1.5t)).$$

$$R'(t) = 0 \implies \sin(1.5t) = 4\cos(1.5t) \implies \tan(1.5t) = 4.$$

By using a table or the $\tan^{-1}$ button on a scientific calculator, we can find that $\tan(1.325817664) = 4$. Thus $\tan\theta = 4$ for $\theta = 1.325817664 + n \cdot \pi$. Therefore $\tan(1.5t) = 4$ for $t = 0.883878442 + n \cdot \frac{\pi}{1.5}$.

$R''(t) = 75 \cdot (4 \cdot \sin(1.5t) \cdot (1.5) + \cos(1.5t) \cdot (1.5))$

$= 112.5 \cdot (4\sin(1.5t) + \cos(1.5t))$.

$R''(0.883878442) = 463.85 > 0$.

$R''\left(0.883878442 + \frac{\pi}{1.5}\right) = -463.85 < 0$.

Thus the maximum value of $R(t)$ occurs when $t = 0.883878442 + (2n + 1) \cdot \frac{\pi}{1.5}$ years. This maximum value is 807 rabbits.

b. The minimum value of $R(t)$ occurs when $t = 0.883878442 + (2n) \cdot \frac{\pi}{1.5}$ years. This minimum value is 419 rabbits.

c. $F(t) = 450 + 125 \cdot \cos(1.5t) - 30\sin(1.5t)$.

$F'(t) = -125 \cdot \sin(1.5t) \cdot (1.5) - 30 \cdot \cos(1.5t) \cdot (1.5)$

$= -7.5 \cdot (25\sin(1.5t) + 6\cos(1.5t))$.

$F'(t) = 0 \implies 25 \cdot \sin(1.5t) = -6 \cdot \cos(1.5t) \implies \tan(1.5t)$

$= -\frac{6}{25} = -0.24$.

By using a table or the tan^{-1} button on a scientific calculator, we find that $tan(-0.23554498) = -0.24$. Thus, $\tan\theta = -0.24$ for $\theta = -0.23554498 + n \cdot \pi$. Therefore $\tan(1.5t) = -0.24$ for $t = 1.937365115 + n \cdot \frac{\pi}{1.5}$.

$F''(t) = -7.5 \cdot (25 \cdot \cos(1.5t) \cdot (1.5) - 6 \cdot \sin(1.5t) \cdot (1.5))$

$= 11.25 \cdot (-25\cos(1.5t) + 6\sin(1.5t))$.

$F''(1.937365115) = 289.24 > 0$.

$F''\left(1.837365115 + \frac{\pi}{1.5}\right) = -289.24 < 0$.

Thus the maximum value of $F(t)$ occurs when

$t = 1.937365115 + (2n + 1) \cdot \frac{\pi}{1.5}$ years. This maximum value is 579 foxes.

d. The minimum value of $F(t)$ occurs when $t = 1.937365115 + (2n) \cdot \frac{\pi}{1.5}$ years.

This minimum value is 321 foxes.

e.

y=R(t)=625−(200sin(1.5t))−(50cos(1.5t))
y=F(t)=450+(125cos(1.5t))−(30sin(1.5t))
900
800
700
600
500
400
300
y
0
1
2
3
4
5
6
7

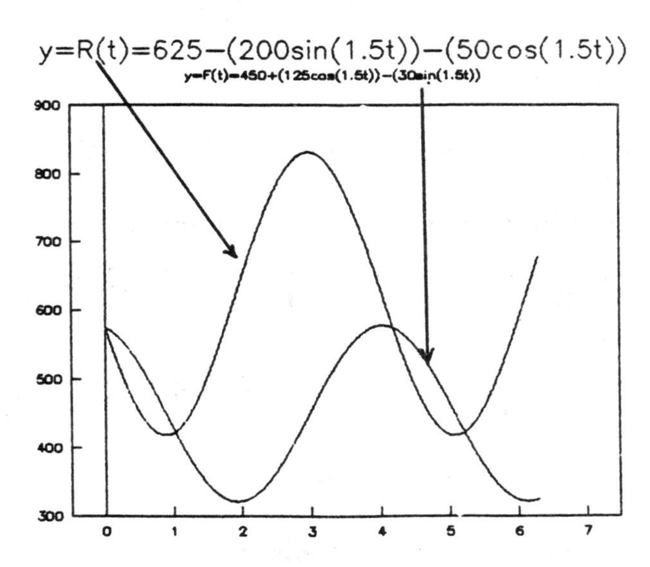

f.

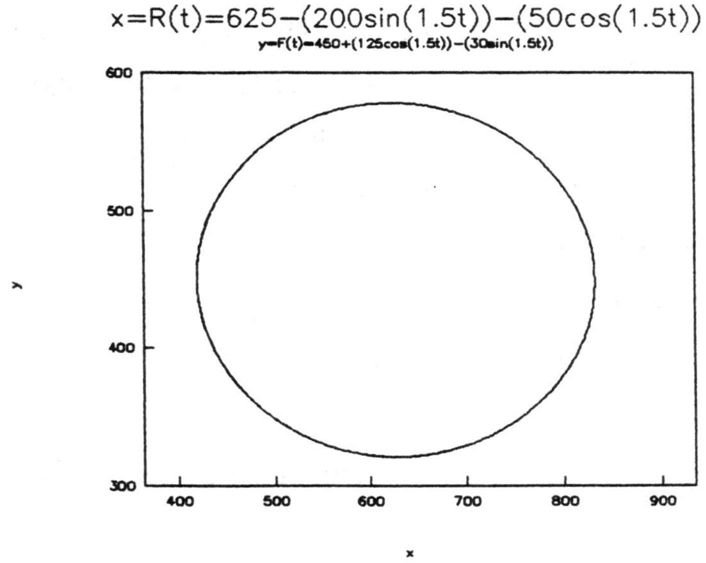

x=R(t)=625−(200sin(1.5t))−(50cos(1.5t))
y=F(t)=450+(125cos(1.5t))−(30sin(1.5t))
600
500
400
300
y
400
500
600
700
800
900
x

CHAPTER 7

Summary of Chapter 7

1. Review of all developed integration rules and the method of substitution.

2. Integration by parts for integrating the product of two functions.

3. Use of integral formulas shown in the front papers in finding antiderivatives for more general types of functions.

4. The definitions for type I and type II improper integrals.

5. Convergence and divergence of improper integrals.

6. The Trapezoidal Rule and Simpson's Rule for approximating definite integrals.

7. Error estimations in Trapezoidal Rule and Simpson's Rule approximations.

Selected Solutions to Exercise Set 7.1

1. $\int \cos x\sqrt{4+\sin x}\, dx$.

   Let $u(x) = 4 + \sin x$. $\qquad du = \cos x\, dx$.

   $\int \cos x\sqrt{4+\sin x}\, dx = \int \sqrt{u}\, du = \frac{2}{3}u^{3/2} + C = \frac{2}{3}(4+\sin x)^{3/2} + C.$

3. $\int_3^4 (x^2-3)\sqrt{x^3-9x}\, dx$.

   Let $u(x) = x^3 - 9x$. $\qquad du = (3x^2-9)dx = 3(x^2-3)dx$.

   $\int (x^2-3)\sqrt{x^3-9x}\, dx = \frac{1}{3}\int \sqrt{u}\, du = \frac{1}{3}\cdot\frac{2}{3}u^{3/2} + C = \frac{2}{9}(x^3-9x)^{3/2} + C.$

   $\int_3^4 (x^2-3)\sqrt{x^3-9x}\, dx = \frac{2}{9}(x^3-9x)^{3/2}\Big]_3^4$

   $= \frac{2}{9}(4^3 - 9\cdot 4)^{3/2} - \frac{2}{9}(3^3 - 9\cdot 3)^{3/2} = \frac{2}{9}\cdot 28^{3/2}.$

5. $\int_0^1 xe^{x^2}\, dx$

   Let $u(x) = x^2$. $\qquad du = 2x\, dx, \qquad x\, dx = \frac{1}{2}\, du.$

   $\int xe^{x^2}\, dx = \frac{1}{2}\int e^u\, du = \frac{1}{2}e^u + C = \frac{1}{2}e^{x^2} + C.$

   $\int_0^1 xe^{x^2}\, dx = \frac{1}{2}e^{x^2}\Big]_0^1 = \frac{1}{2}(e^1 - e^0) = \frac{1}{2}(e-1).$

7. $\int \frac{x^4}{1+x^5}\, dx$.

   Let $u(x) = 1 + x^5$. $\qquad du = 5x^4\, dx$.

   $\int \frac{x^4}{1+x^5}\, dx = \frac{1}{5}\int \frac{1}{u}\, du = \frac{1}{5}\ln|u| + C = \frac{1}{5}\ln|1+x^5| + C.$

9. $\int \frac{e^x}{1+e^x}\, dx$.

   Let $u(x) = 1 + e^x$. $\qquad du = e^x\, dx$.

   $\int \frac{e^x}{1+e^x}\, dx = \int \frac{1}{u}\, du = \ln|u| + C = \ln(1+e^x) + C.$

11. $\int_0^{\pi/4} \tan^3 x\cdot\sec^2 x\, dx$.

    Let $u(x) = \tan x$. $\qquad du = \sec^2 x\, dx$.

    $\int \tan^3 x\cdot \sec^2 x\, dx = \int u^3\, du = \frac{1}{4}u^4 + C = \frac{1}{4}\tan^4 x + C.$

    $\int_0^{\pi/4} \tan^3 x\cdot\sec^2 x\, dx = \frac{1}{4}\tan^4 x\Big]_0^{\pi/4} = \frac{1}{4}\left(\tan^4\frac{\pi}{4} - \tan^4 0\right) = \frac{1}{4}(1^4 - 0) = \frac{1}{4}.$

13. $\int (x+3)(x^2+6x+4)^3\,dx$.

Let $u(x) = x^2+6x+4$. $\quad du = (2x+6)dx = 2(x+3)dx$.

$\int (x+3)(x^2+6x+4)^3\,dx = \frac{1}{2}\int u^3\,du = \frac{1}{8}u^4 + C = \frac{1}{8}(x^2+6x+4)^4 + C$.

15. $\int \sec x \cdot \tan x \cdot e^{\sec x}\,dx$.

Let $u(x) = \sec x$. $\quad du = \sec x \cdot \tan x\,dx$.

$\int \sec x \cdot \tan x \cdot e^{\sec x}\,dx = \int e^u\,du = e^u + C = e^{\sec x} + C$.

17. $\int_1^4 \frac{e^{\sqrt{x}}}{\sqrt{x}}\,dx$.

Let $u(x) = e^{\sqrt{x}}$. $\quad du = \frac{1}{2\sqrt{x}}e^{\sqrt{x}}\,dx$.

$\int \frac{e^{\sqrt{x}}}{\sqrt{x}}\,dx = 2\int du = 2u + C = 2e^{\sqrt{x}} + C$.

$\int_0^4 \frac{e^{\sqrt{x}}}{\sqrt{x}}\,dx = 2\,e^{\sqrt{x}}\Big]_1^4 = 2\left(e^{\sqrt{4}} - e^{\sqrt{1}}\right) = 2(e^2 - e)$.

19. $\int \sin^3 x \cos x\,dx$.

Let $u(x) = \sin x$. $\quad du = \cos x\,dx$.

$\int \sin^3 x \cos x\,dx = \int u^3\,du = \frac{1}{4}u^4 + C = \frac{1}{4}\sin^4 x + C$.

21. $\int \frac{1}{\cos^2 x}\,dx = \int \sec^2 x\,dx = \tan x + C$.

23. $\int_1^2 \frac{dx}{x^2+2x+1} = \int_1^2 \frac{1}{(x+1)^2}dx = -\frac{1}{x+1}\Big]_1^2 = -\frac{1}{2+1} - \left(-\frac{1}{1+1}\right) = \frac{1}{6}$.

25. $\int e^x\left(\tan(1+e^x)\right)dx$

Let $u(x) = 1 + e^x$. $\quad du = e^x\,dx$.

$\int e^x \tan(1+e^x)dx = \int \tan u\,du = \ln|\sec u| + C = \ln|\sec(1+e^x)| + C$.

27. $\int \frac{\cos x}{\sqrt{1-\sin^2 x}}\,dx = \int \frac{\cos x}{\sqrt{\cos^2 x}}\,dx = \int \frac{\cos x}{|\cos x|}\,dx$

$$= \begin{cases} \int 1\,dx & 2n\pi - \frac{\pi}{2} \le x \le 2n\pi + \frac{\pi}{2} \\ -\int 1\,dx & 2n\pi + \frac{\pi}{2} \le x \le 2n\pi + \frac{3\pi}{2} \end{cases}$$

$$= \begin{cases} x + C & 2n\pi - \frac{\pi}{2} \le x \le 2n\pi + \frac{\pi}{2} \\ -x + C & 2n\pi + \frac{\pi}{2} \le x \le 2n\pi + \frac{3\pi}{2} \end{cases},$$

where n is an integer.

29. $\int_e^{e^2} \frac{1}{x\ln x}\,dx$

Let $u(x) = \ln x$. $\quad du = \frac{1}{x}\,dx$.

$\int \frac{1}{x\ln x}\,dx = \int \frac{1}{u}\,du = \ln|u| + C = \ln|\ln x| + C$.

$\int_e^{e^2} \frac{1}{x\ln x}\,dx = \ln\ln x\big]_e^{e^2} = \ln\ln e^2 - \ln\ln e = \ln 2 - \ln 1 = \ln 2$.

31. $\int \frac{1}{\sqrt{x}(1+\sqrt{x})}\,dx$

Let $u(x) = 1 + \sqrt{x}$. $\quad du = \frac{1}{2}\cdot\frac{1}{\sqrt{x}}\,dx$.

$\int \frac{1}{\sqrt{x}(1+\sqrt{x})}\,dx = 2\int \frac{1}{u}\,du = 2\ln|u| + C = 2\ln(1+\sqrt{x}) + C$.

33. $\int_3^5 x\sqrt{x^2-9}\,dx$.

Let $u(x) = x^2 - 9$. $\quad du = 2x\,dx \implies x\,dx = \frac{1}{2}\,du$.

$\int x\sqrt{x^2-9}\,dx = \int u^{1/2}\left(\frac{1}{2}\,du\right) = \frac{1}{2}\cdot\frac{2}{3}u^{3/2} + C = \frac{1}{3}(x^2-9)^{3/2} + C$.

$\int_3^5 x\sqrt{x^2-9}\,dx = \frac{1}{3}(x^2-9)^{3/2}\big]_3^5 = \frac{1}{3}\cdot\left(16^{3/2} - 0\right) = \frac{64}{3}$.

35. $\int_0^2 x\sqrt{1+x^2}\,dx$.

Let $u(x) = 1 + x^2$. $\quad du = 2x\,dx \implies x\,dx = \frac{1}{2}\,du$.

$\int x\sqrt{1+x^2}\,dx = \int u^{1/2}\left(\frac{1}{2}\,du\right) = \frac{1}{2}\cdot\frac{2}{3}u^{3/2} + C = \frac{1}{3}(1+x^2)^{3/2} + C$.

$\int_0^2 x\sqrt{1+x^2}\,dx = \frac{1}{3}(1+x^2)^{3/2}\big]_0^2 = \frac{1}{3}\cdot(5^{3/2} - 1)$.

37. $\int_0^{\sqrt{\pi}/2} t\sin(\pi - t^2)dt$.

Let $u(t) = \pi - t^2$. $\quad du = -2t\,dt \implies t\,dt = -\frac{1}{2}\,du$.

$\int t\sin(\pi - t^2)dt = \int \sin u\left(-\frac{1}{2}\,du\right) = -\frac{1}{2}\cdot(-\cos u) + C = \frac{1}{2}\cos(\pi - t^2) + C$.

$$\int_0^{\sqrt{\pi}/2} t\sin(\pi - t^2)dt = \frac{1}{2}\cos(\pi - t^2)\big]_0^{\sqrt{\pi}/2} = \frac{1}{2}\cdot\left(\cos\frac{3\pi}{4} - \cos\pi\right)$$

$$= \frac{1}{2}\cdot\left(-\frac{\sqrt{2}}{2} + 1\right).$$

39. $\int_0^{\pi/4} \tan^3 x\sec^2 x\,dx$.

Let $u(x) = \tan x$. $\quad du = \sec^2 x\,dx$.

$\int_0^{\pi/4} \tan^3 x\sec^2 x\,dx = \frac{1}{4}\tan^4 x\big]_0^{\pi/4} = \frac{1}{4}$.

41. $\int_1^4 \frac{(4+\sqrt{x})^2}{\sqrt{x}}\,dx.$

Let $u = 4+\sqrt{x}$. $\quad du = \frac{1}{2\sqrt{x}}\,dx \implies \frac{1}{\sqrt{x}}\,dx = 2\,du.$

$\int \frac{(4+\sqrt{x})^2}{\sqrt{x}}\,dx = \int u^2(2\,du) = \frac{2}{3}u^3 + C = \frac{2}{3}\cdot(4+\sqrt{x})^3 + C.$

$\int_1^4 \frac{(4+\sqrt{x})^2}{\sqrt{x}}\,dx = \frac{2}{3}(4+\sqrt{x})^3\Big]_1^4 = \frac{2}{3}\cdot(6^3-5^3) = \frac{182}{3}.$

43. Area of $R = \int_0^{\sqrt{\pi}/2} x\sin x^2\,dx.$

Let $u(x) = x^2$. $\quad du = 2x\,dx.$

$\int x\sin x^2\,dx = \frac{1}{2}\int \sin u\,du = -\frac{1}{2}\cos u + C = -\frac{1}{2}\cos x^2 + C.$

Area of $R = \int_0^{\sqrt{\pi}/2} x\sin x^2\,dx = -\frac{1}{2}\cos x^2\Big]_0^{\sqrt{\pi}/2} = -\frac{1}{2}\left(\cos\left(\frac{\sqrt{\pi}}{2}\right)^2 - \cos 0\right)$

$= -\frac{1}{2}\left(\frac{\sqrt{2}}{2} - 1\right) = -\frac{1}{4}\left(\sqrt{2}-2\right).$

45. Consumers' surplus $= \int_0^{16} \frac{20000}{20+5x}\,dx - (16)(200)$

$= 20000\cdot\frac{1}{5}\ln(20+5x)\Big]_0^{16} - 3200$

$= 4000\left(\ln(20+5\cdot16) - \ln(20+5\cdot0)\right) - 3200$

$= 4000(\ln 100 - \ln 20) - 3200$

$= 4000\ln\left(\frac{100}{20}\right) - 3200 = -3200 + 4000\ln 5.$

47. Let $u(t) = \sin\pi t$. $\quad du = \cos\pi t\cdot\pi\,dt \implies \cos\pi t\,dt = \frac{1}{\pi}\,du.$

$\int R(t)dt = 4,000t + 2,000\cdot\int u\cdot\frac{1}{\pi}\,du = 4,000t + \frac{2,000}{\pi}\cdot\frac{1}{2}u^2 + C$

$= 4,000t + \frac{1,000}{\pi}\cdot\sin^2\pi t + C.$

$\int_0^4 R(t)dt = \left(4,000t + \frac{1,000}{\pi}\cdot\sin^2\pi t\right)\Big]_0^4 = \$16,000.$

49. Let $u(x) = 2+x^4$. $\quad du = 4x^3\,dx \implies x^3\,dx = \frac{1}{4}\,du.$

$\int \frac{x^3}{2+x^4}\,dx = \int \frac{1}{u}\left(\frac{1}{4}\,du\right) = \frac{1}{4}\ln|u| + C = \frac{1}{4}\ln(2+x^4) + C.$

Average value of $f(x)$ on the interval $[0,2]$ is given by

$$\frac{\int_0^2 f(x)dx}{2-0} = \frac{1}{2}\cdot\frac{1}{4}\ln(2+x^4)\Big]_0^2 = \frac{1}{8}\cdot(\ln 18 - \ln 2) = \frac{1}{8}\ln 9.$$

Selected Solutions to Exercise Set 7.2

1. $\int xe^{-x}\,dx$.

   Let $u = x$, $\quad dv = e^{-x}\,dx$.

   Then $du = dx$, $\quad v = -e^{-x}$.

   $\int xe^{-x}\,dx = -xe^{-x} - \int(-e^{-x})dx = -xe^{-x} + (-1)e^{-x} + C = -e^{-x}(x+1) + C$.

3. $\int x\ln x\,dx$.

   Let $u = \ln x$, $\quad dv = x\,dx$. $\quad du = \frac{1}{x}\,dx$, $\quad v = \frac{1}{2}x^2$.

   $\int x\ln x\,dx = \frac{1}{2}x^2\ln x - \int \frac{1}{2}x^2 \cdot \frac{1}{x}\,dx = \frac{1}{2}x^2\ln x - \frac{1}{4}x^2 + C = \frac{1}{4}x^2(2\ln x - 1) + C$.

5. $\int_0^3 x\sqrt{x+1}\,dx$.

   Let $u = x$, $\quad dv = \sqrt{x+1}\,dx$. $\quad du = dx$, $\quad v = \frac{2}{3}(x+1)^{3/2}$.

   $$\int x\sqrt{x+1}\,dx = \tfrac{2}{3}x(x+1)^{3/2} - \int \tfrac{2}{3}(x+1)^{3/2}dx = \tfrac{2}{3}x(x+1)^{3/2} - \tfrac{2}{3}\cdot\tfrac{2}{5}(x+1)^{5/2} + C$$
   $$= \tfrac{2}{15}(x+1)^{3/2}\left[5x - 2(x+1)\right] + C = \tfrac{2}{15}(3x-2)(x+1)^{3/2} + C.$$

   $$\int_0^3 x\sqrt{x+1}\,dx = \tfrac{2}{15}(3x-2)(x+1)^{3/2}\Big]_0^3$$
   $$= \tfrac{2}{15}\left[(3\cdot 3 - 2)(3+1)^{3/2} - (0-2)(0+1)^{3/2}\right] = \tfrac{2}{15}(7\cdot 8 + 2)$$
   $$= \tfrac{116}{15}.$$

7. $\int x\sin x\,dx$.

   Let $u = x$, $\quad dv = \sin x\,dx$. $\quad du = dx$, $\quad v = -\cos x$.

   $\int x\sin x\,dx = -x\cos x - \int(-\cos x)dx = -x\cos x + \sin x + C$.

9. $\int x(2x+1)^5 dx$.

   Let $u = x$, $\quad dv = (2x+1)^5 dx$.

   $du = dx$, $\quad v = \frac{1}{2}\cdot\frac{1}{6}(2x+1)^6 = \frac{1}{12}(2x+1)^6$.

   $$\int x(2x+1)^5 dx = \tfrac{1}{12}x(2x+1)^6 - \int \tfrac{1}{12}(2x+1)^6 dx$$
   $$= \tfrac{1}{12}x(2x+1)^6 - \tfrac{1}{12}\cdot\tfrac{1}{2}\cdot\tfrac{1}{7}(2x+1)^7 + C.$$
   $$= \tfrac{1}{12\cdot 14}(2x+1)^6(14x - 2x - 1) + C = \tfrac{1}{168}(12x-1)(2x+1)^6 + C.$$

11. $\int x^2 e^x \, dx$.

Let $u = x^2$, $dv = e^x \, dx$. $du = 2x \, dx$, $v = e^x$.

$\int x^2 e^x \, dx = x^2 e^x - \int 2x e^x \, dx = x^2 e^x - 2 \int x e^x \, dx$.

In $\int x e^x \, dx$, let $u_1 = x$, $dv_1 = e^x \, dx$. $du_1 = dx$, $v_1 = e^x$.

$\int x e^x \, dx = x e^x - \int e^x \, dx = x e^x - e^x + C_1 = e^x(x-1) + C_1$.

Thus, $\int x^2 e^x \, dx = x^2 e^x - 2\left[e^x(x-1) + C_1\right] = e^x(x^2 - 2x + 2) + C$.

13. $\int_0^{\pi/4} x \sec^2 x \, dx$.

Let $u = x$, $dv = \sec^2 x \, dx$. $du = dx$, $v = \tan x$.

$\int x \sec^2 x \, dx = x \tan x - \int \tan x \, dx = x \tan x - \ln|\sec x| + C$.

$$\int_0^{\pi/4} x \sec^2 x \, dx = (x \tan x - \ln \sec x)\Big]_0^{\pi/4} = \left(\tfrac{\pi}{4} \tan \tfrac{\pi}{4} - \ln \sec \tfrac{\pi}{4}\right) - (0 - \ln \sec 0)$$

$$= \left(\tfrac{\pi}{4} \cdot 1 - \ln \sqrt{2}\right) + \ln 1 = \tfrac{\pi}{4} - \tfrac{1}{2} \ln 2 + 0$$

$$= \tfrac{1}{4}(\pi - 2 \ln 2).$$

15. $\int_0^1 x^2 e^{-x} \, dx$.

Let $u = x^2$, $dv = e^{-x} \, dx$. $du = 2x \, dx$, $v = -e^{-x}$.

$\int x^2 e^{-x} \, dx = -x^2 e^{-x} - \int (-2x e^{-x}) dx = -x^2 e^{-x} + 2 \int x e^{-x} \, dx$.

In $\int x e^{-x} \, dx$, let $u_1 = x$, $dv_1 = e^{-x} \, dx$. $du_1 = dx$, $v_1 = -e^{-x}$.

$\int x e^{-x} dx = -x e^{-x} - \int (-e^{-x}) dx = -x e^{-x} - e^{-x} + C_1$.

Thus $\int x^2 e^{-x} \, dx = -x^2 e^{-x} + 2(-x e^{-x} - e^{-x} + C_1) = -e^{-x}(x^2 + 2x + 2) + C$.

$\int_0^1 x^2 e^{-x} \, dx = -e^{-x}(x^2 + 2x + 2)\Big]_0^1 = -e^{-1}(1 + 2 + 2) + e^0 \cdot 2 = 2 - \frac{5}{e}$.

17. $\int_0^1 \frac{x^3}{\sqrt{1+x^2}} \, dx$.

Let $u = x^2$, $dv = \frac{x}{\sqrt{1+x^2}} \, dx$. $du = 2x \, dx$, $v = \sqrt{1 + x^2}$.

$$\int \frac{x}{\sqrt{1+x^2}} \, dx = x^2 \cdot \sqrt{1+x^2} - \int 2x \cdot \sqrt{1+x^2} \, dx = x^2 \cdot \sqrt{1+x^2} - \tfrac{2}{3}(1+x^2)^{3/2} + C$$

$$= \tfrac{1}{3}\sqrt{1+x^2} \cdot (3x^2 - 2 - 2x^2) + C = \tfrac{1}{3}\sqrt{1+x^2} \cdot (x^2 - 2) + C.$$

$\int_0^1 \frac{x^3}{\sqrt{1+x^2}}\,dx = \frac{1}{3}\sqrt{1+x^2}\,(x^2-2)\Big]_0^1 = \frac{1}{3}\left(-\sqrt{2}+2\right) = 0.1953.$

19. $\int e^{2x}\cos x\,dx.$

Let $u(x) = e^{2x}$, $dv = \cos x\,dx$. $du = 2e^{2x}\,dx$, $v = \sin x$.

$\int e^{2x}\cos x\,dx = e^{2x}\sin x - 2\int e^{2x}\sin x\,dx.$

In $\int e^{2x}\sin x\,dx$, let $u_1 = e^{2x}$, $dv_1 = \sin x\,dx$. $du_1 = 2e^{2x}\,dx$, $v_1 = -\cos x$.

$\int e^{2x}\sin x\,dx = -e^{2x}\cos x + 2\int e^{2x}\cos x\,dx.$

$$\begin{aligned}\int e^{2x}\cos x\,dx &= e^{2x}\sin x - 2\left(-e^{2x}\cos x + 2\int e^{2x}\cos x\,dx\right)\\ &= e^{2x}\sin x + 2e^{2x}\cos x - 4\int e^{2x}\cos x\,dx.\end{aligned}$$

$5\int e^{2x}\cos x\,dx = e^{2x}(\sin x + 2\cos x).$

$\int e^{2x}\cos x\,dx = \frac{1}{5}e^{2x}(\sin x + 2\cos x) + C.$

21. $\int x^3 e^{-x^2}\,dx.$

Let $u(x) = -x^2$. $du = -2x\,dx$, and $x^2 = -u$.

$\int x^3 e^{-x^2}\,dx = \int x^2 e^{-x^2} x\,dx = \frac{1}{2}\int ue^u du.$

In $\int ue^u\,du$, let $w = u$, $dv = e^u\,du$. $dw = du$, $v = e^u$.

$\int ue^u\,du = ue^u - \int e^u\,du = ue^u - e^u + C_1.$

$$\begin{aligned}\text{So } \int x^3 e^{-x^2}\,dx &= \tfrac{1}{2}(ue^u - e^u + C_1) = \tfrac{1}{2}\left(-x^2e^{-x^2} - e^{-x^2} + C_1\right)\\ &= -\tfrac{1}{2}e^{-x^2}(x^2+1) + C.\end{aligned}$$

23. $$\begin{aligned}\text{Area of } R &= \int_0^1 (x - \ln(1+x))\,dx = \int_0^1 x\,dx - \int_0^1 \ln(1+x)dx\\ &= \tfrac{1}{2}x^2\Big]_0^1 - \int_0^1 \ln(1+x)dx = \tfrac{1}{2}(1-0) - (1+x)\cdot(\ln(1+x)-1)\Big]_0^1\\ &= \tfrac{1}{2} - (1+1)\left[\ln(1+1)-1\right] + (1+0)\left[\ln(1+0)-1\right]\\ &= \tfrac{1}{2} - 2(\ln 2 - 1) + (0-1) = \tfrac{3}{2} - 2\ln 2.\end{aligned}$$

25. a. $$\begin{aligned}\text{Present value} &= \int_0^5 (e^{-.10t})\cdot(1000t)dt\\ &= 1000\cdot\left(-\tfrac{1}{.10}te^{-.10t} - \tfrac{1}{.01}e^{-.10t}\right)\Big]_0^5\\ &= 1000\cdot\left(-50e^{-.5} - 100e^{-.5} + 100\right) = \$9{,}020.40.\end{aligned}$$

b. $$\begin{aligned}\text{Value of the total contributions} &= \int_0^5 1000t\,dt.\\ &= 500t^2\Big]_0^5 = \$12{,}500.\end{aligned}$$

27. Average of $f = \frac{1}{e-1}\int_1^e \ln x \, dx = \frac{1}{e-1} \cdot x(\ln x - 1)\Big]_1^e$
$= \frac{1}{e-1}\left[e(\ln e - 1) - 1(\ln 1 - 1)\right] = \frac{1}{e-1}.$

29. Total earnings $= \int_0^5 10,000te^{-.2t}\, dt$
$= 10,000 \cdot \left(-\frac{1}{.2}te^{-.2t} - \frac{1}{.04}e^{-.2t}\right)\Big]_0^5$
$= 10,000 \cdot \left(-\frac{1}{.2} \cdot 5 \cdot e^{-1} - \frac{1}{.04}e^{-1} + \frac{1}{.04}\right)$
$= 10,000 \cdot \left(-50e^{-1} + 25\right) = \$66,060.28.$

31. Distance traveled $= \int_0^{10} v(t)dt = \int_0^{10} 2te^{0.5t}\, dt$
$= 2 \cdot \left(\frac{1}{.5}te^{.5t} - \frac{1}{.25}e^{.5t}\right)\Big]_0^{10} = 2 \cdot \left(16e^5 + 4\right) = 4,757.2211.$

Selected Solutions to Exercise Set 7.3

1. $\int (4+3x)^5\, dx = \frac{1}{3} \cdot \frac{1}{6}(4+3x)^6 + C = \frac{1}{18}(4+3x)^6 + C.$

3. $\int \frac{x}{(5+x)^2}\, dx = \frac{1}{1^2}\left[\ln|5+x| + \frac{5}{5+x}\right] + C = \ln|5+x| + \frac{5}{5+x} + C.$

5. $\int x\sqrt{7+2x}\, dx = -\frac{2(2\cdot 7 - 3\cdot 2x)(7+2x)^{3/2}}{15\cdot 2^2} + C$
$= -\frac{(7-3x)(7+2x)^{3/2}}{15} + C.$

7. $\int \frac{x}{\sqrt{3+2x}}\, dx = -\frac{2(2\cdot 3 - 2x)}{3\cdot 2^2}\sqrt{3+2x} + C = -\frac{(3-x)\sqrt{3+2x}}{3} + C.$

9. $\int \frac{dx}{\sqrt{16+x^2}} = \ln\left|x + \sqrt{16+x^2}\right| + C.$

11. $\int \frac{x\, dx}{(5+x^2)^2} = \int \frac{\frac{1}{2}d(x^2)}{(5+x^2)^2} = \frac{1}{2}\left(-\frac{1}{1(5+x^2)} + C_1\right) = \frac{-1}{2(5+x^2)} + C.$

13. $\int \frac{dx}{9-16x^2} = \frac{1}{16}\int \frac{dx}{\frac{9}{16}-x^2} = \frac{1}{16}\left(\frac{1}{2\cdot\frac{3}{4}}\ln\left|\frac{\frac{3}{4}+x}{\frac{3}{4}-x}\right| + C_1\right)$
$= \frac{1}{24}\ln\left|\frac{3+4x}{3-4x}\right| + C.$

15. $\int \frac{dx}{x\sqrt{x^4+9}} = \int \frac{x}{x^2\sqrt{x^4+9}}\, dx.$ Let $u = x^2$.
$\int \frac{dx}{x\sqrt{x^4+9}} = \frac{1}{2}\int \frac{du}{u\sqrt{u^2+9}} = \frac{1}{2}\left[-\frac{1}{3}\ln\left(\frac{3+\sqrt{u^2+9}}{u}\right) + C_1\right]$
$= -\frac{1}{6}\ln\left(\frac{3+\sqrt{x^4+9}}{x^2}\right) + C.$

17. $\int \frac{4x}{1+e^{x^2}}\, dx.$ Let $u = x^2$. $du = 2x\, dx$.
$\int \frac{4x}{1+e^{x^2}}\, dx = \int \frac{2}{1+e^u}\, du = 2\left[\ln\left(\frac{e^u}{1+e^u}\right) + C_1\right] = 2\ln\left(\frac{e^{x^2}}{1+e^{x^2}}\right) + C.$

19. $\int \frac{dx}{\sqrt{x}(1+e^{\sqrt{x}})}$. Let $u=\sqrt{x}$. $du = \frac{1}{2}\cdot\frac{1}{\sqrt{x}}\,dx$.

$\int \frac{dx}{\sqrt{x}(1+e^{\sqrt{x}})} = 2\int \frac{du}{1+e^u} = 2\left[\ln\left(\frac{e^u}{1+e^u}\right)+C_1\right] = 2\ln\left(\frac{e^{\sqrt{x}}}{1+e^{\sqrt{x}}}\right)+C.$

21. $\int x\sin^3(x^2)dx$. Let $u=x^2$. $du = 2x\,dx$.

$\int x\sin^3(x^2)dx = \frac{1}{2}\int \sin^3 u\,du = \frac{1}{2}\left[-\frac{1}{3}\cos u(\sin^2 u+2)+C_1\right]$

$= -\frac{1}{6}\cos x^2\left(\sin^2(x^2)+2\right)+C.$

23. Let $u=\cos x$. $du = -\sin x\,dx$.

$$\int \frac{\sin x}{\cos x(5-2\cos x)}\,dx = \int \frac{-du}{u\cdot(5-2u)} = \frac{1}{5}\ln\left|\frac{5-2u}{u}\right|+C$$
$$= \frac{1}{5}\ln\left|\frac{5-2\cos x}{\cos x}\right|+C.$$

25. Let $u=\sec x$. $du = \sec x\tan x\,dx$.

$$\int \sqrt{\sec^2 x+9}\sec x\tan x\,dx$$
$$= \int \sqrt{u^2+9}\,du$$
$$= \frac{1}{2}\left[u\sqrt{u^2+9}+9\ln\left(u+\sqrt{u^2+9}\right)\right]+C$$
$$= \frac{1}{2}\left[\sec x\cdot\sqrt{\sec^2 x+9}+9\ln\left(\sec x+\sqrt{\sec^2 x+9}\right)\right]+C.$$

27. Let $u=\sqrt{x}$. $du = \frac{1}{2\sqrt{x}}\,dx \implies \frac{1}{4\sqrt{x}}\,dx = \frac{1}{2}\,du.$

$$\int \frac{1}{4\sqrt{x}\,(1+e^{\sqrt{x}})}\,dx = \frac{1}{2}\int \frac{1}{1+e^u}\,du = \frac{1}{2}\ln\left(\frac{e^u}{1+e^u}\right)+C$$
$$= \frac{1}{2}\ln\left(\frac{e^{\sqrt{x}}}{1+e^{\sqrt{x}}}\right)+C.$$

29. When $p_0=20$, $20 = \frac{100}{1+e^{2x_0}} \implies 20+20e^{2x_0} = 100 \implies e^{2x_0}=4 \implies$

$2x_0 = \ln 4 \implies x_0 = \frac{1}{2}\ln 4 = \ln 4^{1/2} = \ln 2.$

$$\text{Consumer's surplus} = \int_0^{\ln 2} \frac{100}{1+e^{2x}}\,dx - 20\cdot\ln 2$$
$$= 50\ln\left(\frac{e^{2x}}{1+e^{2x}}\right)\Big]_0^{\ln 2} - 20\cdot\ln 2$$
$$= 50\left(\ln\left(\frac{4}{1+4}\right) - \ln\left(\frac{1}{1+1}\right)\right) - 20\cdot\ln 2$$
$$= 50\left(\ln\tfrac{4}{5} - \ln\tfrac{1}{2}\right) - 20\cdot\ln 2$$
$$= 50\ln 1.6 - 20\cdot\ln 2$$
$$= 9.6372.$$

31. $\text{Area} = \int_{-1}^{1} [f(x)-g(x)]\,dx = 2\cdot\int_{-1}^{1}\frac{1}{9-x^2}\,dx = 2\cdot\frac{1}{6}\ln\left|\frac{x+3}{x-3}\right|\Big]_{-1}^{1}$
$= \frac{1}{3}\left(\ln 2 - \ln\frac{1}{2}\right) = \frac{1}{3}\left(\ln 2 - (\ln 1 - \ln 2)\right) = \frac{2}{3}\ln 2 = 0.4621.$

Selected Solutions to Exercise Set 7.4

1. $\int_2^{\infty}\frac{1}{x^2}\,dx = \lim_{t\to\infty}\int_2^t \frac{1}{x^2}\,dx = \lim_{t\to\infty}\left(-\frac{1}{x}\right)\Big]_2^t$
$= \lim_{t\to\infty}\left(-\frac{1}{t}+\frac{1}{2}\right) = \frac{1}{2}$. It converges to $\frac{1}{2}$.

3. $\int_0^{\infty} e^{-x}\,dx = \lim_{t\to\infty}\int_0^t e^{-x}\,dx = \lim_{t\to\infty}(-e^{-x})\Big]_0^t$
$= \lim_{t\to\infty}(-e^{-t}+1) = (-0+1) = 1$. It converges to 1.

5. $\int_1^{\infty}\frac{1}{\sqrt{x}}\,dx = \lim_{t\to\infty}\int_1^t \frac{1}{\sqrt{x}}\,dx = \lim_{t\to\infty}\left(2\sqrt{x}\Big]_1^t\right)$
$= \lim_{t\to\infty}(2\sqrt{t}-2) = +\infty$. Diverges.

7. $\int_0^{\infty}\frac{x}{\sqrt{4+x^2}}\,dx = \lim_{t\to\infty}\int_0^t \frac{x}{\sqrt{4+x^2}}\,dx = \lim_{t\to\infty}\int_0^t \frac{\frac{1}{2}}{\sqrt{4+x^2}}\,dx^2 = \lim_{t\to\infty}\left(\sqrt{4+x^2}\Big]_0^t\right)$
$= \lim_{t\to\infty}(\sqrt{4+t^2}-2) = +\infty$. Diverges.

9. $\int_0^{\infty}\frac{1}{x+2}\,dx = \lim_{t\to\infty}\int_0^t \frac{1}{x+2}\,dx = \lim_{t\to\infty}(\ln|x+2|)]_0^t$
$= \lim_{t\to\infty}(\ln|t+2| - \ln 2) = +\infty$. Diverges.

11. $\int_{-\infty}^{-2} \frac{1}{x^2}\,dx = \lim_{t\to-\infty} \int_t^{-2} \frac{1}{x^2}\,dx = \lim_{t\to-\infty} \left(-\frac{1}{x}\Big]_t^{-2}\right) = \lim_{t\to-\infty} \left(\frac{1}{2}+\frac{1}{t}\right) = \frac{1}{2}.$
It converges to $\frac{1}{2}$.

13. $\int_{-\infty}^{0} \frac{1}{\sqrt{1-x}}\,dx = \lim_{t\to-\infty} \int_t^{0} \frac{1}{\sqrt{1-x}}\,dx = \lim_{t\to-\infty} \left(-2\sqrt{1-x}\right)\Big]_t^0$

$= \lim_{t\to-\infty} \left(-2\cdot\left(1-\sqrt{1-t}\right)\right) = +\infty$. Diverges.

15. $\int_0^1 \frac{1}{\sqrt{1-x}}\,dx = \lim_{t\to1^-} \int_0^t \frac{1}{\sqrt{1-x}}\,dx = \lim_{t\to1^-} \left(-2\sqrt{1-x}\right)\Big]_0^t$

$= \lim_{t\to1^-} \left(-2\left(\sqrt{1-t}-1\right)\right) = 2$. It converges to 2.

17. $\int_0^1 \frac{1}{x^2}\,dx = \lim_{t\to0+} \int_t^1 \frac{1}{x^2}\,dx = \lim_{t\to0+} \left(-\frac{1}{x}\right)\Big]_t^1 = \lim_{t\to0+} \left(-1+\frac{1}{t}\right) = +\infty.$
Diverges.

19. $\int_1^e \frac{1}{x\ln x}\,dx = \lim_{t\to1+} \int_t^e \frac{1}{x\ln x}\,dx = \lim_{t\to1+} \left(\ln|\ln x|\right)\Big]_t^e.$

$= \lim_{t\to1+} \left(\ln\ln e - \ln|\ln t|\right) = +\infty.$ Diverges.

21. $\int_1^3 \frac{1}{x^2-2x+1}\,dx = \lim_{t\to1+} \int_t^3 \frac{dx}{(x-1)^2} = \lim_{t\to1+} \left(-\frac{1}{(x-1)}\right)\Big]_t^3$

$= \lim_{t\to1+} \left(-\frac{1}{2}+\frac{1}{t-1}\right) = +\infty.$ Diverges.

23. $P_0 = \lim_{T\to\infty} \int_0^T 100e^{-0.10t}\,dt = \lim_{T\to\infty} \left(-\frac{100}{0.10}e^{-0.10t}\right)\Big]_0^T$
$= \lim_{T\to\infty} 1000\left(1-e^{-0.10T}\right) = \$1000.$

25. a. $P_0 = \int_0^{10} 1e^{-0.10t}\,dt = -\frac{1}{0.10}e^{-0.10t}\Big]_0^{10} = 10\cdot\left(1-\frac{1}{e}\right)$ million dollars.

b. $P_0 = \lim_{T\to\infty} \int_0^T 1e^{-0.10t}\,dt = \lim_{T\to\infty} \left(-\frac{1}{0.10}e^{-0.10t}\right)\Big]_0^T = \lim_{T\to\infty} 10\left(1-e^{-0.10T}\right)$
$= 10$ million dollars.

27. Volume $= \int_0^\infty \pi(e^{-x})^2 dx = \pi\int_0^\infty e^{-2x}\,dx = \pi\cdot\lim_{t\to\infty} \left(-\frac{1}{2}e^{-2x}\right)\Big]_0^t$
$= \frac{\pi}{2}\lim_{t\to\infty} \left(1-e^{-2t}\right) = \frac{\pi}{2}.$

29. Case 1: $p = 1$.

$\int_t^1 \frac{1}{x}\,dx = \ln x]_t^1 = -\ln t \to \infty$ as $t \to 0^+$.

Case 2: $p \neq 1$.

$\int_t^1 \frac{1}{x^p}\,dx = \int_t^1 x^{-p}\,dx = \frac{1}{-p+1}x^{-p+1}\Big]_t^1 = \frac{1}{-p+1} - \frac{1}{-p+1}t^{-p+1}$.

$\lim_{t\to 0+} \frac{1}{-p+1}t^{-p+1}$ is finite if and only if $-p+1 > 0 \implies p < 1$.

Therefore the improper integral $\int_0^1 \frac{1}{x^p}\,dx$ converges if and only if $p < 1$.

31. $P_0 = \lim_{T\to\infty} \int_0^T 5{,}000e^{-.10T}\,dt = \lim_{T\to\infty} 50{,}000(1 - e^{-.10t}) = \$50{,}000$.

33. $\int_0^2 \frac{1}{x-2}\,dx = \lim_{t\to 2-} \ln|x-2|\Big]_0^t = -\infty$.

Therefore $\int_0^6 \frac{1}{x-2}\,dx$ diverges.

35. $\int_0^2 \frac{1}{(x-2)^{2/3}}\,dx = 3 \cdot \lim_{t\to 2-} (x-2)^{1/3}\Big]_0^t = 3 \cdot (0 - (-2)^{1/3}) = 3 \cdot \sqrt[3]{2}$.

$\int_2^4 \frac{1}{(x-2)^{2/3}}\,dx = 3 \cdot \lim_{t\to 2+} (x-2)^{1/3}\Big]_t^4 = 3 \cdot ((2)^{1/3} - 0) = 3 \cdot \sqrt[3]{2}$.

Therefore the improper integral $\int_0^4 \frac{1}{(x-2)^{2/3}}\,dx$ converges and has the value $3\left(\sqrt[3]{2} + \sqrt[3]{2}\right) = 6\sqrt[3]{2}$.

37. $\int_0^1 \frac{1}{x^2-2x+1}\,dx = \int_0^1 \frac{1}{(x-1)^2}\,dx = \lim_{t\to 1^-} \left(-\frac{1}{x-1}\right)\Big]_0^t = \infty$.

Therefore $\int_0^2 \frac{1}{x^2-2x+1}\,dx$ diverges.

Selected Solutions to Exercise Set 7.5

1. $f(x) = (x^3 - 7x + 4)$. Since $n = 4$, one obtains

$$\int_0^4 f(x)dx \approx \frac{4-0}{2\cdot 4}\left(f(0) + 2f(1) + 2f(2) + 2f(3) + f(4)\right)$$

$$= \tfrac{1}{2}\left(4 + 2(-2) + 2(-2) + 2(10) + 40\right) = \tfrac{1}{2}(56) = 28.$$

3. Put $f(x) = \sin x$. Since $n = 4$, one obtains

$$\int_0^\pi f(x)dx \approx \tfrac{\pi-0}{2\cdot 4}\left(f(0) + 2f\left(\tfrac{\pi}{4}\right) + 2f\left(\tfrac{\pi}{2}\right) + 2f\left(\tfrac{3\pi}{4}\right) + f(\pi)\right)$$
$$= \tfrac{\pi}{8}\left(0 + 2\cdot\tfrac{\sqrt{2}}{2} + 2\cdot 1 + 2\cdot\tfrac{\sqrt{2}}{2} + 0\right)$$
$$= \tfrac{\pi}{8}\left(2 + 2\sqrt{2}\right) = \tfrac{\pi}{4}\left(1 + \sqrt{2}\right) \approx 1.8961.$$

5. Put $f(x) = \frac{4}{1+x^2}$. Since $n = 8$, one obtains

$$\int_0^4 f(x)dx \approx \frac{4-0}{2\cdot 8}\cdot\left(f(0) + 2f\left(\frac{1}{2}\right) + 2f(1) + 2f\left(\frac{3}{2}\right) + 2f(2)\right.$$
$$\left. + 2f\left(\frac{5}{2}\right) + 2f(3) + 2f\left(\frac{7}{2}\right) + f(4)\right)$$
$$= \frac{1}{4}\cdot\left(4 + 2\cdot\frac{16}{5} + 2\cdot 2 + 2\cdot\frac{16}{13} + 2\cdot\frac{4}{5} + 2\cdot\frac{16}{29} + 2\cdot\frac{2}{5} + 2\cdot\frac{16}{53} + \frac{4}{17}\right)$$
$$= \frac{1}{4}\cdot\left(8 + \frac{44}{5} + \frac{32}{13} + \frac{32}{29} + \frac{32}{53} + \frac{4}{17}\right)$$
$$\approx 5.3010.$$

7. Put $f(x) = \cos\left(\frac{\pi x}{4}\right)$. Since $n = 4$, one obtains

$$\int_0^4 f(x)dx \approx \tfrac{4-0}{2\cdot 4}\left(f(0) + 2f(1) + 2f(2) + 2f(3) + f(4)\right)$$
$$= \tfrac{1}{2}\left(1 + 2\cdot\tfrac{\sqrt{2}}{2} + 2\cdot 0 + 2\cdot\left(-\tfrac{\sqrt{2}}{2}\right) + (-1)\right)$$
$$= \tfrac{1}{2}\left(1 + \sqrt{2} - \sqrt{2} - 1\right) = 0.$$

9. Put $f(x) = \frac{1}{1-x^2}$. Since $n = 8$, one obtains

$$\int_2^4 f(x)dx \approx \tfrac{4-2}{2\cdot 8}\left(f(2) + 2f\left(\tfrac{9}{4}\right) + 2f\left(\tfrac{5}{2}\right) + 2f\left(\tfrac{11}{4}\right)\right.$$
$$\left. + 2f(3) + 2f\left(\tfrac{13}{4}\right) + 2f\left(\tfrac{7}{2}\right) + 2f\left(\tfrac{15}{4}\right) + f(4)\right)$$
$$= \tfrac{1}{8}\left(-\tfrac{1}{3} + 2\left(-\tfrac{16}{65}\right) + 2\left(-\tfrac{4}{21}\right) + 2\left(-\tfrac{16}{105}\right) + 2\left(-\tfrac{1}{8}\right)\right.$$
$$\left. + 2\left(-\tfrac{16}{153}\right) + 2\left(-\tfrac{4}{45}\right) + 2\left(-\tfrac{16}{209}\right) + \left(-\tfrac{1}{15}\right)\right) \approx -0.296.$$

11. $\int_0^4 f(x)dx \approx \frac{4-0}{3\cdot 4}\left(f(0)+4f(1)+2f(2)+4f(3)+f(4)\right)$

$= \frac{1}{3}\left(4+4(-2)+2(-2)+4\cdot 10+40\right)$

$= \frac{1}{3}(72) = 24.$

13. $\int_0^\pi f(x)dx \approx \frac{\pi-0}{3\cdot 4}\left(f(0)+4f\left(\frac{\pi}{4}\right)+2f\left(\frac{\pi}{2}\right)+4f\left(\frac{3\pi}{4}\right)+f(\pi)\right)$

$= \frac{\pi}{12}\left(0+4\cdot\frac{\sqrt{2}}{2}+2\cdot 1+4\cdot\frac{\sqrt{2}}{2}+0\right)$

$= \frac{\pi}{12}\left(2+4\sqrt{2}\right) = \frac{\pi}{6}\left(1+2\sqrt{2}\right) \approx 2.005.$

15. $\int_0^4 f(x)dx \approx \frac{4-0}{3\cdot 8}\cdot\left(f(0)+4\cdot f\left(\frac{1}{2}\right)+2\cdot f(1)+4\cdot f\left(\frac{3}{2}\right)+2\cdot f(2)\right.$

$\left.+4\cdot f\left(\frac{5}{2}\right)+2\cdot f(3)+4\cdot f\left(\frac{7}{3}\right)+f(4)\right)$

$= \frac{1}{6}\cdot\left(4+4\cdot\frac{16}{5}+2\cdot 2+4\cdot\frac{16}{13}+2\cdot\frac{4}{5}+4\cdot\frac{16}{29}\right.$

$\left.+2\cdot\frac{2}{5}+4\cdot\frac{16}{53}+\frac{4}{17}\right)$

$= \frac{1}{6}\cdot\left(8+\frac{76}{5}+\frac{64}{13}+\frac{64}{29}+\frac{64}{53}+\frac{4}{17}\right)$

$\approx 5.2955.$

17. $\int_0^4 f(x)dx \approx \frac{4-0}{3\cdot 4}\left(f(0)+4f(1)+2f(2)+4f(3)+f(4)\right)$

$= \frac{1}{3}\left\{1+4\left(\frac{1}{2}\sqrt{2}\right)+2(0)+4\left(-\frac{1}{2}\sqrt{2}\right)+(-1)\right\}$

$= \frac{1}{3}(0) = 0.$

19. $\int_2^4 f(x)dx \approx \frac{4-2}{3\cdot 8}\left(f(2)+4f\left(\frac{9}{4}\right)+2f\left(\frac{10}{4}\right)+4f\left(\frac{11}{4}\right)+2f(3)+4f\left(\frac{13}{4}\right)\right.$

$\left.+2f\left(\frac{14}{4}\right)+4f\left(\frac{15}{4}\right)+f(4)\right)$

$= \frac{1}{12}\left\{-\frac{1}{3}+4\left(-\frac{16}{65}\right)+2\left(-\frac{16}{84}\right)+4\left(-\frac{16}{105}\right)+2\left(-\frac{1}{8}\right)+4\left(-\frac{16}{153}\right)\right.$

$\left.+2\left(-\frac{16}{180}\right)+4\left(-\frac{16}{209}\right)+\left(-\frac{1}{15}\right)\right\}$

$= \frac{1}{12}(-3.5274) \approx -0.2939.$

21. $\int_0^{\pi/4} \frac{2}{1+x^2}\,dx \approx \frac{\frac{\pi}{4}-0}{2\cdot 6}\left(f(0)+2\cdot f\left(\frac{\pi}{24}\right)+2\cdot f\left(\frac{\pi}{12}\right)+2\cdot f\left(\frac{\pi}{8}\right)\right.$

$\left.+\,2\cdot f\left(\frac{\pi}{6}\right)+2\cdot f\left(\frac{5\pi}{24}\right)+f\left(\frac{\pi}{4}\right)\right)$

$= \frac{\pi}{48}\cdot(2+2\cdot(1.966308)+2\cdot(1.871715)+2\cdot(1.732783)$

$+\,2\cdot(1.569667)+2\cdot(1.400200)+1.236973)$

$\approx 1.3298.$

23. Put $f(x)=\sqrt{x^2+2}$. Since $n=6$, one obtains

$\int_1^3 f(x)dx \approx \frac{3-1}{2\cdot 6}\left(f(1)+2f\left(\frac{4}{3}\right)+2f\left(\frac{5}{3}\right)+2f(2)+2f\left(\frac{7}{3}\right)+2f\left(\frac{8}{3}\right)+f(3)\right)$

$= \frac{1}{6}\left\{\sqrt{3}+2\left(\sqrt{\frac{34}{9}}\right)+2\left(\sqrt{\frac{43}{9}}\right)+2\left(\sqrt{6}\right)+2\left(\sqrt{\frac{67}{9}}\right)+2\left(\sqrt{\frac{82}{9}}\right)+\left(\sqrt{11}\right)\right\}$

$= \frac{1}{6}(29.70041) \approx 4.9501.$

25. $f(x)=x\sin\left(\frac{\pi x}{2}\right)$.

$\int_0^2 x\sin\left(\frac{\pi x}{2}\right)dx \approx \frac{2-0}{2\cdot 6}\left(f(0)+2f\left(\frac{1}{3}\right)+2f\left(\frac{2}{3}\right)+2f(1)+2f\left(\frac{4}{3}\right)+2f\left(\frac{5}{3}\right)+f(2)\right)$

$= \frac{1}{6}\left(0+2\left(\frac{1}{3}\right)\left(\frac{1}{2}\right)+2\left(\frac{2}{3}\right)\left(\frac{\sqrt{3}}{2}\right)+2(1)(1)+2\left(\frac{4}{3}\right)\left(\frac{\sqrt{3}}{2}\right)\right.$

$\left.+2\left(\frac{5}{3}\right)\left(\frac{1}{2}\right)+0\right)$

$= \frac{1}{6}\left(\frac{1}{3}+\frac{2}{3}\sqrt{3}+2+\frac{4}{3}\sqrt{3}+\frac{5}{3}\right) = \frac{1}{6}\left(4+2\sqrt{3}\right) \approx 1.2440.$

27. $\int_0^{\pi/4} \frac{2}{1+x^2}\,dx \approx \frac{\frac{\pi}{4}-0}{3\cdot 6}\cdot\left(f(0)+4\cdot f\left(\frac{\pi}{24}\right)+2\cdot f\left(\frac{\pi}{12}\right)+4\cdot f\left(\frac{\pi}{8}\right)\right.$

$\left.+2\cdot f\left(\frac{\pi}{6}\right)+4\cdot f\left(\frac{5\pi}{24}\right)+f\left(\frac{\pi}{4}\right)\right)$

$=\frac{\pi}{72}\,(2+4\cdot(1.966308)+2\cdot(1.871715)+4\cdot(1.732833)$

$+2\cdot(1.569667)+4\cdot(1.400200)+1.236973)$

$\approx 1.3316.$

29. $f(x)=\sqrt{x^2+2}.$

$\int_1^3 f(x)dx \approx \frac{3-1}{3\cdot 6}\left(f(1)+4f\left(\frac{4}{3}\right)+2f\left(\frac{5}{3}\right)+4f(2)+2f\left(\frac{7}{3}\right)+4f\left(\frac{8}{3}\right)+f(3)\right)$

$=\frac{1}{9}\left(\sqrt{3}+4\left(\frac{\sqrt{34}}{3}\right)+2\left(\frac{\sqrt{43}}{3}\right)+4\sqrt{6}+2\left(\frac{\sqrt{67}}{3}\right)+4\left(\frac{\sqrt{82}}{3}\right)+\sqrt{11}\right)$

$\approx 4.9471.$

31. $f(x)=x\sin\left(\frac{\pi x}{2}\right).$

$\int_0^2 f(x)dx \approx \frac{2-0}{3\cdot 6}\left(f(0)+4f\left(\frac{1}{3}\right)+2f\left(\frac{2}{3}\right)+4f(1)+2f\left(\frac{4}{3}\right)+4f\left(\frac{5}{3}\right)+f(2)\right)$

$=\frac{1}{9}\left(0+4\left(\frac{1}{3}\right)\left(\frac{1}{2}\right)+2\left(\frac{2}{3}\right)\left(\frac{\sqrt{3}}{2}\right)+4(1)(1)+2\left(\frac{4}{3}\right)\left(\frac{\sqrt{3}}{2}\right)+4\left(\frac{5}{3}\right)\left(\frac{1}{2}\right)+0\right)$

$=\frac{1}{9}\left(4+\frac{2+6\sqrt{3}+10}{3}\right)=\frac{1}{9}\left(8+2\sqrt{3}\right)\approx 1.27389.$

33. $\int_0^{\pi/4} \frac{1}{1+x^2}\,dx \approx ?$

number of subintervals	approximation to the integral
$n=5$	0.6649153
$n=20$	0.6656965
$n=100$	0.6657707
$n=250$	0.6657731

35. a. $f(x) = \sin\sqrt{x}$.

$$\int_0^3 f(x)dx \approx \frac{3-0}{2\cdot 6}\left(f(0) + 2f\left(\tfrac{1}{2}\right) + 2f(1) + 2f\left(\tfrac{3}{2}\right) + 2f(2) + 2f\left(\tfrac{5}{2}\right) + f(3)\right)$$

$$\approx \tfrac{1}{4}(0 + 2(0.64964) + 2(0.84147) + 2(0.94072) + 2(0.98777)$$

$$+2(0.99995) + 0.98703)$$

$$\approx 2.4565.$$

b. $f(x) = \frac{1}{\sqrt{\cos x}}$.

$$\int_0^{\pi/3} f(x)dx \approx \frac{\frac{\pi}{3}-0}{2\cdot 6}\left(f(0) + 2f\left(\tfrac{\pi}{18}\right) + 2f\left(\tfrac{\pi}{9}\right) + 2f\left(\tfrac{\pi}{6}\right) + 2f\left(\tfrac{2\pi}{9}\right)\right.$$

$$\left. +2f\left(\tfrac{5\pi}{18}\right) + f\left(\tfrac{\pi}{3}\right)\right)$$

$$\approx \tfrac{\pi}{36}(1 + 2(1.00768) + 2(1.03159) + 2(1.07457)$$

$$+2(1.14254) + 2(1.24729) + 1.41421) \approx 1.17125.$$

37. $$\int_{-\Delta x}^{\Delta x} (ax^2 + bx + c)dx = \left(\tfrac{1}{3}ax^3 + \tfrac{1}{2}bx^2 + cx\right)\Big]_{-\Delta x}^{\Delta x}$$

$$= \left\{\tfrac{1}{3}a(\Delta x)^3 + \tfrac{1}{2}b(\Delta x)^2 + c(\Delta x)\right\} - \left\{\tfrac{1}{3}a(-\Delta x)^3 + \tfrac{1}{2}b(-\Delta x)^2 + c(-\Delta x)\right\}$$

$$= \tfrac{\Delta x}{6}\left\{2a(\Delta x)^2 + 3b(\Delta x) + 6c + 2a(\Delta x)^2 - 3b(\Delta x) + 6c\right\}$$

$$= \tfrac{\Delta x}{3}\left\{2a(\Delta x)^2 + 6c\right\}$$

$$= \tfrac{\Delta x}{3}\left\{\left[a(\Delta x)^2 - b(\Delta x) + c\right] + 4c + \left[a(\Delta x)^2 + b(\Delta x) + c\right]\right\}$$

$$= \tfrac{\Delta x}{3}[y_0 + 4y_1 + y_2].$$

39. a. $\ln 5 = \int_1^5 \frac{1}{x}dx \approx \frac{5-1}{3\cdot 8}\left[1+4\left(\frac{2}{3}\right)+2\left(\frac{1}{2}\right)+4\left(\frac{2}{5}\right)+2\left(\frac{1}{3}\right)\right.$

$$\left.+4\left(\frac{2}{7}\right)+2\left(\frac{1}{4}\right)+4\left(\frac{2}{9}\right)+\frac{1}{5}\right]$$

$$= \frac{1}{6}\left[2+\frac{8}{3}+\frac{8}{5}+\frac{2}{3}+\frac{8}{7}+\frac{1}{2}+\frac{8}{9}+\frac{1}{5}\right]$$

$$\approx \frac{1}{6}(9.6650794) \approx 1.6108466.$$

b. $f(x) = \frac{1}{x}$. $f'(x) = -\frac{1}{x^2}$. $f''(x) = \frac{2}{x^3}$. $f'''(x) = -\frac{6}{x^4}$. $f^{iv}(x) = \frac{24}{x^5}$.

The maximum value of $|f^{iv}(x)|$ for $1 \le x \le 5$ is 24. Therefore

$|\text{Error}| \le \frac{(5-1)^5\cdot 24}{180\cdot(8)^4} = \frac{24756}{737280} = 0.03333\ldots.$

c. From the calculator, we get $\ln 5 = 1.6094379$.

$|1.6094379 - 1.6108466| = 0.0014087.$

This is less than the estimated error in part b.

Selected Solutions to the Chapter 7 Review Exercises

1. Let $u = x+5$. $du = dx$.

$\int \sqrt{x+5}\,dx = \int \sqrt{u}\,du = \frac{2}{3}u^{3/2} + C = \frac{2}{3}(x+5)^{3/2} + C.$

3. Let $u = 1+x^2$. $du = 2x\,dx$, $x\,dx = \frac{1}{2}\,du$.

$\int \frac{x}{1+x^2}\,dx = \int \frac{1}{u}\cdot\frac{1}{2}\,du = \frac{1}{2}\ln|u| + C = \frac{1}{2}\ln(1+x^2) + C.$

5. $\int_0^1 (x-2)(x^2+3)dx = \int_0^1 (x^3 - 2x^2 + 3x - 6)dx$

$$= \left(\frac{1}{4}x^4 - \frac{2}{3}x^3 + \frac{3}{2}x^2 - 6x\right)\Big]_0^1$$

$$= \frac{1}{4} - \frac{2}{3} + \frac{3}{2} - 6 = \frac{3-8+18-72}{12} = \frac{-59}{12}.$$

7. Let $u = \sin x$. $du = \cos x\ dx$, $\cos x\ dx = du$.

$$\int_0^{\pi/6} \sin^3 x \cos x\ dx = \int_{x=0}^{x=\pi/6} u^3\ du = \tfrac{1}{4}u^4\Big]_{x=0}^{x=\pi/6} = \tfrac{1}{4}\sin^4 x\Big]_0^{\pi/6}$$

$$= \tfrac{1}{4}\left(\tfrac{1}{16} - 0\right) = \tfrac{1}{64}.$$

9. Let $u = x^2 + 2$. $du = 2x\,dx \implies x\,dx = \tfrac{1}{2}\,du$.

$$\int x\sqrt{x^2+2}\,dx = \int \sqrt{u}\left(\tfrac{1}{2}\,du\right) = \tfrac{1}{2}\cdot\tfrac{2}{3}u^{3/2} + C$$

$$= \tfrac{1}{3}(x^2+2)^{3/2} + C.$$

11. Let $u = x - 3$. $du = dx$.

$$\int_0^2 \frac{dx}{(x-3)^2} = \int_{x=0}^{x=2} u^{-2}\ du = \frac{u^{-1}}{-1}\Big]_{x=0}^{x=2} = -\frac{1}{x-3}\Big]_0^2 = 1 - \tfrac{1}{3} = \tfrac{2}{3}.$$

13. Let $u = \ln x$, $dv = x^2\,dx$. $du = \frac{1}{x}\,dx$, $v = \frac{1}{3}x^3$.

$$\int x^2 \ln x\,dx = (\ln x)\left(\tfrac{1}{3}x^3\right) - \int \left(\tfrac{1}{3}x^3\right)\left(\tfrac{1}{x}\,dx\right)$$

$$= \tfrac{1}{3}x^3 \ln x - \tfrac{1}{3}\int x^2\,dx$$

$$= \tfrac{1}{3}x^3 \ln x - \tfrac{1}{9}x^3 + C.$$

15. $$\int_0^3 \frac{1}{\sqrt[3]{x+1}}\ dx = \tfrac{3}{2}(x+1)^{2/3}\Big]_0^3 = \tfrac{3}{2}\left[(3+1)^{2/3} - 1\right] = \tfrac{3}{2}(4^{2/3} - 1).$$

17. $$\int_0^{1/4} \sec^2 \pi x\ dx = \tfrac{1}{\pi}\tan \pi x\Big]_0^{1/4} = \tfrac{1}{\pi}\left(\tan\tfrac{\pi}{4} - \tan 0\right) = \tfrac{1}{\pi}(1-0) = \tfrac{1}{\pi}.$$

19. Let $u = \ln x$, $dv = x^3\,dx$. $du = \frac{1}{x}\,dx$, $v = \frac{1}{4}x^4$.

$$\int x^3 \ln x\,dx = (\ln x)\left(\tfrac{1}{4}x^4\right) - \int \left(\tfrac{1}{4}x^4\right)\left(\tfrac{1}{x}\,dx\right)$$

$$= \tfrac{1}{4}x^4 \ln x - \tfrac{1}{4}\int x^3\,dx$$

$$= \tfrac{1}{4}x^4 \ln x - \tfrac{1}{16}x^4 + C.$$

21. Let $u = x$, $dv = \csc^2 x\ dx$. $du = dx$, $v = -\cot x$.

$$\int x\csc^2 x\ dx = -x\cot x + \int \cot x\ dx = -x\cot x + \ln|\sin x| + C.$$

23. $$\int_{-\infty}^0 e^x\ dx = \lim_{t\to-\infty}\int_t^0 e^x\ dx = \lim_{t\to-\infty} e^x\Big]_t^0 = \lim_{t\to-\infty}(1 - e^t) = 1.$$

25. Let $u = \ln x$. $du = \frac{1}{x}\, dx$.

$$\int \frac{1}{x \ln x}\, dx = \int \frac{1}{u}\, du = \ln|u| + C = \ln|\ln x| + C.$$

$$\int_{e^2}^{\infty} \frac{dx}{x \ln x} = \lim_{t\to\infty} \int_{e^2}^{t} \frac{1}{x \ln x}\, dx = \lim_{t\to\infty} \ln|\ln x|\Big]_{e^2}^{t}$$

$$= \lim_{t\to\infty} (\ln \ln t - \ln \ln e^2) = +\infty. \quad \text{Diverges.}$$

27. Let $u = 16 - x^2$. $du = -2x\, dx$.

$$\int \frac{x}{\sqrt{16-x^2}}\, dx = -\frac{1}{2}\int \frac{1}{\sqrt{u}}\, du = -\frac{1}{2}\cdot 2u^{1/2} + C = -\sqrt{16-x^2} + C.$$

$$\int_{1}^{4} \frac{x}{\sqrt{16-x^2}}\, dx = \lim_{b\to 4^-} \int_{1}^{b} \frac{x}{\sqrt{16-x^2}}\, dx = \lim_{b\to 4^-} \left(-\sqrt{16-x^2}\right)\Big]_{1}^{b}$$

$$= \lim_{b\to 4^-} \left(-\sqrt{16-b^2} + \sqrt{15}\right) = \sqrt{15}.$$

29. Let $u(x) = x^2 + 4x + 3$. $du = (2x+4)dx$.

$$\int \frac{2x+4}{x^2+4x+3}\, dx = \int \frac{1}{u}\, du = \ln|u| + C = \ln|x^2+4x+3| + C.$$

31. The increase $= \int_2^4 MC(x)dx = \int_2^4 \frac{40x^2}{\sqrt{4+x^3}}\, dx.$

Let $u = 4 + x^3$. $du = 3x^2 dx \implies x^2\, dx = \frac{1}{3}du.$

$$\int \frac{40x^2}{\sqrt{4+x^3}}\, dx = \frac{40}{3}\int (u)^{-1/2} du = \frac{40}{3}\cdot 2u^{1/2} + C = \frac{80}{3}\sqrt{4+x^3} + C.$$

$$\int_2^4 \frac{40x^2}{\sqrt{4+x^3}}\, dx = \frac{80}{3}\sqrt{4+x^3}\,\Big]_2^4 = \frac{80}{3}\left(\sqrt{68} - \sqrt{12}\right) = 127.5229 \text{ cents.}$$

33. Total amount deposited $= \int_0^4 A(t)dt = \int_0^4 400te^{2-t^2}\, dt.$

Let $u = 2 - t^2$. $du = -2t\, dt \implies t\, dt = -\frac{1}{2}\, du.$

$$\int 400te^{2-t^2}\, dt = -200\int e^u\, du = -200e^u + C = -200e^{2-t^2} + C.$$

$$\int_0^4 400te^{2-t^2}\, dt = -200e^{2-t^2}\Big]_0^4 = -200(e^{-14} - e^2)$$

$$= 1477.8111 \text{ ounces.}$$

35. Average value $= \dfrac{\int_0^{\pi/2} 4\cos^3 \pi x\, dx}{\frac{\pi}{2} - 0} = \dfrac{\frac{4}{3\pi}\sin(\pi x)\cdot\left(\cos^2(\pi x)+2\right)\Big]_0^{\pi/2}}{\frac{\pi}{2}}$

$$= \frac{\frac{4}{3\pi}\sin\left(\frac{\pi^2}{2}\right)\cdot\left(\cos^2\left(\frac{\pi^2}{2}\right)+2\right)}{\frac{\pi}{2}}$$

$$= -0.5399.$$

PRACTICE PROBLEMS FOR CHAPTER 7

1. Find each of the following expressions.

a. $\int \frac{x^2\,dx}{\sqrt{x^3+5}}\,dx$ b. $\int \tan^3(2x)\sec(2x)dx$

c. $\int \frac{\ln(x+1)}{2(x+1)}\,dx$ d. $\int e^{(x^3+6x)}\cdot(x^2+2)dx$

e. $\int \frac{1+\cos 2x}{\sin^2 2x}\,dx$ f. $\int x^{1/3}\sqrt{x^{4/3}-1}\,dx$

2. Find each of the following expressions.

a. $\int xe^{(2x+1)}dx$ b. $\int \frac{x}{e^x}\,dx$

c. $\int \frac{x^3}{\sqrt{4-x^2}}\,dx$ d. $\int (x-e^{-x})^2 dx$

e. $\int \sqrt{x}\ln x\,dx$ f. $\int x\cos(2x+1)dx$

3. A printing company estimates that the rate of revenue generated by one of its printing presses at time t will be $80-2t$ thousand dollars per year. Find the present value of this continuous stream of income over the next 4 years at a 10% annual interest rate.

4. Determine each of the following.

a. $\int \sqrt{\left(e^{(2x)}+4\right)^3}\; e^{(2x)}dx$ b. $\int_{-1}^{1} x^4e^{(x^5-1)}dx$

c. $\int x\sqrt{x+1}\,dx$ d. $\int \cos(\ln x)dx$

e. $\int \frac{x}{e^{(x^2)}}\,dx$ f. $\int \frac{x^7}{(x^4+3)^{2/3}}\,dx$

5. Determine which of the following improper integrals converge, and find the value of each convergent one.

a. $\int_1^{\infty} \frac{2x+3}{x^2+3x}\,dx$ b. $\int_0^{\infty} 10e^{(-10x)}dx$

c. $\int_{-\infty}^{-4} \frac{3}{x^4}\,dx$ d. $\int_0^{\infty} xe^{(-x^2)}dx$

e. $\int_1^{\infty} x\ln x\,dx$ f. $\int_0^{\infty} \frac{dx}{1+e^x}$

6. Determine which of the following improper integrals converge, and find the value of each convergent one.

a. $\int_0^{\pi/2} \tan x \, dx$

b. $\int_0^3 \frac{x}{\sqrt{9-x^2}} \, dx$

c. $\int_{-100}^0 \frac{1}{\sqrt{-x}} \, dx$

d. $\int_0^3 \frac{x}{9-x^2} \, dx$

e. $\int_0^1 \frac{dx}{x^{0.999}}$

f. $\int_1^3 (x-1)^{-3/2} dx$

7. The capital value of an asset is sometimes defined as the present value of all future net earnings of the asset. Find the capital value of an asset that at time t is producing income at the rate of $6000e^{-.04t}$ dollars per year, assuming an annual interest rate of 16%.

8. Use the Trapezoidal Rule and Simpson's Rule to approximate each of the following definite integrals with the indicated value of n.

a. $\int_0^{0.5} \frac{1}{\sqrt{2\pi}} e^{-x^2/2} dx$ for $n = 20$.

b. $\int_1^4 \ln(1+x^2) dx$ for $n = 12$.

c. $\int_0^2 \frac{1}{\sqrt{x^4+16}} dx$ for $n = 16$.

d. $\int_1^3 \sqrt{\ln x} \, dx$ for $n = 16$.

e. $\int_0^{\pi} \sqrt{\sin x} \, dx$ for $n = 6$.

f. $\int_0^{\pi} \sin \sqrt{x} \, dx$ for $n = 4$.

9. Let D denote the initial dose of a certain drug administered to a person. Let $r(t)$ denote the person's excretion rate of the drug. The amount of the drug absorbed by the person is the difference between the initial dose and the total amount excreted. If the initial dose is $D = 150$ mg, and if the excretion rate is $r(t) = 25e^{-0.4t}$ mg per hour, find the amount of the drug absorbed by the person.

10. The work W_r required to lift a body of mass m from the surface of a planet of mass M and radius R to a distance r from the center of the planet is given by the formula

$$W_r = \int_R^r \frac{GMm}{x^2}\,dx,$$

where G is a constant determined by the planet.

Let v_0 denote the initial velocity of the body.

a. Find the work required to move the body infinitely far from the planet.

b. Find the initial velocity v_0 so that the initial kinetic energy $\frac{1}{2}mv_0^2$ is equal to the work described in part a.

Solutions to Practice Problems for Chapter 7

1.a. $\int \frac{x^2 dx}{\sqrt{x^3+5}}$.

Let $u = x^3 + 5$. $du = 3x^2 dx$.

$\int \frac{x^2 dx}{\sqrt{x^3+5}} = \int \frac{1}{3} \cdot \frac{1}{\sqrt{u}}\, du = \frac{1}{3} \cdot 2\sqrt{u} + C = \frac{2}{3}\sqrt{x^3+5} + C.$

b. $\int \tan^3(2x)\sec(2x)dx = \int \tan^3(2x) \cdot \tan(2x)\sec(2x)dx$

$= \int \left[\sec^2(2x) - 1\right] \cdot \tan(2x) \cdot \sec(2x)dx.$

Let $u = \sec(2x)$. $du = 2\tan(2x)\sec(2x)dx$.

$\int \tan^3(2x) \cdot \sec(2x)dx = \int (u^2-1)\frac{1}{2}\, du = \frac{1}{2}\left(\frac{1}{3}u^3 - u\right) + C$

$= \frac{1}{6}u(u^2-3) + C = \frac{1}{6}\sec(2x) \cdot \left[\sec^2(2x) - 3\right] + C.$

c. $\int \frac{\ln(x+1)}{2(x+1)}\, dx$.

Let $u = \ln(x+1)$. $du = \frac{1}{x+1}\, dx$.

$\int \frac{\ln(x+1)}{2(x+1)}\, dx = \int \frac{1}{2}u\, du = \frac{1}{4}u^2 + C = \frac{1}{4}\left[\ln(x+1)\right]^2 + C.$

d. $\int e^{(x^3+6x)} \cdot (x^2+2)dx$.

Let $u = x^3 + 6x$. $du = (3x^2+6)dx = 3(x^2+2)dx$.

$\int e^{(x^3+6x)} \cdot (x^2+2)dx = \frac{1}{3}\int e^u du = \frac{1}{3}e^u + C$

$= \frac{1}{3}e^{(x^3+6x)} + C.$

e. $\int \frac{1+\cos 2x}{\sin^2 2x}\, dx = \int \frac{2\cos^2 x}{(2\sin x \cdot \cos x)^2}\, dx = \int \frac{1}{2\sin^2 x}\, dx$

$= \frac{1}{2}\int \csc^2 x\, dx = -\frac{1}{2}\cot x + C.$

f. $\int x^{1/3}\sqrt{x^{4/3}-1}\, dx$.

Let $u = x^{4/3} - 1$. $du = \frac{4}{3}x^{1/3}dx$.

$\int x^{1/3}\sqrt{x^{4/3}-1}\, dx = \frac{3}{4}\int \sqrt{u}\, du = \frac{3}{4} \cdot \frac{2}{3}u^{3/2} + C = \frac{1}{2}\left(x^{4/3}-1\right)^{3/2} + C.$

2.a. $\int x \cdot e^{(2x+1)} dx$.

Let $u = x$, $dv = e^{(2x+1)}dx \implies du = dx$, $v = \frac{1}{2}e^{(2x+1)}$.

$\int x \cdot e^{(2x+1)}dx = \frac{1}{2}xe^{(2x+1)} - \int \frac{1}{2}e^{(2x+1)}dx$

$= \frac{1}{2}xe^{(2x+1)} - \frac{1}{4}e^{(2x+1)} + C = \frac{1}{4}e^{(2x+1)}(2x-1) + C.$

b. $\int \frac{x}{e^x}\, dx = \int x \cdot e^{-x}\, dx$.

Let $u = x$, $dv = e^{-x}dx \implies du = dx$, $v = -e^{-x}$.

$\int \frac{x}{e^x}dx = -xe^{-x} - \int(-e^{-x})dx = -xe^{-x} + (-e^{-x}) + C$

$= -e^{-x}(x+1) + C.$

c. $\int \frac{x^3}{\sqrt{4-x^2}}\, dx = \int x^2 \cdot \frac{x}{\sqrt{4-x^2}}\, dx.$

Let $u = x^2$, $dv = \frac{x}{\sqrt{4-x^2}}\, dx \implies du = 2x\, dx,\ v = -\sqrt{4-x^2}$.

$$\int \frac{x^3}{\sqrt{4-x^2}}\, dx = -x^2\sqrt{4-x^2} - \int(-2x)\sqrt{4-x^2}\, dx$$

$$= -x^2 \cdot \sqrt{4-x^2} - \int \sqrt{u_1}du,$$

where $u_1 = 4 - x^2$

$$= -x^2 \cdot \sqrt{4-x^2} - \tfrac{2}{3}u_1^{3/2} + C$$

$$= -x^2 \cdot \sqrt{4-x^2} - \tfrac{2}{3}(4-x^2)^{3/2} + C.$$

d. $\int(x - e^{-x})^2 dx = \int(x^2 - 2xe^{-x} + e^{-2x})dx$

$$= \int(x^2 + e^{-2x})dx - 2\int xe^{-x}dx.$$

From Problem 2(b): $\int xe^{-x}dx = -e^{-x}(x+1) + C_1$. So

$$\int(x - e^{-x})^2dx = \tfrac{1}{3}x^3 + \left(-\tfrac{1}{2}e^{-2x}\right) + C_0 - 2\left[-e^{-x}(x+1) + C_1\right]$$

$$= \tfrac{1}{3}x^3 - \tfrac{1}{2}e^{-2x} + 2e^{-x}(x+1) + C.$$

e. $\int \sqrt{x}\ln x\, dx$.

Let $u = \ln x$, $dv = \sqrt{x}\, dx \implies du = \frac{1}{x}\, dx$, $v = \frac{2}{3}x^{3/2}$.

$$\int \sqrt{x}\ln x\, dx = \tfrac{2}{3}x^{3/2} \cdot \ln x - \int \tfrac{2}{3}x^{3/2} \cdot \left(\tfrac{1}{x}\right) dx = \tfrac{2}{3}x^{3/2} \cdot \ln x - \tfrac{2}{3}\int x^{1/2}dx$$

$$= \tfrac{2}{3}x^{3/2}\ln x - \tfrac{4}{9}x^{3/2} + C.$$

f. $\int x\cos(2x+1)dx$.

Let $u = x$, $dv = \cos(2x+1)dx \implies du = dx$, $v = \frac{1}{2}\sin(2x+1)$.

$\int x\cos(2x+1)dx = \frac{1}{2}x\sin(2x+1) - \int \frac{1}{2}\sin(2x+1)dx.$

$= \frac{1}{2}x\sin(2x+1) + \frac{1}{4}\cos(2x+1) + C.$

3. The present value of the income stream is

$$P = \int_0^4 (80-2t)e^{-.1t}dt = \int_0^4 80e^{-.1t}dt - 2\int_0^4 te^{-.1t}dt$$

$$= -\frac{80}{0.1}e^{-.1t}\Big]_0^4 - 2\int_0^4 te^{-.1t}dt$$

$$= -800(e^{-.4}-1) - 2\int_0^4 te^{-.1t}dt.$$

Let $u = t$, $dv = e^{-.1t}dt \implies du = dt$, $v = -10e^{-.1t}$.

Then

$$P = -800(e^{-.4}-1) - 2\left[-10te^{-.1t}\Big]_0^4 + \int_0^4 10e^{-.1t}dt\right]$$

$$= -800(e^{-.4}-1) + 20(4e^{-.4}) + 200e^{-.1t}\Big]_0^4$$

$$= -800(e^{-.4}-1) + 80e^{-.4} + 200(e^{-.4}-1)$$

$$= -520 \cdot e^{-.4} + 600 \approx \$251.43.$$

4.a. $\int \sqrt{(e^{2x}+4)^3} \cdot e^{2x}\,dx = \int u^{3/2} \cdot \frac{1}{2}\,du \quad$ where $u = e^{2x}+4$

$= \frac{1}{2} \cdot \frac{2}{5}u^{5/2} + C$

$= \frac{1}{5}(e^{2x}+4)^{5/2} + C.$

b. $\int_{-1}^{1} x^4 e^{(x^5-1)}dx = \frac{1}{5}e^{(x^5-1)}\Big]_{-1}^{1} = \frac{1}{5}(e^0 - e^{-2}) \approx .173.$

c. $\int x\sqrt{x+1}\,dx$.

Let $u = x$, $dv = \sqrt{x+1}\,dx \implies du = dx$, $v = \frac{2}{3}(x+1)^{3/2}$.

So $\int x\sqrt{x+1}\,dx = \frac{2}{3}x(x+1)^{3/2} - \int \frac{2}{3}(x+1)^{3/2}dx$

$$= \frac{2}{3}x(x+1)^{3/2} - \frac{4}{15}(x+1)^{5/2} + C.$$

d. $\int \cos(\ln x)dx$.

Let $u = \cos(\ln x)$, $dv = dx \implies du = \frac{-\sin(\ln x)}{x}\,dx$, $v = x$.

$\int \cos(\ln x)dx = x \cdot \cos(\ln x) + \int x \cdot \frac{\sin(\ln x)}{x}\,dx$

$$= x \cdot \cos(\ln x) + \int \sin(\ln x)dx.$$

For $\int \sin(\ln x)dx$, let $w = \sin(\ln x)$, $dz = dx$

$\implies dw = \frac{\cos(\ln x)}{x}\,dx, \quad z = x.$

So $\int \sin(\ln x)dx = x\sin(\ln x) - \int x \cdot \frac{\cos(\ln x)}{x}\,dx$.

So $\int \cos(\ln x)dx = x \cdot \cos(\ln x) + x \cdot \sin(\ln x) - \int \cos(\ln x)dx$.

Therefore $\int \cos(\ln x)dx = \frac{1}{2}x\,[\cos(\ln x) + \sin(\ln x)] + C$.

e. $\int \frac{x}{e^{(x^2)}}\,dx = \int xe^{-x^2}dx$.

Let $u = e^{-x^2}$. $\quad du = -2xe^{-x^2}dx$.

$\int \frac{x}{e^{(x^2)}}\,dx = -\frac{1}{2}\int du = -\frac{1}{2}u + C = -\frac{1}{2}e^{-x^2} + C$.

f. $\int \frac{x^7}{(x^4+3)^{2/3}}\,dx = \int \frac{x^4 \cdot x^3}{(x^4+3)^{2/3}}\,dx = \int \frac{[(x^4+3)-3]}{(x^4+3)^{2/3}} \cdot x^3\,dx$.

$$= \int \frac{u-3}{u^{2/3}} \cdot \frac{1}{4}\,du \qquad \text{where } u = x^4+3.$$

$$= \int (u^{-1/3} - 3u^{-2/3}) \cdot \frac{1}{4}\,du$$

$$= \frac{1}{4}\left(\frac{3}{2}u^{2/3} - 3 \cdot 3u^{1/3}\right) + C$$

$$= \frac{3}{8}(x^4+3)^{2/3} - \frac{9}{4}(x^4+3)^{1/3} + C.$$

5.a. $\int_1^\infty \frac{2x+3}{x^2+3x}\,dx = \lim_{t\to\infty}\int_1^t \frac{2x+3}{x^2+3x}\,dx$

$$= \lim_{t\to\infty}\int_{x=1}^{x=t} \frac{1}{u}\,du \qquad \text{where } u = x^2+3x$$

$$= \lim_{t\to\infty}\left(\ln|u|\right)\Big]_{x=1}^{x=t} = \lim_{t\to\infty}\left(\ln|x^2+3x|\right)\Big]_1^t$$

$$= \lim_{t\to\infty}\left(\ln|t^2+3t| - \ln 4\right) = +\infty.$$

So the improper integral diverges.

b. $\int_0^\infty 10e^{-10x}dx = \lim_{t\to\infty}\int_0^t 10e^{-10x}dx = \lim_{t\to\infty}\left(-e^{-10x}\right)\Big]_0^t$

$$= \lim_{t\to\infty}\left(1-e^{-10t}\right) = 1.$$

So the improper integral converges to 1.

c. $\int_{-\infty}^{-4} \frac{3}{x^4}\,dx = \lim_{t\to-\infty}\int_t^{-4} \frac{3}{x^4}dx$

$$= \lim_{t\to-\infty}\left(-x^{-3}\right)\Big]_t^{-4} = \lim_{t\to-\infty}\left(t^{-3}-(-4)^{-3}\right) = \frac{1}{64}.$$

So the improper integral converges to $\frac{1}{64}$.

d. $\int_0^\infty xe^{(-x^2)}dx = \lim_{t\to\infty}\int_0^t xe^{(-x^2)}dx$

$$= \lim_{t\to\infty}\left(-\tfrac{1}{2}e^{-x^2}\right)\Big]_0^t \qquad \text{(See Problem 4(e))}$$

$$= \lim_{t\to\infty}\left(\tfrac{1}{2}-\tfrac{1}{2}e^{-t^2}\right) = \tfrac{1}{2}.$$

So the improper integral converges to $\frac{1}{2}$.

e. $\int_1^\infty x\ln x\,dx = \lim_{t\to\infty}\int_1^t x\ln x\,dx.$

Let $u=\ln x$, $dv = x\,dx \implies du = \frac{1}{x}\,dx$, $v=\frac{x^2}{2}$.

Then $\int x\ln x\,dx = \frac{1}{2}x^2\ln x - \frac{1}{2}\int x^2\cdot\frac{1}{x}\,dx$

$$= \tfrac{1}{2}x^2\ln x - \tfrac{1}{4}x^2 + C = \tfrac{1}{4}x^2(2\ln x - 1) + C.$$

So $\int_1^\infty x\ln x\,dx = \lim_{t\to\infty}\frac{1}{4}x^2(2\ln x-1)\Big]_1^t$

$$= \lim_{t\to\infty}\left[\tfrac{1}{4}t^2(2\ln t - 1)+\tfrac{1}{4}\right] = +\infty.$$

So the improper integral diverges.

f. $\int_0^\infty \frac{dx}{1+e^x} = \lim_{t\to\infty} \int_0^t \frac{dx}{1+e^x}.$

Now $\int \frac{1}{1+e^x} dx = \int \frac{1}{1+e^x} \cdot \frac{e^x}{e^x} dx = \int \left(\frac{1}{e^x} - \frac{1}{1+e^x}\right) e^x\,dx$

$$= \int \left(\frac{1}{e^x} - \frac{1}{1+e^x}\right) d(e^x)$$

$$= \ln(e^x) - \ln(1+e^x) + C$$

$$= \ln\left(\frac{e^x}{1+e^x}\right) + C = \ln\left(\frac{1}{e^{-x}+1}\right) + C.$$

So $\int_0^\infty \frac{dx}{1+e^x} = \lim_{t\to\infty} \ln\left(\frac{1}{e^{-x}+1}\right)\Big]_0^t$

$$= \lim_{t\to\infty} \ln\left(\frac{1}{e^{-t}+1}\right) - \ln\left(\tfrac{1}{2}\right) = \ln 2.$$

Thus the improper integral converges to $\ln 2$.

6.a. $\int_0^{\pi/2} \tan x\,dx = \lim_{b\to(\frac{\pi}{2})^-} \int_0^b \tan x\,dx = \lim_{b\to(\frac{\pi}{2})^-} \ln|\sec x|\Big]_0^b$

$$= \lim_{b\to(\frac{\pi}{2})^-} (\ln|\sec b| - \ln 1) = +\infty.$$

So the improper integral diverges.

b. $\int_0^3 \frac{x}{\sqrt{9-x^2}}\,dx = \lim_{b\to3^-} \int_0^b \frac{x}{\sqrt{9-x^2}}\,dx = \lim_{b\to3^-} (-\sqrt{9-x^2})\Big]_0^b$

$$= \lim_{b\to3^-} (3 - \sqrt{9-b^2}) = 3.$$

So the improper integral converges to 3.

c. $\int_{-100}^0 \frac{1}{\sqrt{-x}}\,dx = \lim_{b\to0^-} \int_{-100}^b \frac{1}{\sqrt{-x}}\,dx = \lim_{b\to0^-} (-2\sqrt{-x})\Big]_{-100}^b$

$$= \lim_{b\to0^-} (-2\sqrt{-b} + 20) = 20.$$

So the improper integral converges to 20.

d. $\int_0^3 \frac{x}{9-x^2}\,dx = \lim_{b\to3^-} \int_0^b \frac{x}{9-x^2}\,dx = \lim_{b\to3^-} -\tfrac{1}{2}\ln(9-x^2)\Big]_0^b$

$$= -\tfrac{1}{2} \lim_{b\to3^-} \left[\ln(9-b^2) - \ln 9\right] = +\infty.$$

So the improper integral diverges.

e. $\int_0^1 \frac{dx}{x^{.999}} = \lim_{a\to 0+} \int_a^1 \frac{dx}{x^{.999}} = \lim_{a\to 0+} 1000 \cdot x^{.001}\Big]_a^1$

$= \lim_{a\to 0+} 1000 \cdot (1 - a^{.001}) = 1000.$

So the improper integral converges to 1000.

f. $\int_1^3 (x-1)^{-3/2}\, dx = \lim_{a\to 1+} \int_a^3 (x-1)^{-3/2} dx$

$= \lim_{a\to 1+} -2(x-1)^{-1/2}\Big]_a^3$

$= \lim_{a\to 1+} -2\left[(2^{-1/2}) - (a-1)^{-1/2}\right] = +\infty.$

So the improper integral diverges.

7. The capital value is

$$\int_0^\infty 6000e^{.04t} \cdot e^{-.16t} dt = \int_0^\infty 6000e^{-.12t} dt$$

$$= 6000\left(-\frac{1}{.12}\right) e^{-.12t}\Big]_0^\infty$$

$$= 50,000 \text{ dollars.}$$

8.a. $\int_0^{0.5} \frac{1}{\sqrt{2\pi}} e^{-x^2/2}\, dx$ for $n = 20$.

Let $f(x) = \frac{1}{\sqrt{2\pi}} e^{-x^2/2}$; let $\Delta x = \frac{.5-0}{20} = \frac{1}{40}$, and let

$x_0 = 0$, $x_1 = \frac{1}{40}$, $x_2 = \frac{1}{20}$, ..., $x_{20} = \frac{1}{2} = .5$.

Then by the Trapezoidal Rule:

$$\int_0^{.5} f(x)dx \approx \frac{.5-0}{2(20)}\left[f(0) + 2f\left(\frac{1}{40}\right) + 2f\left(\frac{1}{20}\right) + \cdots + 2f\left(\frac{19}{40}\right) + f\left(\frac{1}{2}\right)\right]$$

$$\approx \frac{1}{80} \cdot \frac{1}{\sqrt{2\pi}}\left[1 + 2e^{-(\frac{1}{40})^2/2} + 2e^{-(\frac{1}{20})^2/2} + 2e^{-(\frac{3}{40})^2/2}\right.$$

$$\left. + 2e^{-(\frac{1}{10})^2/2} + 2e^{-(\frac{1}{8})^2/2} + \cdots + 2e^{-(\frac{19}{40})^2/2} + e^{-(\frac{1}{2})^2/2}\right]$$

$$= \frac{1}{80\sqrt{2\pi}} [1 + 2(.9997 + .9988 + .9972 + .9950 + .9922$$
$$+ .9888 + .9848 + .9802 + .9750 + .9692 + .9629$$
$$+ .9560 + .9486 + .9406 + .9321 + .9231 + .9136$$
$$+ .9037 + .8933) + .8825]$$
$$= \frac{1}{80\sqrt{2\pi}} [1 + 2(18.255) + .8825] = .191454893.$$

By Simpson's Rule:

$$\int_0^{.5} f(x)dx \approx \frac{.5-0}{3(20)} \left[f(0) + 4f\left(\frac{1}{40}\right) + 2f\left(\frac{1}{20}\right) + 4f\left(\frac{3}{40}\right) + 2f\left(\frac{1}{10}\right)\right.$$
$$\left. + \cdots + 2f\left(\frac{18}{40}\right) + 4f\left(\frac{19}{40}\right) + f\left(\frac{1}{2}\right)\right]$$
$$= \frac{1}{120\sqrt{2\pi}} [1 + 4(.9997) + 2(.9988) + 4(.9972) + 2(.9950)$$
$$+ 4(.9922) + 2(.9888) + 4(.9848) + 2(.9802)$$
$$+ 4(.9750) + 2(.9692) + 4(.9629) + 2(.9560)$$
$$+ 4(.9486) + 2(.9406) + 4(.9321) + 2(.9231)$$
$$+ 4(.9136) + 2(.9037) + 4(.8933) + .8825]$$
$$= \frac{1}{120\sqrt{2\pi}}(1 + 55.7084 + .8825)$$
$$= .191462041.$$

b. $\int_1^4 \ln(1+x^2)dx$ for $n = 12$.
Let $f(x) = \ln(1+x^2)$; let $\Delta x = \frac{4-1}{12} = \frac{1}{4}$, and let

$x_0 = 1$, $x_1 = \frac{5}{4}$, $x_2 = \frac{6}{4}$, ..., $x_{11} = \frac{15}{4}$, $x_{12} = 4$.

By the Trapezoidal Rule:

$$\int_1^4 f(x)dx \approx \frac{4-1}{2(12)}\left\{f(1)+2\left[f\left(\frac{5}{4}\right)+f\left(\frac{6}{4}\right)+\cdots+f\left(\frac{15}{4}\right)\right]+f(4)\right\}$$

$$=\frac{1}{8}\left\{\ln 2+2\left[\ln\left(1+\frac{25}{16}\right)+\ln\left(1+\frac{36}{16}\right)+\ln\left(1+\frac{49}{16}\right)\right.\right.$$

$$\left.\left.+\cdots+\ln\left(1+\left(\frac{14}{4}\right)^2\right)+\ln\left(1+\left(\frac{15}{4}\right)^2\right)\right]+\ln(1+4^2)\right\}$$

$$=\frac{1}{8}[.6931+2(.9410+1.1787+1.4018+1.6094$$

$$+1.8021+1.9810+2.1474+2.3026+2.4478$$

$$+2.5840+2.7122)+2.8332]$$

$$=\frac{1}{8}[.6931+2(21.108)+2.8332]=5.7177875.$$

By Simpson's Rule:

$$\int_1^4 f(x)dx \approx \frac{4-1}{3(12)}\left[f(1)+4f\left(\frac{5}{4}\right)+2f\left(\frac{6}{4}\right)+4f\left(\frac{7}{4}\right)+2f\left(\frac{8}{4}\right)\right.$$

$$\left.+\cdots+2f\left(\frac{14}{4}\right)+4f\left(\frac{15}{4}\right)+f(4)\right]$$

$$=\frac{1}{12}[\ln 2+4(.9410)+2(1.1787)+4(1.4018)+2(1.6094)$$

$$+4(1.8021)+2(1.9810)+4(2.1474)+2(2.3026)$$

$$+4(2.4478)+2(2.5840)+4(2.7122)+2.8332]$$

$$\approx\frac{1}{12}(.6931+65.1206+2.8332)=5.720575.$$

c. $\int_0^2 \frac{1}{\sqrt{x^4+16}}\,dx$ for $n=16$.

Let $f(x)=\frac{1}{\sqrt{x^4+16}}$; let $\Delta x=\frac{2-0}{16}=\frac{1}{8}$; and let

$x_0=0,\ x_1=\frac{1}{8},\ x_2=\frac{2}{8},\ \ldots,\ x_{15}=\frac{15}{8},\ x_{16}=2.$

By the Trapezoidal Rule:

$$\int_0^2 f(x)dx \approx \frac{2-0}{2(16)}\left\{f(0)+2\left[f\left(\frac{1}{8}\right)+f\left(\frac{2}{8}\right)+\cdots+f\left(\frac{15}{8}\right)\right]+f(2)\right\}$$

$$= \frac{1}{16}\left\{\frac{1}{4}+2\left[\frac{1}{\sqrt{\left(\frac{1}{8}\right)^4+16}}+\frac{1}{\sqrt{\left(\frac{2}{8}\right)^4+16}}+\frac{1}{\sqrt{\left(\frac{3}{8}\right)^4+16}}\right.\right.$$

$$\left.\left.+\cdots+\frac{1}{\sqrt{\left(\frac{15}{8}\right)^4+16}}\right]+\frac{1}{\sqrt{2^4+16}}\right\}$$

$$= \frac{1}{16}\left[\frac{1}{4}+2(.2500+.24997+.24985+.24951+.24882\right.$$

$$+.24756+.24554+.24254+.23835+.23286$$

$$\left.+.22602+.21789+.20864+.19850+.18778)+.17678\right]$$

$$= \frac{1}{16}[.25+2(3.49383)+.17678] = .4634.$$

By Simpson's Rule:

$$\int_0^2 f(x)dx \approx \frac{2-0}{3(16)}\left[f(0)+4f\left(\frac{1}{8}\right)+2f\left(\frac{2}{8}\right)+4f\left(\frac{3}{8}\right)\right.$$

$$\left.+2f\left(\frac{4}{8}\right)+\cdots+2f\left(\frac{14}{8}\right)+4f\left(\frac{15}{8}\right)+f(2)\right]$$

$$= \frac{1}{24}\left[\frac{1}{4}+\frac{4}{\sqrt{\left(\frac{1}{8}\right)^4+16}}+\frac{2}{\sqrt{\left(\frac{2}{8}\right)^4+16}}+\frac{4}{\sqrt{\left(\frac{3}{8}\right)^4+16}}+\frac{2}{\sqrt{\left(\frac{4}{8}\right)^4+16}}\right.$$

$$\left.+\cdots+\frac{2}{\sqrt{\left(\frac{14}{8}\right)^4+16}}+\frac{4}{\sqrt{\left(\frac{15}{8}\right)^4+16}}+\frac{1}{\sqrt{2^4+16}}\right]$$

$$= \frac{1}{24}[.25 + 4(.2500) + 2(.24997) + 4(.24985) + 2(.24951)$$

$$+ 4(.24882) + 2(.24756) + 4(.24554) + 2(.24254)$$

$$+ 4(.23835) + 2(.23286) + 4(.22602) + 2(.21789)$$

$$+ 4(.20864) + 2(.19850) + 4(.18778) + .17678]$$

$$= \frac{1}{24}(.25 + 10.6977 + .17678) = .46352.$$

d. $\int_1^3 \sqrt{\ln x}\, dx$ for $n = 16$.

Let $f(x) = \sqrt{\ln x}$; let $\Delta x = \frac{3-1}{16} = \frac{1}{8}$; and let

$x_0 = 1$, $x_1 = \frac{9}{8}$, $x_2 = \frac{10}{8}$, $x_3 = \frac{11}{8}$, $\ldots$, $x_{15} = \frac{23}{8}$, $x_{16} = 3$.

By the Trapezoidal Rule:

$$\int_1^3 f(x)dx \approx \frac{\Delta x}{2}\left\{f(1) + 2\left[f\left(\frac{9}{8}\right) + f\left(\frac{10}{8}\right) + \cdots + f\left(\frac{23}{8}\right)\right] + f(3)\right\}$$

$$= \frac{1}{16}\left\{0 + 2\left[\sqrt{\ln\left(\frac{9}{8}\right)} + \sqrt{\ln\left(\frac{10}{8}\right)} + \cdots + \sqrt{\ln\left(\frac{23}{8}\right)}\right] + \sqrt{\ln 3}\right\}$$

$$\approx \frac{1}{16}[2\,(.3432 + .4724 + .5643 + .6368 + .6968$$

$$+ .7481 + .7928 + .8326 + .8682 + .9005$$

$$+ .9301 + .9572 + .9824 + 1.0058 + 1.0276) + 1.0481]$$

$$= \frac{1}{16}[2(11.7587) + 1.0481] = 1.535349279.$$

By Simpson's Rule:

$$\int_1^3 f(x)dx \approx \frac{\Delta x}{3}\left[f(1) + 4f\left(\frac{9}{8}\right) + 2f\left(\frac{10}{8}\right) + \cdots + 2f\left(\frac{22}{8}\right) + 4f\left(\frac{23}{8}\right) + f(3)\right]$$

$$= \frac{1}{24}[0 + 4(.3432) + 2(.4724) + 4(.5643) + 2(.6368)$$

$$+ 4(.6968) + 2(.7481) + 4(.7928) + 2(.8326)$$

$$+ 4(.8682) + 2(.9005) + 4(.9301) + 2(.9572)$$

$$+ 4(.9824) + 2(1.0058) + 4(1.0276) + 1.0481]$$

$$= \frac{1}{24}(35.9284 + 1.0481) = 1.5406875.$$

e. $\int_0^\pi \sqrt{\sin x}\, dx$ for $n = 6$.

Let $f(x) = \sqrt{\sin x}$; let $\Delta x = \frac{\pi - 0}{6} = \frac{\pi}{6}$; and let

$x_0 = 0$, $x_1 = \frac{\pi}{6}$, $x_2 = \frac{\pi}{3}$, $x_3 = \frac{\pi}{2}$, $x_4 = \frac{2\pi}{3}$, $x_5 = \frac{5\pi}{6}$, $x_6 = \pi$.

By the Trapezoidal Rule:

$$\int_0^\pi f(x)dx \approx \frac{\Delta x}{2}\left\{f(0) + 2\left[f\left(\frac{\pi}{6}\right) + f\left(\frac{\pi}{3}\right) + f\left(\frac{\pi}{2}\right) + f\left(\frac{2\pi}{3}\right) + f\left(\frac{5\pi}{6}\right)\right] + f(\pi)\right\}$$

$$= \frac{\pi}{12}\left\{0 + 2\left[\sqrt{\sin\left(\frac{\pi}{6}\right)} + \sqrt{\sin\left(\frac{\pi}{3}\right)} + \sqrt{\sin\left(\frac{\pi}{2}\right)}\right.\right.$$

$$\left.\left. + \sqrt{\sin\left(\frac{2\pi}{3}\right)} + \sqrt{\sin\left(\frac{5\pi}{6}\right)}\right] + 0\right\}$$

$$= \frac{\pi}{12}[2(.7071 + .9306 + 1.0 + .9306 + .7071)]$$

$$= 2.2385942.$$

By Simpson's Rule:

$$\int_0^\pi f(x)dx \approx \frac{\Delta x}{3}\left[f(0) + 4f\left(\frac{\pi}{6}\right) + 2f\left(\frac{\pi}{3}\right) + 4f\left(\frac{\pi}{2}\right) + 2f\left(\frac{2\pi}{3}\right) + 4f\left(\frac{5\pi}{6}\right) + f(\pi)\right]$$

$$= \frac{\pi}{18}[0 + 4(.7071) + 2(.9306) + 4(1.0) + 2(.9306) + 4(.7071) + 0]$$

$$= 2.3351109.$$

f. $\int_0^{\pi} \sin\sqrt{x}\, dx$ for $n = 4$.

Let $f(x) = \sin\sqrt{x}$; let $\Delta x = \frac{\pi - 0}{4} = \frac{\pi}{4}$; and let

$x_0 = 0$, $x_1 = \frac{\pi}{4}$, $x_2 = \frac{\pi}{2}$, $x_3 = \frac{3\pi}{4}$, $x_4 = \pi$.

By the Trapezoidal Rule:

$$\int_0^{\pi} f(x)dx \approx \frac{\Delta x}{2}\left\{f(0) + 2\left[f\left(\frac{\pi}{4}\right) + f\left(\frac{\pi}{2}\right) + f\left(\frac{3\pi}{4}\right)\right] + f(\pi)\right\}$$

$$= \frac{\pi}{8}\{0 + 2(.7747 + .9500 + .9994) + .9797\}$$

$$= 2.52423.$$

By Simpson's Rule:

$$\int_0^{\pi} f(x)dx \approx \frac{\Delta x}{3}\left\{f(0) + 4f\left(\frac{\pi}{4}\right) + 2f\left(\frac{\pi}{2}\right) + 4f\left(\frac{3\pi}{4}\right) + f(\pi)\right\}$$

$$= \frac{\pi}{12}\{0 + 4(.7747) + 2(.9500) + 4(.9994) + .9797\}$$

$$= 2.61174.$$

9. The total amount of the drug excreted by the person is given by

$$\int_0^{\infty} r(t)dt = \int_0^{\infty} 25e^{-0.4t}\, dt$$

$$= \lim_{T\to\infty} 25 \cdot \frac{1}{-0.4}e^{-0.4t}\Big]_0^T$$

$$= -62.5 \cdot \lim_{T\to\infty}\left(e^{-0.4T} - 1\right) = 62.5 \text{ mg}.$$

Therefore the amount of the drug absorbed by the person is

$$150 - 62.5 = 87.5 \text{ mg}.$$

10.a. The required to move the body infinitely far from the planet is given by the improper integral

$$\int_R^\infty \frac{GMm}{x^2}\,dx = \lim_{r\to\infty}\int_R^r \frac{GMm}{x^2}\,dx$$
$$= \lim_{r\to\infty}\left(-\frac{GMm}{x}\right)\Bigg]_R^r$$
$$= \lim_{r\to\infty}\left(-\frac{GMm}{r}+\frac{GMm}{R}\right)$$
$$= \frac{GMm}{R}.$$

b. The initial velocity v_0 so that the initial kinetic energy $\frac{1}{2}mv_0^2$ is equal to the work required to move the body infinitely far from the planet is such that

$$\frac{1}{2}mv_0^2 = \frac{GMm}{R}$$
$$v_0^2 = \frac{2GM}{R}$$
$$v_0 = \sqrt{\frac{2GM}{R}}.$$

CHAPTER 8

SUMMARY OF CHAPTER 8

1. The notion of functions of several independent variables.

2. Three-dimensional coordinate systems.

3. For a function f of two real variables: domain of f, graph of f, level curve for f, continuity of f.

4. For a function $z = f(x, y)$: the definition of the partial derivatives $\frac{\partial f}{\partial x}(x, y)$ and $\frac{\partial f}{\partial y}(x, y)$, other notations for these partial derivatives, a geometric interpretation of these partial derivatives.

5. First order partial derivatives for functions of more than two variables.

6. Competitive and complementary products.

7. Higher order partial derivatives for $z = f(x, y)$; equality of mixed partial derivatives $\frac{\partial^2 f}{\partial x \partial y}$ and $\frac{\partial^2 f}{\partial y \partial x}$ under certain hypotheses.

8. The concepts of a relative maximum and a relative minimum for a function $z = f(x, y)$; the result that if $z_0 = f(x_0, y_0)$ is either a relative maximum or a relative minimum for the function $z = f(x, y)$ and if both partial derivatives $\frac{\partial f}{\partial x}(x_0, y_0)$ and $\frac{\partial f}{\partial y}(x_0, y_0)$ exist, then $\frac{\partial f}{\partial x}(x_0, y_0) = 0$ and $\frac{\partial f}{\partial y}(x_0, y_0) = 0$; the concept of a saddle point on the graph of the function $z = f(x, y)$.

9. The Second Derivative Test for classifying the critical points of the function $z = f(x, y)$ as to producing relative maxima, relative minima, saddle points.

10. The problem of finding where the maximum value or the minimum value of the function $z = f(x, y)$ occurs subject to the constraint $g(x, y) = 0$: the geometry of constraints; solving this problem by first solving the constraint equation for one of the variables x and y in terms of the other, then substituting the expression obtained into $f(x, y)$, and finally using the techniques of Chapter 3 on the resulting function of a single variable; the method of Lagrange Multipliers for solving this problem.

11. Lagrange Multipliers for functions of three variables.

12. For a function $f(x, y)$ of two variables: the linear approximation
$f(x, y) \approx f(x_0, y_0) + \frac{\partial f}{\partial x}(x_0, y_0)(x - x_0) + \frac{\partial f}{\partial y}(x_0 - y_0)(y - y_0)$
to the functional values of f for (x, y) near the point (x_0, y_0), and the resulting approximation
$f(x, y) - f(x_0, y_0) \approx \frac{\partial f}{\partial x}(x_0, y_0)(x - x_0) + \frac{\partial f}{\partial y}(x_0, y_0)(y - y_0)$
to changes in the functional values of f for (x, y) near the point (x_0, y_0); relative and percentage errors for the preceding approximations; the total differential of f given by
$df = \frac{\partial f}{\partial x}(x, y)dx + \frac{\partial f}{\partial y}(x, y)dy$.

13. The Principle of Least Squares for the line $y = mx + b$ that best fits given data points, the formulas for determining m and b; use of the line thus obtained for predicting future values of y.

14. Definition of the double integral $\iint\limits_R f(x,y)dA$; if $f(x,y)$ is continuous and $f(x,y) \geq 0$ for all (x,y) in R, then this double integral is the volume of the solid bounded above by the graph of $z = f(x,y)$ and below by the region R; evaluating $\iint\limits_R f(x,y)dA$ by using an iterated integral.

15. Evaluating a given iterated integral by interchanging the order of integration.

Selected Solutions to Exercise Set 8.1

1. $f(x,y) = 3x + 7y$.

   a. $f(2,3) = 3(2) + 7(3) = 27$.

   b. $f(0,-6) = 3(0) + 7(-6) = -42$.

   c. $f(-1,5) = 3(-1) + 7(5) = 32$.

3. $f(x,y) = \sqrt{xy} - \frac{1}{xy}$.

   a. $f(1,1) = \sqrt{1 \cdot 1} - \frac{1}{1 \cdot 1} = 0$.

   b. $f(-2,-2) = \sqrt{(-2)(-2)} - \frac{1}{(-2)(-2)} = 2 - \frac{1}{4} = \frac{7}{4}$.

   c. $f(1,9) = \sqrt{(1)(9)} - \frac{1}{(1)(9)} = 3 - \frac{1}{9} = \frac{26}{9}$.

5. $f(x,y,z) = x^2 - 2xy + xz^3$.

   a. $f(1,2,-1) = 1^2 - 2(1)(2) + (1)(-1)^3 = -4$.

   b. $f(3,-3,0) = 3^2 - 2(3)(-3) + 3(0)^3 = 27$.

   c. $f(-1,1,4) = (-1)^2 - 2(-1)(1) + (-1)4^3 = -61$.

7. $f(x,y,z) = \sqrt{x^2 + y^2 + z^2}$.

   a. $f(0,1,0) = \sqrt{0^2 + 1^2 + 0^2} = 1$.

   b. $f(-2,1,2) = \sqrt{(-2)^2 + 1^2 + 2^2} = \sqrt{9} = 3$.

   c. $f\left(3,-3,\sqrt{7}\right) = \sqrt{3^2 + (-3)^2 + \left(\sqrt{7}\right)^2} = \sqrt{25} = 5$.

9.

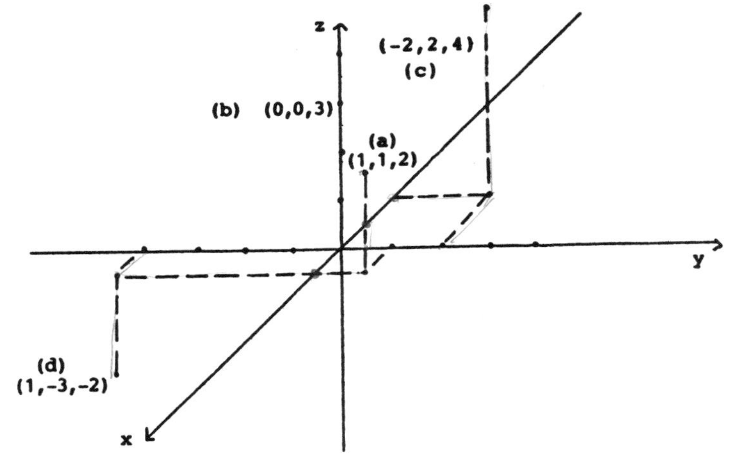

11. $f(x,y) = 3xy^2 - 5x^3y + 7$.

    The domain includes all ordered pairs of real numbers.

13. $f(x,y) = \frac{10x^2}{y^2-x^2}$.

The domain consists of all ordered pairs (x,y) of real numbers such that either $x \neq y$ or $x \neq -y$.

15. $f(x,y) = 4x^{2/3}y^{1/4}$.

The domain consists of all ordered pairs (x,y) of real numbers such that $y \geq 0$.

17. Let x = number of plastic brooms sold

y = number of fiber brooms sold.

Then $z = R(x,y) = 10x + 15y$.

19. $P(x,y) = R(x,y) - C(x,y) = (10x + 15y) - (4x + 6y + 20)$

$= 6x + 9y - 20$.

21. a. $R(x,y) = 0.75x + 0.60y$.

b. $R(200,150) = 0.75(200) + 0.60(150) = \$240$.

23. a. $P(r,t) = 500e^{rt}$.

b. $P(0.03,5) = 500e^{0.03(5)} \approx \$580.92$.

c. $P(0.10,7) = 500e^{0.1(7)} \approx \$1006.88$.

25. $f(K,L) = 100K^{1/4}L^{3/4}$.

a. $f(81,16) = 100(81)^{1/4} \cdot (16)^{3/4} = 100(3)(8) = 2400$.

b. $f(16,81) = 100(16)^{1/4}(81)^{3/4} = 100(2)(27) = 5400$.

27. $E(x,t) = 20x^{3/2}e^{-0.05t}$.

a. $E(16,20) = 20(16)^{3/2}e^{-0.05(20)} = 1280e^{-1}$.

b. $E(64,10) = 20(64)^{3/2}e^{-0.05(10)} = 10240e^{-0.5}$.

Selected Solutions to Exercise Set 8.2

1. $f(xy) = 3x - 6y + 5$.

a. $\frac{\partial f}{\partial x}(x,y) = 3$. b. $\frac{\partial f}{\partial y}(x,y) = -6$.

3. $f(x,y) = 4xy^2 - 3x^3y + y^5$.

a. $\frac{\partial f}{\partial x}(x,y) = 4y^2 - 9x^2y$. b. $\frac{\partial f}{\partial y}(x,y) = 8xy - 3x^3 + 5y^4$.

5. $f(x,y) = \sqrt{x+y} = (x+y)^{1/2}$.

a. $\frac{\partial f}{\partial x}(x,y) = \frac{1}{2}(x+y)^{-1/2} \cdot 1 = \frac{1}{2\sqrt{x+y}}$.

b. $\frac{\partial f}{\partial y}(x,y) = \frac{1}{2}(x+y)^{-1/2} \cdot 1 = \frac{1}{2\sqrt{x+y}}$.

7. $f(x,y) = \frac{x+y}{x-y}$.

a. $\frac{\partial f}{\partial x}(x,y) = \frac{(x-y)-(x+y)}{(x-y)^2} = \frac{-2y}{(x-y)^2}$.

b. $\frac{\partial f}{\partial y}(x,y) = \frac{(x-y)-(x+y)(-1)}{(x-y)^2} = \frac{2x}{(x-y)^2}$.

9. $f(x,y) = \ln\sqrt{x^2+y^2} = \frac{1}{2}\ln(x^2+y^2)$.

a. $\frac{\partial f}{\partial x}(x,y) = \frac{1}{2} \cdot \frac{2x}{x^2+y^2} = \frac{x}{x^2+y^2}$. b. $\frac{\partial f}{\partial y}(x,y) = \frac{1}{2} \cdot \frac{2y}{x^2+y^2} = \frac{y}{x^2+y^2}$.

11. $f(x,y) = x^{2/3}y^{-1/3} + \sqrt{\frac{y}{x}}$.

a. $\frac{\partial f}{\partial x}(x,y) = \frac{2}{3}x^{-1/3}y^{-1/3} - \frac{1}{2}x^{-3/2}y^{1/2}$.

b. $\frac{\partial f}{\partial y}(x,y) = -\frac{1}{3}x^{2/3}y^{-4/3} + \frac{1}{2\sqrt{y}} \cdot \frac{1}{\sqrt{x}} = -\frac{1}{3}x^{2/3}y^{-4/3} + \frac{1}{2\sqrt{xy}}$.

13. $f(x,y) = \tan(x+y) - \sec(x-y)$.

a. $\frac{\partial f}{\partial x}(x,y) = \sec^2(x+y) - \sec(x-y)\tan(x-y)$.

b. $\frac{\partial f}{\partial y}(x,y) = \sec^2(x+y) - \sec(x-y)\tan(x-y)(-1)$

$$= \sec^2(x+y) + \sec(x-y)\tan(x-y).$$

15. $z = \frac{x}{y^2} - \frac{y}{x^2} = xy^{-2} - yx^{-2}$.

$z_x = \frac{1}{y^2} + \frac{2y}{x^3}$. $z_y = x(-2)y^{-3} - x^{-2} = -\frac{2x}{y^3} - \frac{1}{x^2}$.

17. $z = ye^{\sqrt{x}-1}$. $z_x = ye^{\sqrt{x}-1} \cdot \frac{1}{2\sqrt{x}} = \frac{y}{2\sqrt{x}}e^{\sqrt{x}-1}$. $z_y = e^{\sqrt{x}-1}$.

19. $z = \sqrt{\cos^2(xy) + e^{\sin y}}$.

$z_x = \frac{1}{2} \cdot \left(\cos^2(xy) + e^{\sin y}\right)^{-1/2} \cdot (2\cos(xy)\,(-\sin(xy)) \cdot y)$

$= -y\cos(xy)\sin(xy) \cdot \left(\cos^2(xy) + e^{\sin y}\right)^{-1/2}$.

$z_y = \frac{1}{2} \cdot \left(\cos^2(xy) + e^{\sin y}\right)^{-1/2} \cdot \left(2\cos(xy)\,(-\sin(xy)) \cdot x + e^{\sin y} \cdot \cos y\right)$.

21. $f(x,y) = 2x(y-7)$. $\frac{\partial f}{\partial x}(x,y) = 2(y-7)$. $\frac{\partial f}{\partial x}(3,5) = 2(5-7) = -4$.

23. $f(x,y) = \sqrt{y^2+2x}$.

$\frac{\partial f}{\partial y}(x,y) = \frac{2y}{2\sqrt{y^2+2x}} = \frac{y}{\sqrt{y^2+2x}}$. $\frac{\partial f}{\partial y}(4,1) = \frac{1}{\sqrt{1^2+2(4)}} = \frac{1}{3}$.

25. $f(x,y,z) = x\sin(yz) + z\cos^2(xy)$.

$\frac{\partial f}{\partial z}(x,y,z) = x \cdot \cos(yz) \cdot y + \cos^2(xy)$.

$\frac{\partial f}{\partial z}\left(1, \frac{\pi}{4}, -1\right) = 1 \cdot \cos\left(\frac{\pi}{4} \cdot (-1)\right) \cdot \frac{\pi}{4} + \cos^2\left(1 \cdot \frac{\pi}{4}\right) = \frac{\sqrt{2}}{2} \cdot \frac{\pi}{4} + \frac{1}{2}$.

27. $f(x,y,z) = xe^{yz}$.

$\frac{\partial f}{\partial z}(x,y,z) = xye^{yz}$. $\quad \frac{\partial f}{\partial z}(2,0,3) = (2)(0)e^{0\cdot 3} = 0$.

29. $f(x,y) = x^2 - 2xy + y^2$.

a. $\frac{\partial^2 f}{\partial x^2} = \frac{\partial}{\partial x}\left(\frac{\partial f}{\partial x}\right) = \frac{\partial}{\partial x}(2x - 2y) = 2$.

b. $\frac{\partial^2 f}{\partial x \partial y} = \frac{\partial}{\partial x}\left(\frac{\partial f}{\partial y}\right) = \frac{\partial}{\partial x}(-2x + 2y) = -2$.

c. $\frac{\partial^2 f}{\partial y \partial x} = \frac{\partial}{\partial y}\left(\frac{\partial f}{\partial x}\right) = \frac{\partial}{\partial y}(2x - 2y) = -2$.

d. $\frac{\partial^2 f}{\partial y^2} = \frac{\partial}{\partial y}\left(\frac{\partial f}{\partial y}\right) = \frac{\partial}{\partial y}(-2x + 2y) = 2$.

31. $f(x,y) = \ln(x^2 + y^2)$.

a. $\frac{\partial^2 f}{\partial x^2} = \frac{\partial}{\partial x}\left(\frac{2x}{x^2+y^2}\right) = \frac{2(x^2+y^2) - 2x(2x)}{(x^2+y^2)^2} = \frac{-2x^2+2y^2}{(x^2+y^2)^2}$.

b. $\frac{\partial^2 f}{\partial y \partial x} = \frac{\partial}{\partial y}\left(\frac{2x}{x^2+y^2}\right) = \frac{-2x(2y)}{(x^2+y^2)^2} = \frac{-4xy}{(x^2+y^2)^2}$.

c. $\frac{\partial^2 f}{\partial x \partial y} = \frac{\partial}{\partial x}\left(\frac{2y}{x^2+y^2}\right) = \frac{-4xy}{(x^2+y^2)^2}$.

d. $\frac{\partial^2 f}{\partial y^2} = \frac{\partial}{\partial y}\left(\frac{2y}{x^2+y^2}\right) = \frac{2(x^2+y^2) - 2y \cdot 2y}{(x^2+y^2)^2} = \frac{2x^2-2y^2}{(x^2+y^2)^2}$.

33. $\left(\frac{\partial f}{\partial K}\right)(27,8) = \frac{160}{27}$.

35. $u(x,y) = 4x^{2/3} + 6y^{3/2} - xy^2$.

a. $\frac{\partial u}{\partial x} = 4 \cdot \frac{2}{3}x^{-1/3} - y^2 = \frac{8}{3}x^{-1/3} - y^2$

$\Longrightarrow \frac{\partial u}{\partial x}(8,4) = \frac{8}{3} \cdot \frac{1}{2} - 16 = \frac{4}{3} - 16 = -\frac{44}{3}$.

b. $\frac{\partial u}{\partial y} = 6 \cdot \frac{3}{2} \cdot y^{1/2} - 2xy$

$\Longrightarrow \frac{\partial u}{\partial y}(2,4) = 6 \cdot \frac{3}{2} \cdot 2 - 2 \cdot 2 \cdot 4 = 18 - 16 = 2$.

a. means that an increase in slices of pizza from $x = 8$ to $x = 9$ results in a decrease of $\frac{44}{3}$ units of a person's level of satisfaction at $y = 4$, approximately.

b. means that an increase in glasses of soda from $y = 4$ to $y = 5$ results in an increase of 2 units of a person's level of satisfaction at $x = 2$, approximately.

37. $f(K,L) = 100\sqrt{KL}$.

a. $\frac{\partial f}{\partial K} = \frac{50 \cdot L}{\sqrt{KL}} \Longrightarrow \frac{\partial f}{\partial K}(3,27) = \frac{50 \cdot 27}{9} = 150$.

b. $\frac{\partial f}{\partial L} = \frac{50 \cdot K}{\sqrt{KL}} \Longrightarrow \frac{\partial f}{\partial L}(3,27) = \frac{50 \cdot 3}{9} = \frac{50}{3}$.

39. $W(x,y) = 150 - 30x^2 + 20y.$

$B(x,y) = 200 + 40x - 30y^2.$

a. $\frac{\partial W}{\partial x} = -60x \implies \left(\frac{\partial W}{\partial x}\right)(1,1) = -60 < 0.$

$\frac{\partial W}{\partial y} = 20 \implies \left(\frac{\partial W}{\partial y}\right)(1,1) = 20 > 0.$

If the price for white eggs increases from $x = 1$ to $x = 2$ for $y = 1$, then the demand for white eggs decreases approximately by 60 units.

If the price for brown eggs increases from $y = 1$ to $y = 2$ for $x = 1$, then the demand for white eggs increases approximately by 20 units.

b. $\frac{\partial B}{\partial x} = 40 \implies \left(\frac{\partial B}{\partial x}\right)(1,1) = 40 > 0.$

$\frac{\partial B}{\partial y} = -60y \implies \left(\frac{\partial B}{\partial y}\right)(1,1) = -60 < 0.$

If the price for white eggs increases from $x = 1$ to $x = 2$ for $y = 1$, then the demand for brown eggs increases approximately by 40 units.

If the price for brown eggs increases from $y = 1$ to $y = 2$ for $x = 1$, then the demand for brown eggs decreases approximately by 60 units.

41. $f(p,q) = 30 - 4p^2 + 16q \implies \frac{\partial f}{\partial q}(p,q) = 16 > 0.$

$g(p,q) = 80 - 12q^2 + 10p \implies \frac{\partial g}{\partial p}(p,q) = 10 > 0.$

Therefore the products are competitive.

43. $f(p,q) = 25 - 2p^{3/2} + 4\ln q \implies \frac{\partial f}{\partial q}(p,q) = 4 \cdot \frac{1}{q} > 0.$

$g(p,q) = 40 + e^{p/100} - q^{7/3} \implies \frac{\partial g}{\partial p}(p,q) = e^{p/100} \cdot \frac{1}{100} > 0.$

Therefore the products are competitive.

45. $f(p,q) = \frac{10q}{q-p} \implies \frac{\partial f}{\partial q}(p,q) = \frac{(q-p)(10)-(10q)(1)}{(q-p)^2} = \frac{-10p}{(q-p)^2} < 0.$

$q(p,q) = \frac{5p}{p-2q} \implies \frac{\partial g}{\partial p}(p,q) = \frac{(p-2q)(5)-(5p)(1)}{(p-2q)^2} = -\frac{10q}{(p-2q)^2} < 0.$

Therefore the products are complementary.

Selected Solutions to Exercise Set 8.3

1. $f(x,y) = x^2 + y^2 + 4y + 4.$

   $\frac{\partial f}{\partial x} = 2x = 0 \implies x = 0.$ $\frac{\partial f}{\partial y} = 2y + 4 = 0 \implies y = -2.$

   So (0,-2) is a critical point for f.

   $\frac{\partial^2 f}{\partial x^2} = 2,$ $\frac{\partial^2 f}{\partial x \partial y} = 0,$ and $\frac{\partial^2 f}{\partial y^2} = 2.$

   Thus, at the critical point (0,-2) we have

   $$A = 2, \quad B = 0, \quad C = 2, \quad \text{and } D = 0^2 - 2 \cdot 2 = -4.$$

   Since $D < 0$ and $A > 0$, f has a relative minimum value at the critical point (0,-2), and the relative minimum value is $f(0,-2) = 0 + 4 - 8 + 4 = 0.$

3. $f(x,y) = x^2 + 2y^2 - 6x + 4y - 8.$

   $\frac{\partial f}{\partial x} = 2x - 6 = 0 \implies x = 3,$ $\frac{\partial f}{\partial y} = 4y + 4 = 0 \implies y = -1.$

   Thus the critical point is $(3,-1)$.

   $\frac{\partial^2 f}{\partial x^2} = 2,$ $\frac{\partial^2 f}{\partial y^2} = 4,$ $\frac{\partial^2 f}{\partial x \partial y} = 0.$

   $D = 0^2 - 2 \cdot (4) = -8 < 0,$ and $\frac{\partial^2 f}{\partial x^2}(3,-1) > 0.$

   Therefore $f(x,y)$ has a relative minimum value at $(3,-1)$.

5. $f(x,y) = x^2 - y^2 + 6x + 4y + 5.$

   $\frac{\partial f}{\partial x} = 2x + 6 = 0 \implies x = -3,$ $\frac{\partial f}{\partial y} = -2y + 4 = 0 \implies y = 2.$

   Thus (-3,2) is a critical point for f. $\frac{\partial^2 f}{\partial x^2} = 2,$ $\frac{\partial^2 f}{\partial x \partial y} = 0,$ $\frac{\partial^2 f}{\partial y^2} = -2.$

   At the point (-3,2), $A = 2,$ $B = 0,$ $C = -2,$ and $D = 4.$

   Since $D > 0$, $(-3, 2, f(-3,2)) = (-3, 2, 0)$ is a saddle point for f.

7. $f(x,y) = xy + 9.$ $\frac{\partial f}{\partial x} = y = 0,$ $\frac{\partial f}{\partial y} = x = 0.$

   Thus (0,0) is a critical point for f. $\frac{\partial^2 f}{\partial x^2} = 0,$ $\frac{\partial^2 f}{\partial x \partial y} = 1,$ $\frac{\partial^2 f}{\partial y^2} = 0.$

   Since $D = 1 > 0$ at the point (0,0), f has a saddle point at (0,0,9).

9. $f(x,y) = 5x^2 + y^2 - 10x - 6y + 15.$

$\frac{\partial f}{\partial x} = 10x - 10 = 0 \implies x = 1, \quad \frac{\partial f}{\partial y} = 2y - 6 = 0 \implies y = 3.$

$(1,3)$ is a critical point for f. $\frac{\partial^2 f}{\partial x^2} = 10, \quad \frac{\partial^2 f}{\partial x \partial y} = 0, \quad \frac{\partial^2 f}{\partial y^2} = 2.$

Hence, at the point (1,3), $A = 10$, $B = 0$, $C = 2$, and $D = -20$.

Since $A > 0$ and $D < 0$, f has a relative minimum value $f(1,3) = 1$ at the critical point (1,3).

11. $f(x,y) = x^3 - y^3.$

$\frac{\partial f}{\partial x} = 3x^2 = 0 \implies x = 0, \quad \frac{\partial f}{\partial y} = -3y^2 = 0 \implies y = 0.$

So f has a critical point at (0,0). $\frac{\partial^2 f}{\partial x^2} = 6x, \quad \frac{\partial^2 f}{\partial x \partial y} = 0, \quad \frac{\partial^2 f}{\partial y^2} = -6y.$

Since $D = 0$ at the point (0,0), we can't conclude by this test. However, $f(x,0) = x^3$ has neither a relative minimum nor a relative maximum at (0,0); $f(0,y) = -y^3$ has neither a relative minimum nor a relative maximum at (0,0). Therefore, the point (0,0) provides neither a relative maximum nor a relative minimum for f.

13. $f(x,y) = x^3 + y^3 + 4xy.$

$$\frac{\partial f}{\partial x} = 3x^2 + 4y = 0 \qquad (1)$$

$$\frac{\partial f}{\partial y} = 3y^2 + 4x = 0 \qquad (2)$$

From (1) we obtain $y = -3x^2/4$. Thus (2) become $3\left(-\frac{3x^2}{4}\right)^2 + 4x = 0$

$\implies \frac{27}{16}x^4 + 4x = 0 \implies 27x^4 + 64x = 0$

$\implies x(27x^3 + 64) = 0 \implies x = 0, \ -\frac{4}{3}.$

Hence, critical points for f are (0,0) and $\left(-\frac{4}{3}, -\frac{4}{3}\right)$.

$\frac{\partial^2 f}{\partial x^2} = 6x, \quad \frac{\partial^2 f}{\partial x \partial y} = 4, \quad \frac{\partial^2 f}{\partial y^2} = 6y.$

(i) At the point (0,0), $A = 0$, $B = 4$, $C = 0$, and $D = 16$. f has a saddle point at (0,0).

(ii) At the point $\left(-\frac{4}{3}, -\frac{4}{3}\right)$, $A = -8$, $B = 4$, $C = -8$, and $D = -48$. f has a relative maximum value $f\left(-\frac{4}{3}, -\frac{4}{3}\right) = \frac{64}{27}$.

15. $f(x, y) = x^2 + y^2 + x - 2y + xy + 5.$

$$\frac{\partial f}{\partial x} = 2x + 1 + y = 0 \qquad (1)$$
$$\frac{\partial f}{\partial y} = 2y - 2 + x = 0 \qquad (2)$$

$((\mathbf{1}) \times (-2)) + (\mathbf{2}) \implies -3x - 4 = 0 \implies x = -\frac{4}{3}.$

From $(\mathbf{1})$, $y = -2x - 1, \quad y = -2\left(-\frac{4}{3}\right) - 1 = \frac{5}{3}.$

The critical point is $\left(-\frac{4}{3}, \frac{5}{3}\right)$.

$\frac{\partial^2 f}{\partial x^2} = 2, \qquad \frac{\partial^2 f}{\partial x \partial y} = 1, \qquad \frac{\partial^2 f}{\partial y^2} = 2. \qquad D = 1^2 - 2 \cdot 2 = -3 < 0.$

Thus f has a relative minimum value at $\left(-\frac{4}{3}, \frac{5}{3}\right)$.

17. $P(x, y) = 30x + 90y - 0.5x^2 - 2y^2 - xy.$

$$\frac{\partial P}{\partial x} = 30 - x - y = 0 \qquad (1)$$
$$\frac{\partial P}{\partial y} = 90 - 4y - x = 0 \qquad (2)$$

$((\mathbf{1}) \times (-1)) + (\mathbf{2}) \implies -3y + 60 = 0 \implies y = 20.$

From $(\mathbf{1}) \implies x = 30 - y \implies x = 30 - 20 = 10.$

The critical point for P is (10,20).

$\frac{\partial^2 P}{\partial x^2} = -1, \qquad \frac{\partial^2 P}{\partial x \partial y} = -1, \qquad \frac{\partial^2 P}{\partial y^2} = -4.$

$D = (-1)^2 - (-1)(-4) = -3 < 0$, and $A < 0$.

Thus the profit P is maximum when $x = 10$ and $y = 20$.

19. $P(x, y) = 40x + 80y - 2x^2 - 10y^2 - 4xy.$

$$\frac{\partial P}{\partial x} = 40 - 4x - 4y = 0 \qquad (1)$$
$$\frac{\partial P}{\partial y} = 80 - 20y - 4x = 0 \qquad (2)$$

$((\mathbf{1}) \times (-1)) + (\mathbf{2}) \implies 40 - 16y = 0 \implies y = \frac{40}{16} = \frac{5}{2} = 2.5.$

From $(\mathbf{1}) \implies x = 10 - y \implies x = 10 - \left(\frac{5}{2}\right) = \frac{15}{2} = 7.5.$

The critical point for P is (7.5,2.5).

$\frac{\partial^2 P}{\partial x^2} = -4, \qquad \frac{\partial^2 P}{\partial x \partial y} = -4, \qquad \frac{\partial^2 P}{\partial y^2} = -20.$

$D = (-4)^2 - (-4)(-20) = -64 < 0$, and $A < 0$.

Thus the maximum productivity occurs when $(x, y) = (7.5, 2.5)$.

21. The revenue $R(x, y) = p \cdot x + q \cdot y$.

$$R(x, y) = (20 - x)x + \left(46 - \tfrac{5}{2}y\right) y = -x^2 - \tfrac{5}{2}y^2 + 20x + 46y.$$

The profit $P(x, y) = R(x, y) - C(x, y)$

$$= \left(-x^2 - \tfrac{5}{2}y^2 + 20x + 46y\right) - (100 + 4x + 2y + xy)$$
$$= -x^2 - \tfrac{5}{2}y^2 + 16x + 44y - xy - 100.$$

$$\frac{\partial P}{\partial x} = -2x + 16 - y = 0 \qquad (1)$$
$$\frac{\partial P}{\partial y} = -5y + 44 - x = 0 \qquad (2)$$

$(1) + ((2) \times (-2)) \implies -72 + 9y = 0 \implies y = 8.$

From $(1) \implies x = 8 - \frac{y}{2} \implies x = 8 - \frac{1}{2}(8) = 4.$

The critical point for P is (4,8).

$\frac{\partial^2 P}{\partial x^2} = -2, \qquad \frac{\partial^2 P}{\partial x \partial y} = -1, \qquad \frac{\partial^2 P}{\partial y^2} = -5.$

$D = (-1)^2 - (-2)(-5) = -9 < 0$, and $A < 0$.

Thus the maximum profit occurs when $(x, y) = (4, 8)$.

23. Let x be the length of the box in meters, let y be the width of the box in meters, and let z be the height of the box in meters.

Let p be the price in dollars per square meter of the material for the side walls.

Since the volume of the box is to be 64 cubic meters, it follows that

$xyz = 64 \implies z = \frac{64}{xy}.$

Then let $f(x, y)$ be the cost in dollars of the material for the box. Note that

$$\begin{aligned}
f(x,y) &= 2\cdot(xz+yz)\cdot p + 2\cdot xy\cdot 2p \\
&= 2\cdot\left(x\cdot\frac{64}{xy} + y\cdot\frac{64}{xy}\right)\cdot p + 4xyp \\
&= \left(\frac{128}{y} + \frac{128}{x} + 4xy\right)\cdot p \\
&= 4\cdot\left(\frac{32}{y} + \frac{32}{x} + xy\right)\cdot p.
\end{aligned}$$

$$\begin{aligned}
\frac{\partial f}{\partial x} &= 4\cdot\left(-\frac{32}{x^2} + y\right)\cdot p = 0 \implies y = \frac{32}{x^2}. \\
\frac{\partial f}{\partial y} &= 4\cdot\left(-\frac{32}{y^2} + x\right)\cdot p = 0 \implies x = \frac{32}{y^2}.
\end{aligned}$$

Substitute $y = \frac{32}{x^2}$ into the eqution $x = \frac{32}{y^2}$.

$$\begin{aligned}
x &= \frac{32}{\left(\frac{32}{x^2}\right)^2} \implies x = \frac{x^4}{32} \implies x^3 = 32 \implies x = \sqrt[3]{32}. \\
y &= \frac{32}{\left(\sqrt[3]{32}\right)^2} = \sqrt[3]{32}. \\
z &= \frac{64}{\sqrt[3]{32}\cdot\sqrt[3]{32}} = 2\cdot\sqrt[3]{32}. \\
\frac{\partial^2 f}{\partial x^2} &= \frac{256}{x^3}\cdot p, \qquad \frac{\partial^2 f}{\partial y^2} = \frac{256}{y^3}\cdot p, \qquad \frac{\partial^2 f}{\partial x\partial y} = 4p.
\end{aligned}$$

At the critical point $\left(\sqrt[3]{32}, \sqrt[3]{32}\right)$:

$$\begin{aligned}
A &= \frac{\partial^2 f}{\partial x^2}\left(\sqrt[3]{32}, \sqrt[3]{32}\right) = 8p, \qquad C = \frac{\partial^2 f}{\partial y^2}\left(\sqrt[3]{32}, \sqrt[3]{32}\right) = 8p, \\
B &= \frac{\partial^2 f}{\partial x\partial y}\left(\sqrt[3]{32}, \sqrt[3]{32}\right) = 4p. \\
D &= B^2 - AC = (4p)^2 - (8p)(8p) = -48p^2 < 0.
\end{aligned}$$

So therefore the minimum cost occurs when $x = \sqrt[3]{32}$, $y = \sqrt[3]{32}$, $z = 2\cdot\sqrt[3]{32}$.

25. Let $\ell =$ the length of the package

$w =$ the width of the package

$h =$ the height of the package

Then $\ell + 2w + 2h = 84 \implies \ell = 84 - 2w - 2h$.

The volume $V = hw\ell = hw(84 - 2w - 2h)$.

$$V = 84hw - 2hw^2 - 2h^2w \text{ for } h > 0, \quad w > 0.$$

$$\frac{\partial V}{\partial h} = 84w - 2w^2 - 4hw = 0 \implies 42 - w - 2h = 0 \qquad \textbf{(1)}$$

$$\frac{\partial V}{\partial w} = 84h - 4hw - 2h^2 = 0 \implies 42 - h - 2w = 0 \qquad \textbf{(2)}$$

From **(1)**: $w = 42 - 2h$. Then **(2)** becomes

$42 - h - 2(42 - 2h) = 0 \implies 3h - 42 = 0 \implies h = 14.$

Then $w = 42 - 2h = 42 - 2(14) = 14$.

$$\ell = 84 - 2(w + h) = 84 - 2(14 + 14) = 28.$$

$\frac{\partial^2 V}{\partial h^2} = -4w, \qquad \frac{\partial^2 V}{\partial h \partial w} = 84 - 4w - 4h, \qquad \frac{\partial^2 V}{\partial w^2} = -4h.$

At $(14,14), D = (-28)^2 - (-56)(-56) < 0, \quad A = -56 < 0.$

Thus the volume of the package is the largest when $h = w = 14$ inches,

$\ell = 28$ inches.

Selected Solutions to Exercise Set 8.4

1. $f(x, y) = xy. \qquad g(x, y) = x + 4y - 8 = 0.$

$L(x, y, \lambda) = xy + \lambda(x + 4y - 8).$

$$\frac{\partial L}{\partial x} = y + \lambda = 0 \qquad \textbf{(1)}$$

$$\frac{\partial L}{\partial y} = x + 4\lambda = 0 \qquad \textbf{(2)}$$

$$\frac{\partial L}{\partial \lambda} = x + 4y - 8 = 0 \qquad \textbf{(3)}$$

$((\textbf{1}) \times (-4)) + (\textbf{2}) \rightarrow -4y + x = 0 \implies x = 4y.$

Next we substitute $x = 4y$ into equation **(3)**.

$$4y + 4y - 8 = 0 \implies y = 1 \implies x = 4.$$

Note that $f(4,1) = 4 > f(12,-1) = -12$.

Therefore, $f(4,1) = 4$ is the maximum value of f, and f has no minimum value.

3. $f(x,y) = x^2 - 8x + y^2 + 4y - 6. \qquad g(x,y) = 2x - y + 5 = 0.$

   $L(x,y,\lambda) = x^2 - 8x + y^2 + 4y - 6 + \lambda(2x - y + 5).$

$$\frac{\partial L}{\partial x} = 2x - 8 + 2\lambda = 0 \implies x - 4 + \lambda = 0 \qquad \textbf{(1)}$$
$$\frac{\partial L}{\partial y} = 2y + 4 - \lambda = 0 \qquad \textbf{(2)}$$
$$\frac{\partial L}{\partial \lambda} = 2x - y + 5 = 0 \qquad \textbf{(3)}$$
$$\textbf{(1)} + \textbf{(2)} \implies x + 2y = 0 \implies x = -2y \qquad \textbf{(4)}$$

   (3) and **(4)** give $2(-2y) - y + 5 = 0 \implies y = 1.$

   $x = -2y = -2(1) = -2.$

   Since $f(0,5) = 39 > f(-2,1) = 19, \quad f(-2,1) = 19$ is the minimum value of f, and there is no maximum value of f.

5. $f(x,y) = xy. \qquad g(x,y) = x^2 + y^2 - 32 = 0.$

   $L(x,y,\lambda) = xy + \lambda(x^2 + y^2 - 32).$

$$\frac{\partial L}{\partial x} = y + 2x\lambda = 0 \qquad \textbf{(1)}$$
$$\frac{\partial L}{\partial y} = x + 2y\lambda = 0 \qquad \textbf{(2)}$$
$$\frac{\partial L}{\partial \lambda} = x^2 + y^2 - 32 = 0 \qquad \textbf{(3)}$$
$$((\textbf{1}) \cdot (y)) + ((\textbf{2}) \cdot (-x)) \implies y^2 - x^2 = 0 \implies y = \pm x.$$

   For $y = x$: **(3)** gives $x^2 + x^2 - 32 = 0 \implies x = \pm 4.$

   Then critical points are (4,4), (-4,-4).

For $y=-x$: **(3)** gives $x^2+(-x)^2-32=0 \implies x=\pm 4$.

Then critical points are (4,-4), (-4,4).

$f(4,4)=f(-4,-4)=16, \qquad f(4,-4)=f(-4,4)=-16.$

Thus the minimum value of f is -16 and the maximum value is 16.

7. $f(x,y)=x^3-y^3, \qquad g(x,y)=x-y-2=0.$

$L(x,y,\lambda)=x^3-y^3+\lambda(x-y-2).$

$$\frac{\partial L}{\partial x}=3x^2+\lambda=0. \tag{1}$$

$$\frac{\partial L}{\partial y}=-3y^2-\lambda=0. \tag{2}$$

$$\frac{\partial L}{\partial \lambda}=x-y-2=0. \tag{3}$$

$$\textbf{(1)}+\textbf{(2)} \text{ gives } 3(x^2-y^2)=0 \implies x=\pm y \tag{4}$$

From **(3)** and **(4)** we obtain $x=1$ and $y=-1$.

$f(1,-1)=1+1=2$ is the minimum value of f since $f(2,0)=8>f(1,-1)=2$, and there is no maximum value of f.

9. $f(x,y,z)=xyz, \qquad g(x,y,z)=x^2+y^2+z^2-12=0.$

$L(x,y,z,\lambda)=xyz+\lambda(x^2+y^2+z^2-12).$

$$\frac{\partial L}{\partial x}=yz+2\lambda x=0 \tag{1}$$

$$\frac{\partial L}{\partial y}=xz+2\lambda y=0 \tag{2}$$

$$\frac{\partial L}{\partial z}=xy+2\lambda z=0 \tag{3}$$

$$\frac{\partial L}{\partial \lambda}=x^2+y^2+z^2-12=0 \tag{4}$$

From $((\mathbf{1})\cdot y)-((\mathbf{2})\cdot x)$, $(y^2-x^2)z=0 \implies y=x, \quad y=-x$, or $z=0$.

(i) For $z=0$:

(3) and **(4)** become

$$\begin{cases} xy=0 \\ x^2+y^2=12 \end{cases} \implies \begin{cases} x=0 \\ y=\pm 2\sqrt{3} \end{cases} \text{ or } \begin{cases} y=0 \\ x=\pm 2\sqrt{3} \end{cases}$$

Critical points to be checked are $(0, 2\sqrt{3}, 0)$, $(0, -2\sqrt{3}, 0)$, $(2\sqrt{3}, 0, 0)$, and $(-2\sqrt{3}, 0, 0)$.

(ii) For $y = x$:

(2), **(3)**, and **(4)** become

$$x(z + 2\lambda) = 0 \implies x = 0 \quad \text{or} \quad z = -2\lambda$$

$$x^2 + 2\lambda z = 0 \tag{5}$$

$$2x^2 + z^2 - 12 = 0. \tag{6}$$

Substituting $x = 0$ into **(6)**, we obtain $z = \pm 2\sqrt{3}$.

Substituting $z = -2\lambda$ into **(5)** and **(6)** gives

$$x^2 = 4\lambda^2 \tag{7}$$

$$2x^2 + 4\lambda^2 - 12 = 0. \tag{8}$$

From **(7)** and **(8)**, $\lambda = \pm 1$ and so $z = \mp 2$.

Critical points to be checked are $(0, 0, 2\sqrt{3})$, $(0, 0, -2\sqrt{3})$, $(2, 2, -2)$, $(-2, -2, -2)$, $(2, 2, 2)$ and $(-2, -2, 2)$.

(iii) For $y = -x$:

(2), **(3)**, and **(4)** become

$$x(z - 2\lambda) = 0 \implies x = 0 \quad \text{or} \quad z = 2\lambda$$

$$-x^2 + 2\lambda z = 0 \tag{9}$$

$$2x^2 + z^2 - 12 = 0 \tag{10}$$

When $x = 0, \quad z = \pm 2\sqrt{3}$.

When $z = 2\lambda$, $8\lambda^2 + 4\lambda^2 - 12 = 0 \implies \lambda = \pm 1$ and so $z = \pm 2$.

Critical points to be checked are $(0, 0, 2\sqrt{3})$, $(0, 0, -2\sqrt{3})$, $(2, -2, 2)$, $(-2, 2, 2)$,

$(2,-2,-2)$, and $(-2,2,-2)$.

$$f\left(0,2\sqrt{3},0\right)=0 \qquad f\left(0,-2\sqrt{3},0\right)=0$$
$$f\left(2\sqrt{3},0,0\right)=0 \qquad f\left(-2\sqrt{3},0,0\right)=0$$
$$f\left(0,0,2\sqrt{3}\right)=0 \qquad f\left(0,0,-2\sqrt{3}\right)=0$$
$$f(2,2,-2)=-8 \qquad f(-2,-2,-2)=-8$$
$$f(2,2,2)=8 \qquad f(-2,-2,2)=8$$
$$f(2,-2,2)=-8 \qquad f(-2,2,2)=-8$$
$$f(2,-2,-2)=8 \qquad f(-2,2,-2)=8.$$

Hence, the maximum value of f is 8 and the minimum value of f is -8.

11. $f(x,y,z)=x^2+y^2+z^2, \quad g(x,y,z)=x-y+z-1=0.$
$L(x,y,z,\lambda)=x^2+y^2+z^2+\lambda(x-y+z-1).$

$$\frac{\partial L}{\partial x}=2x+\lambda=0 \implies x=-\frac{1}{2}\lambda. \qquad (\mathbf{1})$$
$$\frac{\partial L}{\partial y}=2y-\lambda=0 \implies y=\frac{1}{2}\lambda. \qquad (\mathbf{2})$$
$$\frac{\partial L}{\partial z}=2z+\lambda=0 \implies z=-\frac{1}{2}\lambda. \qquad (\mathbf{3})$$
$$\frac{\partial L}{\partial \lambda}=x-y+z-1=0. \qquad (\mathbf{4})$$

Substituting equations **(1)**, **(2)**, and **(3)** into equation **(4)** gives

$$\left(-\frac{1}{2}\lambda\right)-\left(\frac{1}{2}\lambda\right)+\left(-\frac{1}{2}\lambda\right)-1=0 \implies -\frac{3}{2}\lambda-1=0 \implies \lambda=-\frac{2}{3}.$$

Then

$$x=-\frac{1}{2}\cdot\left(-\frac{2}{3}\right)=\frac{1}{3}, \quad y=\frac{1}{2}\cdot\left(-\frac{2}{3}\right)=-\frac{1}{3}, \quad z=-\frac{1}{2}\cdot\left(-\frac{2}{3}\right)=\frac{1}{3}.$$

Note

$$f\left(\frac{1}{3},-\frac{1}{3},\frac{1}{3}\right)=\frac{1}{3}<f(0,0,1)=1.$$

Therefore, the minimum value of f is $f\left(\frac{1}{3}, -\frac{1}{3}, \frac{1}{3}\right) = \frac{1}{3}$, and f has no maximum value.

13. $P(x, y) = 60x^{1/4}y^{3/4}, \quad g(x, y) = 20x + 10y - 80 = 0.$

$L(x, y, \lambda) = 60x^{1/4}y^{3/4} + \lambda(20x + 10y - 80).$

$$\frac{\partial L}{\partial x} = 15x^{-3/4}y^{3/4} + 20\lambda = 0 \qquad (1)$$

$$\frac{\partial L}{\partial y} = 45y^{-1/4}x^{1/4} + 10\lambda = 0 \qquad (2)$$

$$\frac{\partial L}{\partial \lambda} = 20x + 10y - 80 = 0 \qquad (3)$$

From (1) − ((2) × 2) one obtains

$$15\left(\frac{y}{x}\right)^{3/4} - 90\left(\frac{y}{x}\right)^{-1/4} = 0 \implies \left(\frac{y}{x}\right) - 6 = 0 \implies y = 6x \qquad (4)$$

From (3) and (4), $20x + 60x - 80 = 0 \implies x = 1$ and so $y = 6$.

$P(1, 6) = 60 \cdot 6^{3/4}$. Since $P(4, 0) = 0$, the maximum value of P is $60 \cdot 6^{3/4}$.

15. $P(x, y) = 200x + 100y - 4x^2 - 2y^2.$

$g(x, y) = 20x + 10y - 600 = 0.$

$L(x, y, \lambda) = 200x + 100y - 4x^2 - 2y^2 + \lambda(20x + 10y - 600).$

$$\frac{\partial L}{\partial x} = 200 - 8x + 20\lambda = 0 \qquad (1)$$

$$\frac{\partial L}{\partial y} = 100 - 4y + 10\lambda = 0 \qquad (2)$$

$$\frac{\partial L}{\partial \lambda} = 20x + 10y - 600 = 0 \qquad (3)$$

From (1) − ((2) × 2), $-8x + 8y = 0 \implies y = x.$ (4)

From (3) and (4), $30x = 600 \implies x = 20.$

Thus, the critical point for L is $(20, 20, -2)$.

$P(20, 20) = 4000 + 2000 - 1600 - 800 = 3600.$

Since $P(30,0) = 6000 - 3600 = 2400 < 3600$, P has the maximum value 3600 at (20,20).

Therefore, 20 color sets and 20 black and white sets should be produced in order to maximize profit.

17. $P(x,y) = 20x + 40y - x^2 - y^2$. $\quad g(x,y) = x + y - 40 = 0$.

$L(x,y,\lambda) = 20x + 40y - x^2 - y^2 + \lambda(x + y - 40)$.

$$\frac{\partial L}{\partial x} = 20 - 2x + \lambda = 0 \qquad (1)$$
$$\frac{\partial L}{\partial y} = 40 - 2y + \lambda = 0 \qquad (2)$$
$$\frac{\partial L}{\partial \lambda} = x + y - 40 = 0 \qquad (3)$$

From **(1)** $-$ **(2)**,

$$-2x + 2y - 20 = 0. \qquad (4)$$

From **(3)** and **(4)**, $4y = 100 \implies y = 25$ and $x = 15$.

$$P(15,25) = 300 + 1000 - 225 - 625 = 1300 - 850 = 450.$$

Since $P(20,20) = 400 + 800 - 400 - 400 = 400 < 450$, P has the maximum value 450 when $x = 15$ and $y = 25$.

19. Let x, y, and z be the length, width, and height in inches, respectively, of the rectangular package.

Let $f(x,y,z)$ be the volume in cubic inches of the box.

Note that

$$f(x,y,z) = xyz.$$

We want to maximize $f(x,y,z)$ subject to the constraint

$$x + 2y + 2z = 84.$$

Let

$$L(x,y,z,\lambda) = xyz + \lambda(x+2y+2z-84).$$

$$\frac{\partial L}{\partial x} = yz + \lambda = 0 \quad (1)$$

$$\frac{\partial L}{\partial y} = xz + 2\lambda = 0 \quad (2)$$

$$\frac{\partial L}{\partial z} = xy + 2\lambda = 0 \quad (3)$$

$$\frac{\partial L}{\partial \lambda} = x + 2y + 2z - 84 = 0. \quad (4)$$

$$((\mathbf{1})\cdot 2) - (\mathbf{2}) \to 2yz - xz = 0 \implies z(2y-x) = 0 \implies y = \frac{1}{2}x.$$

$$((\mathbf{1})\cdot 2) - (\mathbf{3}) \to 2yz - xy = 0 \implies y(2z-x) = 0 \implies z = \frac{1}{2}x.$$

Substituting $y = \frac{1}{2}x$ and $z = \frac{1}{2}x$ into equation (4) gives

$$x + 2\cdot\left(\frac{1}{2}x\right) + 2\cdot\left(\frac{1}{2}x\right) - 84 = 0 \implies 3x - 84 = 0 \implies x = 28.$$

Then $y = 14$ and $z = 14$.

Observe next that

$$f(28,14,14) = 5488.$$

Also observe that

$$f(40,20,2) = 1600.$$

Since $f(28,14,14) > f(40,20,2)$, the maximum volume occurs when $x = 28$ inches, $y = 14$ inches, $z = 14$ inches.

21. Let r be the radius of the cylindrical can in cm, and h be the height of the cylindrical can in cm.

Let $f(r,h)$ denote the volume in cubic centimeters of the can.

Note that

$$f(r,h) = \pi r^2 h.$$

The amount of material used to make the can in square cm is given by

$$2\pi rh + 2(\pi r^2).$$

We want to maximize $f(r, h)$ subject to the constraint

$$2\pi rh + 2\pi r^2 = 100.$$

Let

$$(r, h, \lambda) = \pi r^2 h + \lambda(2\pi rh + 2\pi r^2 - 100).$$

$$\frac{\partial L}{\partial r} = 2\pi rh + 2\pi h\lambda + 4\pi r\lambda = 0 \qquad \textbf{(1)}$$

$$\frac{\partial L}{\partial h} = \pi r^2 + 2\pi r\lambda = 0 \qquad \textbf{(2)}$$

$$\frac{\partial L}{\partial \lambda} = 2\pi rh + 2\pi r^2 - 100 = 0 \qquad \textbf{(3)}$$

$$\textbf{(2)} \implies \lambda = -\frac{r}{2}.$$

Substituting $\lambda = -\frac{r}{2}$ into equation **(1)** gives

$$2\pi rh + 2\pi h \cdot \left(-\frac{r}{2}\right) + 4\pi r \cdot \left(-\frac{r}{2}\right) = 0 \implies \pi rh - 2\pi r^2 = 0 \implies$$
$$\pi r(h - 2r) = 0 \implies h = 2r.$$

Substituting $h = 2r$ into equation **(3)** gives

$$2\pi r \cdot 2r + 2\pi r^2 - 100 = 0 \implies 6\pi r^2 = 100 \implies r^2 = \frac{100}{6\pi} \implies r = \frac{10}{\sqrt{6\pi}}.$$

Then $h = \frac{20}{\sqrt{6\pi}}$.

Observe that

$$f\left(\frac{10}{\sqrt{6\pi}}, \frac{20}{\sqrt{6\pi}}\right) = \pi \cdot \frac{100}{6\pi} \cdot \frac{20}{\sqrt{6\pi}} = \frac{1000}{3\sqrt{6\pi}} > f\left(2, \frac{25 - 2\pi}{\pi}\right).$$

Therefore, the maximum volume occurs when $r = \frac{10}{\sqrt{6\pi}}$ cm and $h = \frac{20}{\sqrt{6\pi}}$ cm.

23. Let r and h be the radius and height in centimeters, respectively, of the cylindrical jar.

Let $f(r, h)$ denote the exterior surface area of the jar in square centimeters. Note that

$$f(r, h) = 2\pi r h + 2\pi r^2.$$

The volume of the jar is given by

$$\pi r^2 h = 2000.$$

We want to minimize $f(r, h)$ subject to the constraint

$$\pi r^2 h - 2000 = 0.$$

Let

$$L(r, h, \lambda) = 2\pi r h + 2\pi r^2 + \lambda(\pi r^2 h - 2000).$$

$$\frac{\partial L}{\partial r} = 2\pi h + 4\pi r + 2\pi r h \lambda = 0 \quad \textbf{(1)}$$

$$\frac{\partial L}{\partial h} = 2\pi r + \pi r^2 \lambda = 0 \quad \textbf{(2)}$$

$$\frac{\partial L}{\partial \lambda} = \pi r^2 h - 2000 = 0 \quad \textbf{(3)}$$

$$((\mathbf{1}) \cdot r) - ((\mathbf{2}) \cdot 2h) \to (2\pi r h + 4\pi r^2 + 2\pi r^2 h \lambda) - (4\pi r h + 2\pi r^2 h \lambda) = 0$$

$$\implies -2\pi r h + 4\pi r^2 = 0 \implies 2\pi r(-h + 2r) = 0 \implies h = 2r.$$

Substituting $h = 2r$ into equation **(3)** gives

$$\pi r^2 \cdot 2r - 2000 = 0 \implies r^3 = \frac{1000}{\pi} \implies r = \frac{10}{\pi^{1/3}}.$$

Then $h = \frac{20}{\pi^{1/3}}$.

Observe that

$$f\left(\frac{10}{\pi^{1/3}}, \frac{20}{\pi^{1/3}}\right) = 2\pi \cdot \frac{10}{\pi^{1/3}} \cdot \frac{20}{\pi^{1/3}} + 2\pi \left(\frac{10}{\pi^{1/3}}\right)^2 = 400\pi^{1/3} + 200\pi^{1/3}$$
$$= 600\pi^{1/3}.$$

Also observe that

$$f\left(10, \frac{20}{\pi}\right) = 2\pi \cdot 10 \cdot \frac{20}{\pi} + 2\pi(10)^2 = 400 + 200\pi.$$

Since $f\left(\frac{10}{\pi^{1/3}}, \frac{20}{\pi^{1/3}}\right) < f\left(10, \frac{20}{\pi}\right)$, the minimum exterior surface area of the jar is obtained when $r = \frac{10}{\pi^{1/3}}$ cm and $h = \frac{20}{\pi^{1/3}}$ cm.

Selected Solutions to Exercise Set 8.5

1. $f(x,y) = 6x^2 + 4y^3$.

   $\frac{\partial f}{\partial x} = 12x, \qquad \frac{\partial f}{\partial y} = 12y^2.$

   $\frac{\partial f}{\partial x}(1,4) = 12, \qquad \frac{\partial f}{\partial y}(1,4) = 192.$

   $$\begin{aligned} f(1.2, 3.9) &\approx f(1,4) + \frac{\partial f}{\partial x}(1,4)(1.2-1) + \frac{\partial f}{\partial y}(1,4)(3.9-4) \\ &= 262 + 12 \cdot (0.2) + 192 \cdot (-0.1) \\ &= 245.2. \end{aligned}$$

3. $f(x,y) = 3x^2e^{-y}$.

   $\frac{\partial f}{\partial x} = 6xe^{-y}, \qquad \frac{\partial f}{\partial y} = -3x^2e^{-y}.$

   $\frac{\partial f}{\partial x}(4,-2) = 24e^2, \qquad \frac{\partial f}{\partial y}(4,-2) = -48e^2.$

   $$\begin{aligned} f(4.1,-2) &\approx f(4,-2) + \frac{\partial f}{\partial x}(4,-2) \cdot (4.1-4) + \frac{\partial f}{\partial y}(4,-2) \cdot (-2-(-2)) \\ &= 48e^2 + 24e^2(0.1) + (-48e^2)(0) \\ &= 50.4e^2 = 372.4084. \end{aligned}$$

5. $f(x,y) = \sqrt{x}\sin y$.

   $\frac{\partial f}{\partial x} = \frac{1}{2\sqrt{x}}\sin y, \qquad \frac{\partial f}{\partial y} = \sqrt{x} \cdot \cos y.$

   $\frac{\partial f}{\partial x}\left(4, \frac{\pi}{2}\right) = \frac{1}{4}, \qquad \frac{\partial f}{\partial y}\left(4, \frac{\pi}{2}\right) = 0.$

   $$\begin{aligned} f\left(4.15, \tfrac{13\pi}{24}\right) &\approx f\left(4, \tfrac{\pi}{2}\right) + \frac{\partial f}{\partial x}\left(4, \tfrac{\pi}{2}\right) \cdot (4.15-4) + \frac{\partial f}{\partial y}\left(4, \tfrac{\pi}{2}\right) \cdot \left(\tfrac{13\pi}{24} - \tfrac{\pi}{2}\right) \\ &= 2 + \tfrac{1}{4} \cdot 0.15 + 0 \cdot \tfrac{\pi}{24} \\ &= 2.0375. \end{aligned}$$

7. $f(x,y) = \frac{x}{x+y}$.

$\frac{\partial f}{\partial x} = \frac{(x+y)\cdot(1)-(x)\cdot(1)}{(x+y)^2} = \frac{y}{(x+y)^2}$, $\qquad \frac{\partial f}{\partial y} = \frac{(x+y)\cdot(0)-(x)\cdot(1)}{(x+y)^2} = \frac{-x}{(x+y)^2}$.

$\frac{\partial f}{\partial x}(6,2) = \frac{2}{64} = \frac{1}{32}$, $\qquad \frac{\partial f}{\partial y}(6,2) = \frac{-6}{64} = -\frac{3}{32}$.

$$\begin{aligned} f(6.1, 2.05) &\approx f(6,2) + \tfrac{\partial f}{\partial x}(6,2)\cdot(6.1-6) + \tfrac{\partial f}{\partial y}(6,2)\cdot(2.05-2) \\ &= \tfrac{3}{4} + \tfrac{1}{32}\cdot(0.1) + \left(-\tfrac{3}{32}\right)\cdot(0.05) \\ &= 0.7484. \end{aligned}$$

9. $f(x,y,z) = xy + yz + xz$.

$\frac{\partial f}{\partial x} = y + z$, $\qquad \frac{\partial f}{\partial y} = x + z$, $\qquad \frac{\partial f}{\partial z} = y + x$.

$\frac{\partial f}{\partial x}(1,3,7) = 10$, $\qquad \frac{\partial f}{\partial y}(1,3,7) = 8$, $\qquad \frac{\partial f}{\partial z}(1,3,7) = 4$.

$$\begin{aligned} f(0.95, 3.10, 7.05) &\approx f(1,3,7) + \tfrac{\partial f}{\partial x}(1,3,7)\cdot(0.95-1) + \tfrac{\partial f}{\partial y}(1,3,7)\cdot(3.10-3) \\ &\quad + \tfrac{\partial f}{\partial z}(1,3,7)\cdot(7.05-7) \\ &= 31 + 10\cdot(-0.05) + 8\cdot(0.10) + 4\cdot(0.05) \\ &= 31.5. \end{aligned}$$

11. $f(x,y) = 3xy^2 + 4x^3y$.

$\frac{\partial f}{\partial x} = 3y^2 + 12x^2y$, $\qquad \frac{\partial f}{\partial y} = 6xy + 4x^3$.

$\frac{\partial f}{\partial x}(1,-1) = -9$, $\qquad \frac{\partial f}{\partial y}(1,-1) = -2$.

$df = \frac{\partial f}{\partial x}(1,-1)dx + \frac{\partial f}{\partial y}(1,-1)dy = -9\,dx - 2\,dy$.

13. $f(x,y) = \sin(xy)$.

$\frac{\partial f}{\partial x} = \cos(xy)\cdot(y)$, $\qquad \frac{\partial f}{\partial y} = \cos(xy)\cdot x$.

$\frac{\partial f}{\partial x}\left(\frac{\pi}{4},1\right) = \frac{\sqrt{2}}{2}$, $\qquad \frac{\partial f}{\partial y}\left(\frac{\pi}{4},1\right) = \frac{\pi}{4}\cdot\frac{\sqrt{2}}{2}$.

$df = \frac{\partial f}{\partial x}\left(\frac{\pi}{4},1\right)dx + \frac{\partial f}{\partial y}\left(\frac{\pi}{4},1\right)dy = \frac{\sqrt{2}}{2}\,dx + \frac{\pi}{4}\cdot\frac{\sqrt{2}}{2}\,dy = \frac{\sqrt{2}}{2}\,dx + \frac{\pi\sqrt{2}}{8}\,dy$.

15. $f(x,y) = \ln(x^2 + y^2)$.

$\frac{\partial f}{\partial x} = \frac{2x}{x^2+y^2}$, $\qquad \frac{\partial f}{\partial y} = \frac{2y}{x^2+y^2}$.

$\frac{\partial f}{\partial x}(0,e) = 0$, $\qquad \frac{\partial f}{\partial y}(0,e) = \frac{2e}{e^2} = 2e^{-1}$.

$df = \frac{\partial f}{\partial x}(0,e)dx + \frac{\partial f}{\partial y}(0,e)dy = 0\,dx + 2e^{-1}\,dy = \frac{2}{e}\,dy$.

17. $f(x,y,z) = xyz + xe^{-z}$.

$\frac{\partial f}{\partial x} = yz + e^{-z}$, $\quad \frac{\partial f}{\partial y} = xz$, $\quad \frac{\partial f}{\partial z} = xy - xe^{-z}$.

$\frac{\partial f}{\partial x}(1,1,0) = 1$, $\quad \frac{\partial f}{\partial y}(1,1,0) = 0$, $\quad \frac{\partial f}{\partial z}(1,1,0) = 0$.

$$df = \frac{\partial f}{\partial x}(1,1,0)dx + \frac{\partial f}{\partial y}(1,1,0)dy + \frac{\partial f}{\partial z}(1,1,0)dz$$
$$= dx + 0 + 0 = dx.$$

19. $M = C + D$.

$dM = \frac{\partial M}{\partial C} \cdot dC + \frac{\partial M}{\partial D} \cdot dD = 1 \cdot (0.10C) + 1 \cdot (-0.05D)$.

$\frac{dM}{M} = \frac{0.10C - 0.05D}{C+D}$.

(percentage change in the money supply) $\approx \frac{0.10C-0.05D}{C+D} \cdot 100$.

21. $\frac{\partial f}{\partial x} = 300x^{-1/4}y^{1/4}$, $\quad \frac{\partial f}{\partial y} = 100x^{3/4}y^{-3/4}$.

$\frac{\partial f}{\partial x}(16,256) = 600$, $\quad \frac{\partial f}{\partial y}(15,256) = 12.5$.

$$\text{(Approximate change in output)} = df = \frac{\partial f}{\partial x}(16,256) \cdot dx + \frac{\partial f}{\partial y}(16,256) \cdot dy$$
$$= 600 \cdot (-2) + 12.5 \cdot 24$$
$$= -900.$$

23. $V(T,P) = \frac{2600T}{P}$.

$V(300,780) = 1000$, $\quad V(310,775) = 1040$.

The actual change in the volume is

$$V(310,775) - V(300,780) = 40 \text{ cm}^3.$$

25. a. $S(K,L) = 4KL - L^2$.

$S(16,20) = 4(16)(20) - (20)^2 = 880$.

b. $\frac{\partial S}{\partial K} = 4L$, $\quad \frac{\partial S}{\partial L} = 4K - 2L$.

$\frac{\partial S}{\partial K}(16,20) = 80$, $\quad \frac{\partial S}{\partial L}(16,20) = 24$.

The approximate change in sales is

$$dS = \frac{\partial S}{\partial K}(16,20) \cdot dK + \frac{\partial S}{\partial L}(16,20) \cdot dL$$
$$= 80 \cdot dK + 24 \cdot dL$$
$$= 80 \cdot 2 + 24 \cdot 0$$
$$= 160.$$

Selected Solutions to Exercise Set 8.6

1.

j	x_j	y_j	x_j^2	$x_j y_j$
1	4	9	16	36
2	3	6	9	18
3	5	7	25	35
4	5	8	25	40
Σ	17	30	75	129

$m = \frac{4(129)-(17)(30)}{4(75)-(17)^2} = \frac{6}{11} \approx 0.545.$

$b = \frac{(75)(30)-(17)(129)}{4(75)-(17)^2} = \frac{57}{11} \approx 5.182.$

The least squares regression line is $y = 0.545x + 5.182$.

3.

j	x_j	y_j	x_j^2	$x_j y_j$
1	52	74	2704	3848
2	46	66	2116	3036
3	69	94	4761	6486
4	54	91	2916	4914
5	61	84	3721	5124
6	48	80	2304	3840
Σ	330	489	18522	27248

$m = \frac{6(27248)-330(489)}{6(18522)-(330)^2} = \frac{2118}{2232} \approx 0.949.$

$b = \frac{18522(489)-330(27248)}{6(18522)-(330)^2} = \frac{65418}{2232} \approx 29.309.$

The least squares regression line is $y = 0.949x + 29.309$.

5.

j	x_j	y_j	x_j^2	x_jy_j
1	0	1	0	0
2	1	0	1	0
3	1	2	1	2
4	2	1	4	2
5	2	2	4	4
6	3	2	9	6
7	3	1	9	3
8	3	3	9	9
9	4	2	16	8
10	4	3	16	12
Σ	23	17	69	46

$m = \frac{10(46)-23(17)}{10(69)-(23)^2} = \frac{69}{161} \approx 0.429.$

$b = \frac{69(17)-23(46)}{10(69)-(23)^2} = \frac{115}{161} \approx 0.714.$

The least squares regression line is $y = 0.429x + 0.714$.

7.

j	x_j	y_j	x_j^2	x_jy_j
1	0	10	0	0
2	1	14	1	14
3	2	8	4	16
4	3	4	9	12
5	4	2	16	8
6	5	3	25	15
Σ	15	41	55	65

Let $x =$ the number of years after 1988, and let $y =$ the number of orders for new railroad cars received by the manufacturer.

a. $y = mx + b$.

$m = \frac{6(65)-15(41)}{6(55)-(15)^2} = -\frac{225}{105} = -2.143.$

$b = \frac{55(41)-15(65)}{6(55)-(15)^2} = \frac{1280}{105} = 12.19.$

The least squares regression line is $y = -2.143x + 12.19$.

b. For the year 1994, $x = 6$. The predicted number of orders the company would receive in 1994 is

$$y = -2.143(6) + 12.19 = -0.668 \text{ (668 cars returned).}$$

9.

j	x_j	y_j	x_j^2	$x_j y_j$
1	0	3.21	0	0
2	1	3.29	1	3.29
3	2	3.08	4	6.16
4	3	3.15	9	9.45
5	4	3.23	16	12.92
Σ	10	15.96	30	31.82

Let x = the number of years after 1989, and let y = the corresponding overall grade point average for undergraduates.

a. $y = mx + b$.

$m = \frac{5(31.82)-10(15.96)}{5(30)-(10)^2} = -\frac{0.5}{50} = -.01.$

$b = \frac{30(15.96)-10(31.82)}{5(30)-(10)^2} = \frac{160.6}{50} = 3.212.$

The least squares regression line is $y = -0.01x + 3.212$.

b. For the year 1995, $x = 6$. The predicted overall grade point average for undergraduates for the year 1995 is

$$y = -0.01(6) + 3.212 = 3.152.$$

c. This least squares regression line does not suggest grade inflation since the slope is negative.

11.

j	x_j	y_j	x_j^2	$x_j y_j$
1	0	20	0	0
2	1	24	1	24
3	2	25	4	50
4	3	32	9	96
5	4	34	16	136
Σ	10	135	30	306

$$m = \frac{5(306) - 10(135)}{5(30) - (10)^2} = \frac{180}{50} = 3.6.$$

$$b = \frac{30(135) - 10(306)}{5(30) - (10)^2} = \frac{990}{50} = 19.8.$$

a. For the year 1994, $x = 5$. The predicted $y = 3.6(5) + 19.8 = 37.8\%$.

b. For the year 2000, $x = 8$. The predicted $y = 3.6(8) + 19.8 = 48.6\%$.

13.

j	x_j	y_j	x_j^2	x_jy_j
1	2	5	4	10
2	0	7	0	0
3	4	2	16	8
4	6	4	36	24
5	8	3	64	24
Σ	20	21	120	66

$$m = \frac{5(66) - 20(21)}{5(120) - (20)^2} = -\frac{90}{200} = -0.45.$$

$$b = \frac{120(21) - 20(66)}{5(120) - (20)^2} = \frac{1200}{200} = 6.$$

The regression line is $y = -0.45x + 6$.

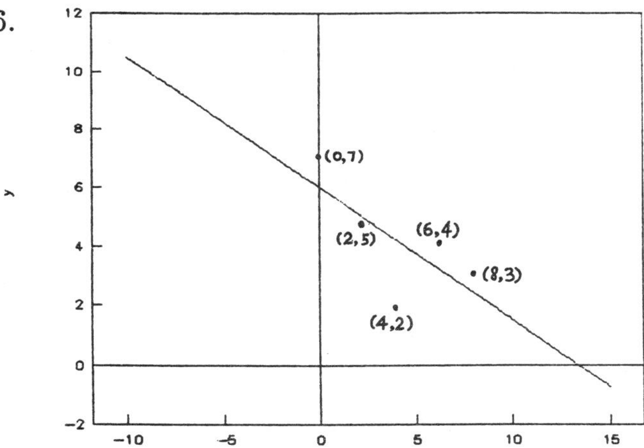

15.

j	x_j	y_j	x_j^2	x_jy_j
1	4	−2	16	−8
2	2	6	4	12
3	0	3	0	0
4	−2	8	4	−16
Σ	4	15	24	−12

$$m = \frac{4(-12) - 4(15)}{4(24) - (4)^2} = -\frac{108}{80} = -1.35.$$

$$b = \frac{24(15) - 4(-12)}{4(24) - (4)^2} = \frac{408}{80} = 5.1.$$

The regression line is $y = -1.35x + 5.1$.

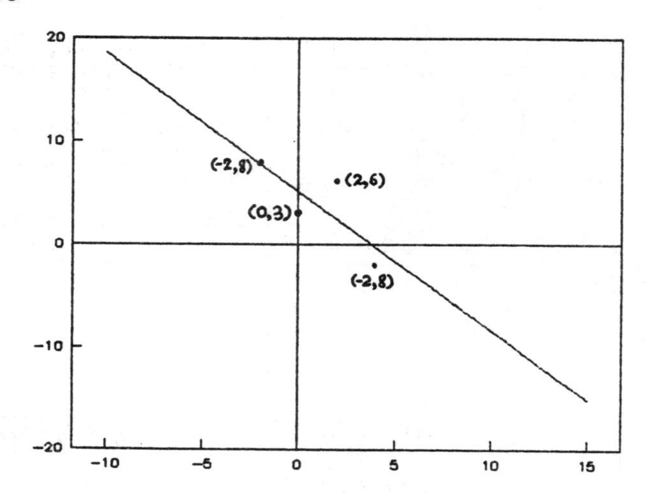

17.

j	x_j	y_j	x_j^2	$x_j y_j$
1	2	4	4	8
2	3	5	9	15
3	3.5	3	12.25	10.5
4	4.2	4	17.64	16.8
5	5	6	25	30
Σ	17.7	22	67.89	80.3

$$m = \frac{5(80.3) - 17.7(22)}{5(67.89) - (17.7)^2} = \frac{12.1}{26.16} = 0.463.$$

$$b = \frac{(67.89)(22) - 17.7(80.3)}{5(67.89) - (17.7)^2} = \frac{72.27}{26.16} = 2.763.$$

a. The least squares regression line is $y = 0.463x + 2.763$.

b. The size of the litter increases with age since the slope of the regression line is positive.

19. Here let p denote the price of an ice cream cone in cents, and let r denote the corresponding daily revenue in cents.

j	p_j	r_j	p_j^2	$p_j r_j$
1	40	2600	1600	104000
2	50	3500	2500	175000
3	60	3300	3600	198000
4	70	3640	4900	254800
5	80	3680	6400	294400
Σ	300	16720	19000	1,026,200

$$m = \frac{5(1,025,200) - 300(16,720)}{5(19,000) - (300)^2} = \frac{115,000}{5,000} = 23.$$
$$b = \frac{19,000(16,720) - 300(1,026,200)}{5(19,000) - (300)^2} = \frac{9,820,000}{5,000} = 1964.$$

a. The least squares regression line for revenue as a function of price is given by

$$r = mp + b = 23p + 1964.$$

b. According to this regression model, the daily revenue increases with increases in price since the slope of the regression line is positive.

Selected Solutions to Exercise Set 8.7

1. $\int_0^2 \int_0^1 (x-y)dx\,dy = \int_0^2 \left[\left(\frac{x^2}{2} - xy \right) \Big]_0^1 \right] dy = \int_0^2 \left(\frac{1}{2} - y \right) dy$

$$= \left(\tfrac{1}{2}y - \tfrac{1}{2}y^2 \right) \Big]_0^2 = \tfrac{1}{2}(2-4) = -1.$$

3. $\int_{-1}^1 \int_0^2 x^2 y\,dy\,dx = \int_{-1}^1 \left[\tfrac{1}{2}x^2y^2 \Big]_0^2 \right] dx$

$$= \int_{-1}^1 \tfrac{1}{2}x^2 \cdot 4\,dx = \tfrac{2}{3}x^3 \Big]_{-1}^1 = \tfrac{4}{3}.$$

5. $\int_0^2 \int_0^{x^2} (x-2y)dy\,dx = \int_0^2 \left[(xy - y^2) \Big]_0^{x^2} \right] dx = \int_0^2 (x^3 - x^4)dx$

$$= \left(\tfrac{1}{4}x^4 - \tfrac{1}{5}x^5 \right) \Big]_0^2 = \tfrac{1}{4} \cdot 16 - \tfrac{1}{5} \cdot 32 = -\tfrac{48}{20} = -\tfrac{12}{5}.$$

7. $\int_0^1 \int_0^1 ye^{x-y^2}\,dy\,dx.$

Let $u(y) = x - y^2$. $\quad du = -2y\,dy$.

$$\int_0^1 ye^{x-y^2}\,dy = -\tfrac{1}{2} \int_{y=0}^{y=1} e^u\,du = -\tfrac{1}{2}e^u \Big]_{y=0}^{y=1} = -\tfrac{1}{2}e^{x-y^2} \Big]_0^1 = -\tfrac{1}{2}\left(e^{x-1} - e^x\right).$$

$$\int_0^1 \int_0^1 ye^{x-y^2}\,dy\,dx = \int_0^1 \left[\int_0^1 ye^{x-y^2}\,dy \right] dx = \int_0^1 \left(-\tfrac{1}{2}(e^{x-1} - e^x) \right) dx$$

$$= -\tfrac{1}{2}(e^{x-1} - e^x) \Big]_0^1 = -\tfrac{1}{2}(e^0 - e^1 - e^{-1} + e^0) = \tfrac{1}{2}\left(e + \tfrac{1}{e} - 2\right).$$

9. $\int_0^1 \int_{-x}^{\sqrt{x}} \frac{y}{1+x}\, dydx = \int_0^1 \left[\frac{y^2}{2(1+x)} \right]_{-x}^{\sqrt{x}} dx = \int_0^1 \frac{x-x^2}{2(1+x)}\, dx = \int_0^1 \frac{(x-x^2+2)-2}{2(1+x)}\, dx$

$= \frac{1}{2} \int_0^1 \left(2 - x - \frac{2}{1+x}\right) dx = \frac{1}{2}\left[2x - \frac{1}{2}x^2 - 2\ln(1+x)\right]\Big]_0^1$

$= \frac{1}{2}\left(2 - \frac{1}{2} - 2\ln 2\right) = \frac{3}{4} - \ln 2.$

11. $\int_0^2 \int_0^{\pi/2} x^2 \cos y\, dydx = \int_0^2 \left[x^2 \sin y\Big]_0^{\pi/2} \right] dx$

$= \int_0^2 x^2\, dx = \frac{1}{3}x^3\Big]_0^2 = \frac{8}{3}.$

13. $\iint_R (x+y^2)dA = \int_0^2 \int_0^1 (x+y^2)dxdy = \int_0^2 \left[\left(\frac{1}{2}x^2 + xy^2\right)\Big]_0^1 \right] dy$

$= \int_0^2 \left(\frac{1}{2} + y^2\right) dy = \left(\frac{1}{2}y + \frac{1}{3}y^3\right)\Big]_0^2 = \frac{1}{2}\cdot 2 + \frac{1}{3}\cdot 8 = \frac{11}{3}.$

15. $\iint_R \frac{xy}{\sqrt{x^2+y^2}}\, dA = \int_1^2 \int_1^2 \frac{xy}{\sqrt{x^2+y^2}}\, dxdy = \int_1^2 \left[\int_1^2 \frac{y}{\sqrt{x^2+y^2}} \frac{1}{2} d(x^2+y^2) \right] dy$

$= \int_1^2 \left[\frac{1}{2}y \cdot 2\sqrt{x^2+y^2}\Big]_1^2 \right] dy = \int_1^2 y\left(\sqrt{4+y^2} - \sqrt{1+y^2}\right) dy$

$= \int_1^2 \left(\sqrt{4+y^2} - \sqrt{1+y^2}\right) \frac{1}{2} d(y^2) = \frac{1}{2}\left(\frac{2}{3}(4+y^2)^{3/2} - \frac{2}{3}(1+y^2)^{3/2}\right)\Big]_1^2$

$= \frac{1}{3}\left[(4+4)^{3/2} - (1+4)^{3/2}\right] - \frac{1}{3}\left[(4+1)^{3/2} - (1+1)^{3/2}\right]$

$= \frac{1}{3}8^{3/2} - \frac{2}{3}5^{3/2} + \frac{1}{3}2^{3/2}.$

17. $\iint_R xy\; dA = \int_0^1 \int_y^{\sqrt{y}} xy\; dxdy = \int_0^1 \left[\frac{1}{2}x^2 y\Big]_y^{\sqrt{y}} \right] dy$

$= \int_0^1 \left[\frac{1}{2}\left(\sqrt{y}\right)^2 y - \frac{1}{2}y^2 y \right] dy = \frac{1}{2}\int_0^1 (y^2 - y^3)dy = \frac{1}{2}\left(\frac{1}{3}y^3 - \frac{1}{4}y^4\right)\Big]_0^1$

$= \frac{1}{2}\left(\frac{1}{3} - \frac{1}{4}\right) = \frac{1}{24}.$

19. $\int_{-1}^{1}\int_{0}^{x+1}(x+y)dydx = \int_{0}^{2}\int_{y-1}^{1}(x+y)dxdy$

$= \int_{0}^{2}\left[\left(\tfrac{1}{2}x^2+xy\right)\Big]_{y-1}^{1}\right]dy$

$= \int_{0}^{2}\left\{\left(\tfrac{1}{2}+y\right)-\left[\tfrac{1}{2}(y-1)^2+(y-1)y\right]\right\}dy$

$= \int_{0}^{2}\left(-\tfrac{3}{2}y^2+3y\right)dy = \left(-\tfrac{1}{2}y^3+\tfrac{3}{2}y^2\right)\Big]_{0}^{2}$

$= -\tfrac{1}{2}\cdot 8+\tfrac{3}{2}\cdot 4 = 2.$

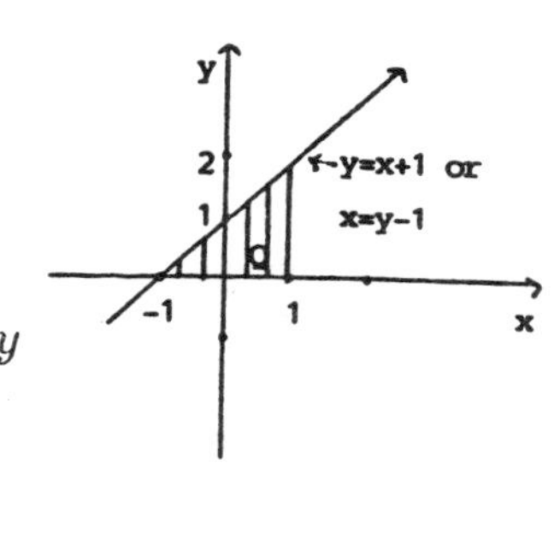

21. $\int_{0}^{1}\int_{0}^{y}xy^2\ dxdy = \int_{0}^{1}\int_{x}^{1}xy^2\ dydx$

$= \int_{0}^{1}\left[\tfrac{1}{3}xy^3\Big]_{x}^{1}\right]dx = \tfrac{1}{3}\int_{0}^{1}(x-x^4)dx$

$= \tfrac{1}{3}\left(\tfrac{1}{2}x^2-\tfrac{1}{5}x^5\right)\Big]_{0}^{1} = \tfrac{1}{3}\left(\tfrac{1}{2}-\tfrac{1}{5}\right) = \tfrac{1}{10}.$

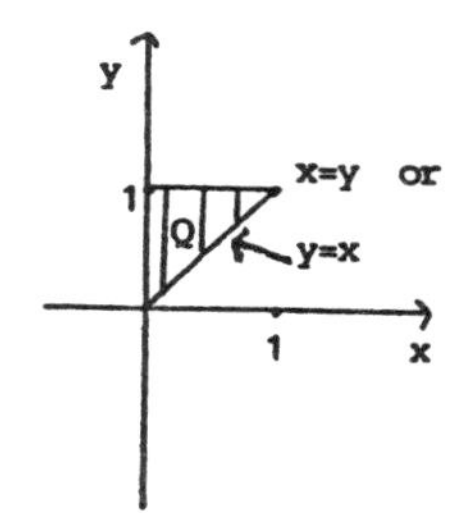

23. Volume $= \int_{0}^{3}\int_{0}^{3}(6-x-y)dxdy = \int_{0}^{3}\left[\left(6x-\tfrac{1}{2}x^2-xy\right)\Big]_{0}^{3}\right]dy$

$= \int_{0}^{3}\left(18-\tfrac{9}{2}-3y\right)dy = \left(18y-\tfrac{9}{2}y-\tfrac{3}{2}y^2\right)\Big]_{0}^{3} = 54-\tfrac{27}{2}-\tfrac{3}{2}\cdot 9 = 27.$

25. Volume $= \int_{0}^{2}\int_{0}^{4-x^2}4\ dydx = 4\int_{0}^{2}\left(y\Big]_{0}^{4-x^2}\right)dx = 4\int_{0}^{2}(4-x^2)dx$

$= 4\left(4x-\tfrac{1}{3}x^3\right)\Big]_{0}^{2} = 4\left(8-\tfrac{8}{3}\right) = \tfrac{64}{3}.$

27. Volume $= \iint_R(12-4x-2y)dA = \int_{0}^{2}\int_{0}^{1}(12-4x-2y)dydx$

$= \int_{0}^{2}\left\{\left[(12-4x)y-y^2\right]\Big]_{0}^{1}\right\}dx = \int_{0}^{2}(12-4x-1)dx$

$= \int_{0}^{2}(11-4x)dx = \left(11x-2x^2\right)\Big]_{0}^{2} = 11\cdot(2)-2\cdot(4) = 14.$

29. Volume $= \iint_R f(x,y)dA = \int_0^2 \int_0^{3-\frac{3}{2}x} (4+x+y)dydx$

$= \int_0^2 \left\{ \left[(4+x)y + \frac{1}{2}y^2\right]\Big]_0^{3-\frac{3}{2}x} \right\} dx$

$= \int_0^2 \left[(4+x)\left(3-\frac{3}{2}x\right) + \frac{1}{2}\left(3-\frac{3}{2}x\right)^2\right] dx$

$= \int_0^2 \frac{3}{8}(44-20x-x^2)dx = \frac{3}{8}\left(44x - 10x^2 - \frac{1}{3}x^3\right)\Big]_0^2$

$= \frac{3}{8}\left(88 - 40 - \frac{1}{3}\cdot 8\right) = 17.$

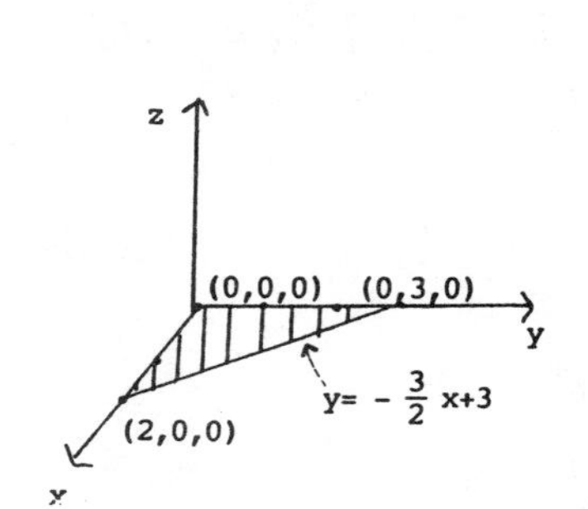

31. Area $= \iint_R dA = \int_0^1 \int_{x^3}^{x^2} dydx$

$= \int_0^1 \left(y\Big]_{x^3}^{x^2}\right) dx = \int_0^1 (x^2 - x^3)dx$

$= \left(\frac{1}{3}x^3 - \frac{1}{4}x^4\right)\Big]_0^1$

$= \frac{1}{3} - \frac{1}{4} = \frac{1}{12}.$

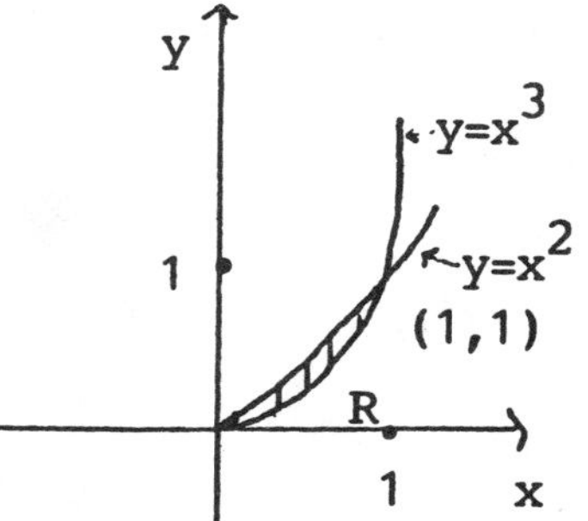

33. This formula for the average value of f over R follows because

$$\text{Area of } R = \iint_R 1 \cdot dA.$$

35. The area of the region R is 6.

$\iint_R f(x,y)dA = \int_0^4 \int_0^{-\frac{3}{4}x+3} 12x\, dydx = \int_0^4 12xy\Big]_0^{-\frac{3}{4}x+3} dx$

$= \int_0^4 (-9x^2 + 36x)dx = (-3x^3 + 18x^2)\Big]_0^4 = 96.$

The average value $= \frac{96}{6} = 16.$

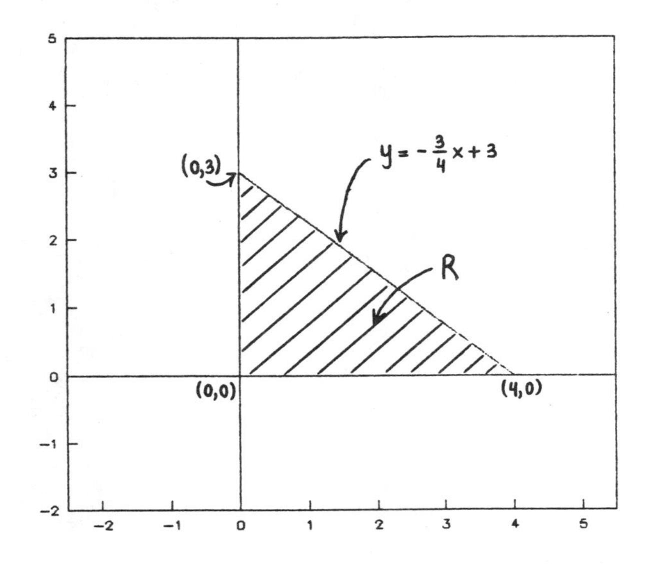

Selected Solutions to Chapter 8 - Review Exercises

1. $f(x,y) = \sqrt{25 - x^2 - y^2}$.

   The domain of f is the set of ordered pairs (x, y) of real numbers such that $x^2 + y^2 \leq 25$.

3. $f(x,y) = \frac{y-x}{y+x}$.

   a. $f(2,-1) = \frac{-1-2}{-1+2} = -3$.

   b. $f(1,-2) = \frac{-2-1}{-2+1} = 3$.

5. $f(x,y) = \frac{1+x^2}{1-xy}$.

   $f(x,y)$ is discontinuous on the set of points $\{(x,y) \mid xy = 1\}$.

7. $f(x,y) = (x-y)^3 + \ln xy$.

   $\frac{\partial f}{\partial x} = 3(x-y)^2 + \frac{y}{xy} = 3(x-y)^2 + \frac{1}{x}$.

   $\frac{\partial f}{\partial y} = 3(x-y)^2(-1) + \frac{x}{xy} = -3(x-y)^2 + \frac{1}{y}$.

9. $f(x,y) = \frac{xy}{x+y}$.

   $\frac{\partial f}{\partial x} = \frac{y(x+y)-xy}{(x+y)^2} = \left(\frac{y}{x+y}\right)^2$.

   $\frac{\partial f}{\partial y} = \frac{x(x+y)-xy}{(x+y)^2} = \left(\frac{x}{x+y}\right)^2$.

11. $f(x,y) = x^{2/3}y^{1/3} - \frac{x^{3/4}}{y^{1/4}}$.

    $\frac{\partial f}{\partial x} = \frac{2}{3}x^{-1/3}y^{1/3} - \frac{3x^{-1/4}}{4y^{1/4}} = \frac{2y^{1/3}}{3x^{1/3}} - \frac{3}{4x^{1/4}y^{1/4}}$.

    $\frac{\partial f}{\partial y} = \frac{1}{3}x^{2/3}y^{-2/3} - \left(-\frac{1}{4}\right)y^{-5/4}x^{3/4} = \frac{x^{2/3}}{3y^{2/3}} + \frac{x^{3/4}}{4y^{5/4}}$.

13. $f(x,y) = xy\sqrt{y^2 - x^2}$.

    $\frac{\partial f}{\partial x} = y\sqrt{y^2 - x^2} + xy \cdot \frac{-2x}{2\sqrt{y^2-x^2}} = y\sqrt{y^2 - x^2} - \frac{x^2y}{\sqrt{y^2-x^2}}$.

    $\frac{\partial f}{\partial y} = x\sqrt{y^2 - x^2} + xy \cdot \frac{2y}{2\sqrt{y^2-x^2}} = x\sqrt{y^2 - x^2} + \frac{xy^2}{\sqrt{y^2-x^2}}$.

15. $f(x,y) = (xy^2 - x^2y)^{2/3}$.

    $\frac{\partial f}{\partial x} = \frac{2}{3}(xy^2 - x^2y)^{-1/3}(y^2 - 2xy) = \frac{2}{3}y(y-2x)(xy^2 - x^2y)^{-1/3}$.

    $\frac{\partial f}{\partial y} = \frac{2}{3}(xy^2 - x^2y)^{-1/3}(2xy - x^2) = \frac{2}{3}x(2y-x)(xy^2 - x^2y)^{-1/3}$.

17. $f(x,y,z) = \ln\sqrt{x^2+4y^2+2z}$.

$\frac{\partial f}{\partial x} = \frac{1}{\sqrt{x^2+4y^2+2z}} \cdot \frac{1}{2}\left(x^2+4y^2+2z\right)^{-1/2} \cdot 2x = \frac{x}{x^2+4y^2+2z}$.

$\frac{\partial f}{\partial y} = \frac{1}{\sqrt{x^2+4y^2+2z}} \cdot \frac{1}{2}\left(x^2+4y^2+2z\right)^{-1/2} \cdot 8y = \frac{4y}{x^2+4y^2+2z}$.

$\frac{\partial f}{\partial z} = \frac{1}{\sqrt{x^2+4y^2+2z}} \cdot \frac{1}{2}\left(x^2+4y^2+2z\right)^{-1/2} \cdot 2 = \frac{1}{x^2+4y^2+2z}$.

19. $f(x,y) = \sqrt{y^2-x^2}$.

$\frac{\partial f}{\partial x} = \frac{-2x}{2\sqrt{y^2-x^2}} = -x(y^2-x^2)^{-1/2}$. $\quad \frac{\partial f}{\partial y} = \frac{2y}{2\sqrt{y^2-x^2}} = y(y^2-x^2)^{-1/2}$.

$\frac{\partial^2 f}{\partial x^2} = -(y^2-x^2)^{-1/2} - x\left(-\frac{1}{2}\right)(y^2-x^2)^{-3/2}(-2x)$

$= -(y^2-x^2)^{-1/2} - x^2(y^2-x^2)^{-3/2} = \frac{-(y^2-x^2+x^2)}{(y^2-x^2)^{3/2}} = \frac{-y^2}{(y^2-x^2)^{3/2}}$.

$\frac{\partial^2 f}{\partial x \partial y} = -x\left(-\frac{1}{2}\right)(y^2-x^2)^{-3/2}(2y) = \frac{xy}{(y^2-x^2)^{3/2}}$.

$\frac{\partial^2 f}{\partial y^2} = (y^2-x^2)^{-1/2} + y\left(-\frac{1}{2}\right)(y^2-x^2)^{-3/2}(2y)$

$= \frac{1}{\sqrt{y^2-x^2}} - \frac{y^2}{(y^2-x^2)^{3/2}} = \frac{y^2-x^2-y^2}{(y^2-x^2)^{3/2}} = \frac{-x^2}{(y^2-x^2)^{3/2}}$.

21. $f(x,y) = 3x^2 + xy - 6y^2$.

$$\frac{\partial f}{\partial x} = 6x + y = 0 \qquad \textbf{(1)}$$

$$\frac{\partial f}{\partial y} = x - 12y = 0 \qquad \textbf{(2)}$$

$$\textbf{(1)} + (\textbf{(2)} \times (-6)) \implies 73y = 0 \implies y = 0 \qquad \textbf{(3)}$$

From **(2)** and **(3)**: $x = 12y = 12 \cdot (0) = 0$.

So the critical point is (0,0).

$\frac{\partial^2 f}{\partial x^2} = 6, \quad \frac{\partial^2 f}{\partial x \partial y} = 1, \quad \frac{\partial^2 f}{\partial y^2} = -12.$

At (0,0), $D = 1 - 6(-12) = 73 > 0$.

$f(x,y)$ has a saddle point at (0,0,0).

23. $f(x,y) = e^{x^2-4xy}$.

$$\frac{\partial f}{\partial x} = e^{x^2-4xy}(2x-4y) = 0 \implies 2x - 4y = 0 \qquad \textbf{(1)}$$

$$\frac{\partial f}{\partial y} = e^{x^2-4xy}(-4x) = 0 \implies 4x = 0 \qquad \textbf{(2)}$$

From **(2)** : $4x = 0 \implies x = 0.$ **(3)**

From **(1)** and **(3)** : $2 \cdot (0) - 4y = 0 \implies y = 0.$

So (0,0) is the critical point.

$$\frac{\partial^2 f}{\partial x^2} = e^{x^2-4xy}(2x-4y)^2 + 2e^{x^2-4xy},$$
$$\frac{\partial^2 f}{\partial x \partial y} = e^{x^2-4xy}(2x-4y)(-4x) + (-4)e^{x^2-4xy},$$
$$\frac{\partial^2 f}{\partial y^2} = e^{x^2-4xy} \cdot 16x^2.$$

At (0,0): $A = 2, \quad B = -4, \quad C = 0. \quad D = 16 > 0.$

Thus (0,0,1) is a saddle point of f.

25. $f(x,y) = e^{1+x^2-y^2}.$

$$\frac{\partial f}{\partial x} = e^{1+x^2-y^2}(2x) = 0 \implies x = 0.$$
$$\frac{\partial f}{\partial y} = e^{1+x^2-y^2}(-2y) = 0 \implies y = 0.$$

So (0,0) is the critical point.

$$\frac{\partial^2 f}{\partial x^2} = e^{1+x^2-y^2}(4x^2) + 2e^{1+x^2-y^2},$$
$$\frac{\partial^2 f}{\partial x \partial y} = e^{1+x^2-y^2}(-4xy),$$
$$\frac{\partial^2 f}{\partial y^2} = e^{1+x^2-y^2}(4y^2) - 2e^{1+x^2-y^2}.$$

At (0,0): $A = 2, \quad B = 0, \quad C = -2.$ Then $D = 4 > 0.$

So (0,0) $\to$ a saddle point.

27. $\int_{-1}^{1}\int_{0}^{2}(2x+3y)dxdy = \int_{-1}^{1}\left[(x^2+3xy)\right]_0^2 dy$

$= \int_{-1}^{1}(4+6y)dy = (4y+3y^2)\Big]_{-1}^{1} = 4+3-(-4+3) = 8.$

29. $\int_0^1\int_0^x xy\sqrt{x^2+y^2}\;dydx = \int_0^1\left[\int_0^x x\sqrt{x^2+y^2}\;\frac{1}{2}\;d(y^2)\right]dx$

$= \int_0^1\left[\frac{1}{2}x\cdot\frac{2}{3}(x^2+y^2)^{3/2}\Big]_0^x\right]dx = \int_0^1 \frac{1}{3}x\left[(x^2+x^2)^{3/2}-(x^2+0)^{3/2}\right]dx$

$= \frac{1}{3}(2^{3/2}-1)\int_0^1 x^4\;dx = \frac{1}{3}(2^{3/2}-1)\cdot\frac{1}{5}x^5\Big]_0^1 = \frac{1}{15}(2^{3/2}-1).$

31. $\int_0^1\int_0^x xy(7+y^2)dydx = \int_0^1\left[\int_0^x x(7+y^2)\frac{1}{2}d(y^2)\right]dx = \int_0^1\left[\frac{1}{4}x(7+y^2)^2\Big]_0^x\right]dx$

$= \frac{1}{4}\int_0^1 x\left[(7+x^2)^2-7^2\right]dx = \frac{1}{4}\int_0^1\left[(7+x^2)^2-49\right]\cdot\frac{1}{2}d(x^2)$

$= \frac{1}{8}\left[\frac{1}{3}(7+x^2)^3-49x^2\right]\Big]_0^1 = \frac{1}{8}\left(\frac{1}{3}\cdot 8^3-49-\frac{1}{3}\cdot 7^3\right) = \frac{1}{8}\cdot\frac{22}{3} = \frac{11}{12}.$

33. a. $R(x,y) = xp+yq$

$= x(80-2x)+y(120-4y)$

$= -2x^2-4y^2+80x+120y.$

b. $P(x,y) = R(x,y)-C(x,y)$

$= (-2x^2-4y^2+80x+120y)-(1000+20x+20y)$

$= -2x^2-4y^2+60x+100y-1000.$

c. $\frac{\partial R}{\partial x} = -4x+80 = 0 \implies x = 20$

$\frac{\partial R}{\partial y} = -8y+120 = 0 \implies y = 15$

$\frac{\partial^2 R}{\partial x^2} = -4,\quad \frac{\partial^2 R}{\partial x\partial y} = 0,\quad \frac{\partial^2 R}{\partial y^2} = -8.$

At (20,15), $D = -32 < 0$ and $A < 0$,

thus the revenue is a maximum when $(x,y) = (20,15)$.

d. $\frac{\partial P}{\partial x} = -4x+60 = 0 \implies x = 15$

$\frac{\partial P}{\partial y} = -8y+100 = 0 \implies y = 12.5$

$\frac{\partial^2 P}{\partial x^2} = -4,\quad \frac{\partial^2 P}{\partial x\partial y} = 0,\quad \frac{\partial^2 P}{\partial y^2} = -8.$

At (15,12.5), $D = -32 < 0$ and $A < 0$.

Thus the profit is a maximum when $(x,y) = (15,12.5)$.

35. a. $P(4,16) = 200\sqrt{4\cdot 16} = 1600.$

b. $\frac{\partial P}{\partial x} = 200\cdot\frac{y}{2\sqrt{xy}} = 100\sqrt{\frac{y}{x}}.$

When $y = 16$, the increase in $P \approx \frac{\partial P}{\partial x}(4,16)\cdot(5-4) = 100\sqrt{\frac{16}{4}} = 200.$

37. $P(x,y) = 25x + 5y - x^2 - xy$.

$$\frac{\partial P}{\partial x} = 25 - 2x - y = 0 \qquad (1)$$

$$\frac{\partial P}{\partial y} = 5 - x = 0 \qquad (2)$$

From **(2)** $\implies x = 5$.

From **(1)** $\implies y = 25 - 2x \implies y = 25 - 2(5) = 15$.

$$\frac{\partial^2 P}{\partial x^2} = -2, \quad \frac{\partial^2 P}{\partial x \partial y} = -1, \quad \frac{\partial^2 P}{\partial y^2} = 0.$$

At (5,15): $D = 1 > 0$. $(5,15) \to$ a saddle point of P.

Note that $P(2.5, 20) > P(5, 15)$. Thus there is no maximum profit.

39. $f(x,y,z) = x + 2y - 3z + 1, \qquad g(x,y,z) = x^2 + y^2 + z^2 - 14 = 0.$

$L(x,y,z,\lambda) = f(x,y,z) + \lambda g(x,y,z)$.

$$\frac{\partial L}{\partial x} = 1 + 2x\lambda = 0 \qquad (1)$$

$$\frac{\partial L}{\partial y} = 2 + 2y\lambda = 0 \qquad (2)$$

$$\frac{\partial L}{\partial z} = -3 + 2z\lambda = 0 \qquad (3)$$

$$\frac{\partial L}{\partial \lambda} = x^2 + y^2 + z^2 - 14 = 0 \qquad (4)$$

From **(1)**, **(2)**, and **(3)** $\implies x = -\dfrac{1}{2\lambda}, \quad y = -\dfrac{1}{\lambda}, \quad z = \dfrac{3}{2\lambda}$. **(5)**

(4) and **(5)** give $\frac{1}{4\lambda^2} + \frac{1}{\lambda^2} + \frac{9}{4\lambda^2} = 14 \implies 1 + 4 + 9 = 14 \cdot (4\lambda^2)$

$\implies \lambda^2 = \frac{1}{4} \implies \lambda = \pm\frac{1}{2}$.

For $\lambda = \frac{1}{2}$: $x = -1, \quad y = -2, \quad z = 3$.

For $\lambda = -\frac{1}{2}$: $x = 1, \quad y = 2, \quad z = -3$.

Since $f(-1,-2,3) = -13, \quad f(1,2,-3) = 15$,

thus the maximum value of f is $f(1,2,-3) = 15$.

41. We are to maximize $S = 10xy + 2x + 4y$ subject to the constraint $x + y = 10$.

Let $L(x, y, \lambda) = (10xy + 2x + 4y) + \lambda \cdot (x + y - 10)$.

$\frac{\partial L}{\partial x} = 10y + 2 + \lambda(1) = 0.$

$\frac{\partial L}{\partial y} = 10x + 4 + \lambda(1) = 0.$

$\frac{\partial L}{\partial \lambda} = x + y - 10 = 0.$

$\frac{\partial L}{\partial x} = 0 \implies \lambda = -10y - 2.$

$\frac{\partial L}{\partial y} = 0 \implies \lambda = -10x - 4.$

$$-10y - 2 = -10x - 4.$$

$$10x - 10y = -2.$$

$$x + y = 10.$$

$$10x - 10(10 - x) = -2 \implies 20x = 98 \implies x = 4.9 \implies y = 5.1.$$

43. The area of the region is given by

$$\int_0^1 \left\{ \int_{x^2}^{\sqrt{x}} 1\ dy \right\} dx = \int_0^1 \{\sqrt{x} - x^2\}\, dx$$

$$= \left(\frac{2}{3} x^{3/2} - \frac{1}{3} x^3 \right) \Big]_0^1 = \left(\frac{2}{3} - \frac{1}{3} \right) = \frac{1}{3}.$$

45. Revenue $R = (1200 - 40x + 20y) \cdot x + (2650 + 10x - 30y) \cdot y$

$$= -40x^2 + 30xy - 30y^2 + 1200x + 2650y.$$

Cost $C = (1200 - 40x + 20y) \cdot 65 + (2650 + 10x - 30y) \cdot 80$

$$= 78000 - 2600x + 1300y + 212000 + 800x - 2400y$$

$$= -1800x - 1100y + 290,000.$$

Profit $P = R - C$

$$= -40x^2 + 30xy - 30y^2 + 3000x + 3750y - 290,000.$$

$\frac{\partial P}{\partial x} = -80x + 30y + 3000.$

$\frac{\partial P}{\partial y} = 30x - 60y + 3750.$

$\frac{\partial P}{\partial x} = 0$ and $\frac{\partial P}{\partial y} = 0$:

$$\begin{cases} -80x + 30y + 3000 = 0 \\ 30x - 60y + 3750 = 0 \end{cases}$$

$$\begin{cases} -160x + 60y + 6000 = 0 \\ 30x - 60y + 3750 = 0 \end{cases}$$

$$-130x + 9750 = 0 \implies x = 75.$$

$$-80 \cdot 75 + 30y + 3000 = 0 \implies 30y = 3000 \implies y = 100.$$

$\frac{\partial^2 P}{\partial x^2} = -80, \quad \frac{\partial^2 P}{\partial y^2} = -60, \quad \frac{\partial P}{\partial x \partial y} = 30.$

$D = (30^2) - (-80)(-60) = -3900 < 0.$

$A = -80 < 0.$

Therefore the maximum profit P occurs when $x = 75$, $y = 100$.

Practice Problems for Chapter 8

1. Find the indicated values of the following functions.
   a. $f(x, y) = \frac{x\sqrt{y}}{1+x}$; $f(2, 9)$, $f(5, 1)$, and $f(0, 0)$.
   b. $f(x, y) = xe^{y/x}$; $f(1, 0)$, $f(3, 12)$, and $f\left(\frac{1}{2}, \ln 4\right)$.
   c. $f(x, y, z) = x^2 e^{\sqrt{y^2+z^2}}$; $f(1, -1, 0)$, $f(2, 5, -2)$, and $f(3, 1, 3)$.
   d. $f(r, s, t, u) = \ln(ru)$; $f(1, 5, 3, 1)$ and $f(e, \ln 2, -3, e^{-3})$.
   e. $f(x, y) = x^2 \sin y + y^2 \cos x$; $f\left(\frac{\pi}{2}, \frac{5\pi}{3}\right)$ and $f\left(-\frac{\pi}{4}, -\frac{\pi}{2}\right)$.
   f. $f(x, y) = \frac{x^2+3xy+3y^2}{x+y}$; show that $f(2a, 2b) = 2f(a, b)$.
2. Find the domain of each of the following functions.
   a. $f(x, y) = \frac{1}{\sqrt{x^2-4}}$.
   b. $f(x, y, z) = \frac{2^{xyz}}{z^{1/4}}$.
   c. $f(x, y, z) = \sqrt{\ln\left[3 - (x^2 + y^2 + z^2)\right]}$.
   d. $f(x, y) = e^{y/x}$.
   e. $f(x, y) = 3x^{4/5} \cdot y^{3/4} - x^{-3/2} \cdot y^{2/3}$.
   f. $f(x, y) = \frac{2 \sin x}{1 - 2\cos(xy)}$.

3. Find all the first-order partial derivatives for each of the following functions.

a. $f(x,y) = 3x^2 + xy + 5y^2$.

b. $f(x,y) = e^{x/y}$.

c. $f(x,y) = \sqrt[3]{x^2 + y^2}$.

d. $f(x,y,z) = \frac{x}{y-z}$.

e. $f(x,y,z) = x^2y^3z^5 - 3x^2y^4z^3 + 5xz$.

f. $f(x,y) = x^2e^{3xy}\ln\left(\frac{2}{y}\right)$.

g. $f(x,y,z) = ze^{z/(xy)}$.

h. $f(x,y) = \sqrt{x}\,\sin^2\left(\frac{y}{x}\right)$.

4. Find the indicated partial derivative(s) at the given point.

a. $f(x,y) = xye^{2x-y}$; $\frac{\partial f}{\partial x}(2,-3)$.

b. $f(x,y,z) = xy^2z + 5$; $\frac{\partial f}{\partial y}(2,-1,3)$.

c. $f(x,y,z) = \frac{x}{y-z}$; $\frac{\partial f}{\partial x}(2,-1,3)$ and $\frac{\partial f}{\partial z}(2,-1,3)$.

d. $f(x,y,z) = e^{xz} \cdot \sqrt{y+2z}$; $\frac{\partial f}{\partial z}(0,1,4)$.

e. $f(r,s,t,u) = (s^2 + tu) \cdot \ln(2r + 7st)$; $\frac{\partial f}{\partial s}(1,1,0,1)$.

5. Find all second-order partial derivatives for each of the following functions.

a. $f(x,y) = xe^{y/x}$.

b. $f(x,y) = \ln\sqrt{x^2 + y^2}$.

c. $f(x,y) = (x^2 + xy + y^2)(x^2 + xy + 1)$.

d. $f(x,y) = \cos(x+y) + \sin(x-y)$.

6. For each of the following functions, find all critical points for f, and determine whether these points correspond to a relative maximum, a relative minimum, or a saddle point.

a. $f(x,y) = x^2 - 3y^2 + 4x + 6y + 8$.

b. $f(x,y) = -2x^2 + 2xy - y^2 + 4x - 6y + 5$.

c. $f(x,y) = 2x^2 + 3xy + 5y^2$.

d. $f(x,y) = xy - \frac{1}{x} - \frac{1}{y}$.

e. $f(x,y) = \frac{1}{3}(x^3 + 8y^3) - 2(x^2 + y^2) + 1$.

f. $f(x,y) = x^3 + y^3 - 3xy$.

g. $f(x,y) = e^{(7x^2 - 5xy + y^2 + x - y + 6)}$.

h. $f(x,y) = x^3 - y^2 - 3x + 4y + 11$.

7. In the following exercises, use the method of Lagrange Multipliers to find where the minimum and maximum values of f occur subject to the given constraint.

   a. $f(x, y) = x^2 - y^2$ subject to $2x + y = 3$.

   b. $f(x, y) = x^2 + xy + y^2 - 2x - 5y + 10$ subject to $1 - x + y = 0$.

   c. $f(x, y, z) = 3x + 2y + z$ subject to $xyz = 36$.

   d. $f(x, y, z) = xyz^2$ subject to $x^2 + y^2 + z^2 = 20$.

8. Evaluate each of the following iterated integrals.

   a. $\int_1^3 \int_1^2 (x^2 - y)dxdy$.

   b. $\int_0^3 \int_0^x (x^2 + y^2)dydx$.

   c. $\int_1^{\ln 8} \int_0^{\ln y} e^{x+y}\, dxdy$.

   d. $\int_0^2 \int_0^{\sqrt{4-y^2}} x\, dxdy$.

   e. $\int_0^3 \int_{y^2}^{3y} x\, dxdy$.

   f. $\int_{-2}^1 \int_{x^2+4x}^{3x+2} dydx$.

   g. $\int_0^2 \int_y^2 e^{x^2} dxdy$.

   h. $\int_0^\pi \int_x^\pi \frac{\sin y}{y}\, dydx$.

9. Evaluate the following double integrals.

   a. $\iint_R (x^2 + 2xy - 3y^2)dA$, $R = \{(x, y)\,|\,0 \le x \le 1,\ x^2 \le y \le \sqrt{x}\}$.

   b. $\iint_R x^2y^2\, dA$, $R = \{(x, y)\,|\,0 \le x \le y,\ 1 \le y \le 2\}$.

   c. $\iint_R x \sin y\, dA$, $R = \{(x, y)\,|\,0 \le x \le \pi,\ 0 \le y \le x\}$.

   d. $\iint_R (x^3 - xy)dA$, $R = \{(x, y)\,|\,0 \le x \le 1,\ x^3 \le y \le x^2\}$.

10. The productivity of a country is given by

$$f(x, y) = 300x^{2/3} \cdot y^{1/3},$$

where x and y are the amounts of labor and capital, respectively.

   a. Find the marginal productivity of labor and capital when $x = 125$ and $y = 64$.

   b. What would be the approximate effect of utilizing 125 units of labor and cutting back to 62 units of capital?

11. The base of a rectangular box costs three times as much per square foot as the sides and top. Find the relative dimensions for the most economical box for a given volume.

12. A monopolist manufactures and sells two competing products, call them I and II, that cost $\$P_I$ and $\$P_{II}$ per unit, respectively, to produce. Let $R(x, y)$ be the revenue from marketing x units of product I and y units of product II. Show that if the monopolist's profit is maximized when $x = a$, $y = b$, then

$$\frac{\partial R}{\partial x}(a, b) = P_I \quad \text{and} \quad \frac{\partial R}{\partial y}(a, b) = P_{II}.$$

13. The production function for a firm is

$$f(x, y) = 64x^{3/4} \cdot y^{1/4},$$

where x and y are the numbers of units of labor and capital utilized. Suppose that labor costs \$96 per unit and capital costs \$162 per unit, and that the firm decides to produce 3456 units of goods.

a. Determine the amounts of labor and capital that should be utilized in order to minimize the cost.

b. Find the value of λ at the optimal level of production.

14. a. By the method of least squares find the linear regression line for the following data. Refer to 1985 as year $x = 1$.

Overseas Shipments of Dooflets by Klas Company

(in millions)

Year	1985	1986	1987	1988	1989
Quantity	35	31	26	24	26

b. For the data in part (a), refer to 1985 as year $x = -2$, 1986 as year $x = -1$, 1987 as year $x = 0$, etc. Then $\sum_{i=1}^{5} x_i = 0$. Fit a least squares line and observe how the calculation is simplified.

15. a. Find the volume of the solid bounded above by $f(x, y) = x + y + 1$ and below by the triangle $R = \{(x, y) \mid 0 \leq x \leq 1,\ 0 \leq y \leq x\}$.

b. Find the volume of the solid bounded above by $f(x, y) = x \cdot \sqrt{x^2 + y^2}$ and below by the region $R = \{(x, y) \mid 0 \leq x \leq \sqrt{4 - y^2},\ -2 \leq y \leq 2\}$.

c. Find the area of the region R bounded by the graphs of the parabola $x = y - y^2$ and of the line $x + y = 0$.

16. Use a linear approximation to calculate the value of $f(x, y) = xy\sqrt{x^2 + y^2}$ at $x = 2.98$, $y = 4.04$.

17. Find a linear approximation for values of the function $f(x, y) = (2 + xy)\cos y$ near the point $(2, \frac{\pi}{2})$.

18. Find those points of the curve $x^2 + 24xy + 8y^2 = 1700$ that are closest to the origin.

19. Consider two products A and B. Let p represent the unit selling price of product A, and let q represent the unit selling price of product B. Let $f(p, q)$ represent the demand for product A, and let $g(p, q)$ represent the demand for product B. For each of the following cases determine whether A and B are competitive, complementary, or neither.

a. $f(p, q) = 1000 - 50p + 2q$, $g(p, q) = 500 + 4p - 20q$.

b. $f(p, q) = \frac{100}{p\sqrt{q}}$, $g(p, q) = \frac{500}{q\sqrt[3]{p}}$.

20. As in Problem 19, we consider two product A and B. Let p represent the unit selling price of product A, and let q represent the unit selling price of product B. Let the demand for product A be given by

$$f(p,q) = 1000 - 50p + 2q,$$

and let the demand for product B be given by

$$g(p,q) = 500 + 4p - 20q.$$

Let $R(p,q)$ denote the corresponding revenue. Find the values of p and q which maximize $R(p,q)$.

SOLUTIONS TO PRACTICE PROBLEMS FOR CHAPTER 8

1.a. $f(x,y) = \frac{x\sqrt{y}}{1+x}$.

$f(2,9) = \frac{2\sqrt{9}}{1+2} = 2, \quad f(5,1) = \frac{5\sqrt{1}}{1+5} = \frac{5}{6}, \quad f(0,0) = 0.$

b. $f(x,y) = xe^{y/x}$.

$f(1,0) = 1 \cdot e^{0/1} = 1, \quad f(3,12) = 3e^{12/3} = 3e^4$, and

$f\left(\frac{1}{2}, \ln 4\right) = \frac{1}{2}e^{(\ln 4)/\frac{1}{2}} = \frac{1}{2}e^{2\ln 4} = \frac{1}{2}e^{\ln 4^2} = \frac{1}{2}(4^2) = 8.$

c. $f(x,y,z) = x^2 \cdot e^{\sqrt{y^2+z^2}}$.

$f(1,-1,0) = 1^2 \cdot e^{\sqrt{(-1)^2+0^2}} = e.$

$f(2,5,-2) = 2^2 \cdot e^{\sqrt{5^2+(-2)^2}} = 4e^{\sqrt{29}}.$

$f(3,1,3) = 3^2 \cdot e^{\sqrt{1^2+3^2}} = 9e^{\sqrt{10}}.$

d. $f(r,s,t,u) = \ln(ru)$.

$f(1,5,3,1) = \ln(1 \cdot 1) = 0.$

$f(e, \ln 2, -3, e^{-3}) = \ln(e \cdot e^{-3}) = \ln(e^{-2}) = -2.$

e. $f(x,y) = x^2 \sin y + y^2 \cos x$.

$$f\left(\frac{\pi}{2}, \frac{5\pi}{3}\right) = \left(\frac{\pi}{2}\right)^2 \sin\left(\frac{5\pi}{3}\right) + \left(\frac{5\pi}{3}\right)^2 \cos\left(\frac{\pi}{2}\right)$$
$$= \frac{\pi^2}{4}\left(-\frac{\sqrt{3}}{2}\right) + \left(\frac{5\pi}{3}\right)^2 \cdot 0 = -\frac{\pi^2\sqrt{3}}{8}.$$

$$f\left(-\frac{\pi}{4}, -\frac{\pi}{2}\right) = \left(-\frac{\pi}{4}\right)^2 \sin\left(-\frac{\pi}{2}\right) + \left(-\frac{\pi}{2}\right)^2 \cos\left(-\frac{\pi}{4}\right)$$
$$= \frac{\pi^2}{16}(-1) + \frac{\pi^2}{4} \cdot \frac{\sqrt{2}}{2} = \frac{\pi^2}{16}\left(2\sqrt{2} - 1\right).$$

f. $f(x,y) = \frac{x^2+3xy+3y^2}{x+y}$.

$$f(2a,2b) = \frac{(2a)^2+3(2a)(2b)+3(2b)^2}{2a+2b}$$
$$= \frac{4(a^2+3ab+3b^2)}{2(a+b)} = 2 \cdot \frac{a^2+3ab+3b^2}{a+b}$$
$$= 2f(a,b).$$

2.a. $f(x, y) = \frac{1}{\sqrt{x^2-4}}$.

The function is defined only when $x^2 - 4 > 0$

$\Longrightarrow x^2 > 4 \Longrightarrow |x| > 2 \Longrightarrow x > 2$ or $x < -2$.

Thus the domain is the set of all (x, y) such that $x > 2$ or $x < -2$, and all real numbers y.

b. $f(x, y, z) = \frac{2^{xyz}}{z^{1/4}}$.

The domain of f is the set of all (x, y, z) except those points with $z \leq 0$ since the function f is not defined if the denominator $z^{1/4} = 0$ and z underneath the even radical sign is negative.

c. $f(x, y, z) = \sqrt{\ln [3 - (x^2 + y^2 + z^2)]}$.

The function f is defined only when

$\ln \left[3 - (x^2 + y^2 + z^2)\right] \geq 0 \Longrightarrow 3 - (x^2 + y^2 + z^2) \geq 1$

$\Longrightarrow x^2 + y^2 + z^2 \leq 2$ or $\sqrt{x^2 + y^2 + z^2} \leq \sqrt{2}$.

Thus the domain is the set of all (x, y, z) on or inside the sphere $x^2 + y^2 + z^2 = 2$.

d. $f(x, y) = e^{y/x}$.

The domain is the set of (x, y) except the points such that $x = 0$.

e. $f(x, y) = 3x^{4/5} \cdot y^{3/4} - x^{-3/2} \cdot y^{2/3}$.

In the expression of $f(x, y)$, $y^{3/4}$ is defined only when $y \geq 0$, and $x^{-3/2}$ is defined only when $x > 0$. Thus, the domain is the set of (x, y) such that $x > 0$ and $y \geq 0$.

f. $f(x, y) = \frac{2 \sin x}{1 - 2\cos(x, y)}$.

The function f is defined only when $1 - 2\cos(xy) \neq 0$.

$\Longrightarrow \cos(xy) \neq \frac{1}{2} \Longrightarrow xy \neq \frac{\pi}{3} + 2n\pi$ and $xy \neq \frac{5\pi}{3} + 2n\pi$,

$n = 0, \pm 1, \pm 2, \ldots$. Thus the domain is the set of (x, y) such that $xy \neq \frac{\pi}{3} + 2n\pi$ and $xy \neq \frac{5\pi}{3} + 2n\pi$, $(n = 0, \pm 1, \pm 2, \ldots)$.

3.a. $f(x,y) = 3x^2 + xy + 5y^2$.

$\frac{\partial f}{\partial x} = 6x + y, \qquad \frac{\partial f}{\partial y} = x + 10y.$

b. $f(x,y) = e^{x/y}$.

$\frac{\partial f}{\partial x} = e^{x/y} \cdot \left(\frac{1}{y}\right) = \frac{1}{y} e^{x/y}, \qquad \frac{\partial f}{\partial y} = e^{x/y} \cdot \left(-\frac{x}{y^2}\right) = -\frac{x}{y^2} \cdot e^{x/y}.$

c. $f(x,y) = \sqrt[3]{x^2 + y^2}$.

$\frac{\partial f}{\partial x} = \frac{1}{3}(x^2 + y^2)^{-2/3} \cdot (2x) = \frac{2}{3}x \cdot (x^2 + y^2)^{-2/3}.$

$\frac{\partial f}{\partial y} = \frac{1}{3}(x^2 + y^2)^{-2/3} \cdot (2y) = \frac{2}{3}y \cdot (x^2 + y^2)^{-2/3}.$

d. $f(x,y,z) = \frac{x}{y-z}$.

$\frac{\partial f}{\partial x} = \frac{1}{y-z}, \qquad \frac{\partial f}{\partial y} = -\frac{x}{(y-z)^2}, \qquad \frac{\partial f}{\partial z} = \frac{x}{(y-z)^2}.$

e. $f(x,y,z) = x^2y^3z^5 - 3x^2y^4z^3 + 5xz$.

$\frac{\partial f}{\partial x} = 2xy^3z^5 - 6xy^4z^3 + 5z.$

$\frac{\partial f}{\partial y} = 3x^2y^2z^5 - 12x^2y^3z^3.$

$\frac{\partial f}{\partial z} = 5x^2y^3z^4 - 9x^2y^4z^2 + 5x.$

f. $f(x,y) = x^2 \cdot e^{3xy} \cdot \ln\left(\frac{2}{y}\right) = -x^2e^{3xy}\ln\left(\frac{y}{2}\right)$.

$\frac{\partial f}{\partial x} = \left(-2x \cdot e^{3xy} - x^2e^{3xy} \cdot 3y\right)\ln\left(\frac{y}{2}\right)$

$= xe^{3xy}(2 + 3xy)\ln\left(\frac{2}{y}\right).$

$\frac{\partial f}{\partial y} = -x^2\left(e^{3xy} \cdot 3x \cdot \ln\left(\frac{y}{2}\right) + e^{3xy} \cdot \frac{2}{y} \cdot \frac{1}{2}\right)$

$= x^2e^{3xy}\left(3x\ln\left(\frac{2}{y}\right) - \frac{1}{y}\right).$

g. $f(x,y,z) = ze^{z/(xy)}$.

$\frac{\partial f}{\partial x} = ze^{z/(xy)}\left(-\frac{z}{x^2y}\right) = -\frac{z^2}{x^2y}e^{z/(xy)}.$

$\frac{\partial f}{\partial y} = ze^{z/(xy)}\left(-\frac{z}{xy^2}\right) = -\frac{z^2}{xy^2}e^{z/(xy)}.$

$\frac{\partial f}{\partial z} = e^{z/(xy)} + z \cdot e^{z/(xy)} \cdot \left(\frac{1}{xy}\right) = e^{z/(xy)} \cdot \left(1 + \frac{z}{xy}\right).$

h. $f(x,y) = \sqrt{x}\sin^2\left(\frac{y}{x}\right)$.

$\frac{\partial f}{\partial x} = \frac{1}{2\sqrt{x}}\sin^2\left(\frac{y}{x}\right) + \sqrt{x} \cdot 2\sin\left(\frac{y}{x}\right) \cdot \cos\left(\frac{y}{x}\right) \cdot \left(-\frac{y}{x^2}\right)$

$= \frac{1}{2}x^{-1/2}\sin^2\left(\frac{y}{x}\right) - x^{-3/2}y\sin\left(\frac{2y}{x}\right).$

$\frac{\partial f}{\partial y} = \sqrt{x} \cdot 2\sin\left(\frac{y}{x}\right) \cdot \cos\left(\frac{y}{x}\right)\left(\frac{1}{x}\right) = \frac{\sin\left(\frac{2y}{x}\right)}{\sqrt{x}}.$

4.a. $f(x,y) = xye^{2x-y}$.

$\frac{\partial f}{\partial x} = ye^{2x-y} + xye^{2x-y}(2) = ye^{2x-y}(1+2x)$.

$\frac{\partial f}{\partial x}(2,-3) = (-3)e^{2(2)-(-3)} \cdot (1+2(2)) = -15e^7$.

b. $f(x,y,z) = xy^2z + 5$.

$\frac{\partial f}{\partial y} = 2xyz$. $\qquad \frac{\partial f}{\partial y}(2,-1,3) = 2(2)(-1)(3) = -12$.

c. $f(x,y,z) = \frac{x}{y-z}$.

$\frac{\partial f}{\partial x} = \frac{1}{y-z}$, $\quad \frac{\partial f}{\partial z} = \frac{-x}{(y-z)^2} \cdot (-1) = \frac{x}{(y-z)^2}$.

$\frac{\partial f}{\partial x}(2,-1,3) = \frac{1}{(-1)-(3)} = -\frac{1}{4}$.

$\frac{\partial f}{\partial z}(2,-1,3) = \frac{2}{(-1-3)^2} = \frac{2}{16} = \frac{1}{8}$.

d. $f(x,y,z) = e^{xz} \cdot \sqrt{y+2z}$.

$$\frac{\partial f}{\partial z} = e^{xz} \cdot x \cdot \sqrt{y+2z} + e^{xz} \cdot \tfrac{1}{2}(y+2z)^{-1/2} \cdot 2$$

$$= e^{xz}\left(x \cdot \sqrt{y+2z} + \frac{1}{\sqrt{y+2z}}\right).$$

$$\frac{\partial f}{\partial z}(0,1,4) = e^{(0)\cdot(4)}\left(0 \cdot \sqrt{1+(2)(4)} + \frac{1}{\sqrt{1+(2)(4)}}\right) = \frac{1}{3}.$$

e. $f(r,s,t,u) = (s^2+tu) \cdot \ln(2r+7st)$.

$\frac{\partial f}{\partial s} = 2s \cdot \ln(2r+7st) + (s^2+tu) \cdot \frac{1}{2r+7st} \cdot (7t)$.

$\frac{\partial f}{\partial s}(1,1,0,1) = 2\ln(2) + 1 \cdot \frac{1}{2} \cdot 0 = 2\ln 2$.

5.a. $f(x,y) = xe^{y/x}$.

$\frac{\partial f}{\partial x} = e^{y/x} + xe^{y/x} \cdot \left(-\frac{y}{x^2}\right) = e^{y/x} \cdot \left(1 - \frac{y}{x}\right)$.

$\frac{\partial f}{\partial y} = xe^{y/x} \cdot \frac{1}{x} = e^{y/x}$.

$\frac{\partial^2 f}{\partial x^2} = \frac{y}{x^2} \cdot e^{y/x} + \left(1 - \frac{y}{x}\right)e^{y/x} \cdot \left(-\frac{y}{x^2}\right) = \frac{y^2}{x^3}e^{y/x}$.

$\frac{\partial^2 f}{\partial y^2} = \frac{1}{x}e^{y/x}$.

$$\frac{\partial^2 f}{\partial y \partial x} = -\frac{1}{x}e^{y/x} + \left(1 - \frac{y}{x}\right) \cdot e^{y/x} \cdot \left(\frac{1}{x}\right)$$

$$= \frac{1}{x}e^{y/x}\left(-1 + 1 - \frac{y}{x}\right) = -\frac{y}{x^2} \cdot e^{y/x}.$$

$$\frac{\partial^2 f}{\partial x \partial y} = \frac{\partial}{\partial x}\left(e^{y/x}\right) = e^{y/x}\left(-\frac{y}{x^2}\right) = -\frac{y}{x^2} \cdot e^{y/x}.$$

b. $f(x,y) = \ln\sqrt{x^2+y^2}$.

$\frac{\partial f}{\partial x} = \frac{1}{\sqrt{x^2+y^2}} \cdot \frac{1}{2\sqrt{x^2+y^2}} \cdot 2x = \frac{x}{x^2+y^2}$.

$\frac{\partial f}{\partial y} = \frac{1}{\sqrt{x^2+y^2}} \cdot \frac{2y}{2\sqrt{x^2+y^2}} = \frac{y}{x^2+y^2}$.

$\frac{\partial^2 f}{\partial x^2} = \frac{\partial}{\partial x}\left(\frac{x}{x^2+y^2}\right) = \frac{1\cdot(x^2+y^2)-x\cdot(2x)}{(x^2+y^2)^2} = \frac{y^2-x^2}{(x^2+y^2)^2}$.

$\frac{\partial^2 f}{\partial y^2} = \frac{\partial}{\partial y}\left(\frac{y}{x^2+y^2}\right) = \frac{1\cdot(x^2+y^2)-y\cdot(2y)}{(x^2+y^2)^2} = \frac{x^2-y^2}{(x^2+y^2)^2}$.

$\frac{\partial^2 f}{\partial y\partial x} = \frac{\partial}{\partial y}\left(\frac{x}{x^2+y^2}\right) = \frac{-x\cdot 2y}{(x^2+y^2)^2} = -\frac{2xy}{(x^2+y^2)^2}$.

$\frac{\partial^2 f}{\partial x\partial y} = \frac{\partial}{\partial x}\left(\frac{y}{x^2+y^2}\right) = \frac{-y\cdot 2x}{(x^2+y^2)^2} = -\frac{2xy}{(x^2+y^2)^2}$.

c. $f(x,y) = (x^2+xy+y^2)(x^2+xy+1)$.

$\frac{\partial f}{\partial x} = (2x+y)(x^2+xy+1) + (x^2+xy+y^2)(2x+y)$

$= (2x+y)(2x^2+2xy+y^2+1)$.

$\frac{\partial f}{\partial y} = (x+2y)(x^2+xy+1) + (x^2+xy+y^2)(x)$

$= (x+2y)(x^2+xy+1) + (x^3+x^2y+xy^2)$.

$\frac{\partial^2 f}{\partial x^2} = 2(2x^2+2xy+y^2+1) + (2x+y)(4x+2y)$

$= 2(6x^2+6xy+2y^2+1)$.

$\frac{\partial^2 f}{\partial y^2} = 2(x^2+xy+1) + (x+2y)\cdot x + (x^2+2xy)$

$= 4x^2+6xy+2$.

$\frac{\partial^2 f}{\partial y\partial x} = \frac{\partial}{\partial y}\left[(2x+y)(2x^2+2xy+y^2+1)\right]$

$= 1\cdot(2x^2+2xy+y^2+1) + (2x+y)(2x+2y)$

$= 6x^2+8xy+3y^2+1$.

$\frac{\partial^2 f}{\partial x\partial y} = \frac{\partial}{\partial x}\left[(x+2y)(x^2+xy+1) + (x^3+x^2y+xy^2)\right]$

$= 1\cdot(x^2+xy+1) + (x+2y)(2x+y) + (3x^2+2xy+y^2)$

$= 6x^2+8xy+3y^2+1$.

d. $f(x,y) = \cos(x+y) + \sin(x-y)$.

$\frac{\partial f}{\partial x} = -\sin(x+y) + \cos(x-y)$.

$\frac{\partial f}{\partial y} = -\sin(x+y) + \cos(x-y)\cdot(-1) = -\sin(x+y) - \cos(x-y)$.

$\frac{\partial^2 f}{\partial x^2} = -\cos(x+y) - \sin(x-y)$.

$\frac{\partial^2 f}{\partial y^2} = -\cos(x+y) + \sin(x-y)\cdot(-1) = -\cos(x+y) - \sin(x-y)$.

$$\frac{\partial^2 f}{\partial y \partial x} = \frac{\partial}{\partial y}\left[-\sin(x+y)+\cos(x-y)\right]$$
$$= -\cos(x+y) - \sin(x-y)\cdot(-1) = -\cos(x+y)+\sin(x-y).$$

$$\frac{\partial^2 f}{\partial x \partial y} = \frac{\partial}{\partial x}\left[-\sin(x+y)-\cos(x-y)\right]$$
$$= -\cos(x+y)+\sin(x-y).$$

6.a. $f(x,y) = x^2 - 3y^2 + 4x + 6y + 8.$

$\frac{\partial f}{\partial x} = 2x + 4 = 0 \implies x = -2.$

$\frac{\partial f}{\partial y} = -6y + 6 = 0 \implies y = 1.$

So the critical point of f is $(-2,1)$.

Now, $A = \frac{\partial^2 f}{\partial x^2}(-2,1) = 2 > 0$, $B = \frac{\partial^2 f}{\partial x \partial y}(-2,1) = 0$, $C = \frac{\partial^2 f}{\partial y^2}(-2,1) = -6$.

$D = B^2 - AC = 12 > 0.$

Thus the point $(-2, 1, f(-2,1))$ is a saddle point.

b. $f(x,y) = -2x^2 + 2xy - y^2 + 4x - 6y + 5.$

$$\frac{\partial f}{\partial x} = -4x + 2y + 4 = 0 \implies -2x + y = -2 \qquad (1)$$
$$\frac{\partial f}{\partial y} = 2x - 2y - 6 = 0 \implies x - y = 3 \qquad (2)$$

(1) + (2) gives: $-x = 1 \implies x = -1.$

Substituting $x = -1$ into (2) gives: $(-1) - y = 3 \implies y = -4.$

So the critical point of f is (-1,-4).

Now, $A = \frac{\partial^2 f}{\partial x^2}(-1,-4) = -4 < 0,\ B = \frac{\partial^2 f}{\partial x \partial y}(-1,-4) = 2,$

$C = \frac{\partial^2 f}{\partial y^2}(-1,-4) = -2 \implies D = B^2 - AC = -4 < 0,\ A < 0.$

Thus f has a relative maximum at (-1,-4).

c. $f(x,y) = 2x^2 + 3xy + 5y^2.$

$$\frac{\partial f}{\partial x} = 4x + 3y = 0 \qquad (1)$$
$$\frac{\partial f}{\partial y} = 3x + 10y = 0 \qquad (2)$$

Solving (1) and (2) gives $x = 0,\ y = 0.$

So the critical point of f is (0,0).

Now $A = \frac{\partial^2 f}{\partial x^2}(0,0) = 4$, $B = \frac{\partial^2 f}{\partial x \partial y}(0,0) = 3$, $C = \frac{\partial^2 f}{\partial y^2}(0,0) = 10$

$\Longrightarrow D = B^2 - AC = -31 < 0$ and $A > 0$.

Thus f has a relative minimum of 0 at the point (0,0).

d. $f(x,y) = xy - \frac{1}{x} - \frac{1}{y}$.

$$\frac{\partial f}{\partial x} = y + \frac{1}{x^2} = 0 \Longrightarrow x^2 y = -1, \ (x \neq 0) \quad (1)$$

$$\frac{\partial f}{\partial y} = x + \frac{1}{y^2} = 0 \Longrightarrow xy^2 = -1, \ (y \neq 0) \quad (2)$$

$$(1) - (2) \text{ gives } x^2 y - xy^2 = 0$$

$$xy(x-y) = 0 \Longrightarrow x = y \quad (3)$$

From (3) and (1) we have $x = y = -1$.

So the critical point of f is $(-1,-1)$.

$\frac{\partial^2 f}{\partial x^2} = -\frac{2}{x^3}$, $\frac{\partial^2 f}{\partial x \partial y} = 1$, $\frac{\partial^2 f}{\partial y^2} = -\frac{2}{y^3}$.

$A = \frac{\partial^2 f}{\partial x^2}(-1,-1) = 2$, $B = \frac{\partial^2 f}{\partial x \partial y}(-1,-1) = 1$, $C = \frac{\partial^2 f}{\partial y^2}(-1,-1) = 2$.

$\Longrightarrow D = B^2 - AC = -3 < 0$, and $A > 0$.

Thus f has a relative minimum of 3 at (-1,-1).

e. $f(x,y) = \frac{1}{3}(x^3 + 8y^3) - 2(x^2 + y^2) + 1$.

$\frac{\partial f}{\partial x} = x^2 - 4x = 0 \Longrightarrow x(x-4) = 0 \Longrightarrow x = 0, x = 4$.

$\frac{\partial f}{\partial y} = 8y^2 - 4y = 0 \Longrightarrow 4y(2y-1) = 0 \Longrightarrow y = 0, y = \frac{1}{2}$.

So the critical points of f are (0,0), $\left(0, \frac{1}{2}\right)$, (4,0), and $\left(4, \frac{1}{2}\right)$.

$\frac{\partial^2 f}{\partial x^2} = 2x - 4$, $\frac{\partial^2 f}{\partial x \partial y} = 0$, $\frac{\partial^2 f}{\partial y^2} = 16y - 4$.

At (0,0): $A = -4, B = 0, C = -4 \Longrightarrow D = B^2 - AC = -16 < 0, A < 0$.

Thus $f(0,0) = 1$ is a relative maximum.

At $\left(0, \frac{1}{2}\right)$: $A = -4, B = 0, C = 4 \Longrightarrow D = 16 > 0$.

At $(4,0)$: $A = 4, B = 0, C = -4 \Longrightarrow D = 16 > 0$.

At $\left(4, \frac{1}{2}\right)$: $A = 4, B = 0, C = 4 \Longrightarrow D = -16 < 0$, and $A > 0$.

Thus $f\left(4, \frac{1}{2}\right)$ is a relative minimum; both $\left(0, \frac{1}{2}, f\left(0, \frac{1}{2}\right)\right)$ and $(4, 0, f(4,0))$ are the saddle points.

f. $f(x,y) = x^3 + y^3 - 3xy$.

$$\frac{\partial f}{\partial x} = 3x^2 - 3y = 0 \implies y = x^2 \tag{1}$$

$$\frac{\partial f}{\partial y} = 3y^2 - 3x = 0 \implies x = y^2 \tag{2}$$

Substituting $x = y^2$ from (2) into (1) gives:

$$y = y^4 \implies y^4 - y = 0 \implies y(y^3 - 1) = 0$$

$$\implies y = 0 \text{ or } y^3 = 1$$

$$\implies y = 0 \text{ or } y = 1.$$

Then by (2) we obtain (0,0) and (1,1) as the critical points for f.

$$\frac{\partial^2 f}{\partial x^2} = 6x, \quad \frac{\partial^2 f}{\partial x \partial y} = -3, \quad \frac{\partial^2 f}{\partial y^2} = 6y.$$

At (0,0): $A = 0$, $B = -3$, $C = 0 \implies D = 9 > 0$.

At (1,1): $A = 6$, $B = -3$, $C = 6 \implies D = -27 < 0$, $A > 0$.

Thus $f(1,1) = -1$ is the relative minimum, and (0,0,0) is the saddle point.

g. $f(x,y) = e^{(7x^2 - 5xy + y^2 + x - y + 6)}$.

$$\frac{\partial f}{\partial x} = e^{(7x^2 - 5xy + y^2 + x - y + 6)} \cdot (14x - 5y + 1) = 0$$

$$\implies 14x - 5y = -1 \tag{1}$$

$$\frac{\partial f}{\partial y} = e^{(7x^2 - 5xy + y^2 + x - y + 6)} \cdot (-5x + 2y - 1) = 0$$

$$\implies -5x + 2y = 1 \tag{2}$$

$$\text{From (2): } 2y = 1 + 5x \implies y = \frac{1 + 5x}{2} \tag{3}$$

Substituting (3) into (1) gives:

$$14x - 5\left(\frac{1 + 5x}{2}\right) = -1$$

$$28x - 5 - 25x = -2$$

$$3x = 3 \implies x = 1.$$

Then $y = \frac{1+5(1)}{2} = 3$.

So the critical point of f is (1,3).

$$\frac{\partial^2 f}{\partial x^2} = e^{(7x^2-5xy+y^2+x-y+6)} \cdot (14x - 5y + 1)^2$$
$$+ e^{(7x^2-5xy+y^2+x-y+6)} \cdot (14)$$
$$= e^{(7x^2-5xy+y^2+x-y+6)} \cdot \left[(14x - 5y + 1)^2 + 14\right].$$

$$\frac{\partial^2 f}{\partial x \partial y} = e^{(7x^2-5xy+y^2+x-y+6)} \cdot (-5x + 2y - 1)(14x - 5y + 1)$$
$$+ e^{(7x^2-5xy+y^2+x-y+6)} \cdot (-5).$$
$$= e^{(7x^2-5xy+y^2+x-y+6)} \cdot [(-5x + 2y - 1)(14x - 5y + 1) - 5].$$

$$\frac{\partial^2 f}{\partial y^2} = e^{(7x^2-5xy+y^2+x-y+6)} \cdot \left[(-5x + 2y - 1)^2 + 2\right].$$

At (1,3):

$A = \frac{\partial^2 f}{\partial x^2}(1,3) = 14e^5$, $B = -5e^5$, $C = 2e^5$

$\implies D = B^2 - AC = e^{10}(25 - 14 \cdot 2) = -3e^{10} < 0$, $A > 0$.

Thus $f(1,3) = e^5$ is a relative minimum.

h. $f(x,y) = x^3 - y^2 - 3x + 4y + 11$.

$\frac{\partial f}{\partial x} = 3x^2 - 3 = 0 \implies x^2 = 1 \implies x = -1,\ x = 1$.

$\frac{\partial f}{\partial y} = -2y + 4 = 0 \implies y = 2$.

So the critical points of f are (-1,2) and (1,2).

$\frac{\partial^2 f}{\partial x^2} = 6x$, $\frac{\partial^2 f}{\partial x \partial y} = 0$, $\frac{\partial^2 f}{\partial y^2} = -2$.

At (-1,2): $A = -6$, $B = 0$, $C = -2 \implies D = -12 < 0$, $A < 0$.

At (1,2): $A = 6$, $B = 0$, $C = -2 \implies D = 12 > 0$.

Thus, $f(-1,2)$ is the relative maximum, and the point $(1,2,f(1,2))$ is the saddle point.

7.a. $f(x,y) = x^2 - y^2$ subject to $2x + y = 3$.

Let $F(x,y,\lambda) = f(x,y) + \lambda(2x + y - 3)$.

$$\frac{\partial F}{\partial x} = f_x(x,y) + 2\lambda = 2x + 2\lambda = 0 \qquad (1)$$
$$\frac{\partial F}{\partial y} = f_y(x,y) + \lambda = -2y + \lambda = 0 \qquad (2)$$
$$\frac{\partial F}{\partial \lambda} = 2x + y - 3 = 0 \qquad (3)$$

From (1) and (2) $\Longrightarrow$ $x = -\lambda$ and $y = \frac{1}{2}\lambda$.

Substituting them into (3) gives:

$$2(-\lambda) + \frac{1}{2}\lambda - 3 = 0 \Longrightarrow -\frac{3}{2}\lambda = 3 \Longrightarrow \lambda = -2.$$

So $x = 2$ and $y = -1$.

Since $f(0,3) = -9 < f(2,-1) = 3$, $f(2,-1) = 3$ is the maximum value of f subject to the given constraint.

b. $f(x,y) = x^2 + xy + y^2 - 2x - 5y + 10$ subject to $1 - x + y = 0$.

Let $F(x,y,\lambda) = f(x,y) + \lambda(1 - x + y)$.

$$\frac{\partial F}{\partial x} = 2x + y - 2 - \lambda = 0 \qquad (1)$$
$$\frac{\partial F}{\partial y} = x + 2y - 5 + \lambda = 0 \qquad (2)$$
$$\frac{\partial F}{\partial \lambda} = 1 - x + y = 0 \qquad (3)$$
$$(1) + (2) \text{ gives: } 3x + 3y - 7 = 0 \qquad (4)$$
$$((3) \times 3) + (4) \text{ gives: } 6y - 4 = 0 \Longrightarrow y = \frac{2}{3}.$$

Substituting $y = \frac{2}{3}$ into (3) gives $1 - x + \frac{2}{3} = 0 \Longrightarrow x = \frac{5}{3}$.

So the critical point is $\left(\frac{5}{3}, \frac{2}{3}\right)$.

Since $f(1,1) = 6 > f\left(\frac{5}{3}, \frac{2}{3}\right) = \frac{69}{9}$, $f\left(\frac{5}{3}, \frac{2}{3}\right) = \frac{69}{9}$ must be the minimum value.

c. $f(x, y, z) = 3x + 2y + z$ subject to $xyz = 36$.

Let $F(x, y, z, \lambda) = f(x, y, z) + \lambda(xyz - 36)$.

$$\frac{\partial F}{\partial x} = 3 + yz\lambda = 0 \implies yz\lambda = -3 \qquad (1)$$
$$\frac{\partial F}{\partial y} = 2 + xz\lambda = 0 \implies xz\lambda = -2 \qquad (2)$$
$$\frac{\partial F}{\partial z} = 1 + xy\lambda = 0 \implies xy\lambda = -1 \qquad (3)$$
$$\frac{\partial F}{\partial \lambda} = xyz - 36 = 0 \implies xyz = 36 \qquad (4)$$

Dividing each side of Eq. (1) by the corresponding side of Eq. (2) (the division is valid since $xyz \neq 0$), we obtain

$$\frac{yz\lambda}{xz\lambda} = \frac{-3}{-2} \implies \frac{y}{x} = \frac{3}{2} \implies y = \frac{3}{2}x.$$

Similarly from Eq. (1) and Eq. (3) we obtain

$$\frac{yz\lambda}{xy\lambda} = \frac{-3}{-1} \implies z = 3x.$$

Substituting the results into Eq. (4) gives

$$x \cdot \left(\frac{3}{2}x\right) \cdot (3x) = 36$$
$$\frac{9}{2}x^3 = 36$$
$$x^3 = \frac{2 \times 36}{9} = 8 \implies x = 2.$$

Thus $y = 3$ and $z = 6$.

So the critical point is (2,3,6).

Since $f(1, 6, 6) = 21 > f(2, 3, 6) = 18$, f has the minimum value of 18 subject to the given constraint.

d. $f(x, y, z) = xyz^2$ subject to $x^2 + y^2 + z^2 = 20$.

Let $F(x, y, z, \lambda) = f(x, y, z) + \lambda(x^2 + y^2 + z^2 - 20)$.

$$\frac{\partial F}{\partial x} = yz^2 + 2x\lambda = 0 \quad (1)$$
$$\frac{\partial F}{\partial y} = xz^2 + 2y\lambda = 0 \quad (2)$$
$$\frac{\partial F}{\partial z} = 2xyz + 2z\lambda = 0 \quad (3)$$
$$\frac{\partial F}{\partial \lambda} = x^2 + y^2 + z^2 - 20 = 0 \quad (4)$$

$(y \times \text{Eq. }(1)) + ((-x) \times \text{Eq. }(2))$ gives

$$y^2z^2 - x^2z^2 = 0$$
$$z^2(y^2 - x^2) = 0 \implies y^2 = x^2 \qquad (z \neq 0).$$

$(z \times \text{Eq. }(1)) + ((-x) \times \text{Eq. }(3))$ gives

$$yz^3 - 2x^2yz = 0$$
$$yz(z^2 - 2x^2) = 0 \implies z^2 = 2x^2 \qquad (yz \neq 0).$$

Substituting into Eq. (4) gives

$$x^2 + x^2 + 2x^2 = 20$$
$$4x^2 = 20$$
$$x = \sqrt{5}, -\sqrt{5}.$$

Thus there are eight critical points:
$(\sqrt{5}, \sqrt{5}, \sqrt{10})$, $(\sqrt{5}, \sqrt{5}, -\sqrt{10})$, $(\sqrt{5}, -\sqrt{5}, \sqrt{10})$, $(\sqrt{5}, -\sqrt{5}, -\sqrt{10})$,
$(-\sqrt{5}, \sqrt{5}, \sqrt{10})$, $(-\sqrt{5}, \sqrt{5}, -\sqrt{10})$, $(-\sqrt{5}, -\sqrt{5}, \sqrt{10})$, $(-\sqrt{5}, -\sqrt{5}, -\sqrt{10})$.
Since
$f(\sqrt{5}, \sqrt{5}, \sqrt{10}) = f(\sqrt{5}, \sqrt{5}, -\sqrt{10}) = f(-\sqrt{5}, -\sqrt{5}, \sqrt{10})$

$$= f\left(-\sqrt{5}, -\sqrt{5}, -\sqrt{10}\right) = 50,$$

$$f\left(-\sqrt{5}, \sqrt{5}, \sqrt{10}\right) = f\left(-\sqrt{5}, \sqrt{5}, -\sqrt{10}\right) = f\left(\sqrt{5}, -\sqrt{5}, \sqrt{10}\right)$$
$$= f\left(\sqrt{5}, -\sqrt{5}, -\sqrt{10}\right) = -50,$$

the maximum value of f is 50, and the minimum value is -50.

8.a. $\int_1^3 \int_1^2 (x^2 - y)dxdy = \int_1^3 \left[\left(\frac{x^3}{3} - yx\right)\right]_{x=1}^{x=2} dy$

$= \int_1^3 \left(\frac{8}{3} - 2y - \frac{1}{3} + y\right) dy = \int_1^3 \left(\frac{7}{3} - y\right) dy = \left(\frac{7}{3}y - \frac{y^2}{2}\right)\Big]_1^3$

$= \frac{7}{3}(3-1) - \frac{1}{2}(9-1) = \frac{14}{3} - 4 = \frac{2}{3}.$

b. $\int_0^3 \int_0^x (x^2 + y^2)dydx = \int_0^3 \left[\left(x^2y + \frac{1}{3}y^3\right)\right]_{y=0}^{y=x} dx$

$= \int_0^3 \left(x^3 + \frac{1}{3}x^3 - 0\right) dx = \int_0^3 \frac{4}{3}x^3 dx = \frac{1}{3}x^4\Big]_0^3 = 27.$

c. $\int_1^{\ln 8} \int_0^{\ln y} e^{x+y} dxdy = \int_1^{\ln 8} \left(e^{x+y}\Big]_{x=0}^{x=\ln y}\right) dy$

$= \int_1^{\ln 8} (e^{y+\ln y} - e^y)dy = \int_1^{\ln 8} (e^y \cdot e^{\ln y} - e^y)dy$

$= \int_1^{\ln 8} (e^y \cdot y - e^y)dy = \int_1^{\ln 8} (y-1)e^y \, dy.$

Let $u = y - 1$, $dv = e^y dy \implies du = dy$, $v = e^y$. Then

$\int_1^{\ln 8} \int_0^{\ln y} e^{x+y} dxdy = \int_1^{\ln 8} (y-1)e^y dy = (y-1)e^y\Big]_1^{\ln 8} - \int_1^{\ln 8} e^y \, dy$

$= [(y-1)e^y - e^y]\Big]_1^{\ln 8} = (y-2)e^y\Big]_1^{\ln 8} = (\ln 8 - 2)e^{\ln 8} - (-1)e$

$= 8(\ln 8 - 2) + e.$

d. $\int_0^2 \int_0^{\sqrt{4-y^2}} x \, dxdy = \int_0^2 \left[\frac{1}{2}x^2\Big]_0^{\sqrt{4-y^2}}\right] dy = \int_0^2 \frac{1}{2}(4 - y^2)dy$

$= \left(2y - \frac{1}{6}y^3\right)\Big]_0^2 = 4 - \frac{8}{6} = \frac{8}{3}.$

e. $\int_0^3 \int_{y^2}^{3y} x \, dxdy = \int_0^3 \left(\frac{1}{2}x^2\Big]_{x=y^2}^{x=3y}\right) dy = \frac{1}{2}\int_0^3 (9y^2 - y^4)dy$

$= \frac{1}{2}\left(3y^3 - \frac{1}{5}y^5\right)\Big]_0^3 = \frac{1}{2}\left(3 \cdot 27 - \frac{1}{5} \cdot 3^5\right) = \frac{81}{5}.$

f. $\int_{-2}^1 \int_{x^2+4x}^{3x+2} dydx = \int_{-2}^1 \left(y\Big]_{x^2+4x}^{3x+2}\right) dx = \int_{-2}^1 (3x + 2 - x^2 - 4x)dx$

$= \int_{-2}^1 (2 - x - x^2)dx = \left(2x - \frac{x^2}{2} - \frac{x^3}{3}\right)\Big]_{-2}^1 = \left(2 - \frac{1}{2} - \frac{1}{3}\right) - \left(-4 - \frac{4}{2} + \frac{8}{3}\right) = \frac{9}{2}.$

g. $\int_0^2 \int_y^2 e^{x^2}\, dx\,dy$.

By interchanging the order of integration we have

$$\int_0^2 \int_y^2 e^{x^2}\, dx\,dy = \int_0^2 \int_0^x e^{x^2}\, dy\,dx$$

$$= \int_0^2 \left[\left(ye^{x^2}\right)\Big]_0^x\right] dx$$

$$= \int_0^2 xe^{x^2}\, dx = \int_{x=0}^{x=2} \tfrac{1}{2}e^u du$$

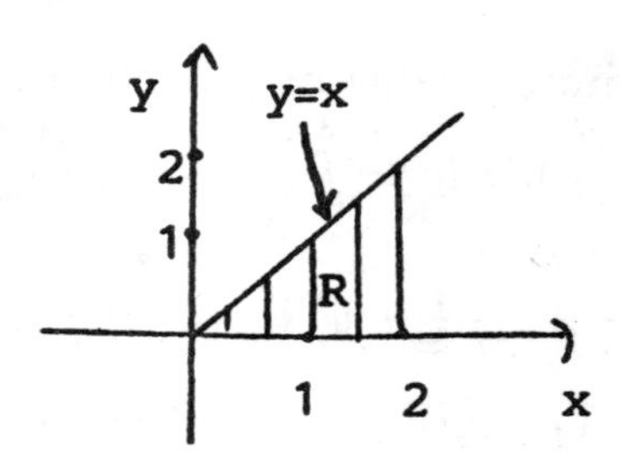

where $u = x^2$

$$= \tfrac{1}{2}e^u\Big]_{x=0}^{x=2} = \tfrac{1}{2}e^{x^2}\Big]_0^2 = \tfrac{1}{2}\left(e^4 - 1\right).$$

h. $\int_0^\pi \int_x^\pi \frac{\sin y}{y}\, dy\,dx$.

By interchanging the order of integration we have

$$\int_0^\pi \int_x^\pi \frac{\sin y}{y}\, dy\,dx = \int_0^\pi \int_0^y \frac{\sin y}{y}\, dx\,dy$$

$$= \int_0^\pi \left[\frac{\sin y}{y} \cdot x\right]_{x=0}^{x=y} dy = \int_0^\pi \frac{\sin y}{y}(y-0)dy$$

$$= \int_0^\pi \sin y\, dy = -\cos y]_0^\pi = -\cos\pi + \cos 0 = 2.$$

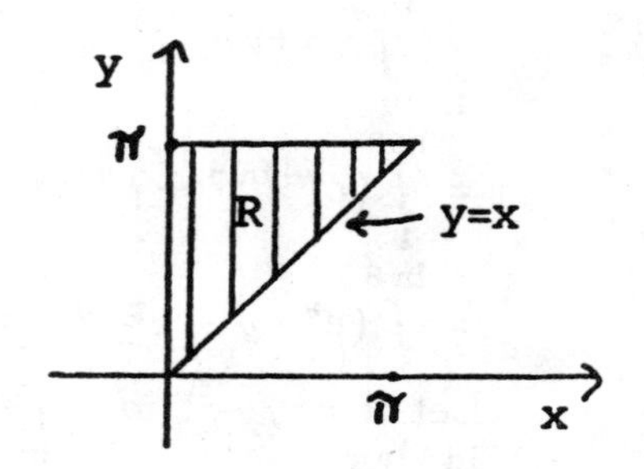

9.a. $\iint_R (x^2 + 2xy - 3y^2)dA = \int_0^1 \int_{x^2}^{\sqrt{x}} (x^2 + 2xy - 3y^2)dy\,dx$

$$= \int_0^1 \left[\left(x^2y + xy^2 - y^3\right)\Big]_{y=x^2}^{y=\sqrt{x}}\right] dx$$

$$= \int_0^1 \left[\left(x^2 \cdot \sqrt{x} + x \cdot (\sqrt{x})^2 - (\sqrt{x})^3\right) - \left(x^2 \cdot x^2 + x \cdot (x^2)^2 - (x^2)^3\right)\right] dx$$

$$= \int_0^1 \left[\left(x^{5/2} + x^2 - x^{3/2}\right) - \left(x^4 + x^5 - x^6\right)\right] dx$$

$$= \left(\tfrac{2}{7}x^{7/2} + \tfrac{1}{3}x^3 - \tfrac{2}{5}x^{5/2} - \tfrac{1}{5}x^5 - \tfrac{1}{6}x^6 + \tfrac{1}{7}x^7\right)\Big]_0^1$$

$$= \tfrac{2}{7} + \tfrac{1}{3} - \tfrac{2}{5} - \tfrac{1}{5} - \tfrac{1}{6} + \tfrac{1}{7} = \tfrac{3}{7} + \tfrac{1}{6} - \tfrac{3}{5}$$

$$= \tfrac{90}{210} + \tfrac{35}{210} - \tfrac{126}{210} = -\tfrac{1}{210}.$$

b. $\iint_R x^2y^2\,dA = \int_1^2\int_0^y x^2y^2\,dx\,dy = \int_1^2 \left(\frac{1}{3}x^3y^2\Big]_{x=0}^{x=y}\right) dy$

$= \frac{1}{3}\int_1^2 y^5\,dy = \frac{1}{18}y^6\Big]_1^2 = \frac{1}{18}(2^6-1) = \frac{63}{18} = \frac{7}{2}.$

c. $\iint_R x\sin y\;dA = \int_0^\pi\int_0^x x\sin y\;dy\,dx = \int_0^\pi \left[x(-\cos y)\Big]_{y=0}^{y=x}\right] dx$

$= -\int_0^\pi x(\cos x-1)dx = -\int_0^\pi x\cos x\;dx + \frac{x^2}{2}\Big]_0^\pi.$

Let $u = x$, $dv = \cos x\;dx \implies du = dx$, $v = \sin x$. Then

$\int x\cos x\;dx = x\sin x - \int \sin x\;dx = x\sin x + \cos x + C.$

Thus

$\iint_R x\sin y\;dA = \left(-x\sin x - \cos x + \frac{1}{2}x^2\right)\Big]_0^\pi = 2 + \frac{1}{2}\pi^2.$

d. $\iint_R (x^3 - xy)dA = \int_0^1\int_{x^3}^{x^2}(x^3-xy)dy\,dx = \int_0^1 \left[\left(x^3y - \frac{1}{2}xy^2\right)\Big]_{y=x^3}^{y=x^2}\right] dx$

$= \int_0^1 \left[\left(x^5 - \frac{1}{2}x^5\right) - \left(x^6 - \frac{1}{2}x^7\right)\right] dx = \int_0^1 \left(\frac{1}{2}x^7 - x^6 + \frac{1}{2}x^5\right) dx$

$= \left(\frac{1}{16}x^8 - \frac{1}{7}x^7 + \frac{1}{12}x^6\right)\Big]_0^1 = \frac{1}{16} - \frac{1}{7} + \frac{1}{12} = \frac{1}{336}.$

10. $f(x,y) = 300x^{2/3}\cdot y^{1/3}.$

a. The marginal productivity of labor and capital are $\frac{\partial f}{\partial x}$ and $\frac{\partial f}{\partial y}$, respectively.

$\frac{\partial f}{\partial x} = 200x^{-1/3}\cdot y^{1/3}$, $\frac{\partial f}{\partial y} = 100x^{2/3}\cdot y^{-2/3}.$

Then

$\frac{\partial f}{\partial x}(125,64) = 200\cdot(125)^{-1/3}\cdot(64)^{1/3} = 160.$

$\frac{\partial f}{\partial y}(125,64) = 100\cdot(125)^{2/3}\cdot(64)^{-2/3} = \frac{625}{4} = 156.25.$

b. From part (a) we know that if we increase capital by 1 unit and hold the amount of labor fixed at $x = 125$, production will increase approximately by 156.25 units of goods. Therefore, utilizing $x = 125$ units of labor and cutting back to 62 units of capital will cause approximately $156.25\cdot(2) = 312.50$ fewer units of goods to be produced.

11. Let x = width of the box,

y = length of the box,

z = height of the box.

Let V_0 = given volume of the box.

Let C = the cost of the box.

Let a = cost per square foot for the sides and top of the box.

Then

$$C = (2xz + 2yz + xy) \cdot a + xy \cdot (3a)$$
$$= 2a(2xy + xz + yz).$$

But $V_0 = xyz \implies z = V_0/xy \qquad (xyz \neq 0).$

So

$$C = 2a\left(2xy + \frac{V_0}{y} + \frac{V_0}{x}\right).$$

$$\frac{\partial C}{\partial x} = 2a\left(2y - \frac{V_0}{x^2}\right) = 0 \tag{1}$$

$$\frac{\partial C}{\partial y} = 2a\left(2x - \frac{V_0}{y^2}\right) = 0 \tag{2}$$

From Eq. (1) $\implies y = V_0/2x^2$. Substituting into (2) gives

$$2x - \frac{V_0}{\left(\frac{V_0}{2x^2}\right)^2} = 0 \implies 2x - \frac{4x^4}{V_0} = 0$$
$$\implies x(V_0 - 2x^3) = 0$$
$$\implies x^3 = V_0/2 \implies x = \sqrt[3]{\frac{V_0}{2}}.$$

Then $y = \frac{V_0}{2\left(\sqrt[3]{\frac{V_0}{2}}\right)^2} = \sqrt[3]{\frac{V_0}{2}}$, and $z = \frac{V_0}{\sqrt[3]{\frac{V_0}{2}} \cdot \sqrt[3]{\frac{V_0}{2}}} = \sqrt[3]{4V_0}$.

$\frac{\partial^2 C}{\partial x^2} = \frac{4aV_0}{x^3}$, $\frac{\partial^2 C}{\partial x \partial y} = 4a$, $\frac{\partial^2 C}{\partial y^2} = \frac{4aV_0}{y^3}$.

At $x = y = \sqrt[3]{\frac{V_0}{2}}$:

$$A = 4aV_0 \Big/ \left(\sqrt[3]{\tfrac{V_0}{2}}\right)^3 = 8a > 0,\ B = 4a,\ C = 8a$$

$$\Longrightarrow D = B^2 - AC = -48a^2 < 0,\ A > 0.$$

Thus the dimensions of the box that minimize the cost are $x = y = \sqrt[3]{\frac{V_0}{2}}$ and $z = \sqrt[3]{4V_0} = 2x$.

12. Let $C(x, y)$ be the costs for producing and selling the two products I and II.

Then $C(x, y) = xP_{\text{I}} + yP_{\text{II}}$.

The profit is then

$$P(x,y) = R(x,y) - C(x,y) = R(x,y) - xP_{\text{I}} - yP_{\text{II}}.$$

$$\frac{\partial P}{\partial x} = \frac{\partial R}{\partial x} - P_{\text{I}} \quad \text{and} \quad \frac{\partial P}{\partial y} = \frac{\partial R}{\partial y} - P_{\text{II}}.$$

Since the profit $P(x, y)$ is maximized when $x = a$, $y = b$, we must have

$$\frac{\partial P}{\partial x}(a,b) = 0 \quad \text{and} \quad \frac{\partial P}{\partial y}(a,b) = 0$$

$$\Longrightarrow \begin{cases} \frac{\partial P}{\partial x}(a,b) = \frac{\partial R}{\partial x}(a,b) - P_{\text{I}} = 0 \\ \frac{\partial P}{\partial y}(a,b) = \frac{\partial R}{\partial y}(a,b) - P_{\text{II}} = 0 \end{cases}$$

$$\Longrightarrow \frac{\partial R}{\partial x}(a,b) = P_{\text{I}} \quad \text{and} \quad \frac{\partial R}{\partial y}(a,b) = P_{\text{II}}.$$

13.a. The cost $C(x, y) = 96x + 162y$.

We need to minimize the cost $C(x, y)$ subject to the constraint $64x^{3/4} \cdot y^{1/4} = 3456$.

Let $F(x, y, \lambda) = C(x, y) + \lambda(64x^{3/4} \cdot y^{1/4} - 3456)$.

$$\frac{\partial F}{\partial x} = 96 + 48x^{-1/4}y^{1/4}\lambda = 0 \quad \Longrightarrow \left(\frac{y}{x}\right)^{1/4} = -\frac{2}{\lambda} \tag{1}$$

$$\frac{\partial F}{\partial y} = 162 + 16x^{3/4}y^{-3/4}\lambda = 0 \quad \Longrightarrow \left(\frac{x}{y}\right)^{3/4} = -\frac{81}{8\lambda} \tag{2}$$

$$\frac{\partial F}{\partial \lambda} = 64x^{3/4} \cdot y^{1/4} - 3456 = 0 \quad \Longrightarrow x^{3/4} \cdot y^{1/4} = 54 \tag{3}$$

Dividing each side of Eq. (1) by the corresponding side of Eq. (2) gives

$$\frac{y}{x} = \frac{16}{81} \Longrightarrow y = \frac{16}{81}x.$$

Substituting into Eq. (3) gives

$$x^{3/4} \cdot \left(\frac{16}{81}x\right)^{1/4} = 54$$

$$\frac{2}{3}x = 54 \Longrightarrow x = 81.$$

Then $y = \frac{16}{81} \cdot 81 = 16$.

Thus 81 units of labor and 16 units of capital should be utilized in order to minimize the cost.

b. From Eq. (1) in part (a) we have

$$\lambda = -2\left(\frac{x}{y}\right)^{1/4}.$$

So at the optimal level of production $x = 81$, $y = 16$

$$\lambda = -2\left(\frac{81}{16}\right)^{1/4} = -3.$$

14.a. We obtain the following table for the data.

j	x_j	y_j	x_j^2	$x_j y_j$
1	1	35	1	35
2	2	31	4	62
3	3	26	9	78
4	4	24	16	96
5	5	26	25	130
Σ	15	142	55	401

Then

$$m = \frac{5 \cdot (401) - 15 \cdot (142)}{5 \cdot (55) - (15)^2} = \frac{-125}{50} = -2.5$$

and

$$b = \frac{55 \cdot (142) - 15 \cdot (401)}{5 \cdot (55) - (15)^2} = \frac{1795}{50} = 35.9.$$

The least squares regression line therefore is $y = -2.5x + 35.9$.

b. We have the corresponding table for the data.

j	x_j	y_j	x_j^2	$x_j y_j$
1	−2	35	4	−70
2	−1	31	1	−31
3	0	26	0	0
4	1	24	1	24
5	2	26	4	52
Σ	0	142	10	−25

Then

$$m = \frac{5 \cdot (-25) - 0 \cdot (142)}{5 \cdot (10) - (0)^2} = -2.5$$

and

$$b = \frac{10 \cdot (142) - 0 \cdot (-25)}{5 \cdot (10) - (0)^2} = 28.4.$$

The least squares regression line is then

$$y = -2.5x + 28.4.$$

15.a. The volume $V = \iint_R f(x,y)dA = \int_0^1 \int_0^x (x + y + 1)dydx$

$$= \int_0^1 \left[\left(xy + \tfrac{1}{2}y^2 + y\right)\Big]_0^x \right] dx$$

$$= \int_0^1 \left(x^2 + \tfrac{1}{2}x^2 + x\right) dx$$

$$= \left(\tfrac{1}{2}x^3 + \tfrac{1}{2}x^2\right)\Big]_0^1 = \tfrac{1}{2} + \tfrac{1}{2} = 1.$$

b. The volume of the solid $= \iint\limits_R f(x,y)dA$

$$= \int_{-2}^{2} \int_{0}^{\sqrt{4-y^2}} x\sqrt{x^2+y^2}\, dxdy$$

$$= \int_{-2}^{2} \int_{0}^{\sqrt{4-y^2}} \tfrac{1}{2}\sqrt{u+y^2}\, dudy \quad \text{where } u = x^2$$

$$= \int_{-2}^{2} \left[\tfrac{1}{2} \cdot \tfrac{2}{3} \cdot (u+y^2)^{3/2} \Big]_{x=0}^{x=\sqrt{4-y^2}} \right] dy$$

$$= \int_{-2}^{2} \tfrac{1}{3} \left(x^2+y^2\right)^{3/2} \Big]_{x=0}^{x=\sqrt{4-y^2}} dy$$

$$= \int_{-2}^{2} \tfrac{1}{3} \left\{ \left[\left(\sqrt{4-y^2}\right)^2 + y^2 \right]^{3/2} - (0+y^2)^{3/2} \right\} dy$$

$$= \tfrac{1}{3} \int_{-2}^{2} (8-y^3)dy = \tfrac{1}{3}\left(8y - \tfrac{1}{4}y^4\right)\Big]_{-2}^{2}$$

$$= \tfrac{1}{3}\left[(16-4)-(-16-4)\right] = \tfrac{32}{3}.$$

c. The points of intersection of the graphs of

$x = y - y^2$ and $x + y = 0$ are (0,0) and (-2,2).

Then the area of $R = \iint\limits_R 1 \cdot dA$

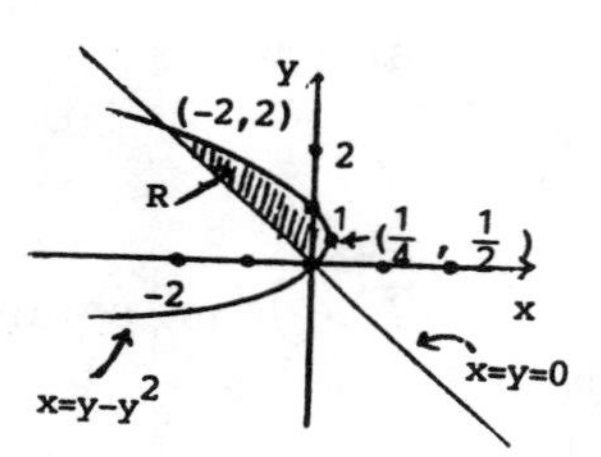

$$= \int_{0}^{2} \int_{-y}^{y-y^2} dxdy = \int_{0}^{2} \left(x\Big]_{x=-y}^{x=y-y^2} \right) dy = \int_{0}^{2} \left[y - y^2 - (-y)\right] dy$$

$$= \int_{0}^{2} (2y-y^2)dy = \left(y^2 - \tfrac{1}{3}y^3\right)\Big]_0^2 = 4 - \tfrac{1}{3}\cdot 8 = \tfrac{4}{3}.$$

16. $f(x,y) = xy \cdot (x^2+y^2)^{1/2}$.

$$\tfrac{\partial f}{\partial x}(x,y) = y \cdot (x^2+y^2)^{1/2} + xy \cdot \tfrac{1}{2}\left(x^2+y^2\right)^{-1/2} \cdot 2x$$

$$= y \cdot (x^2+y^2)^{-1/2} \cdot \left((x^2+y^2)+x^2\right)$$

$$= y \cdot (x^2+y^2)^{-1/2} \cdot (2x^2+y^2).$$

$$\tfrac{\partial f}{\partial y}(x,y) = x \cdot (x^2+y^2)^{1/2} + xy \cdot \tfrac{1}{2}(x^2+y^2)^{-1/2} \cdot 2y$$

$$= x \cdot (x^2+y^2)^{-1/2} \cdot \left((x^2+y^2)+y^2\right)$$

$$= x \cdot (x^2+y^2)^{-1/2} \cdot (x^2+2y^2).$$

$f(3,4) = 3 \cdot 4 \cdot (3^2+4^2)^{1/2} = 60.$

$\tfrac{\partial f}{\partial x}(3,4) = 4 \cdot (3^2+4^2)^{-1/2} \cdot (2 \cdot 3^2 + 4^2) = \tfrac{136}{5} = 27.2$

$\frac{\partial f}{\partial y}(3,4) = 3 \cdot (3^2 + 4^2)^{-1/2} \cdot (3^2 + 2 \cdot 4^2) = \frac{123}{5} = 24.6.$

$f(2.98, 4.04) \approx f(3,4) + \frac{\partial f}{\partial x}(3,4) \cdot (2.98 - 3) + \frac{\partial f}{\partial y}(3,4) \cdot (4.04 - 4)$

$= 60 + 27.2 \cdot (-.02) + 24.6 \cdot (.04) = 60.44.$

17. $f(x,y) = (2 + xy) \cdot \cos y.$

$\frac{\partial f}{\partial x}(x,y) = y \cdot \cos y,$

$\frac{\partial f}{\partial y}(x,y) = -2 \sin y + x \cdot \cos y - xy \cdot \sin y.$

$f\left(2, \frac{\pi}{2}\right) = 0, \quad \frac{\partial f}{\partial x}\left(2, \frac{\pi}{2}\right) = 0, \quad \frac{\partial f}{\partial y}\left(2, \frac{\pi}{2}\right) = -2 - \pi.$

The desired linear approximation is

$$\begin{aligned} f(x,y) &\approx f\left(2, \frac{\pi}{2}\right) + \frac{\partial f}{\partial x}\left(2, \frac{\pi}{2}\right) \cdot (x - 2) + \frac{\partial f}{\partial y}\left(2, \frac{\pi}{2}\right) \cdot \left(y - \frac{\pi}{2}\right) \\ &= (-2 - \pi) \cdot \left(y - \frac{\pi}{2}\right). \end{aligned}$$

18. The distance between a point (x,y) and the origin is $\sqrt{x^2 + y^2}$. The problem here is equivalent to finding where the minimum value of

$$f(x,y) = x^2 + y^2$$

occurs subject to the constraint

$$g(x,y) = x^2 + 24xy + 8y^2 - 1700 = 0.$$

Let

$$\begin{aligned} L(x,y,\lambda) &= f(x,y) + \lambda \cdot g(x,y) \\ &= x^2 + y^2 + \lambda \cdot (x^2 + 24xy + 8y^2 - 1700). \end{aligned}$$

$$\frac{\partial L}{\partial x} = 2x + \lambda \cdot (2x + 24y) = 0. \qquad (1)$$

$$\frac{\partial L}{\partial y} = 2y + \lambda \cdot (24x + 16y) = 0. \qquad (2)$$

(1) $\Longrightarrow \lambda = \frac{-2x}{2x + 24y}.$

(2) $\Longrightarrow \lambda = -\frac{2y}{24x + 16y}.$

$$\frac{2x}{2x+24y} = \frac{2y}{24x+16y}.$$

$$48x^2 + 32xy = 4xy + 48y^2.$$

$$48x^2 + 28xy - 48y^2 = 0.$$

$$12x^2 + 7xy - 12y^2 = 0.$$

$$(4x - 3y)(3x + 4y) = 0.$$

Therefore, $y = \frac{4}{3}x$ or $y = -\frac{3}{4}x$.

Case 1: $y = \frac{4}{3}x$.

$$\frac{\partial L}{\partial \lambda} = x^2 + 24xy + 8y^2 - 1700 = 0 \implies$$

$$x^2 + 24x \cdot \left(\frac{4}{3}x\right) + 8 \cdot \left(\frac{4}{3}x\right)^2 - 1700 = 0.$$

$$x^2 + 32x^2 + \frac{128}{9}x^2 - 1700 = 0.$$

$$9x^2 + 288x^2 + 128x^2 - 15300 = 0.$$

$$425x^2 = 15300.$$

$$x^2 = 36 \implies x = 6 \quad \text{or} \quad x = -6.$$

Therefore, we have the candidate points $(6, 8)$ and $(-6, -8)$.

Case 2: $y = -\frac{3}{4}x$.

$$\frac{\partial L}{\partial \lambda} = 0 \implies x^2 + 24x \cdot \left(-\frac{3}{4}x\right) + 8 \cdot \left(-\frac{3}{4}x\right)^2 - 1700 = 0.$$

$$x^2 - 18x^2 + \frac{9}{2}x^2 - 1700 = 0.$$

$$2x^2 - 36x^2 + 9x^2 - 3400 = 0.$$

$$-25x^2 = 3400, \quad \text{contradiction.}$$

$$f(6, 8) = f(-6, -8) = 100.$$

Therefore the minimum distance occurs at the points $(6, 8)$ and $(-6, -8)$.

19. a. $f(p,q) = 1000 - 50p + 2q, \quad g(p,q) = 500 + 4p - 20q.$

$\frac{\partial f}{\partial q} = 2 > 0, \quad \frac{\partial g}{\partial p} = 4 > 0.$

Hence, A and B are competitive.

b. $f(p,q) = 100p^{-1}q^{-1/2}, \quad g(p,q) = 500q^{-1}p^{-1/3}.$

$\frac{\partial f}{\partial q} = 100p^{-1}\left(-\frac{1}{2}q^{-3/2}\right) < 0, \quad \frac{\partial g}{\partial p} = 500q^{-1}\left(-\frac{1}{3}p^{-4/3}\right) < 0.$

Therefore A and B are complementary.

20. The revenue

$$\begin{aligned} R(p,q) &= (1000 - 50p + 2q)\cdot p + (500 + 4p - 20q)\cdot q \\ &= 1000p - 50p^2 + 2pq + 500q + 4pq - 20q^2 \\ &= -50p^2 + 6pq - 20q^2 + 1000p + 500q. \end{aligned}$$

$\frac{\partial R}{\partial p} = -100p + 6q + 1000, \quad \frac{\partial R}{\partial q} = 6p - 40q + 500.$

$\frac{\partial R}{\partial p} = 0$ and $\frac{\partial R}{\partial q} = 0$:

$$\begin{cases} -100p + 6q + 1000 = 0 & \cdot 40 \implies -4000p + 240q + 40,000 = 0 \\ 6p - 40q + 500 = 0 & \cdot 6 \implies 36p - 240q + 3000 = 0 \quad + \\ & \overline{-3964p + 43000 = 0} \\ & p = \frac{43000}{3964} = 10.85. \end{cases}$$

$$\begin{aligned} 6\cdot(10.85) - 40q + 500 &= 0 \\ -40q &= -6\cdot(10.85) - 500 \\ -40q &= -5651 \\ q &= \frac{5651}{40} = 14.13. \end{aligned}$$

$$\frac{\partial^2 R}{\partial p^2} = -100, \quad \frac{\partial^2 R}{\partial q \partial p} = 6, \quad \frac{\partial^2 R}{\partial q^2} = -40.$$

$$\begin{aligned} D &= \left[\frac{\partial^2 R}{\partial q \partial p}\right]^2 - \left[\frac{\partial^2 R}{\partial p^2}\right]\cdot\left[\frac{\partial^2 R}{\partial q^2}\right] \\ &= [6]^2 - [-100]\cdot[-40] < 0. \end{aligned}$$

$$A = \frac{\partial^2 R}{\partial p^2} < 0.$$

Therefore the maximum revenue occurs when $p = \$10.85$ and $q = \$14.13$.

CHAPTER 9

Summary of Chapter 9

1. The definition of an infinite sequence of real numbers; limits of sequences, convergence and divergence of sequences; properties of limits of sequences; a definite integral as the limit of a sequence of approximating sums; lot sampling for defectives.

2. The concept of an infinite series of real numbers; partial sums of an infinite series, convergence of an infinite series to a sum S; divergence of an infinite series; properties of infinite series.

3. The geometric series $\sum_{k=0}^{\infty} x^k$, convergence of this series to the sum $\frac{1}{1-x}$ if $|x| < 1$.

4. The necessary condition that if an infinite series $\sum a_k$ converges then $\lim_{k\to\infty} a_k = 0$.

5. The Integral Test for the convergence or divergence of an infinite series of nonnegative real numbers; the consequence that the p-series $\sum_{k=1}^{\infty} \frac{1}{k^p}$, $p > 0$, converges if and only if $p > 1$.

6. The Comparison Test for the convergence or divergence of an infinite series of nonnegative real numbers; the Limit Comparison Test for the convergence or divergence of an infinite series of positive real numbers.

7. The Ratio Test for the convergence or divergence of an infinite series of positive real numbers.

8. The concept of an alternating infinite series, the Alternating Series Test for the convergence of such an infinite series under certain hypotheses.

9. The definition of absolute convergence for an infinite series of real numbers; the theorem that an absolutely convergent infinite series of real numbers converges; the concept of conditional convergence for an infinite series of real numbers.

10. The nth Taylor polynomial $P_n(x)$ for a function $f(x)$ expanded about $x = a$, use of $P_n(x)$ to approximate functional values $f(x)$ for x near a, Taylor's Theorem for determining an upper bound on the absolute value of the error in this approximation.

11. The concept of the Taylor series for a function $f(x)$ expanded about $x = a$; determining the interval of convergence and the radius of convergence of a Taylor series; differentiation and integration of Taylor series; the observation that for any function f discussed in Chapter 9, if the Taylor series under consideration converges at x, then it converges to the value $f(x)$.

12. More general power series.

13. Newton's Method for approximating zeros of functions.

Selected Solutions to Exercise Set 9.1

1. $\{2n+3\}$.

$$a_1 = 2(1)+3 = 5, \quad a_2 = 2(2)+3 = 7, \quad a_3 = 2(3)+3 = 9,$$

$$a_4 = 2(4)+3 = 11, \quad a_5 = 2(5)+3 = 13.$$

3. $\left\{\frac{2n+1}{n+5}\right\}$.

$$a_1 = \frac{2\cdot1+1}{1+5} = \frac{1}{2}, \quad a_2 = \frac{2\cdot2+1}{2+5} = \frac{5}{7}, \quad a_3 = \frac{2\cdot3+1}{3+5} = \frac{7}{8},$$

$$a_4 = \frac{2\cdot4+1}{4+5} = 1, \quad a_5 = \frac{2\cdot5+1}{5+5} = \frac{11}{10}.$$

5. $\left\{\frac{(-1)^n}{e^n}\right\}$.

$$a_1 = \frac{(-1)}{e^1} = -\frac{1}{e}, \quad a_2 = \frac{(-1)^2}{e^2} = \frac{1}{e^2}, \quad a_3 = \frac{(-1)^3}{e^3} = -\frac{1}{e^3},$$

$$a_4 = \frac{(-1)^4}{e^4} = \frac{1}{e^4}, \quad a_5 = \frac{(-1)^5}{e^5} = -\frac{1}{e^5}.$$

7. $\left\{\frac{(n-1)(n+1)}{2n^2+2n+2}\right\} = \left\{\frac{n^2-1}{2(n^2+n+1)}\right\}$.

$$a_1 = \frac{1^2-1}{2(1^2+1+1)} = 0, \quad a_2 = \frac{2^2-1}{2(2^2+2+1)} = \frac{3}{14}, \quad a_3 = \frac{3^2-1}{2(3^2+3+1)} = \frac{8}{26} = \frac{4}{13},$$

$$a_4 = \frac{4^2-1}{2(4^2+4+1)} = \frac{15}{42} = \frac{5}{14}, \quad a_5 = \frac{5^2-1}{2(5^2+5+1)} = \frac{24}{62} = \frac{12}{31}.$$

9. $\left\{(-1)^{n+1}\sin\left(\frac{n\pi}{4}\right)\right\}$.

$$a_1 = (-1)^{1+1}\sin\left(\frac{1\cdot\pi}{4}\right) = \frac{\sqrt{2}}{2}, \quad a_2 = (-1)^{2+1}\sin\left(\frac{2\cdot\pi}{4}\right) = -1,$$

$$a_3 = (-1)^{3+1}\sin\left(\frac{3\cdot\pi}{4}\right) = \frac{\sqrt{2}}{2},$$

$$a_4 = (-1)^{4+1}\sin\left(\frac{4\cdot\pi}{4}\right) = 0,$$

$$a_5 = (-1)^{5+1}\sin\left(\frac{5\cdot\pi}{4}\right) = -\frac{\sqrt{2}}{2}.$$

11. $\left\{\frac{n}{3n+6}\right\}$.

$$\lim_{n\to\infty} \frac{n}{3n+6} = \lim_{n\to\infty} \frac{\frac{n}{n}}{\frac{3n}{n}+\frac{6}{n}} = \lim_{n\to\infty} \frac{1}{3+\frac{6}{n}} = \frac{1}{3}.$$

13. $\{\sqrt{5+n}\}$.

$\lim_{n\to\infty} \sqrt{5+n} = \infty \implies$ the sequence diverges.

15. $\left\{\frac{e^n+3}{e^n}\right\}$.

$\lim_{n\to\infty} \frac{e^n+3}{e^n} = \lim_{n\to\infty} \left(1+\frac{3}{e^n}\right) = 1$ since $\lim_{n\to\infty} \frac{3}{e^n} = 0$.

17. $\left\{\frac{n^2-2n+1}{3+n^2}\right\}$.

$\lim_{n\to\infty} \frac{n^2-2n+1}{3+n^2} = \lim_{n\to\infty} \frac{\frac{(n^2-2n+1)}{n^2}}{\frac{(3+n^2)}{n^2}} = \lim_{n\to\infty} \frac{1-\frac{2}{n}+\frac{1}{n^2}}{\frac{3}{n^2}+1} = 1.$

19. $\{\ln(n+1)-\ln(n)\}$.

$\lim_{n\to\infty} [\ln(n+1)-\ln(n)] = \lim_{n\to\infty} \ln\left(\frac{n+1}{n}\right) = \lim_{n\to\infty} \ln\left(1+\frac{1}{n}\right) = \ln 1 = 0.$

21. Since $\sin n\pi = 0$ for all $n \geq 1$, $\lim_{n\to\infty} \sin n\pi = 0$.

23. $\lim_{n\to\infty} \cos \frac{n-1}{n^2} = \lim_{n\to\infty} \cos \frac{\frac{1}{n}-\frac{1}{n^2}}{1} = \cos 0 = 1.$

25. $\lim_{n\to\infty} \left\{\frac{1}{n} - \frac{1}{n+1}\right\} = \lim_{n\to\infty} \frac{(n+1)-n}{n(n+1)} = \lim_{n\to\infty} \frac{1}{n(n+1)} = 0.$

27. $\lim_{n\to\infty} \left(1+\frac{1}{n}\right)^n = e$ as developed in Chapter 4.

29. $\lim_{n\to\infty} \frac{\sqrt{n}}{1+\sqrt{n}} = \lim_{n\to\infty} \frac{\frac{\sqrt{n}}{\sqrt{n}}}{\frac{(1+\sqrt{n})}{\sqrt{n}}} = \lim_{n\to\infty} \frac{1}{\frac{1}{\sqrt{n}}+1} = 1.$

31. $\left\{\cos\left(\frac{n\pi}{4}\right)\right\}$.

The terms of this sequence oscillate as $n \to \infty$.

Therefore this sequence diverges.

33. $N(n) = \frac{A}{B+Ce^{-n}}$.

$\lim_{n\to\infty} N(n) = \frac{A}{B}$ since $\lim_{n\to\infty} e^{-n} = 0$.

35. $a_n = 1000 \cdot (1.08)^n$.

a. $a_1 = 1000 \cdot 1.08 = \$1080$.

b. $a_2 = 1000 \cdot (1.08)^2 = \$1166.40$.

c. $a_5 = 1000 \cdot (1.08)^5 = \$1469.33$.

37. Assume that the fleet average fuel consumption in miles per gallon increases by 10 percent at the end of each year.

    a. $a_1 = 22 + 22 \cdot .10 = 22 \cdot (1.10) = 24.2.$

    b. $a_2 = 22 \cdot (1.10) + 22 \cdot (1.10) \cdot .10 = 22 \cdot (1.10) \cdot (1.10) = 22 \cdot (1.10)^2 = 26.62.$

    c. $a_n = 22 \cdot (1.10)^n.$

39. $a_1 = 50.$

    $a_2 = 50 \cdot \frac{1}{2} + 50 = 50 \cdot \left(1 + \frac{1}{2}\right).$

    $a_3 = 50 \cdot \left(1 + \frac{1}{2}\right) \cdot \frac{1}{2} + 50 = 50 \cdot \left(1 + \frac{1}{2} + \frac{1}{4}\right).$

    $a_4 = 50 \cdot \left(1 + \frac{1}{2} + \frac{1}{4}\right) \cdot 12 + 50 = 50 \cdot \left(1 + \frac{1}{2} + \frac{1}{4} + \frac{1}{8}\right).$

    etc.

    $a_n = 50 \cdot \left(1 + \frac{1}{2} + \frac{1}{4} + \frac{1}{8} + \cdots + \frac{1}{2^{n-1}}\right).$

41. $P(n) = 1 - \left(\frac{100-10}{100}\right)^n = 1 - \left(\frac{9}{10}\right)^n$ or $100 \cdot \left[1 - \left(\frac{9}{10}\right)^n\right]$ percent.

    $P(1) = 1 - \frac{9}{10} = .1$ or 10 percent.

    $P(2) = 1 - \left(\frac{9}{10}\right)^2 = 1 - \frac{81}{100} = .19$ or 19 percent.

    $P(3) = 1 - \left(\frac{9}{10}\right)^3 = 1 - \frac{729}{1000} = .271$ or 27.1 percent.

Selected Solutions to Exercise Set 9.2

1. $\sum_{k=1}^{\infty} (2k+3).$

    $a_1 = 2(1)+3 = 5, \quad a_2 = 2(2)+3 = 7, \quad a_3 = 2(3)+3 = 9, \quad a_4 = 2(4)+3 = 11,$

    $S_4 = \sum_{k=1}^{4} a_k = 5 + 7 + 9 + 11.$

3. $\sum_{k=0}^{\infty} \frac{1}{5^k}.$

    $S_4 = \frac{1}{5^0} + \frac{1}{5^1} + \frac{1}{5^2} + \frac{1}{5^3} = 1 + \frac{1}{5} + \frac{1}{25} + \frac{1}{125}.$

5. $\sum_{k=1}^{\infty} \frac{(-1)^k}{1+k^2}$.

$S_4 = \sum_{k=1}^{4} \frac{(-1)^k}{1+k^2} = \frac{(-1)^1}{1+1^2} + \frac{(-1)^2}{1+2^2} + \frac{(-1)^3}{1+3^2} + \frac{(-1)^4}{1+4^2} = -\frac{1}{2} + \frac{1}{5} - \frac{1}{10} + \frac{1}{17}$.

7. $\sum_{k=1}^{\infty} \frac{\cos(\pi k)}{1+k}$.

$$S_4 = \sum_{k=1}^{4} \frac{\cos(\pi k)}{1+k} = \frac{\cos(\pi\cdot 1)}{1+1} + \frac{\cos(\pi\cdot 2)}{1+2} + \frac{\cos(\pi\cdot 3)}{1+3} + \frac{\cos(\pi\cdot 4)}{1+4}$$
$$= -\frac{1}{2} + \frac{1}{3} - \frac{1}{4} + \frac{1}{5}.$$

9. $\sum_{k=0}^{\infty} \frac{1}{4^k} = \sum_{k=0}^{\infty} \left(\frac{1}{4}\right)^k = \frac{1}{1-\frac{1}{4}} = \frac{4}{3}$ since $\left|\frac{1}{4}\right| < 1$.

11. $\sum_{k=0}^{\infty} \frac{3^k}{7^k} = \sum_{k=0}^{\infty} \left(\frac{3}{7}\right)^k = \frac{1}{1-\frac{3}{7}} = \frac{7}{4}$ since $\left|\frac{3}{7}\right| < 1$.

13. $\sum_{k=1}^{\infty} \frac{1}{e^{-k}} = \sum_{k=1}^{\infty} e^k$.

This is a divergent geometric series since $e > 1$.

15. $\sum_{k=1}^{\infty} \frac{7^k}{5^k} = \sum_{k=0}^{\infty} \left(\frac{7}{5}\right)^k - 1$ diverges since $\frac{7}{5} > 1$.

17. $\sum_{k=0}^{\infty} \frac{1+3^k}{8^k} = \sum_{k=0}^{\infty} \left(\frac{1}{8}\right)^k + \sum_{k=0}^{\infty} \left(\frac{3}{8}\right)^k = \frac{1}{1-\frac{1}{8}} + \frac{1}{1-\frac{3}{8}} = \frac{1}{7/8} + \frac{1}{5/8}$

$$= \frac{8}{7} + \frac{8}{5} = \frac{40}{35} + \frac{56}{35} = \frac{96}{35}.$$

19. $\sum_{k=1}^{\infty} \frac{2^k-3^k}{4^k} = \sum_{k=0}^{\infty} \left[\left(\frac{2}{4}\right)^k - \left(\frac{3}{4}\right)^k\right] - \frac{2^0-3^0}{4^0} = \sum_{k=0}^{\infty} \left(\left(\frac{1}{2}\right)^k + (-1)\left(\frac{3}{4}\right)^k\right)$

$$= \sum_{k=0}^{\infty} \left(\frac{1}{2}\right)^k + \sum_{k=0}^{\infty}(-1)\left(\frac{3}{4}\right)^k = \sum_{k=0}^{\infty} \left(\frac{1}{2}\right)^k + (-1)\sum_{k=0}^{\infty} \left(\frac{3}{4}\right)^k = \frac{1}{1-\frac{1}{2}} - \frac{1}{1-\frac{3}{4}} = -2.$$

21. $\sum_{k=1}^{\infty} \frac{5\cdot 3^k - 3\cdot 7^k}{5^k} = \sum_{k=0}^{\infty} \left[5\left(\frac{3}{5}\right)^k - 3\cdot\left(\frac{7}{5}\right)^k\right] - \frac{5\cdot 3^0 - 3\cdot 7^0}{5^0}$

The infinite series $\sum_{k=0}^{\infty} \left(\frac{3}{5}\right)^k$ is a convergent geometric series, but the infinite series $\sum_{k=0}^{\infty} \left(\frac{7}{5}\right)^k$ is a divergent geometric series. Therefore the given infinite series diverges.

23. $0.333\bar{3} = 3\sum_{k=1}^{\infty} \left(\frac{1}{10}\right)^k = 3\left[\sum_{k=0}^{\infty} \left(\frac{1}{10}\right)^k - \left(\frac{1}{10}\right)^0\right]$

$$= 3\left(\frac{1}{1-\frac{1}{10}} - 1\right) = 3\left(\frac{10}{9} - 1\right) = 3\cdot\frac{1}{9} = \frac{1}{3}.$$

25. a. Total amount in the first five years

$$= 1000\left(1 + \tfrac{1}{2} + \tfrac{1}{2^2} + \tfrac{1}{2^3} + \tfrac{1}{2^4}\right) = 1000\left(\frac{1-\frac{1}{2^5}}{1-\frac{1}{2}}\right) = 1000 \cdot \tfrac{31}{16} = \$1937.50.$$

b. Total amount received in perpetuity $= 1000\left(\frac{1}{1-\frac{1}{2}}\right) = \$2000.$

27. $1 + 0.75 + (0.75)^2 + (0.75)^3 + \cdots = \frac{1}{1-0.75} = 4.$ Total is \$4 million.

29. a. $S(n) = 1000 \cdot e^{-r} + 1000 \cdot e^{-2r} + \cdots + 1000 \cdot e^{-nr} = \sum_{k=1}^{n} 1000 \cdot e^{-kr}.$

b. $S(n) = \frac{1000 \cdot e^{-r}(1-e^{-nr})}{1-e^{-r}}.$

c. $S = \lim_{n\to\infty} S(n) = \sum_{k=1}^{\infty} 1000 \cdot e^{-rk}.$

d. $e^{-r} < 1$ for $r > 0$, so the infinite series in (c) converges.

e. $S = \frac{1000 \cdot e^{-r}}{1-e^{-r}} = \frac{1000}{e^r - 1}.$

31. $20 + (0.75) \cdot 20 + (0.75)^2 \cdot 20 + \cdots = \frac{20}{1-0.75} = \$80.$

33. Let a_n denote the number of people who have heard the rumor by the end of the nth day.

$a_1 = 3.$

$a_2 = 3 + 3 \cdot 2 = 9.$

$a_3 = 9 + 9 \cdot 2 = 27.$

$a_4 = 27 + 27 \cdot 2 = 81.$

In general, $a_n = 3^n$ which is analogous to the compound interest formula. Note that $a_n \to \infty$ as $n \to \infty$.

35. Let a_0 be the initial amount of the pollutant.

Let a_n be the amount of the pollutant that remains after n days.

$a_1 = .40 \cdot a_0.$

$a_2 = .40 \cdot a_0 - (.40 \cdot a_0) \cdot (.60) = .16a_0$

$a_3 = .16a_0 - (.16a_0) \cdot (.60) = .064a_0.$

$a_4 = .064a_0 - (.064a_0) \cdot (.60) = .0256a_0.$

In general, $a_n = .4^n a_0.$

$\lim_{n\to\infty} a_n = 0.$

Selected Solutions to Exercise Set 9.3

1. $\int_1^\infty \frac{1}{x+2}\,dx = \lim_{t\to\infty}\int_1^t \frac{1}{x+2}\,dx = \lim_{t\to\infty} \ln(x+2)\Big]_1^t$
$= \lim_{t\to\infty}[\ln(t+2)-\ln 3] = \infty$. So $\sum_{k=1}^{\infty}\frac{1}{k+2}$ diverges.

3. Put $f(x) = \frac{x}{(x^2+1)^{3/2}}$. $f'(x) = \frac{(x^2+1)^{3/2} - x\cdot\frac{3}{2}(x^2+1)^{1/2}2x}{(x^2+1)^3} = \frac{1-2x^2}{(x^2+1)^{5/2}} < 0$ for $x \ge 1$. Thus $f(x)$ is decreasing.
$\int_1^\infty \frac{x}{(x^2+1)^{3/2}}\,dx = \lim_{t\to\infty}\int_1^t \frac{x}{(x^2+1)^{3/2}}\,dx = \lim_{t\to\infty}\int_1^t \frac{1}{(x^2+1)^{3/2}}\cdot\frac{1}{2}\,d(x^2+1)$
$= \lim_{t\to\infty}\frac{1}{2}(-2)(x^2+1)^{-1/2}\Big]_1^t = \lim_{t\to\infty}\left[-(t^2+1)^{-1/2}+2^{-1/2}\right] = \frac{1}{\sqrt{2}}$.
So $\sum \frac{k}{(k^2+1)^{3/2}}$ converges.

5. $\int_2^\infty \frac{1}{x\ln x}\,dx = \lim_{t\to\infty}\int_2^t \frac{1}{x\ln x}\,dx = \lim_{t\to\infty}\int_2^t \frac{1}{\ln x}\,d(\ln x)$
$= \lim_{t\to\infty}(\ln\ln x)\Big]_2^t = \lim_{t\to\infty}(\ln\ln t - \ln\ln 2) = \infty$. So $\sum\frac{1}{k\ln k}$ diverges.

7. Put $f(x) = x^2e^{-x^3}$.
$f'(x) = 2xe^{-x^3} + x^2e^{-x^3}(-3x^2) = xe^{-x^3}(2-3x^3) < 0$ for $x \ge 1$.
Thus $f(x)$ is decreasing.
$\int_1^\infty f(x)\,dx = \lim_{t\to\infty}\int_1^t x^2e^{-x^3}dx = \lim_{t\to\infty}\int_1^t e^{-x^3}\left(\frac{1}{3}\right)d(x^3)$
$= \lim_{t\to\infty}\left(-\frac{1}{3}e^{-x^3}\right)\Big]_1^t = \lim_{t\to\infty}\left(-\frac{1}{3}e^{-t^3}+\frac{1}{3}e^{-1}\right) = \frac{1}{3e}$.
So $\sum k^2e^{-k^3}$ converges.

9. Since $\frac{3}{k} > \frac{1}{k}$ and $\sum\frac{1}{k}$ diverges, $\sum\frac{3}{k}$ diverges.

11. Since $\frac{1}{2+k^{3/2}} < \frac{1}{k^{3/2}}$ and $\sum\frac{1}{k^{3/2}}$ converges, $\sum\frac{1}{2+k^{3/2}}$ converges.

13. Since $\frac{3+k}{k^2} > \frac{k}{k^2} = \frac{1}{k}$ and $\sum\frac{1}{k}$ diverges, $\sum\frac{3+k}{k^2}$ diverges.

15. Since $\frac{2k}{1+k^3} < \frac{2k}{k^3} = \frac{2}{k^2}$ and $\sum\frac{2}{k^2}$ converges, $\sum\frac{2k}{1+k^3}$ converges.

17. $\lim_{k\to 0}\frac{\frac{k+3}{2k^2+1}}{\frac{1}{k}} = \lim_{k\to\infty}\frac{k^2+3k}{2k^2+1} = \frac{1}{2}$. Since $\sum\frac{1}{k}$ diverges, $\sum\frac{k+3}{2k^2+1}$ diverges.

19. $\lim_{k\to\infty}\frac{\frac{k+\sqrt{k}}{k+k^3}}{\frac{1}{k^2}} = \lim_{k\to\infty}\frac{k^3+k^2\sqrt{k}}{k+k^3} = 1$. Since $\sum\frac{1}{k^2}$ converges, $\sum\frac{k+\sqrt{k}}{k+k^3}$ converges.

21. Comparison Test.
Since $\frac{k^2+5}{3k^2} > \frac{k^2}{3k^2} = \frac{1}{3}$ and $\sum\frac{1}{3}$ diverges, $\sum\frac{k^2+5}{3k^2}$ diverges.

23. Comparison Test.
Since $\frac{k}{4+k^3} < \frac{k}{k^3} = \frac{1}{k^2}$ and $\sum\frac{1}{k^2}$ converges, $\sum\frac{k}{4+k^3}$ converges.

25. Comparison Test.

Since $\frac{\cos^2 \pi k}{k^2} \le \frac{1}{k^2}$ and $\sum \frac{1}{k^2}$ converges, $\sum \frac{\cos^2 \pi k}{k^2}$ converges.

27. Comparison Test.

Note that $e^{\sqrt{k}} > 1$ for $k > 0$.

$\frac{e^{\sqrt{k}}}{\sqrt{k}} > \frac{1}{\sqrt{k}}$ and $\sum \frac{1}{\sqrt{k}}$ diverges. So $\sum \frac{e^{\sqrt{k}}}{\sqrt{k}}$ diverges.

29. Use the Limit Comparison Test.

Since $\lim\limits_{k\to\infty} \frac{\frac{2k^2-2k}{k^4+3k}}{\frac{1}{k^2}} = \lim\limits_{k\to\infty} \frac{2k^4-2k^3}{k^4+3k} = 2$ and $\sum \frac{1}{k^2}$ converges, $\sum \frac{2k^2-2k}{k^4+3k}$ converges.

31. Limit Comparison Test.

Since $\lim\limits_{k\to\infty} \frac{\frac{k}{\sqrt{1+k^2}}}{1} = 1$ and $\sum 1$ diverges, $\sum \frac{k}{\sqrt{1+k^2}}$ diverges.

33. Comparison Test.

$\frac{1}{\sqrt{4k(k+1)}} \ge \frac{1}{\sqrt{4k(k+k)}} = \frac{1}{\sqrt{8}\,k}$ for $k \ge 1$.

Since $\sum \frac{1}{\sqrt{8}\,k} = \frac{1}{\sqrt{8}} \sum \frac{1}{k}$ diverges, $\sum \frac{1}{\sqrt{4k(k+1)}}$ diverges.

35. Comparison Test.

$\frac{2^k}{5+3^{k+1}} < \frac{2^k}{3^{k+1}} = \frac{1}{3}\left(\frac{2}{3}\right)^k$ and $\sum \frac{1}{3}\left(\frac{2}{3}\right)^k = \frac{1}{3}\sum\left(\frac{2}{3}\right)^k$ converges.

So $\sum \frac{2^k}{5+3^{k+1}}$ converges.

37. Integral Test.

Put $f(x) = \frac{\ln x}{x}$. $\quad f'(x) = \frac{\frac{1}{x}\cdot x - \ln x}{x^2} = \frac{1-\ln x}{x^2} < 0$ for $x > e$.

Thus $f(x)$ is decreasing for $x > e$.

$\int_3^\infty f(x)dx = \lim\limits_{t\to\infty} \int_3^t \frac{\ln x}{x}\, dx = \lim\limits_{t\to\infty} \int_3^t \ln x\; d(\ln x) = \lim\limits_{t\to\infty} \frac{1}{2}(\ln x)^2\Big]_3^t$

$= \lim\limits_{t\to\infty} \left[\frac{1}{2}(\ln t)^2 - \frac{1}{2}(\ln 3)^2\right] = \infty$. So $\sum \frac{\ln k}{k}$ diverges.

39. Use the Limit Comparison Test.

$\lim\limits_{k\to\infty} \frac{\frac{7+k}{1+k^2}}{\frac{1}{k}} = \lim\limits_{k\to\infty} \frac{7k+k^2}{1+k^2} = \lim\limits_{k\to\infty} \frac{\frac{7}{k}+1}{\frac{1}{k^2}+1} = \frac{0+1}{0+1} = 1$.

$\sum \frac{1}{k}$ diverges. Therefore $\sum \frac{7+k}{1+k^2}$ diverges.

41. Comparison Test.

Since $\frac{2^k}{k+5} > \frac{1}{k+5}$ for $k > 0$ and $\sum \frac{1}{k+5}$ diverges, $\sum \frac{2^k}{k+5}$ diverges.

43. Limit Comparison Test.

Since $\lim\limits_{k\to\infty} \frac{\frac{k}{(k+2)2^k}}{\frac{1}{2^k}} = 1$ and $\sum \frac{1}{2^k}$ converges, $\sum \frac{k}{(k+2)2^k}$ converges.

45. Limit Comparison Test.

Since $\lim\limits_{k\to\infty} \frac{\frac{2k+2}{\sqrt{k^3+2}}}{\frac{1}{\sqrt{k}}} = 2$ and $\sum \frac{1}{\sqrt{k}}$ diverges, $\sum \frac{2k+2}{\sqrt{k^3+2}}$ diverges.

47. Comparison Test.
$\frac{k}{2+\ln k} > \frac{k}{2+k}$, so $\sum \frac{k}{2+\ln k}$ diverges.
Here $\sum \frac{k}{2+k}$ diverges by Theorem 5.

Selected Solutions to Exercise Set 9.4

1. $a_k = \frac{2^k}{k+2}$.

$\rho = \lim_{k\to\infty} \frac{a_{k+1}}{a_k} = \lim_{k\to\infty} \frac{2^{k+1}}{(k+1)+2} \cdot \frac{k+2}{2^k} = \lim_{k\to\infty} \frac{2(k+2)}{k+3} = 2 > 1.$

Thus $\sum \frac{2^k}{k+2}$ diverges.

3. $a_k = k^2e^{-k}$.

$\rho = \lim_{k\to\infty} \frac{a_{k+1}}{a_k} = \lim_{k\to\infty} \frac{(k+1)^2e^{-k-1}}{k^2e^{-k}} = \lim_{k\to\infty} \left(\frac{k+1}{k}\right)^2 \cdot e^{-1}$

$= \lim_{k\to\infty} \frac{1}{e}\left(1+\frac{1}{k}\right)^2 = \frac{1}{e} < 1$. So $\sum k^2e^{-k}$ converges.

5. $a_k = \frac{\ln k}{ke^k}$.

$a_k < \frac{k}{ke^k} = \frac{1}{e^k}$ for integers $k \geq 1$.

The infinite series $\sum \frac{1}{e^k}$ converges. Therefore $\sum \frac{\ln k}{ke^k}$ converges.

7. $a_k = \frac{k+2}{1+k^3}$.

$\rho = \lim_{k\to\infty} \frac{a_{k+1}}{a_k} = \lim_{k\to\infty} \frac{k+3}{1+(k+1)^3} \cdot \frac{1+k^3}{k+2} = \lim_{k\to\infty} \frac{k+3}{k+2} \cdot \frac{k^3+1}{k^3+3k^2+3k+2} = 1.$

Therefore there is no conclusion from the Ratio Test.

We use the Limit Comparison Test.

$\lim_{k\to\infty} \frac{\frac{k+2}{1+k^3}}{\frac{1}{k^2}} = \lim_{k\to\infty} \frac{k^3+2k^2}{1+k^3} = 1$. Since $\sum \frac{1}{k^2}$ converges, the infinite series $\sum \frac{k+2}{1+k^3}$ converges.

9. $a_k = \frac{k!}{ke^k}$.

$$\rho = \lim_{k\to\infty} \frac{a_{k+1}}{a_k} = \lim_{k\to\infty} \frac{(k+1)!}{(k+1)e^{k+1}} \cdot \frac{ke^k}{k!} = \lim_{k\to\infty} \frac{k}{k+1} \cdot \frac{k+1}{e} = \infty > 1.$$

So $\sum \frac{k!}{ke^k}$ diverges.

11. $a_k = \frac{e^{5k}}{k!}$.

$$\rho = \lim_{k\to\infty} \frac{a_{k+1}}{a_k} = \lim_{k\to\infty} \frac{e^{5(k+1)}}{(k+1)!} \cdot \frac{k!}{e^{5k}} = \lim_{k\to\infty} \frac{e^5}{k+1} = 0 < 1.$$

Therefore $\sum \frac{e^{5k}}{k!}$ converges.

13. $a_k = \frac{1}{k+2}$.

Since $a_k = \frac{1}{k+2} > \frac{1}{k+3} = a_{k+1}$ and $\lim\limits_{k\to\infty} a_k = \lim\limits_{k\to\infty} \frac{1}{k+2} = 0$,

$\sum\limits_{k=0}^{\infty} \frac{(-1)^k}{k+2}$ converges.

15. $\sum\limits_{k=1}^{\infty} \frac{\cos \pi k}{k} = -1 + \frac{1}{2} - \frac{1}{3} + \frac{1}{4} - \cdots = \sum\limits_{k=1}^{\infty} \frac{(-1)^k}{k}$ where $a_k = \frac{1}{k}$.

Since $a_k = \frac{1}{k} > \frac{1}{k+1} = a_{k+1}$ and $\lim\limits_{k\to\infty} a_k = \lim\limits_{k\to\infty} \frac{1}{k} = 0$,

$\sum\limits_{k=1}^{\infty} \frac{\cos \pi k}{k}$ converges.

17. $a_k = \frac{k^2}{\sqrt{k+1}}$.

Since $\lim\limits_{k\to\infty} a_k = \lim\limits_{k\to\infty} \frac{k^2}{\sqrt{k+1}} = \infty \neq 0$, the condition (ii) of the alternating series test fails. $\sum\limits_{k=1}^{\infty} \frac{(-1)^k k^2}{\sqrt{k+1}}$ diverges by Theorem 5.

19. $\sum\limits_{k=1}^{\infty} \frac{(-1)^k}{k^2}$.

$\sum\limits_{k=1}^{\infty} \left| \frac{(-1)^k}{k^2} \right| = \sum\limits_{k=1}^{\infty} \frac{1}{k^2}$ converges, so $\sum \frac{(-1)^k}{k^2}$ converges absolutely.

21. $\sum_{k=1}^{\infty} \frac{(-1)^k \cdot k}{(k+1)!}$.

$a_k = \left| \frac{(-1)^k \cdot k}{(k+1)!} \right| = \frac{k}{(k+1)!}$

$$\lim_{k\to\infty} \frac{a_{k+1}}{a_k} = \lim_{k\to\infty} \frac{\frac{k+1}{(k+2)!}}{\frac{k}{(k+1)!}} = \lim_{k\to\infty} \frac{(k+1)(k+1)!}{k(k+2)!}$$

$$= \lim_{k\to\infty} \frac{k+1}{k(k+2)} = 0.$$

Therefore $\sum_{k=1}^{\infty} \frac{(-1)^k \cdot k}{(k+1)!}$ converges absolutely by the ratio test.

23. $\sum_{k=1}^{\infty} \frac{(-1)^k}{1+\sqrt{k}}$.

$$\lim_{k\to\infty} \frac{\frac{1}{1+\sqrt{k}}}{\frac{1}{\sqrt{k}}} = \lim_{k\to\infty} \frac{\sqrt{k}}{1+\sqrt{k}} = \lim_{k\to\infty} \frac{1}{\frac{1}{\sqrt{k}}+1} = 1.$$

$\sum_{k=1}^{\infty} \frac{1}{\sqrt{k}}$ diverges. Therefore $\sum_{k=1}^{\infty} \frac{(-1)^k}{1+\sqrt{k}}$ does not converge absolutely. But this series does converge by the alternating series test.

25. $\sum_{k=1}^{\infty} \frac{1}{k} \sin \frac{k\pi}{2} = \sum_{k=1}^{\infty} (-1)^{k+1} \cdot \frac{1}{2k-1}$ converges by the alternating series test, but does not converge absolutely since $\sum_{k=1}^{\infty} \frac{1}{2k-1} \geq \sum_{k=1}^{\infty} \frac{1}{2k} = \frac{1}{2} \sum_{k=1}^{\infty} \frac{1}{k}$.

27. $\sum_{k=1}^{\infty} \frac{(-1)^k \sqrt{k}}{(1+k^2)}$.

$a_k = \left| \frac{(-1)^k \sqrt{k}}{(1+k^2)} \right| = \frac{\sqrt{k}}{(1+k^2)} < \frac{\sqrt{k}}{k^2} = \frac{1}{k^{3/2}}$.

Therefore, $\sum_{k=1}^{\infty} \frac{(-1)^k \sqrt{k}}{(1+k^2)}$ converges absolutely by the comparison test.

29. Note that $3^{20} < 20!$.

Therefore $3^k < k!$ for $k \geq 20$, so that $\frac{k!}{3^k} > 1$ for $k \geq 20$.

Hence $\frac{(-1)^{k+1} k!}{3^k}$ does not approach 0 as k approaches ∞.

Thus the given infinite series does not converge.

31. $\sum (-1)^k a_k$ with $a_k > a_{k+1}$ for all k.

Since $\lim_{k\to\infty} a_k = L \neq 0$, by Theorem 5, it does not converge.

Selected Solutions to Exercise Set 9.5

1. $f(x) = e^{-x}$. $\quad f(0) = 1$

   $f'(x) = -e^{-x}, \quad f'(0) = -1$

   $f''(x) = e^{-x}, \quad f''(0) = 1$

   $f'''(x) = -e^{-x}, \quad f'''(0) = -1$

   $P_3(x) = 1 - x + \frac{x^2}{2} - \frac{x^3}{3!}$.

3. $f(x) = \cos x$. $\quad f(0) = 1$

   $f'(x) = -\sin x, \quad f'(0) = 0$

   $f''(x) = -\cos x, \quad f''(0) = -1$

   $f'''(x) = \sin x, \quad f'''(0) = 0$

   $f^{(4)}(x) = \cos x, \quad f^{(4)}(0) = 1$

   $P_4(x) = 1 - \frac{x^2}{2} + \frac{x^4}{4!}$.

5. $f(x) = \frac{1}{1+x}$. $\quad f(0) = 1$.

   $f'(x) = \frac{-1}{(1+x)^2}, \quad f'(0) = -1$

   $f''(x) = \frac{2}{(1+x)^3}, \quad f''(0) = 2$

   $P_2(x) = 1 - x + x^2$.

7. $f(x) = (x+2)^{1/2}$. $\quad f(0) = \sqrt{2}$.

   $f'(x) = \frac{1}{2}(x+2)^{-1/2}, \quad f'(0) = \frac{1}{2\sqrt{2}}$

   $f''(x) = -\frac{1}{4}(x+2)^{-3/2}, \quad f''(0) = -\frac{1}{4 \cdot 2^{3/2}} = -\frac{1}{8\sqrt{2}}$

   $f'''(x) = \frac{3}{8}(x+2)^{-5/2}, \quad f'''(0) = \frac{3}{8 \cdot 2^{5/2}} = \frac{3}{32\sqrt{2}}$.

   $P_3(x) = \sqrt{2} + \frac{1}{2\sqrt{2}}x - \frac{1}{16\sqrt{2}}x^2 + \frac{1}{64\sqrt{2}}x^3$.

9. $f(x) = \sin x$. $\quad f\left(\frac{\pi}{4}\right) = \frac{\sqrt{2}}{2}$

   $f'(x) = \cos x, \quad f'\left(\frac{\pi}{4}\right) = \frac{\sqrt{2}}{2}$

   $f''(x) = -\sin x, \quad f''\left(\frac{\pi}{4}\right) = -\frac{\sqrt{2}}{2}$

   $f'''(x) = -\cos x, \quad f'''\left(\frac{\pi}{4}\right) = -\frac{\sqrt{2}}{2}$.

   $f^{(4)}(x) = \sin x, \quad f^{(4)}\left(\frac{\pi}{4}\right) = \frac{\sqrt{2}}{2}$.

   $P_4(x) = \frac{\sqrt{2}}{2} + \frac{\sqrt{2}}{2}\left(x - \frac{\pi}{4}\right) - \frac{\sqrt{2}}{4}\left(x - \frac{\pi}{4}\right)^2 - \frac{\sqrt{2}}{12}\left(x - \frac{\pi}{4}\right)^3 + \frac{\sqrt{2}}{48}\left(x - \frac{\pi}{4}\right)^4$.

11. $f(x) = e^x.$ $\quad f(1) = e$

$f'(x) = e^x,$ $\quad f'(1) = e$

$f''(x) = e^x,$ $\quad f''(1) = e$

$f'''(x) = e^x,$ $\quad f'''(1) = e$

$f^{(4)}(x) = e^x,$ $\quad f^{(4)}(1) = e$

$$P_4(x) = e + e(x-1) + \tfrac{e}{2}(x-1)^2 + \tfrac{e}{3!}(x-1)^3 + \tfrac{e}{4!}(x-1)^4$$
$$= e\left[1 + (x-1) + \tfrac{1}{2}(x-1)^2 + \tfrac{1}{3!}(x-1)^3 + \tfrac{1}{4!}(x-1)^4\right].$$

13. $f(x) = \sqrt{x}.$ $\quad f(9) = 3$

$f'(x) = \frac{1}{2\sqrt{x}},$ $\quad f'(9) = \frac{1}{6}$

$f''(x) = \frac{-1}{4x\sqrt{x}},$ $\quad f''(9) = \frac{-1}{108}$

$f'''(x) = \frac{3}{8x^2\sqrt{x}},$ $\quad f'''(9) = \frac{1}{648}$

$$P_3(x) = 3 + \tfrac{1}{6}(x-9) - \tfrac{1}{216}(x-9)^2 + \tfrac{1}{3888}(x-9)^3.$$

15. $f(x) = e^{x^2}.$ $\quad f(0) = 1$

$f'(x) = e^{x^2} \cdot 2x,$ $\quad f'(0) = 0$

$f''(x) = 2 \cdot e^{x^2} + 4x^2 \cdot e^{x^2},$ $\quad f''(0) = 2$

$P_2(x) = 1 + x^2.$

17. $f(x) = x \cdot \cos x.$ $\quad f\left(\frac{\pi}{4}\right) = \frac{\sqrt{2}}{8}\pi$

$f'(x) = \cos x - x\sin x,$ $\quad f'\left(\frac{\pi}{4}\right) = \frac{\sqrt{2}}{2} - \frac{\sqrt{2}}{8}\pi$

$f''(x) = -2\sin x - x \cdot \cos x,$ $\quad f''\left(\frac{\pi}{4}\right) = -\sqrt{2} - \frac{\sqrt{2}}{8}\pi$

$$P_2(x) = \tfrac{\sqrt{2}}{8}\pi + \left(\tfrac{\sqrt{2}}{2} - \tfrac{\sqrt{2}}{8}\pi\right)\left(x - \tfrac{\pi}{4}\right) + \left(-\tfrac{\sqrt{2}}{2} - \tfrac{\sqrt{2}}{16}\pi\right)\left(x - \tfrac{\pi}{4}\right)^2.$$

19. $f(x) = e^{-x}.$ $\quad f(0) = 1$

$f'(x) = -e^{-x},$ $\quad f'(0) = -1$

$f''(x) = e^{-x},$ $\quad f''(0) = 1$

$f'''(x) = -e^{-x},$ $\quad f'''(0) = -1$

$P_3(x) = 1 - x + \frac{1}{2}x^2 - \frac{1}{3!}x^3.$

$e^{-1} = f(1) \approx P_3(1) = 1 - 1 + \frac{1}{2} - \frac{1}{6} = \frac{1}{3} = 0.3333.$

21. $f(x) = \sqrt{x}$.

$P_3(x) = 3 + \frac{1}{6}(x-9) - \frac{1}{216}(x-9)^2 + \frac{1}{3888}(x-9)^3$ by Problem 13.

$\sqrt{10} = f(10) \approx P_3(10) = 3 + \frac{1}{6} - \frac{1}{216} + \frac{1}{3888} = \frac{12295}{3888}$.

23. $f(x) = e^{-x}$.

$P_n(x) = 1 - x + \frac{x^2}{2!} - \frac{x^3}{3!} + \cdots + (-1)^n \frac{x^n}{n!}$.

$P_n'(x) = -1 + x - \frac{x^2}{2} + \cdots + (-1)^n \frac{x^{n-1}}{(n-1)!}$

$= -\left(1 - x + \frac{x^2}{2} + \cdots + (-1)^{n-1} \frac{x^{n-1}}{(n-1)!}\right)$

$= -P_{n-1}(x)$.

25. $P_4(x) = x - \frac{x^3}{3!}$.

$Q_4(x) = 1 - \frac{x^2}{2!} + \frac{x^4}{4!}$.

$Q_4'(x) = -x + \frac{x^3}{3!} = -\left(x - \frac{x^3}{3!}\right) = -P_4(x)$.

Selected Solutions to Exercise Set 9.6

1. Put $f(x) = \ln x$. $\quad f(1) = 0$.

$f'(x) = \frac{1}{x}, \quad f''(x) = -\frac{1}{x^2}, \quad f'''(x) = \frac{2}{x^3}$.

$f'(1) = 1, \quad f''(1) = -1, \quad f'''(1) = 2$.

$P_3(x) = f(1) + (x-1)f'(1) + \frac{1}{2}f''(1)(x-1)^2 + \frac{1}{3!}f'''(1)(x-1)^3$.

$\ln(1.4) \approx P_3(1.4) = 0 + (1.4-1) + \frac{1}{2}(-1)(1.4-1)^2 + \frac{1}{6}\cdot 2(1.4-1)^3$

$= 0 + 0.4 - \frac{1}{2}\cdot(0.16) + \frac{1}{3}(0.064) = 0.3413$.

$|f^{(4)}(x)| = \left|-\frac{6}{x^4}\right| = \frac{6}{x^4} \le 6$ in [1,1.4]. So $M = 6$.

$|R_3(1.4)| \le \frac{M}{4!}(1.4-1)^4 = \frac{6}{4\cdot 3\cdot 2}0.4^4 = 0.0064$.

3. Put $f(x) = \sqrt{x}$.

$f(4) = 2. \quad f'(x) = \frac{1}{2\sqrt{x}}, \quad f''(x) = -\frac{1}{4}x^{-3/2}$.

$f'(4) = \frac{1}{4}, \quad f''(4) = -\frac{1}{4}2^{-3} = -\frac{1}{32}$.

$\sqrt{3.91} \approx P_2(3.91) = f(4) + f'(4)(3.91-4) + \frac{1}{2}f''(4)(3.91-4)^2$

$= 2 - \frac{1}{4}(0.09) - \frac{1}{64}(0.09)^2 = 1.9774$.

$|f'''(x)| = \left|\frac{3}{8}x^{-5/2}\right| \le \frac{3}{8}(3.91)^{-5/2} \approx 0.0124$ for x in $[3.91,4]$, so $M = 0.0124$.

$|R_2(3.91)| \le \frac{M}{3!}|3.91 - 4|^3 = \frac{0.0124}{6} \cdot (0.09)^3 \approx 1.506 \cdot 10^{-6}$.

5. Put $f(x) = \sin x$. $\quad f\left(\frac{\pi}{4}\right) = \frac{\sqrt{2}}{2}$.

$f'(x) = \cos x, \quad f''(x) = -\sin x, \quad f'''(x) = -\cos x, \quad f^{(4)}(x) = \sin x$.

$48° = \frac{48\cdot\pi}{180} = \frac{4\pi}{15}$.

$$\begin{aligned}\sin 48° &= \sin\left(\tfrac{4\pi}{15}\right) \approx P_4\left(\tfrac{4\pi}{15}\right) = f\left(\tfrac{\pi}{4}\right) + f'\left(\tfrac{\pi}{4}\right)\left(\tfrac{4\pi}{15} - \tfrac{\pi}{4}\right) \\ &\quad + \tfrac{1}{2}f''\left(\tfrac{\pi}{4}\right)\left(\tfrac{4\pi}{15} - \tfrac{\pi}{4}\right)^2 + \tfrac{1}{3!}f'''\left(\tfrac{\pi}{4}\right)\left(\tfrac{4\pi}{15} - \tfrac{\pi}{4}\right)^3 + \tfrac{1}{4!}f^{(4)}\left(\tfrac{\pi}{4}\right)\left(\tfrac{4\pi}{15} - \tfrac{\pi}{4}\right)^4 \\ &= \tfrac{\sqrt{2}}{2} + \tfrac{\sqrt{2}}{2}\left(\tfrac{\pi}{60}\right) - \tfrac{\sqrt{2}}{4}\left(\tfrac{\pi}{60}\right)^2 - \tfrac{\sqrt{2}}{12}\left(\tfrac{\pi}{60}\right)^3 + \tfrac{\sqrt{2}}{48}\left(\tfrac{\pi}{60}\right)^4 \\ &= \tfrac{\sqrt{2}}{2}\left[1 + \tfrac{\pi}{60} - \tfrac{1}{2}\left(\tfrac{\pi}{60}\right)^2 - \tfrac{1}{6}\left(\tfrac{\pi}{60}\right)^3 + \tfrac{1}{24}\left(\tfrac{\pi}{60}\right)^4\right] \\ &= \tfrac{\sqrt{2}}{2}\left(1 + 0.05236 - 0.00137 - 0.00002 + 3 \times 10^{-7}\right) \approx 0.7431.\end{aligned}$$

$|f^{(5)}(x)| = |\cos x| \le 1$ in $\left[\frac{\pi}{4}, \frac{4\pi}{15}\right]$, so $M = 1$.

$|R_4(48°)| \le \frac{M}{5!}\left(\frac{4\pi}{15} - \frac{\pi}{4}\right)^5 = \frac{1}{5!}\left(\frac{\pi}{60}\right)^5 = 3 \times 10^{-9}$.

7. Put $f(x) = \tan x$.

$f'(x) = \sec^2 x, \quad f''(x) = 2\sec x \cdot \sec x \tan x = 2\sec^2 x \tan x$,

$f'''(x) = 4\sec x \cdot \sec x \cdot \tan x \cdot \tan x + 2\sec^2 x \cdot \sec^2 x = 4\sec^2 x \tan^2 x + 2\sec^4 x$.

$f(0) = 0, \quad f'(0) = 1, \quad f''(0) = 0, \quad f'''(0) = 2$.

$$\begin{aligned}\tan\tfrac{\pi}{12} &\approx P_3\left(\tfrac{\pi}{12}\right) = f(0) + f'(0)\cdot\tfrac{\pi}{12} + \tfrac{1}{2}f''(0)\left(\tfrac{\pi}{12}\right)^2 + \tfrac{1}{3!}f'''(0)\left(\tfrac{\pi}{12}\right)^3 \\ &= \tfrac{\pi}{12} + \tfrac{1}{3}\left(\tfrac{\pi}{12}\right)^3 \approx 0.2618 + 0.00598 = 0.2678.\end{aligned}$$

$$\begin{aligned}f^{(4)}(x) &= 8\sec x \cdot \sec x \tan x \cdot \tan^2 x + 4\sec^2 x \cdot 2\tan x \cdot \sec^2 x \\ &\quad + 8\sec^3 x \sec x \tan x \\ &= 8\sec^2 x \tan^3 x + 16\sec^4 x \cdot \tan x.\end{aligned}$$

$$\begin{aligned}|f^{(4)}(x)| &\le 8\sec^2\left(\tfrac{\pi}{4}\right)\tan^3\left(\tfrac{\pi}{4}\right) + 16\sec^4\left(\tfrac{\pi}{4}\right)\cdot\tan\left(\tfrac{\pi}{4}\right) \\ &= 8\left(\sqrt{2}\right)^2\cdot 1 + 16\cdot\left(\sqrt{2}\right)^4 = 16 + 64 = 80 = M \text{ in } \left[0, \tfrac{\pi}{12}\right].\end{aligned}$$

$\left|R_3\left(\frac{\pi}{12}\right)\right| \le \frac{M}{4!}\left(\frac{\pi}{12}\right)^4 = \frac{80}{24}\left(\frac{\pi}{12}\right)^4 \approx 0.0157$.

9. $e^x \approx 1 + x + \frac{x^2}{2} = P_2(x)$.

$|f'''(x)| = e^x \le e < 3$ in $[0,1]$.

$|R_2(x)| \le \frac{M}{3!}x^3 \le \frac{3}{3!}\cdot 1^3 = \frac{1}{2}$.

11. The error associated with the approximation $\sin x \approx P_n(x)$ satisfies

$|R_n(x)| \leq \frac{1}{(n+1)!}|x|^{n+1}$ since $\left|\frac{d^n}{dx^n}\sin x\right| \leq 1$ for all $n > 0$

$\leq \frac{1}{(n+1)!}(0.2)^{n+1}$ since $|x| < 0.2$.

Therefore to get the accuracy $|R_n(x)| \leq 0.001$, it is enough to get $\frac{1}{(n+1)!}(0.2)^{n+1} \leq 0.001$.

If $n = 2$: $\frac{1}{3!}(0.2)^3 \leq 0.001$?

$0.00133 \not\leq 0.001$

If $n = 3$: $\frac{1}{4!}(0.2)^4 \leq 0.001$?

$0.000067 < 0.001$.

Thus the integer n can be ≥ 3 in order to get the desired accuracy.

13. a. Define $h(x) = g(x) - f(x)$.

b. Since $g(x) \geq f(x)$ for $a \leq x \leq b$,

$h(x) = g(x) - f(x) \geq 0$ for $a \leq x \leq b$.

c. Then $g(x) = f(x) + h(x)$ and

$\int_a^b g(x)dx = \int_a^b [f(x) + h(x)]\,dx.$

Since $f(x)$ and $g(x)$ are continuous on $[a, b]$, $h(x)$ is continuous on $[a, b]$. Hence $h(x)$ is also integrable on $[a, b]$, and by the Property (I1) in Section 5.5

$$\int_a^b [f(x) + h(x)]\,dx = \int_a^b f(x)dx + \int_a^b h(x)dx.$$

So

$$\int_a^b g(x)dx = \int_a^b f(x)dx + \int_a^b h(x)dx.$$

d. Since $h(x) \geq 0$ for $a \leq x \leq b$, the graph of $h(x)$ is above the x-axis. Geometrically $\int_a^b h(x)dx$ is the area of the region bounded by the graph of $y = h(x)$ and the x-axis for $a \leq x \leq b$. Therefore $\int_a^b h(x)dx \geq 0$.

e. From (c) and (d) we have

$$\int_a^b g(x)dx = \int_a^b f(x)dx + \int_a^b h(x) \geq \int_a^b f(x)dx$$

since $\int_a^b h(x)dx \geq 0$. That is,

$$\int_a^b f(x)dx \leq \int_a^b g(x)dx.$$

Selected Solutions to Exercise Set 9.7

1. $f(x) = e^{2x}, \qquad a = 0.$

   $f(0) = 1, \quad f'(x) = 2e^{2x}, \quad f''(x) = 4e^{2x}, \cdots, f^{(k)}(x) = 2^k e^{2x}$ for all k.

   $f'(0) = 2, \quad f''(0) = 4, \cdots, f^{(k)}(0) = 2^k$ for all k.

   The Taylor series is
   $\sum_{k=0}^{\infty} \frac{f^{(k)}(0)}{k!}x^k = \sum_{k=0}^{\infty} \frac{2^k}{k!}x^k = 1 + 2x + \frac{4}{2}x^2 + \frac{8}{3!}x^3 + \cdots + \frac{2^k}{k!}x^k + \cdots.$

3. $f(x) = \frac{1}{1+x}, \qquad a = 0.$

   $f(0) = 1, \quad f'(x) = \frac{-1}{(1+x)^2}, \quad f''(x) = 2(1+x)^{-3}, \quad f'''(x) = -6(1+x)^{-4}, \cdots.$

   $f^{(k)}(x) = (-1)^k k!(1+x)^{-k-1}$ for all $k = 0, 1, 2, \cdots$.

   $f'(0) = -1, \quad f''(0) = 2, \cdots, f^{(k)}(0) = (-1)^k k!$ for all $k = 0, 1, 2, \cdots$.

   The Taylor series is
   $\sum_{k=0}^{\infty} \frac{f^{(k)}(0)}{k!}x^k = \sum_{k=0}^{\infty} \frac{(-1)^k k!}{k!}x^k = \sum_{k=0}^{\infty}(-1)^k x^k = 1-x+x^2-x^3+\cdots+(-1)^k x^k+\cdots.$

5. $f(x) = \ln x$, $a = 1$.

   $f'(x) = \frac{1}{x}, \quad f''(x) = -x^{-2}, \quad f'''(x) = 2x^{-3}, \quad \cdots,$

   $f^{(k)}(x) = (-1)^{k-1}(k-1)!x^{-k}$ for $k = 1, 2, 3, \cdots$.

   $f(1) = 0, \quad f'(1) = 1, \quad f''(1) = -1, \quad f'''(1) = 2, \quad \cdots,$

$f^{(k)}(1) = (-1)^{k-1}(k-1)!$ for $k = 1, 2, 3, \cdots$.

The Taylor series is

$$\sum_{k=0}^{\infty} \frac{f^{(k)}(1)}{k!}(x-1)^k = (x-1) - \frac{1}{2}(x-1)^2 + \frac{2}{3!}(x-1)^3 + \cdots + \frac{(-1)^{k-1}(k-1)!}{k!}(x-1)^k + \cdots$$

$$= \sum_{k=1}^{\infty} \frac{(-1)^{k-1}}{k}(x-1)^k.$$

7. $f(x) = \sin x, \qquad a = \frac{\pi}{2}.$

$f\left(\frac{\pi}{2}\right) = 1, \quad f'(x) = \cos x, \quad f''(x) = -\sin x, \quad f'''(x) = -\cos x,$
$f^{(4)}(x) = \sin x, \cdots.$

$$f'\left(\tfrac{\pi}{2}\right) = 0, \quad f''\left(\tfrac{\pi}{2}\right) = -1, \quad f'''\left(\tfrac{\pi}{2}\right) = 0, \quad f^{(4)}\left(\tfrac{\pi}{2}\right) = 1, \cdots.$$

$\Longrightarrow f^{(2k)}\left(\frac{\pi}{2}\right) = (-1)^k$ and $f^{(2k+1)}\left(\frac{\pi}{2}\right) = 0$ for $k = 0, 1, 2, \cdots$.

The Taylor series is

$$\sum_{k=0}^{\infty} \frac{f^{(k)}\left(\frac{\pi}{2}\right)}{k!}\left(x - \frac{\pi}{2}\right)^k = 1 - \frac{1}{2!}\left(x - \frac{\pi}{2}\right)^2 + \frac{1}{4!}\left(x - \frac{\pi}{2}\right)^4 + \cdots$$

$$+ \frac{(-1)^k}{(2k)!}\left(x - \frac{\pi}{2}\right)^{2k} + \cdots.$$

9. $f(x) = \sin x, \quad a = \frac{\pi}{3}.$

$f\left(\frac{\pi}{3}\right) = \frac{\sqrt{3}}{2}, \quad f'(x) = \cos x, \quad f''(x) = -\sin x, \quad f'''(x) = -\cos x,$

$f^{(4)}(x) = \sin x, \cdots.$

$f'\left(\frac{\pi}{3}\right) = \frac{1}{2}, \quad f''\left(\frac{\pi}{3}\right) = -\frac{\sqrt{3}}{2}, \quad f'''\left(\frac{\pi}{3}\right) = -\frac{1}{2}, \quad f^{(4)}\left(\frac{\pi}{3}\right) = \frac{\sqrt{3}}{2}, \cdots.$

The Taylor series is

$$\sum_{k=0}^{\infty} \frac{f^{(k)}\left(\frac{\pi}{3}\right)}{k!}\left(x - \frac{\pi}{3}\right)^k = \frac{\sqrt{3}}{2} + \frac{1}{2}\left(x - \frac{\pi}{3}\right) - \frac{\sqrt{3}}{2}\frac{\left(x - \frac{\pi}{3}\right)^2}{2!}$$

$$- \frac{1}{2}\frac{\left(x - \frac{\pi}{3}\right)^3}{3!} + \cdots.$$

11. $a_k = \frac{1}{k!}(2x)^k.$

$\lim_{k\to\infty} \frac{|a_{k+1}|}{|a_k|} = \lim_{k\to\infty} \frac{\frac{|2x|^{k+1}}{(k+1)!}}{\frac{|2x|^k}{k!}} = \lim_{k\to\infty} \frac{2|x|}{k+1} = 0$ for all real numbers $x \neq 0$.

By the ratio test, the Taylor series $\sum_{k=0}^{\infty} \frac{(2x)^k}{k!}$ converges absolutely for all real numbers x. Therefore, the interval of convergence is $(-\infty, \infty)$.

13. $\sum_{k=1}^{\infty} \frac{(-1)^{k-1}}{k}(x-1)^k$. $a_k = \frac{(-1)^{k-1}}{k}(x-1)^k$.

$\lim_{k\to\infty} \frac{|a_{k+1}|}{|a_k|} = \lim_{k\to\infty} \frac{\frac{|x-1|^{k+1}}{k+1}}{\frac{|x-1|^k}{k}} = \lim_{k\to\infty} \frac{k}{k+1}|x-1| = |x-1|$.

By the Ratio Test, the Taylor series converges absolutely when $|x-1| < 1$, i.e., $0 < x < 2$.

Case $|x-1| > 1$: The Taylor series diverges since

$$\lim_{k\to\infty} \frac{|x-1|^k}{k} = \infty \text{ if } |x-1| > 1.$$

Case $x = 2$: $\sum_{k=1}^{\infty} \frac{(-1)^{k-1}}{k} = 1 - \frac{1}{2} + \frac{1}{3} - \frac{1}{4} + \cdots$ converges by the Alternating Series Test.

Case $x = 0$: $\sum_{k=1}^{\infty} \frac{(-1)^{k-1}}{k}(-1)^k = \sum_{k=1}^{\infty} \frac{(-1)^{2k-1}}{k} = -\sum_{k=1}^{\infty} \frac{1}{k}$ diverges.

Hence, the interval of convergence for the Taylor series is $(0, 2]$.

15. $\sin x = \sum_{k=0}^{\infty} \frac{(-1)^k}{(2k)!}\left(x - \frac{\pi}{2}\right)^{2k}$

$$\Longrightarrow (\sin x)' = \left(\sum_{k=0}^{\infty} \frac{(-1)^k}{(2k)!}\left(x - \frac{\pi}{2}\right)^{2k}\right)'$$

$$\Longrightarrow \cos x = \sum_{k=1}^{\infty} \frac{(-1)^k}{(2k-1)!}\left(x - \frac{\pi}{2}\right)^{2k-1}$$

$$= -\left(x - \frac{\pi}{2}\right) + \frac{\left(x-\frac{\pi}{2}\right)^3}{3!} - \frac{\left(x-\frac{\pi}{2}\right)^5}{5!} + \cdots.$$

17. $\frac{1}{1-x} = 1 + x + x^2 + x^3 + x^4 + \cdots$ for $|x| < 1$.

$\frac{1}{(1-x)^2} = \frac{d}{dx}\frac{1}{1-x} = 1 + 2x + 3x^2 + 4x^3 + 5x^4 + 6x^5 + \cdots$.

$\frac{1}{(1-x)^3} = \frac{d}{dx}\frac{1}{(1-x)^2} = 2 + 6x + 12x^2 + 20x^3 + 30x^4 + \cdots$.

19. $\int \frac{1}{1-x}\,dx = -\ln(1-x) + C$.

$\frac{1}{1-x} = 1 + x + x^2 + x^3 + x^4 + \cdots$ for $|x| < 1$.

$x + \frac{1}{2}x^2 + \frac{1}{3}x^3 + \frac{1}{4}x^4 + \frac{1}{5}x^5 \cdots = -\ln(1-x) + C$.

$C = 0$, by considering $x = 0$ in the preceding formula.

$$\ln(1-x) = -x - \frac{1}{2}x^2 - \frac{1}{3}x^3 - \frac{1}{4}x^4 - \frac{1}{5}x^5 + \cdots.$$

21.

$$\lim_{k\to\infty} \frac{\left|\frac{x^{k+1}}{k+3}\right|}{\left|\frac{x^k}{k+2}\right|} = |x| \cdot \lim_{k\to\infty} \frac{k+2}{k+3} = |x| \cdot 1 = |x|.$$

The series converges absolutely for $|x| < 1 \implies -1 < x < 1$.

Test $x = -1$: $\frac{1}{2} - \frac{1}{3} + \frac{1}{4} - \frac{1}{5} + \cdots$, which converges by the alternating series test.

Test $x = 1$: $\frac{1}{2} + \frac{1}{3} + \frac{1}{4} + \frac{1}{5} + \cdots$, which diverges by the integral test.

The interval of convergence is $[-1, 1)$.

The radius of convergence is 1.

23.

$$\lim_{k\to\infty} \left| \frac{\frac{(-1)^{k+2}}{(k+1)!} x^{k+1}}{\frac{(-1)^{k+1}}{k!} x^k} \right| = |x| \cdot \lim_{k\to\infty} \frac{1}{k+1} = 0.$$

The series converges absolutely for all x.

25.

$$\lim_{k\to\infty} \frac{\left|\frac{(k+1)^2+1}{(k+1)!} \cdot x^{k+1}\right|}{\left|\frac{k^2+1}{k^2} \cdot x^k\right|} = \lim_{k\to\infty} |x| \cdot \frac{k^2+2k+2}{k^2+1} \cdot \frac{1}{k+1}$$
$$= |x| \cdot 1 \cdot 0 = 0.$$

The series converges absolutely for all x.

27. Since $\lim_{k\to\infty} \frac{x^k}{k} = +\infty$ for $x > 1$, $\lim_{k\to\infty} |a_k| = \lim_{k\to\infty} \frac{x^k}{k} = \infty$ for all $x > 1$. Hence $\sum_{k=1}^{\infty} a_k = \sum_{k=1}^{\infty} (-1)^{k+1} \frac{x^k}{k}$ does not converge by Theorem 5.

29. $f(x) = \frac{1}{1-x} \implies f(0) = 1.$

$f'(x) = \frac{1}{(1-x)^2} \implies f'(0) = 1$

$f''(x) = \frac{2!}{(1-x)^3} \implies f''(0) = 2!$

$f'''(x) = \frac{3!}{(1-x)^4} \implies f'''(0) = 3!$

$\vdots$

$f^{(k)}(x) = \frac{k!}{(1-x)^{k+1}} \implies f^{(k)}(0) = k!$

Hence $f(x) = \sum_{k=0}^{\infty} \frac{f^{(k)}(0)}{k!} x^k = \sum_{k=0}^{\infty} \frac{k!}{k!} x^k = \sum_{k=0}^{\infty} x^k.$

31. $f(x) = \cos x \implies f(0) = 1$

$f'(x) = -\sin x \implies f'(0) = 0$

$f''(x) = -\cos x \implies f''(0) = -1$

$f'''(x) = \sin x \implies f'''(0) = 0$

$f^{(4)}(x) = \cos x \implies f^{(4)}(0) = 1$, etc.

Hence $f^{(2k+1)}(0) = 0$ and $f^{(2k)}(0) = (-1)^k$ for $k = 0, 1, 2, \cdots$. The Taylor series for $\cos x$ is

$$\sum_{k=0}^{\infty} \frac{f^{(k)}(0)}{k!} x^k = \sum_{k=0}^{\infty} \frac{f^{(2k)}(0)}{(2k)!} x^{2k} = \sum_{k=0}^{\infty} \frac{(-1)^k}{(2k)!} x^{2k}.$$

Selected Solutions to Exercise Set 9.8

1. $f(x) = x^2 - \frac{7}{2}x + \frac{3}{4}, \quad a = 0,\ b = 2.$

Since $f(0) = \frac{3}{4}$ and $f(2) = -\frac{9}{4}$, a zero of $f(x)$ must lie within [0,2].

The approximation scheme for Newton's Method is

$$x_{n+1} = x_n - \frac{f(x_n)}{f'(x_n)} = x_n - \frac{x_n^2 - \frac{7}{2}x_n + \frac{3}{4}}{2x_n - \frac{7}{2}}.$$

Using $x_1 = 0$ as a first approximation, we get results in the table:

n	x_n	$f(x_n)$	$f'(x_n)$	x_{n+1}
1	0.0	0.75	−3.5	0.214286
2	0.214286	0.045917	−3.071428	0.229236
3	0.229236	0.000223	−3.041528	0.229309
4	0.229309	0.000001	−3.041382	0.229309

Thus $x_4 = 0.229309$ is an approximation to the zero.

3. $f(x) = 1 - x - x^2, \quad a = 0, \quad b = 1.$

Since $f(0) = 1$ and $f(1) = -1$ there must exist a zero in [0,1]. The approximation scheme for Newton's Method is

$$x_{n+1} = x_n - \frac{1 - x_n - x_n^2}{-1 - 2x_n} = x_n + \frac{1 - x_n - x_n^2}{1 + 2x_n}.$$

Using $x_1 = 1$ as a first approximation, we obtain the results:

n	x_n	$f(x_n)$	$f'(x_n)$	x_{n+1}
1	1.0	−1.0	−3.0	0.666667
2	0.666667	−0.111112	−2.333334	0.619048
3	0.619048	−0.002268	−2.238096	0.618035
4	0.618035	−0.000002	−2.236070	0.618034

Thus $x_5 = 0.618034$ is a good approximation to the zero.

5. $f(x) = \sqrt{x+3} - x, \quad a = 1, \quad b = 3.$

Since $f(1) = 1$ and $f(3) = \sqrt{6} - 3 < 0$ there must exist a zero in [1,3]. The approximation scheme for Newton's method is

$$x_{n+1} = x_n - \frac{\sqrt{x_n+3} - x_n}{\frac{1}{2\sqrt{x_n+3}} - 1} = x_n + \frac{2(x_n+3) - 2x_n\sqrt{x_n+3}}{2\sqrt{x_n+3} - 1}$$

$$x_{n+1} = \frac{2x_n\sqrt{x_n+3} - x_n + 2(x_n+3) - 2x_n\sqrt{x_n+3}}{2\sqrt{x_n+3} - 1}$$

$$= \frac{x_n + 6}{2\sqrt{x_n+3} - 1}.$$

Using $x_1 = 1$, we obtain the following results.

n	x_n	$f(x_n)$	$f'(x_n)$	x_{n+1}
1	1.0	1.0	−0.75	2.333333
2	2.333333	−0.023932	−0.783494	2.302788
3	2.302788	−0.000010	−0.782871	2.302775
4	2.302775	0.0000005	−0.782871	2.302775

Thus $x_4 = 2.302775$ is a good approximation to the zero.

7. $f(x) = x^4 - 5, \quad a = -2, \quad b = -1.$

Since $f(-2) = 11$ and $f(-1) = -4$ there must be a zero in [-2,-1]. The approximation scheme for Newton's method is

$$x_{n+1} = x_n - \frac{x_n^4 - 5}{4x_n^3}.$$

Using $x_1 = -2$, we have the following results.

n	x_n	$f(x_n)$	$f'(x_n)$	x_{n+1}
1	−2.0	11.0	−32.0	−1.65625
2	−1.65625	2.524949	−18.173462	−1.517314
3	−1.517314	0.300317	−13.972895	−1.495821
4	−1.495821	0.006319	−13.387481	−1.495349
5	−1.495349	0.000003	−13.374812	−1.495349

Thus $x_5 = -1.495349$ is a good approximation to the zero.

9. Finding a number for which $\sqrt{x+3} = x$ is equivalent to finding a zero of the function $h(x) = \sqrt{x+3} - x$. From the results of Exercise 5, $h(x) = \sqrt{x+3} - x$ has a zero in [1,3] approximately equal to 2.302776. Thus a point of intersection between the graphs of $f(x)$ and $g(x)$ is approximately (2.302776, 2.302776).

11. If the first guess x_1 is precisely the desired zero the results of every iteration will be the same number as the first guess. In this case $f(x_1) = 0$, hence $x_2 = x_1 - \frac{f(x_1)}{f'(x_1)} = x_1$, and, in consequence of the Newton's Method, $x_{n+1} = x_n = \cdots = x_1$ for all $n \geq 1$.

13. $f(x) = (x-2)^{1/3}$.

Use Newton's Method.

$$x_{n+1} = x_n - \frac{(x_n-2)^{1/3}}{\frac{1}{3}(x_n-2)^{-2/3}} = x_n - 3(x_n - 2) = 6 - 2x_n, \quad x_1 = 3.$$

n	x_n	$f(x_n)$	$f'(x_n)$	x_{n+1}
1	3.0	1.0	$\frac{1}{3}$	0.0
2	0.0	−1.259921	0.209987	6.0
3	6.0	1.587401	0.132283	−6.0
4	−6.0	−2.0	0.083333	18.0

The results show that this approximation does not converge to the zero of $f(x)$ but moves away.

Selected Solutions to Chapter 9 - Review Exercises

1. $\left\{\sin \frac{n\pi}{2}\right\}$ does not converge since

$$\left\{\sin \frac{n\pi}{2}\right\} = \left\{\sin \frac{\pi}{2},\ \sin \pi,\ \sin \frac{3\pi}{2},\ \sin 2\pi,\ \sin \frac{5\pi}{2}, \cdots\right\}$$
$$= \{1, 0, -1, 0, 1, 0, -1, \cdots\}.$$

These terms oscillate among 1, 0 and -1 and do not approach any single number as $n \to \infty$.

3. $\lim_{n\to\infty} (\ln \sqrt{n} - \ln n) = \lim_{n\to\infty} \ln\left(\frac{\sqrt{n}}{n}\right) = \lim_{n\to\infty} \ln\left(\frac{1}{\sqrt{n}}\right) = -\infty.$
Thus $\{\ln \sqrt{n} - \ln n\}$ diverges.

5. $\lim_{n\to\infty} \frac{n^2+n-5}{4+2n^2} = \lim_{n\to\infty} \frac{1+\frac{1}{n}-\frac{5}{n^2}}{\frac{4}{n^2}+2} = \frac{1+0-0}{0+2} = \frac{1}{2}.$

7. $\lim_{n\to\infty} \frac{3^n+4^n}{5^n} = \lim_{n\to\infty} \left[\left(\frac{3}{5}\right)^n + \left(\frac{4}{5}\right)^n\right] = \lim_{n\to\infty} \left(\frac{3}{5}\right)^n + \lim_{n\to\infty} \left(\frac{4}{5}\right)^n = 0 + 0 = 0.$

9. $\sum_{k=0}^{\infty} \frac{1}{3^k} = \sum_{k=0}^{\infty} \left(\frac{1}{3}\right)^k = \frac{1}{1-\frac{1}{3}} = \frac{3}{2}.$

11. $\sum_{k=1}^{\infty} \frac{3^k}{2^k} = \sum_{k=1}^{\infty} \left(\frac{3}{2}\right)^k$ is a divergent geometric series.

13. $\sum_{k=1}^{\infty} \frac{1}{1+k^2}$ converges by comparison to the known convergent series $\sum_{k=1}^{\infty} \frac{1}{k^2}$.

15. $\sum_{k=1}^{\infty} \frac{3^k}{k^2}$.
By the Ratio Test, since

$$\lim_{k\to\infty} \frac{a_{k+1}}{a_k} = \lim_{k\to\infty} \frac{3^{k+1}}{(k+1)^2} \cdot \frac{k^2}{3^k} = \lim_{k\to\infty} \left(\frac{k}{k+1}\right)^2 \cdot 3 = 3 > 1,\ \sum_{k=1}^{\infty} \frac{3^k}{k^2} \text{ diverges.}$$

17. $\sum_{k=1}^{\infty} \frac{k^2}{k!}$.

$$\lim_{k\to\infty} \frac{a_{k+1}}{a_k} = \lim_{k\to\infty} \frac{(k+1)^2}{(k+1)!} \cdot \frac{k!}{k^2} = \lim_{k\to\infty} \frac{1}{k+1} \cdot \left(\frac{k+1}{k}\right)^2 = 0.$$

Thus $\sum_{k=1}^{\infty} \frac{k^2}{k!}$ converges by the Ratio Test.

19. $\sum_{k=0}^{\infty} \frac{k-1}{k+1}$. Since $\lim_{k\to\infty} \frac{k-1}{k+1} = 1 \neq 0$, by Theorem 5 in Section 9.3 $\sum_{k=0}^{\infty} \frac{k-1}{k+1}$ diverges.

21. $\sum_{k=1}^{\infty} \frac{1}{\sqrt{k+1}}$.
By the Limit Comparison Test:

$$\lim_{k\to\infty} \frac{1}{\sqrt{k+1}} \bigg/ \frac{1}{\sqrt{k}} = \lim_{k\to\infty} \sqrt{\frac{k}{k+1}} = 1 > 0,$$ and $\sum_{k=1}^{\infty} \frac{1}{\sqrt{k}}$, a p-series with $p = \frac{1}{2} < 1$, diverges. Thus $\sum_{k=1}^{\infty} \frac{1}{\sqrt{k+1}}$ diverges.

23. $\sum_{k=1}^{\infty} \frac{1}{k^2+1}$.

Since $\frac{1}{k^2+1} < \frac{1}{k^2}$ and $\sum_{k=1}^{\infty} \frac{1}{k^2}$ converges, by the Comparison Test $\sum_{k=1}^{\infty} \frac{1}{k^2+1}$ converges.

25. $\sum_{k=0}^{\infty} \frac{(-1)^k k}{\sqrt{k+1}}$. $\qquad a_k = \frac{k}{\sqrt{k+1}}$.

Since $\lim_{k\to\infty} a_k = \lim_{k\to\infty} \frac{k}{\sqrt{k+1}} = \infty$ the condition (ii) in the Alternating Series Test fails. By the necessary condition for convergence of series, Theorem 5 in Section 9.3, $\sum_{k=0}^{\infty} \frac{(-1)^k k}{\sqrt{k+1}}$ diverges.

27. $\sum_{k=1}^{\infty} \frac{k\cdot 2^k}{k!}$.

$\lim_{k\to\infty} \frac{a_{k+1}}{a_k} = \lim_{k\to\infty} \frac{(k+1)2^{k+1}}{(k+1)!} \cdot \frac{k!}{k\cdot 2^k} = \lim_{k\to\infty} \frac{2(k+1)}{(k+1)\cdot k} = 0 < 1.$

By the Ratio Test $\sum_{k=1}^{\infty} \frac{k\cdot 2^k}{k!}$ converges.

29. $\sum_{k=0}^{\infty} \frac{1}{(k+2)^{3/2}}$.

Since $\frac{1}{(k+2)^{3/2}} < \frac{1}{k^{3/2}}$ for all $k > 0$ and $\sum_{k=1}^{\infty} \frac{1}{k^{3/2}}$, a p-series with $p = \frac{3}{2} > 1$, converges, $\sum_{k=0}^{\infty} \frac{1}{(k+2)^{3/2}}$ converges.

31. $f(x) = e^{-3x}, \quad a = 0, \quad n = 3.$

$f'(x) = -3e^{-3x}, \quad f''(x) = 9e^{-3x}, \quad f'''(x) = -27e^{-3x}.$

$$P_3(x) = f(0) + f'(0)x + \tfrac{1}{2!}f''(0)x^2 + \tfrac{1}{3!}f'''(0)x^3$$
$$= 1 - 3x + \tfrac{9}{2}x^2 - \tfrac{27}{3!}x^3 = 1 - 3x + \tfrac{9}{2}x^2 - \tfrac{9}{2}x^3.$$

33. $f(x) = \tan x, \quad a = 0, \quad n = 2.$

$f'(x) = \sec^2 x, \quad f''(x) = 2\sec x \cdot \sec x \tan x = 2\sec^2 x \cdot \tan x.$

$f(0) = 0, \quad f'(0) = 1, \quad f''(0) = 0.$

$P_2(x) = f(0) + f'(0)x + \tfrac{1}{2!}f''(0)x^2 = x.$

35. $f(x) = \ln(1 + x^2), \quad a = 0, \quad n = 2.$

$f'(x) = \frac{2x}{1+x^2}, \quad f''(x) = \frac{2(1+x^2)-2x\cdot 2x}{(1+x^2)^2} = \frac{2(1-x^2)}{(1+x^2)^2}.$

$f(0) = 0, \quad f'(0) = 0, \quad f''(0) = 2.$

$P_2(x) = f(0) + f'(0)x + \tfrac{1}{2!}f''(0)x^2 = x^2.$

37. $f(x) = \sqrt{x}, \quad a = 9, \quad n = 3.$

$f'(x) = \frac{1}{2\sqrt{x}}, \quad f''(x) = -\frac{1}{4}x^{-3/2}, \quad f'''(x) = \frac{3}{8}x^{-5/2}.$

$f(9) = 3, \quad f'(9) = \frac{1}{6}, \quad f''(9) = -\frac{1}{4 \cdot 27}, \quad f'''(9) = \frac{1}{8 \cdot 81}.$

$P_3(x) = f(9) + f'(9)(x-9) + \frac{1}{2!}f''(9)(x-9)^2 + \frac{1}{3!}f'''(9)(x-9)^3$

$= 3 + \frac{1}{6}(x-9) - \frac{1}{216}(x-9)^2 + \frac{1}{3888}(x-9)^3.$

39. $f(x) = xe^x, \quad a = 0.$

$f'(x) = e^x(1+x), \quad f''(x) = e^x(2+x), \cdots, f^{(k)}(x) = e^x(k+x)$ for all $k \geq 0$.

$f^{(k)}(0) = k, \quad k \geq 0.$

The Taylor series for $f(x)$ is

$\sum_{k=0}^{\infty} \frac{f^{(k)}(0)}{k!}x^k = x + \frac{2}{2!}x^2 + \frac{3}{3!}x^3 + \cdots + \frac{k}{k!}x^k + \cdots$

$= x + x^2 + \frac{1}{2!}x^3 + \cdots + \frac{1}{(k-1)!}x^k + \cdots.$

41. $f(x) = \frac{1}{1-x} - 1, \quad a = 0.$

$f'(x) = \frac{1}{(1-x)^2}, \quad f''(x) = \frac{2!}{(1-x)^3}, \quad f'''(x) = \frac{3!}{(1-x)^4}, \cdots,$

$f^{(k)}(x) = \frac{k!}{(1-x)^{k+1}}$ for all $k \geq 1$.

$f^{(k)}(0) = k!$ for all $k \geq 1$.

The Taylor series for $f(x)$ is

$\sum_{k=0}^{\infty} \frac{f^{(k)}(0)}{k!}x^k = \sum_{k=1}^{\infty} \frac{k!}{k!}x^k = \sum_{k=1}^{\infty} x^k.$

43. $f(x) = x^3 + 2x^2 - x + 3.$

$f'(x) = 3x^2 + 4x - 1, \quad f''(x) = 6x + 4, \quad f'''(x) = 6, \quad f^{(k)}(x) = 0$ for all $k \geq 4$.

a. $f'(0) = -1, \quad f''(0) = 4, \quad f'''(0) = 6, \quad f^{(k)}(0) = 0$ for $k \geq 4$.

The Taylor Series for $f(x)$ is

$$\sum_{k=0}^{\infty} \frac{f^{(k)}(0)}{k!}x^k = 3 - x + \frac{4}{2!}x^2 + \frac{6}{3!}x^3 = 3 - x + 2x^2 + x^3$$

$$= x^3 + 2x^2 - x + 3.$$

b. $f(1) = 5$, $f'(1) = 6$, $f''(1) = 10$, $f'''(1) = 6$, $f^{(k)}(1) = 0$ for $k \geq 4$.

The Taylor Series for $f(x)$ is

$$\begin{aligned}\sum_{k=0}^{\infty} \frac{f^{(k)}(1)}{k!}(x-1)^k &= 5 + 6(x-1) + \frac{10}{2!}(x-1)^2 + \frac{6}{3!}(x-1)^3 \\ &= 5 + 6(x-1) + 5(x-1)^2 + (x-1)^3 \\ &= x^3 + 2x^2 - x + 3.\end{aligned}$$

45. Let $S(n)$ = the total amount paid by the annuity in n years. Then

$$\begin{aligned} S(1) &= 500 \\ S(2) &= 500 + 500\left(\frac{1}{3}\right) = 500\left(1 + \frac{1}{3}\right) \\ S(3) &= 500 + 500\left(\frac{1}{3}\right) + 500\left(\frac{1}{3}\right)^2 = 500\left(1 + \frac{1}{3} + \frac{1}{3^2}\right) \\ &\vdots \\ S(n) &= 500 + 500\left(\frac{1}{3}\right) + \cdots + 500\left(\frac{1}{3}\right)^{n-1} = 500\sum_{k=0}^{n-1} \frac{1}{3^k}. \end{aligned}$$

Thus the total amount paid by the annuity in perpetuity is

$$\begin{aligned} S = \lim_{n\to\infty} S_n &= \lim_{n\to\infty} 500\sum_{k=0}^{n-1}\frac{1}{3^k} = 500\sum_{k=0}^{\infty}\frac{1}{3^k} \\ &= 500\left(\frac{1}{1-\frac{1}{3}}\right) = \$750. \end{aligned}$$

47. $\cot x = x \implies f(x) = \cot x - x = 0.$

Since $f(\frac{\pi}{4}) = 1 - \frac{\pi}{4} > 0$ and $f\left(\frac{\pi}{2}\right) = -\frac{\pi}{2} < 0$ there must exist a zero of $f(x)$ in $\left[0, \frac{\pi}{2}\right]$. The approximation scheme is

$$x_{n+1} = x_n - \frac{\cot x_n - x_n}{-\csc^2 x_n - 1}.$$

Using $x_1 = \frac{\pi}{2}$, we obtain the following information:

n	x_n	$f(x_n)$	$f'(x_n)$	x_{n+1}
1	$\frac{\pi}{2}$	$-\frac{\pi}{2}$	−2.0	0.785398
2	0.785398	0.214602	−3.000001	0.856932
3	0.856932	0.009338	−2.750424	0.860327
4	0.860327	0.000018	−2.740194	0.860334
5	0.860334	0.000001	−2.740173	0.860334

The solution of the equation in $\left[0, \frac{\pi}{2}\right]$ is approximately $x_5 = 0.860334$.

49. Note that at a relative maximum point x of f, $f'(x) = 0$, i.e., $\sin x + x\cos x = 0$. Let $F(x) = \sin x + x\cos x$. We then need to find a zero of $F(x)$ in [0,3]. Since $F(1) = 1.381773$ and $F(3) = -2.828857$ there must exist a zero of $F(x)$ in [1,3]. By Newton's Method

$$x_{n+1} = x_n - \frac{\sin x_n + x_n \cos x_n}{2\cos x_n - x_n \sin x_n},$$

and using $x_1 = 3$ we have

n	x_n	$F(x_n)$	$F'(x_n)$	x_{n+1}
1	3.0	−2.828857	−2.403345	1.822950
2	1.822950	0.513569	−2.264284	2.049763
3	2.049763	−0.057187	−2.740831	2.028898
4	2.028898	−0.000379	−2.704198	2.028758
5	2.028758	−0.0000004377	−2.703947	2.028758

Note that $F(x) > 0$ for all $0 < x \leq \frac{\pi}{2}$. Also note that $F'(x) < 0$ for all $\frac{\pi}{2} < x < 3$. Therefore there is exactly one point x in the open interval (0,3) such that $F(x) = 0$. Hence $f(x)$ has the unique critical value 2.028758 in the open interval (0,3). Since $f(0) = 0$, $f(3) = 3\sin 3 = 0.42336$, and $f(2.028758) = 1.8197$, the maximum value of $f(x)$ for x in [0,3] is $f(2.028758) = 1.8197$.

51. a. The total distance traveled by the ball when it has struck the ground 3 times is

$$10 + 2 \cdot (10) \cdot \frac{5}{6} + 2 \cdot (10) \left(\frac{5}{6}\right)^2 = 10 + \frac{50}{3} + \frac{125}{9} = \frac{365}{9} \text{ feet.}$$

b. The total distance traveled by the ball when it has come to rest is

$$10 + 2 \cdot 10 \cdot \frac{5}{6} \cdot \sum_{n=0}^{\infty} \left(\frac{5}{6}\right)^n = 10 + \frac{50}{3} \cdot \frac{1}{1 - \frac{5}{6}} = 160 \text{ feet.}$$

PRACTICE PROBLEMS FOR CHAPTER 9

1. Write the first five terms of each of the following sequences.

a. $\{n^2 + 4\}$
b. $\left\{\frac{n}{2^n}\right\}$
c. $\{(-1)^{n+1}(2n)^2\}$
d. $\left\{e^{\frac{2n+1}{2n-1}}\right\}$
e. $\left\{\frac{\ln n}{n}\right\}$
f. $\left\{1 + \frac{(-1)^n}{n}\right\}$

2. Determine whether each of the following sequences converges or diverges, and find the limit of each convergent one.

a. $\left\{\frac{3n-1}{2n}\right\}$
b. $\left\{2 + \frac{n^3-2}{n^2}\right\}$
c. $\left\{\sin\left(\frac{n-2}{2n}\pi\right)\right\}$
d. $\left\{\frac{(100)^n}{(101)^n}\right\}$
e. $\left\{\frac{1+(-1)^n}{n}\right\}$
f. $\left\{\frac{e^n}{e^{2n}+1}\right\}$
g. $\{\ln(3^n + 5^n) - \ln(5^n)\}$
h. $\left\{\left(1 + \frac{1}{2n}\right)^n\right\}$

3. Write the first five terms of each of the following infinite series.

a. $\sum_{k=0}^{\infty} (k^2 - 2k)$
b. $2 \sum_{k=0}^{\infty} 2^k$
c. $\sum_{k=1}^{\infty} \left[(-1)^{k+1}\left(1 + \frac{1}{k}\right)\right]$
d. $\sum_{k=1}^{\infty} \frac{1}{k^2 \cdot \sqrt{k+2}}$
e. $\sum_{k=0}^{\infty} \frac{1-e^k}{e^{3k}+1}$
f. $\sum_{k=1}^{\infty} \ln\left(\frac{1}{k(k+1)}\right)$

4. Determine which of the following series converges, and find the sum of each convergent one.

   a. $\sum_{k=0}^{\infty} \left(-\frac{2}{3}\right)^k$

   b. $\frac{1}{(0.1)} + \frac{1}{(0.1)^2} + \frac{1}{(0.1)^3} + \ldots$

   c. $0.02 + 0.002 + 0.0002 + \ldots$

   d. $\sum_{k=0}^{\infty} \frac{4^k}{3^{k+1}}$

   e. $\sum_{k=0}^{\infty} \frac{2 \cdot 3^k - 2^k}{6^k}$

   f. $\sum_{k=1}^{\infty} \frac{1}{k(k+1)}$

5. Determine which of the following infinite series converges, and state the test used.

   a. $\sum_{k=1}^{\infty} \frac{k+1}{\sqrt{k(k+2)}}$

   b. $\sum_{k=0}^{\infty} \frac{\sin\left(\frac{\pi}{k+1}\right)}{2^k}$

   c. $\sum_{k=0}^{\infty} \frac{3^k}{(k+3)!}$

   d. $\sum_{k=0}^{\infty} \frac{\sqrt{k}}{k^2+k+1}$

   e. $\sum_{k=2}^{\infty} \frac{1}{k(\ln k)^{3/2}}$

   f. $\sum_{k=1}^{\infty} \frac{8k^2-7}{e^k(k+1)^2}$

   g. $\sum_{k=1}^{\infty} (-1)^{k+1} \left(\frac{2k+3}{3k+2}\right)$

   h. $\sum_{k=0}^{\infty} (-1)^{k-1} \cdot \frac{1}{\sqrt{k^2+1}}$

   i. $\sum_{k=1}^{\infty} \frac{e^k}{k+\ln k}$

   j. $\sum_{k=0}^{\infty} \frac{(k+3)!}{3!k!3^k}$

6. For each of the following functions find the n-th Taylor polynomial expanded about $x = a$.

   a. $f(x) = \frac{1}{x^2+1}$; $a = 0$, $n = 3$.

   b. $f(x) = \sqrt{x^2+9}$; $a = -4$, $n = 3$.

   c. $f(x) = x \cdot e^{-2x}$; $a = 0$, $n = 4$.

   d. $f(x) = \tan x$; $a = \frac{\pi}{4}$, $n = 3$.

   e. $f(x) = \ln(\cos x)$; $a = \frac{\pi}{3}$, $n = 3$.

7. For each of the given function values, use Taylor's Theorem to find an approximation using a Taylor polynomial of degree n expanded about $x = a$, and estimate the accuracy of the approximation.

   a. $\ln 0.99$; $a = 1$, $n = 4$.

   b. $e^{-0.1}$; $a = 0$, $n = 4$.

c. $\sqrt[3]{26}$; $a = 27$, $n = 2$.

d. $\sin 5°$; $a = 0$, $n = 3$.

8. The quantity $\sqrt{e} = e^{0.5}$ is to be computed using a Taylor polynomial

$$e^x = 1 + x + \frac{x^2}{2!} + \cdots + \frac{x^n}{n!}.$$

How large should one choose n to guarantee that

$$|R_n(0.5)| < 0.0005?$$

9. Find a Taylor series for each of the following functions expanded about $x = a$, and determine the interval of convergence.

   a. $f(x) = x \cdot \ln x$; $a = 1$.

   b. $f(x) = \cos x$; $a = \frac{\pi}{3}$.

   c. $f(x) = e^{5-2x}$; $a = \frac{5}{2}$.

   d. $f(x) = \sqrt[3]{6-x}$; $a = -2$.

10. Find a Taylor series for $f(x) = \frac{1}{(1-2x)^3}$ expanded about $a = 0$ by twice differentiating the Taylor series for $g(x) = \frac{1}{1-2x}$.

11. Consider an annuity which has a payment of \$5,000 due at the end of each year. Suppose the annual interest rate is 8 percent compounded annually.

   a. Find the present value of the annuity if 7 payments are to be made.

   b. Find the present value of the annuity if the payments are to be continued to perpetuity.

12. An alumna decides to donate a permanent scholarship of \$1200 per year. How much money should be deposited in the bank at 10% annual interest compounded annually in order to be able to supply the money for the scholarship at the end of each year?

13. Show that to establish an annuity paying in perpetuity R dollars at the end of each interest period, it requires a deposit of R/i dollars, where i is the interest rate per interest period.

14. a. Use Newton's Method to approximate the solution of $x^4 + x - 3 = 0$ lying in the interval [1,2].

    b. Use Newton's Method to approximate the positive solution of $2x = 3\sin x$.

    c. Use Newton's Method to approximate $\sqrt[4]{80}$.

15. The equation

$$\tan y = x$$

defines implictly $y = f(x)$ where $f(x)$ is a differentiable function defined for all real numbers x such that $f(0) = 0$. This function is the principal inverse tangent function $\tan^{-1} x$ to be found on a scientific calculator.

    a. Use implicit differentiation to show that $f'(x) = \frac{1}{1+x^2}$ for all real numbers x.

    b. Explain why the power series $1 - x^2 + x^4 - x^6 + x^8 - x^{10} + \cdots$ converges to $\frac{1}{1+x^2}$ for all real numbers x such that $-1 < x < 1$.

    c. Explain why the Maclaurin series for $f(x)$ is given by

$$\begin{aligned} x - \frac{1}{3}x^3 + \frac{1}{5}x^5 - \frac{1}{7}x^7 + \frac{1}{9}x^9 - \frac{1}{11}x^11 + \cdots \\ = \sum_{k=0}^{\infty} (-1)^k \cdot \frac{1}{2k+1} \cdot x^{2k+1}. \end{aligned}$$

    d. For each nonnegative integer n let

$$S_n(x) = \sum_{k=0}^{n} (-1)^k \cdot \frac{1}{2k+1} \cdot x^{2k+1}.$$

    Find $S_n\left(\frac{1}{\sqrt{3}}\right)$ for $n = 2, 4, 6, 8, 10, 12$. What is $\lim_{n\to\infty} S_n\left(\frac{1}{\sqrt{3}}\right)$?

    e. Explain why $f(x) = \frac{\pi}{2} - f\left(\frac{1}{x}\right)$ for $x > 1$. Explain why $f(-x) = -f(x)$.

16. Suppose you put A_0 dollars into an account on December 1, 1990. Assume a constant annual interest rate of 6 percent with k equal compounding interest periods per year. On each December 1 beginning with December 1, 1995 you will withdraw \$1,000 from the account at the start of the day.

a. Let $V(m, k)$ denote the value of the account on December 1 m years after December 1, 1995 right after the withdrawal is made. Find the formula for $V(m, k)$.

b. Find a formula $A_0 = F(m, k)$ so that the account will run out on December 1 m years after December 1, 1995 right after the withdrawal is made.

c. Find $A(k) = \lim_{m \to \infty} F(m, k)$.

d. Find $\lim_{k \to \infty} A(k)$.

17. Put A_0 dollars into an account on December 1, 1990. Assume a constant annual interest rate of 6 percent with k equal compounding interest periods per year. At the beginning of the first day of each interest period starting with December 1, 1995 you will withdraw $\frac{1000}{k}$ dollars from the account.

a. Let $V(j, k)$ denote the value of the account on the first day of the jth of these interest periods starting with December 1, 1995 right after the withdrawal is made. Find the formula for $V(j, k)$.

b. Find a formula for $A_0 = F(m, k)$ so that the account will run out on December 1 m years after December 1, 1995 right after the withdrawal is made.

c. Find $A(k) = \lim_{m \to \infty} F(m, k)$.

d. Find $\lim_{k \to \infty} A(k)$.

18. Show that the Maclaurin series for $\cos x$ converges to $\cos x$ for all x.

Solutions to Practice Problems for Chapter 9

1. a. $\{5, 8, 13, 20, 29\}$.

   b. $\{\frac{1}{2}, \frac{1}{2}, \frac{3}{8}, \frac{1}{4}, \frac{5}{32}\}$.

   c. $\{4, -16, 36, -64, 100\}$.

   d. $\{e^3, e^{5/3}, e^{7/5}, e^{9/7}, e^{11/9}\}$.

   e. $\{0, \frac{\ln 2}{2}, \frac{\ln 3}{3}, \frac{\ln 4}{4}, \frac{\ln 5}{5}\}$.

   f. $\{0, \frac{3}{2}, \frac{2}{3}, \frac{5}{4}, \frac{4}{5}\}$.

2. a. $\lim\limits_{n\to\infty} \frac{3n-1}{2n} = \lim\limits_{n\to\infty} \frac{3-\frac{1}{n}}{2} = \frac{3}{2}$, so it converges to $\frac{3}{2}$.

   b. $\lim\limits_{n\to\infty} (2 + n - \frac{2}{n^2}) = \infty$, so it diverges.

   c. $\lim\limits_{n\to\infty} \sin\left(\frac{n-2}{2n}\pi\right) = \sin\frac{\pi}{2} = 1$, so it converges to 1.

   d. $\lim\limits_{n\to\infty} \frac{(100)^n}{(101)^n} = \lim\limits_{n\to\infty} \left(\frac{100}{101}\right)^n = 0$, so it converges to 0.

   e. $\lim\limits_{n\to\infty} \frac{1+(-1)^n}{n} = 0$, so it converges to 0.

   f. $\lim\limits_{n\to\infty} \frac{e^n}{e^{2n}+1} = \lim\limits_{n\to\infty} \frac{\frac{1}{e^n}}{1+\frac{1}{e^{2n}}} = 0$, so it converges to 0.

   g. $\lim\limits_{n\to\infty} (\ln(3^n + 5^n) - \ln(5^n)) = \lim\limits_{n\to\infty} \ln \frac{3^n+5^n}{5^n} = \lim\limits_{n\to\infty} \ln\left(\left(\frac{3}{5}\right)^n + 1\right)$
   $= \ln 1 = 0$, so it converges to 0.

   h. $\lim\limits_{n\to\infty} \left(1 + \frac{1}{2n}\right)^n = \lim\limits_{n\to\infty} \left(\left(1 + \frac{1}{2n}\right)^{2n}\right)^{1/2} = e^{1/2}$, so it converges to $e^{1/2}$.

3. a. $0 + (-1) + 0 + 3 + 8$.

   b. $2(1 + 2 + 4 + 8 + 16)$.

   c. $2 + (-\frac{3}{2}) + \frac{4}{3} + (-\frac{5}{4}) + \frac{6}{5}$.

   d. $\left(\frac{1}{\sqrt{3}} + \frac{1}{8} + \frac{1}{9\sqrt{5}} + \frac{1}{16\sqrt{6}} + \frac{1}{25\sqrt{7}}\right)$.

   e. $0 + \left(\frac{1-e}{e^3+1}\right) + \left(\frac{1-e^2}{e^6+1}\right) + \left(\frac{1-e^3}{e^9+1}\right) + \left(\frac{1-e^4}{e^{12}+1}\right)$.

   f. $\ln\frac{1}{2} + \ln\frac{1}{6} + \ln\frac{1}{12} + \ln\frac{1}{20} + \ln\frac{1}{30}$.

4. a. It converges to $\frac{1}{1-\left(-\frac{2}{3}\right)} = \frac{3}{5}$.

b. Since the n-th term of the series is 10^n, it diverges.

c. The n-th term of the series is $0.02 \cdot \left(\frac{1}{10}\right)^n$, so it converges to $\frac{0.02}{1-\frac{1}{10}} = 0.02 \cdot \frac{10}{9} = \frac{0.2}{9} = \frac{1}{45}$.

d. The k-th term is $\frac{1}{3}\left(\frac{4}{3}\right)^k$. Since $\frac{4}{3} > 1$, the series diverges.

e. The k-th term is $2\cdot\left(\frac{1}{2}\right)^k - \left(\frac{1}{3}\right)^k$. It converges to $2\cdot\frac{1}{1-\frac{1}{2}} - \frac{1}{1-\frac{1}{3}} = 4 - \frac{3}{2} = \frac{5}{2}$.

f. Since $\frac{1}{k(k+1)} = \frac{1}{k} - \frac{1}{k+1}$, $\sum_{k=1}^{n} \frac{1}{k(k+1)} = \left(1-\frac{1}{2}\right) + \left(\frac{1}{2}-\frac{1}{3}\right) + \dots$
$+\left(\frac{1}{n-1}-\frac{1}{n}\right) + \left(\frac{1}{n}-\frac{1}{n+1}\right) = 1 - \frac{1}{n+1}$.
Thus $\sum_{k=1}^{\infty} \frac{1}{k(k+1)} = \lim_{n\to\infty} \sum_{k=1}^{n} \frac{1}{k(k+1)} = \lim_{n\to\infty} \left(1-\frac{1}{n+1}\right) = 1$.

5. a. Since $\lim_{k\to\infty} \frac{k+1}{\sqrt{k(k+2)}} = 1$, the series diverges by Theorem 5.

b. Since $\left|\frac{\sin\left(\frac{\pi}{k+1}\right)}{2^k}\right| \le \frac{1}{2^k}$ and $\sum_{k=0}^{\infty} \frac{1}{2^k} = 2$, the series converges by the comparison test.

c. By the ratio test, the series converges since $\lim_{k\to\infty} \frac{\frac{3^{k+1}}{(k+4)!}}{\frac{3^k}{(k+3)!}} = \lim_{k\to\infty} \frac{3}{k+4} = 0 < 1$.

d. By the comparison test, the series converges since $\frac{\sqrt{k}}{k^2+k+1} \le \frac{\sqrt{k}}{k^2} = \frac{1}{k^{3/2}}$ and $\sum k^{-3/2}$ converges.

e.
$$\int_2^{\infty} \frac{1}{x(\ln x)^{3/2}}dx = \int_{x=2}^{x=\infty} y^{-3/2}dy, \text{ where } y = \ln x,$$
$$= -2\cdot\left(y^{-1/2}\right)\Big|_{x=2}^{x=\infty}$$
$$= -2\cdot(\ln x)^{-1/2}\Big|_{x=2}^{x=\infty}$$
$$= -2\cdot\left(0-(\ln 2)^{-1/2}\right) = 2\cdot(\ln 2)^{-1/2}.$$

By the integral test, the series converges.

f. $\lim_{k\to\infty} \frac{\frac{8k^2-7}{e^k(k+1)^2}}{\frac{1}{e^k}} = \lim_{k\to\infty} \frac{8k^2-7}{(k+1)^2} = 8$, and $\sum e^{-k}$ converges.
By the limit comparison test, the series converges.

g. $\lim_{k\to\infty} (-1)^{k+1}\left(\frac{2k+3}{3k+2}\right)$ does not exist, so by Theorem 5 the series diverges.

h. It converges by the alternating series test.

i. $\lim_{k\to\infty} \frac{e^k}{k+\ln k} = \infty$ since e^k approaches ∞ faster than $k + \ln k$. Thus, it diverges by Theorem 5.

j. $\lim_{k\to\infty} \frac{(k+4)!/\left(3!(k+1)!3^{k+1}\right)}{(k+3)!/(3!k!3^k)} = \lim_{k\to\infty} \frac{k+4}{3(k+1)} = \frac{1}{3} < 1$, so by the ratio test the series converges.

6. a. $f(x) = \frac{1}{x^2+1} \implies f(0) = 1.$

$f'(x) = \frac{-2x}{(x^2+1)^2} \implies f'(0) = 0.$

$f''(x) = \frac{-2\cdot(x^2+1)^2+2x\cdot 2(x^2+1)\cdot 2x}{(x^2+1)^4}$

$= \frac{-2+6x^2}{(x^2+1)^3} \implies f''(0) = -2.$

$f'''(x) = \frac{12x(x^2+1)^3-(-2+6x^2)\cdot 3\cdot(x^2+1)^2\cdot 2x}{(x^2+1)^6}$

$= \frac{12x(x^2+1)-6x(-2+6x^2)}{(x^2+1)^4}$

$= \frac{-24x^3+24x}{(x^2+1)^4} \implies f'''(0) = 0.$

$P_3(x) = 1 + \frac{-2}{2!}x^2 = 1 - x^2.$

b. $f(x) = \sqrt{x^2+9} \implies f(-4) = 5.$

$f'(x) = \frac{2x}{2\sqrt{x^2+9}} = \frac{x}{\sqrt{x^2+9}} \implies f'(-4) = -\frac{4}{5}.$

$f''(x) = \frac{\sqrt{x^2+9}-x\cdot\frac{1}{2}\cdot(x^2+9)^{-1/2}\cdot 2x}{x^2+9}$

$= \frac{\sqrt{x^2+9}-x^2(x^2+9)^{-1/2}}{x^2+9} \implies f''(-4) = \frac{9}{125}$

$= (x^2+9)^{-1/2} - x^2\cdot(x^2+9)^{-3/2}$

$= (x^2+9)^{-3/2}(x^2+9-x^2) = 9(x^2+9)^{-3/2}.$

$f'''(x) = 9\cdot\left(-\frac{3}{2}\right)(x^2+9)^{-5/2}\cdot 2x = -27x(x^2+9)^{-5/2}$

$\implies f'''(-4) = -27\cdot(-4)\cdot 5^{-5} = \frac{108}{3125}.$

$P_3(x) = 5 - \frac{4}{5}(x+4) + \frac{9}{250}(x+4)^2 + \frac{108}{3125}\cdot\frac{1}{3!}(x+4)^3.$

c. $f(x) = x\cdot e^{-2x} \implies f(0) = 0.$

$f'(x) = e^{-2x} + x\cdot e^{-2x}\cdot(-2) = (1-2x)e^{-2x} \implies f'(0) = 1.$

$f''(x) = -2\cdot e^{-2x} + (1-2x)\cdot e^{-2x}\cdot(-2)$

$= (-2-2+4x)\cdot e^{-2x} \implies f''(0) = -4.$

$f'''(x) = 4 \cdot e^{-2x} + (-4+4x) \cdot e^{-2x} \cdot (-2)$

$= e^{-2x}(4+8-8x)$

$= (12-8x) \cdot e^{-2x} \implies f'''(0) = 12.$

$f^{(4)}(x) = -8 \cdot e^{-2x} + (12-8x) \cdot e^{-2x} \cdot (-2)$

$= e^{-2x}(-8-24+16x)$

$= (-32+16x)e^{-2x} \implies f^{(4)}(0) = -32.$

$P_4(x) = x - 2x^2 + 2x^3 - \frac{4}{3}x^4.$

d. $f(x) = \tan x \implies f\left(\frac{\pi}{4}\right) = 1.$

$f'(x) = \sec^2 x \implies f'\left(\frac{\pi}{4}\right) = 2.$

$f''(x) = 2 \cdot \sec^2 x \cdot \tan x \implies f''\left(\frac{\pi}{4}\right) = 4.$

$f'''(x) = 2 \cdot (2 \cdot \sec^2 x \cdot \tan^2 x + \sec^4 x) \implies f'''\left(\frac{\pi}{4}\right) = 16.$

$P_3(x) = 1 + 2\left(x - \frac{\pi}{4}\right) + 2\left(x - \frac{\pi}{4}\right)^2 + \frac{8}{3}\left(x - \frac{\pi}{4}\right)^3.$

e. $f(x) = \ln(\cos x) \implies f\left(\frac{\pi}{3}\right) = \ln\left(\frac{1}{2}\right) = -\ln 2.$

$f'(x) = \frac{-\sin x}{\cos x} = -\tan x \implies f'\left(\frac{\pi}{3}\right) = -\sqrt{3}.$

$f''(x) = -\sec^2 x \implies f''\left(\frac{\pi}{3}\right) = -4.$

$f'''(x) = -2 \cdot \sec^2 x \cdot \tan x \implies f'''\left(\frac{\pi}{3}\right) = -2 \cdot 4 \cdot \sqrt{3} = -8\sqrt{3}.$

$P_3(x) = -\ln 2 - \sqrt{3}\left(x - \frac{\pi}{3}\right) - 2 \cdot \left(x - \frac{\pi}{3}\right)^2 - \frac{4\sqrt{3}}{3} \cdot \left(x - \frac{\pi}{3}\right)^3.$

7. a. $\ln 0.99 = \ln(1 - 0.01).$

Let $f(x) = \ln x$. Then $f(1) = 0$.

$f'(x) = \frac{1}{x} \implies f'(1) = 1.$

$f''(x) = \frac{-1}{x^2} \implies f''(1) = -1.$

$f'''(x) = \frac{2}{x^3} \implies f'''(1) = 2.$

$f^{(4)}(x) = \frac{-6}{x^4} \implies f^{(4)}(1) = -6.$

$P_4(x) = (x-1) - \frac{1}{2}(x-1)^2 + \frac{1}{3}(x-1)^3 - \frac{1}{4}(x-1)^4.$

$P_4(0.99) = -0.01 - \frac{1}{2}(-0.01)^2 + \frac{1}{3}(-0.01)^3 - \frac{1}{4}(-0.01)^4.$

$= -0.01 - \frac{1}{2} \cdot 10^{-4} - \frac{1}{3} \cdot 10^{-6} - \frac{1}{4} \cdot 10^{-8}$

$= -0.01 - 0.00005 - 0.000000333\ldots - 0.0000000025$

$= -0.010050335\ldots.$

$|f^{(5)}(x)| = \left|\frac{24}{x^5}\right| \le \frac{24}{(0.99)^5} \le 25.24$ for $0.99 \le x \le 1$.

$|R_4(0.99)| \le \frac{25.24}{5!}|0.99 - 1|^5 \le 2.11 \cdot 10^{-11}$.

b. Let $f(x) = e^x$. Then $f(0) = 1$.

$P_4(x) = 1 + x + \frac{x^2}{2} + \frac{x^3}{3!} + \frac{x^4}{4!}$.

$P_4(-0.1) = 1 - 0.1 + 0.005 - 0.000166667 + 0.000004167 = 0.9048375\cdots$.

$|R_4(-0.1)| \le \frac{e^0}{5!}(0.1)^5 \le 8.3 \cdot 10^{-8}$.

c. Let $f(x) = \sqrt[3]{x}$. Then $f(27) = 3$.

$f'(x) = \frac{1}{3}x^{-2/3} \implies f'(27) = \frac{1}{3} \cdot \frac{1}{9} = \frac{1}{27}$.

$f''(x) = -\frac{2}{9}x^{-5/3} \implies f''(27) = -\frac{2}{9} \cdot \frac{1}{3^5} = -\frac{2}{2187}$.

$P_2(x) = 3 + \frac{1}{27}(x - 27) - \frac{1}{2187}(x - 27)^2$.

$P_2(26) = 3 + \frac{1}{27}(-1) - \frac{1}{2187} = 2.962505716$.

$|f'''(x)| = \left|\frac{10}{27}x^{-8/3}\right| \le \frac{10}{27} \cdot \frac{1}{26^{8/3}} < \frac{10}{27} \cdot \frac{1}{(24.389)^{8/3}} = \frac{10}{27} \cdot \frac{1}{(2.9)^8}$

≤ 0.000075 for $26 \le x \le 27$.

$|R_2(26)| \le \frac{0.000075}{3!} \cdot |26 - 27|^3 \le 0.000013$.

d. $\sin 5° = \sin \frac{5}{180}\pi = \sin(0.08727)$.

Let $f(x) = \sin x$. Then $f(0) = 0$.

$P_3(x) = x - \frac{x^3}{3!}$.

$P_3\left(\frac{5}{180} \cdot \pi\right) = 0.087266462 - 0.000110762 = 0.0871557$.

$\left|R_3\left(\frac{5}{180} \cdot \pi\right)\right| \le \frac{1}{4!}\left(\frac{5}{180}\pi\right)^4 \le 0.000003$.

8. $|R_n(x)| \le \frac{3}{(n+1)!}|x|^{n+1}$ for $x \le 1$. Since $e < 3$

$\implies |R_n(0.5)| \le \frac{3}{(n+1)!} \cdot |0.5|^{n+1}$.

For $n = 2$, $|R_2(0.5)| \le 0.0625$.

For $n = 3$, $|R_3(0.5)| \le 0.0078125$.

For $n = 4$, $|R_4(0.5)| \le 0.00078125$.

For $n = 5$, $|R_5(0.5)| \le 0.00006511 < 0.0005$.

Hence if $n \ge 5$, $|R_n(0.5)| < 0.0005$.

9. a. $f(x) = x \cdot \ln x \implies f(1) = 0.$

$f'(x) = \ln x + 1 \implies f'(1) = 1.$

$f''(x) = \frac{1}{x} \implies f''(1) = 1.$

$f'''(x) = -\frac{1}{x^2} \implies f'''(1) = -1.$

$f^{(4)}(x) = \frac{2}{x^3} \implies f^{(4)}(1) = 2.$

$f^{(5)}(x) = \frac{-6}{x^4} \implies f^{(5)}(1) = -6.$

etc.

$$\implies f(x) \sim (x-1) + \tfrac{1}{2}(x-1)^2 - \tfrac{1}{3!}(x-1)^3 + \tfrac{2}{4!}(x-1)^4 - \tfrac{6}{5!}(x-1)^5 + \ldots.$$
$$= (x-1) + \tfrac{1}{2}(x-1)^2 + \sum_{n=3}^{\infty}(-1)^n \tfrac{(n-2)!}{n!}(x-1)^n.$$

By the Ratio Test

$$\lim_{n\to\infty} \frac{\frac{(n-1)!}{(n+1)!}|x-1|^{n+1}}{\frac{(n-2)!}{n!}|x-1|^n}$$
$$= \lim_{n\to\infty}\left(\frac{n-1}{n+1}\right)|x-1| = |x-1|.$$

If $|x-1| < 1$, i.e., if $0 < x < 2$, the series converges absolutely.

If $x = 0$,

$$-1 + \sum_{n=2}^{\infty}(-1)^n \frac{(n-2)!}{n!}(-1)^n = -1 + \sum_{n=2}^{\infty} \frac{(n-2)!}{n!} = -1 + \sum_{n=2}^{\infty} \frac{1}{n(n-1)}$$
$$= -1 + \sum_{n=2}^{\infty}\left(\frac{1}{n-1} - \frac{1}{n}\right) = -1 + \lim_{k\to\infty}\left(1 - \frac{1}{k}\right) = 0$$

$\therefore$ the series converges for $x = 0$.

If $x < 0$, $\lim_{n\to\infty} \frac{(1-x)^n}{n(n-1)} = \infty$ and hence the series diverges.

If $x > 2$, $\lim_{n\to\infty} \frac{(x-1)^n}{n(n-1)} = \infty$ and hence the series diverges.

If $x = 2$, $1 + \sum_{n=2}^{\infty}(-1)^n \frac{(n-2)!}{n!} = 1 + \sum_{n=2}^{\infty}(-1)^n \frac{1}{n(n-1)}$ converges by the Alternating Series Test.

Therefore the interval of convergence is [0,2].

b. $f(x) = \cos x \implies f\left(\frac{\pi}{3}\right) = \frac{1}{2}$.

$f'(x) = -\sin x \implies f'\left(\frac{\pi}{3}\right) = -\frac{\sqrt{3}}{2}$.

$f''(x) = -\cos x \implies f''\left(\frac{\pi}{3}\right) = -\frac{1}{2}$.

$f'''(x) = \sin x \implies f'''\left(\frac{\pi}{3}\right) = \frac{\sqrt{3}}{2}$.

$f^{(4)}(x) = \cos x \implies f^{(4)}\left(\frac{\pi}{3}\right) = \frac{1}{2}$.

etc.

$$\implies f(x) \sim \frac{1}{2}\left(1 - \sqrt{3}\left(x - \frac{\pi}{3}\right) - \frac{1}{2}\left(x - \frac{\pi}{3}\right)^2 + \frac{\sqrt{3}}{3!}\left(x - \frac{\pi}{3}\right)^3 + \frac{1}{4!}\left(x - \frac{\pi}{3}\right)^4 + \ldots\right)$$

$$= \frac{1}{2}\left\{1 - \frac{1}{2}\left(x - \frac{\pi}{3}\right)^2 + \frac{1}{4!}\left(x - \frac{\pi}{3}\right)^4 + \ldots\right\} + \frac{\sqrt{3}}{2}\left\{-\left(x - \frac{\pi}{3}\right) + \frac{1}{3!}\left(x - \frac{\pi}{3}\right)^3 - \frac{1}{5!}\left(x - \frac{\pi}{3}\right)^5 + \ldots\right\}.$$

By the Ratio test, each series converges for all x.

$\therefore$ The Taylor series converges for all x.

c. $f(x) = e^{5-2x} \implies f\left(\frac{5}{2}\right) = 1$.

$f'(x) = -2 \cdot e^{5-2x} \implies f'\left(\frac{5}{2}\right) = -2$.

$f''(x) = (-2)^2 \cdot e^{5-2x} \implies f''\left(\frac{5}{2}\right) = (-2)^2$.

$\vdots \qquad \vdots$

$f^{(n)}(x) = (-2)^n \cdot e^{5-2x} \implies f^{(n)}\left(\frac{5}{2}\right) = (-2)^n$

$\implies f(x) \sim \sum_{n=0}^{\infty} \frac{(-2)^n\left(x-\frac{5}{2}\right)^n}{n!}$.

Since $\lim_{n\to\infty} \frac{\left|\frac{(-2)^{n+1}}{(n+1)!}\right| \cdot \left|x-\frac{5}{2}\right|^{n+1}}{\left|\frac{(-2)^n}{n!}\right| \left|x-\frac{5}{2}\right|^n} = \lim_{n\to\infty} \frac{2\cdot\left|x-\frac{5}{2}\right|}{(n+1)} = 0$ for each $x \neq \frac{5}{2}$, by the Ratio test the Taylor series converges for all x. Hence, the interval of convergence is all real numbers.

d. $f(x) = \sqrt[3]{6-x} \implies f(-2) = 2.$

$f'(x) = -\frac{1}{3}(6-x)^{-2/3} \implies f'(-2) = -\frac{1}{3}\left(\frac{1}{2}\right)^2.$

$f''(x) = -\frac{2}{9}(6-x)^{-5/3} \implies f''(-2) = -\frac{2}{9}\left(\frac{1}{2}\right)^5.$

$f'''(x) = -\frac{2\cdot 5}{27}(6-x)^{-8/3} \implies f'''(-2) = -\frac{2\cdot 5}{27}\left(\frac{1}{2}\right)^8.$

$\vdots \qquad\qquad\qquad\qquad \vdots$

$f^{(n)}(x) = -\frac{2\cdot 5\ldots(3n-4)}{3^n}(6-x)^{-\frac{3n-1}{3}}$

$\implies f^{(n)}(-2) = -\frac{2\cdot 5\ldots(3n-4)}{3^n}(2)^{-(3n-1)}$ for $n \geq 2$.

$$\implies f(x) \sim 2 - \left(\frac{1}{2}\right)^2\left(\frac{1}{3}(x+2) + \frac{2}{9}\left(\frac{1}{2}\right)^3 \cdot \frac{(x+2)^2}{2!} + \ldots + \frac{2\cdot 5\ldots(3n-4)}{3^n}\cdot\left(\frac{1}{2}\right)^{(3n-3)}\frac{(x+2)^n}{n!} + \ldots\right).$$

By the Ratio Test,

$$\lim_{n\to\infty} \frac{\frac{2\cdot 5\cdot 8\ldots(3n-1)}{3^{n+1}}\left(\frac{1}{2}\right)^{3n}\cdot\frac{|x+2|^{n+1}}{(n+1)!}}{\frac{2\cdot 5\cdot 8\ldots(3n-4)}{3^n}\left(\frac{1}{2}\right)^{3n-3}\cdot\frac{|x+2|^n}{n!}}$$

$$= |x+2|\cdot\frac{1}{3}\cdot\left(\frac{1}{2}\right)^3\cdot\lim_{n\to\infty}\frac{3n-1}{n+1} = |x+2|\cdot\frac{1}{3}\cdot\frac{1}{8}\cdot 3$$

$$= \frac{1}{8}\cdot|x+2|.$$

The series converges absolutely for $\frac{1}{8}|x+2| < 1 \implies |x+2| < 8 \implies -10 < x < 6$. The series diverges when $x < -10$ or $x > 6$. For $x = -10$, the series is

$$2 + \frac{2}{3} - \left(\frac{1}{2}\right)^2\cdot\sum_{n=2}^{\infty}\frac{2\cdot 5\cdot 8\ldots(3n-4)}{3^n}\left(\frac{1}{2}\right)^{3n-3}\cdot\frac{(-8)^n}{n!}$$

$$= 2 + \frac{2}{3} - \sum_{n=2}^{\infty}\frac{2\cdot 5\cdot 8\ldots(3n-4)}{3^n\cdot 2^{3n-1}}(-1)^n\cdot\frac{2^{3n}}{n!}$$

$$= 2 + \frac{2}{3} + 2\cdot\sum_{n=2}^{\infty}(-1)^{n+1}\frac{2\cdot 5\cdot 8\ldots(3n-4)}{3\cdot 6\cdot 9\ldots(3n)}.$$

For $x = 6$, the series is

$$2 - \frac{2}{3} - 2 \cdot \sum_{n=2}^{\infty} \frac{2 \cdot 5 \cdot 8 \ldots (3n-4)}{3 \cdot 6 \cdot 9 \ldots (3n)}.$$

It can be shown by more advanced methods that

$$\frac{2 \cdot 5 \cdot 8 \ldots (3n-4)}{3 \cdot 6 \cdot 9 \ldots (3n)} \leq \frac{1}{9}\left(\frac{2}{n}\right)^{(4/3)} \qquad \text{for } n \geq 2.$$

Therefore, the series also converges for $x = -10$ and $x = 6$.

10. $g(x) = \frac{1}{1-2x} \implies g(0) = 1.$

$g'(x) = \frac{2}{(1-2x)^2} \implies g'(0) = 2.$

$g''(x) = \frac{2^2 \cdot 2!}{(1-2x)^3} \implies g''(0) = 8$

$g'''(x) = \frac{2^3 \cdot 3!}{(1-2x)^4} \implies g'''(0) = 2^3 \cdot 3!$

$\vdots \qquad \vdots$

$g^{(n)}(x) = \frac{2^n \cdot n!}{(1-2x)^n} \implies g^{(n)}(0) = 2^n \cdot n!$

$$g(x) = \frac{1}{1-2x} = 1 + 2x + 4x^2 + 8x^3 + \cdots + 2^n x^n + \ldots$$

$$g'(x) = \frac{2}{(1-2x)^2} = 2 + 8x + 24x^2 + \cdots + n \cdot 2^n \cdot x^{n-1} + \ldots$$

$$g''(x) = \frac{8}{(1-2x)^3} = 8 + 48x + 192x^2 + \cdots + n(n-1)2^n x^{n-2} + \ldots, \ n \geq 2$$

$$\implies \frac{1}{(1-2x)^3} = 1 + 6x + 24x^2 + \cdots + n(n-1)2^{n-3} \cdot x^{n-2} + \ldots, \ n \geq 2.$$

11. a. $5000(1.08)^{-1} + 5000(1.08)^{-2} + \cdots + 5000(1.08)^{-7}$

$= 5000 \cdot \frac{(1.08)^{-1}(1-(1.08)^{-7})}{1-(1.08)^{-1}} \doteqdot 5000 \frac{0.92593(0.41651)}{0.07407} \doteqdot \$26033.42.$

b. $\lim_{n\to\infty} 5000 \frac{(1.08)^{-1}(1-(1.08)^{-n})}{1-(1.08)^{-1}} = 5000 \cdot \frac{(1.08)^{-1}}{1-(1.08)^{-1}} = \$62500.$

12. $1200(1.1)^{-1} + 1200(1.1)^{-2} + 1200(1.1)^{-3} + \ldots$

$= \lim_{n\to\infty} 1200 \frac{(1.1)^{-1}(1-(1.1)^{-n})}{1-(1.1)^{-1}} = 1200 \cdot \frac{(1.1)^{-1}}{1-(1.1)^{-1}} = \$12000.$

13. $R(1+i)^{-1} + R(1+i)^{-2} + \cdots = \lim_{n\to\infty} \frac{R(1+i)^{-1}(1-(1+i)^{-n})}{1-(1+i)^{-1}}$

$= R \cdot \frac{(1+i)^{-1}}{1-(1+i)^{-1}} = R \cdot \frac{1}{(1+i)-1} = \frac{R}{i}.$

14. a. Let $f(x) = x^4 + x - 3$.

Since $f(1) = -1 < 0$ and $f(2) = 15 > 0$, $f(x)$ must equal zero for some $x \in (1, 2)$. Start with $x_1 = 1$.

$f'(x) = 4x^3 + 1$.

$x_2 = x_1 - \frac{f(x_1)}{f'(x_1)} = 1 - \frac{-1}{5} = \frac{6}{5} = 1.2$.

$x_3 = 1.2 - \frac{0.2736}{7.912} \doteq 1.1654$.

$x_4 = 1.1654 - \frac{0.009991}{7.3312} \doteq 1.1640$.

$x_5 = 1.1640 + \frac{0.000257}{7.3084} \doteq 1.1640$.

$\therefore$ The approximate solution to $f(x) = 0$ between 1 and 2 is 1.1640.

b. Let $f(x) = 2x - 3\sin x \implies f'(x) = 2 - 3\cos x$.

$f(1) < 0$ and $f(2) > 0$.

Start with $x_1 = 2$.

$x_2 = x_1 - \frac{f(x_1)}{f'(x_1)} = 2 - \frac{1.2721}{3.2484} = 1.6084$.

$x_3 = 1.6084 - \frac{0.21892}{2.11278} = 1.5048$.

$x_4 = 1.5048 - \frac{0.01613}{1.8022} = 1.49585$.

$x_5 = 1.49585 - \frac{0.000121484}{1.77537} = 1.495782$.

$\therefore$ The approximate solution to $f(x) = 0$ is 1.4958.

c. Since $\sqrt[4]{80}$ is a solution to $x^4 = 80$, let $f(x) = x^4 - 80$.

Then $f'(x) = 4x^3$. Start with $x_1 = 2$.

$x_2 = x_1 - \frac{f(x_1)}{f'(x_1)} = 2 - \frac{-64}{32} = 4$.

$x_3 = 4 - \frac{176}{256} = 3.3125$.

$x_4 = 3.3125 - \frac{40.39992}{145.3877} \doteq 3.0346$.

$x_5 = 3.0346 - \frac{4.8019451}{111.78006} \doteq 2.9916$.

$x_6 = 2.9916 - \frac{0.0966031}{107.09534} \doteq 2.9907$.

$x_7 = 2.9907 - \frac{0.0002608}{106.99871} \doteq 2.9907$.

$\therefore \sqrt[4]{80} \doteq 2.9907$.

15. a. $\frac{d}{dx}\tan y = \frac{d}{dx}x \implies \sec^2 y \cdot \frac{dy}{dx} = 1 \implies \frac{dy}{dx} = \frac{1}{\sec^2 y}$.

$\sec^2 y = 1 + \tan^2 y = 1 + x^2$.

$f'(x) = \frac{dy}{dx} = \frac{1}{1+x^2}$.

b. The power series

$$1 - x^2 + x^4 - x^6 + x^8 - x^{10} + \cdots = \sum_{k=0}^{\infty}(-x^2)^k$$

is a geometric series with ratio $r = -x^2$. Thus this power series converges to $\frac{1}{1-(-x^2)} = \frac{1}{1+x^2}$ for all real numbers x such that $x^2 < 1 \implies -1 < x < 1$.

c. The Maclaurin series for $f(x)$ is given by

$$\int \frac{1}{1+x^2}\,dx = \int \left(\sum_{k=0}^{\infty}(-1)^k x^{2k}\right) dx$$

$$= \sum_{k=0}^{\infty}(-1)^k \cdot \frac{1}{2k+1}x^{2k+1} + C.$$

Since $f(0) = 0$, $C = 0$.

d. From the computer we find that

$$S_2\left(\frac{1}{\sqrt{3}}\right) = .52603 \qquad S_8\left(\frac{1}{\sqrt{3}}\right) = .5236$$

$$S_4\left(\frac{1}{\sqrt{3}}\right) = .523767 \qquad S_{10}\left(\frac{1}{\sqrt{3}}\right) = .523599$$

$$S_6\left(\frac{1}{\sqrt{3}}\right) = .523612 \qquad S_{12}\left(\frac{1}{\sqrt{3}}\right) = .523599$$

$\lim_{n\to\infty} S_n\left(\frac{1}{\sqrt{3}}\right) = \frac{\pi}{6}$.

e. Note that $-\frac{\pi}{2} < f(x) < \frac{\pi}{2}$ for all x.

Also note that $0 < f(x) < \frac{\pi}{2}$ for all $x > 0$.

For $x > 1$:

$$\tan\left(\frac{\pi}{2} - f\left(\frac{1}{x}\right)\right) = \cot f\left(\frac{1}{x}\right) = \frac{1}{\tan f\left(\frac{1}{x}\right)} = \frac{1}{\frac{1}{x}} = x.$$

Therefore, $\tan\left(\frac{\pi}{2} - f\left(\frac{1}{x}\right)\right) = \tan f(x)$.

Note that $0 < \frac{\pi}{2} - f\left(\frac{1}{x}\right) < \frac{\pi}{2}$ and $0 < f(x) < \frac{\pi}{2}$.

Therefore, $\frac{\pi}{2} - f\left(\frac{1}{x}\right) = f(x)$.

For any x:

$\tan f(x) = x$ and $\tan(-f(x)) = -\tan f(x) = -x$.

Therefore, $\tan(-f(x)) = \tan f(-x)$.

Note that $-\frac{\pi}{2} < -f(x) < \frac{\pi}{2}$ and $-\frac{\pi}{2} < f(-x) < \frac{\pi}{2}$.

Therefore, $-f(x) = f(-x)$.

16. a. The value of the account on December 1, 1995 right after the withdrawal is made would be

$$A_0 \cdot \left(1 + \frac{.06}{k}\right)^{5k} - 1000.$$

The value of the account on December 1, 1996 right after the withdrawal is made would be

$$\left[A_0 \cdot \left(1 + \frac{.06}{k}\right)^{5k} - 1000\right] \cdot \left(1 + \frac{.06}{k}\right)^{k} - 1000$$

$$= A_0 \cdot \left(1 + \frac{.06}{k}\right)^{6k} - 1000 \cdot \left[1 + \left(1 + \frac{.06}{k}\right)^{k}\right].$$

The value of the account on December 1, 1997 right after the withdrawal is made would be

$$\left\{A_0 \cdot \left(1 + \frac{.06}{k}\right)^{6k} - 1000 \cdot \left[1 + \left(1 + \frac{.06}{k}\right)^{k}\right]\right\} \cdot \left(1 + \frac{.06}{k}\right)^{k} - 1000$$

$$= A_0 \cdot \left(1 + \frac{.06}{k}\right)^{7k} - 1000 \cdot \left[1 + \left(1 + \frac{.06}{k}\right)^{k} + \left(1 + \frac{.06}{k}\right)^{2k}\right].$$

Continuing in this manner we find that the value of the account on December 1 m years after December 1, 1995 right after the withdrawal is made is given by

$$V(m,k) = A_0 \cdot \left(1 + \frac{.06}{k}\right)^{(5+m)k} - 1000 \cdot \sum_{i=0}^{m} \left[\left(1 + \frac{.06}{k}\right)^{k}\right]^{i}$$
$$= A_0 \cdot \left(1 + \frac{.06}{k}\right)^{(5+m)k} - 1000 \cdot \frac{1 - \left[\left(1 + \frac{.06}{k}\right)^{k}\right]^{m+1}}{1 - \left(1 + \frac{.06}{k}\right)^{k}}.$$

b. Suppose the account runs out on December 1 m years after December 1, 1995 right after the withdrawal is made. Then:

$$A_0 \cdot \left(1 + \frac{.06}{k}\right)^{5k+mk} = 1000 \cdot \frac{1 - \left(1 + \frac{.06}{k}\right)^{k+mk}}{1 - \left(1 + \frac{.06}{k}\right)^{k}}.$$

$$A_0 = \frac{1000}{1 - \left(1 + \frac{.06}{k}\right)^{k}} \cdot \frac{1 - \left(1 + \frac{.06}{k}\right)^{k+mk}}{\left(1 + \frac{.06}{k}\right)^{5k+mk}}.$$

$$A_0 = \frac{1000}{\left(1 + \frac{.06}{k}\right)^{k} - 1} \cdot \frac{\left(1 + \frac{.06}{k}\right)^{k+mk} - 1}{\left(1 + \frac{.06}{k}\right)^{5k+mk}}.$$

$$A_0 = \frac{1000}{\left(1 + \frac{.06}{k}\right)^{k} - 1} \cdot \left[\frac{\left(1 + \frac{.06}{k}\right)^{k+mk}}{\left(1 + \frac{.06}{k}\right)^{5k+mk}} - \frac{1}{\left(1 + \frac{.06}{k}\right)^{5k+mk}}\right].$$

Therefore,

$$A_0 = F(m,k) = \frac{1000}{\left(1 + \frac{.06}{k}\right)^{k} - 1} \cdot \left[\frac{1}{\left(1 + \frac{.06}{k}\right)^{4k}} - \frac{1}{\left(1 + \frac{.06}{k}\right)^{5k+mk}}\right].$$

c. As $m \to \infty$, $A_0 = F(m,k) \to \frac{1000}{(1+\frac{.06}{k})^{k}-1} \cdot \frac{1}{(1+\frac{.06}{k})^{4k}} = A(k)$.

d. Note that

$$\left(1 + \frac{.06}{k}\right)^{k} = \left(\left(1 + \frac{1}{\left(\frac{k}{.06}\right)}\right)^{\left(\frac{k}{.06}\right)}\right)^{.06} \to e^{.06} \quad \text{as } k \to \infty.$$

Therefore

$$\lim_{k\to\infty} A(k) = \frac{1000}{e^{.06} - 1} \cdot \frac{1}{(e^{.06})^4} = \frac{1000}{e^{.3} - e^{.24}}$$
$$= \$12,721.08\ .$$

17. a. The value of the account on the first day of the first of these interest periods starting with December 1, 1995 right after the withdrawal is made would be

$$A_0 \cdot \left(1 + \frac{.06}{k}\right)^{5k} - \frac{1000}{k} .$$

The value of the account on the first day of the second of these interest periods starting with December 1, 1995 right after the withdrawal is made would be

$$\left[A_0 \cdot \left(1 + \frac{.06}{k}\right)^{5k} - \frac{1000}{k}\right] \cdot \left(1 + \frac{.06}{k}\right) - \frac{1000}{k}$$

$$= A_0 \cdot \left(1 + \frac{.06}{k}\right)^{5k+1} - \frac{1000}{k} \cdot \left[1 + \left(1 + \frac{.06}{k}\right)\right] .$$

The value of the account on the first day of the third of these interest periods starting with December 1, 1995 right after the withdrawal is made would be

$$\left\{A_0 \cdot \left(1 + \frac{.06}{k}\right)^{5k+1} - \frac{1000}{k} \cdot \left[1 + \left(1 + \frac{.06}{k}\right)\right]\right\} \cdot \left(1 + \frac{.06}{k}\right) - \frac{1000}{k}$$

$$= A_0 \cdot \left(1 + \frac{.06}{k}\right)^{5k+2} - \frac{1000}{k} \cdot \left[1 + \left(1 + \frac{.06}{k}\right) + \left(1 + \frac{.06}{k}\right)^2\right] .$$

Continuing in this manner we find that the value of the account on the first day of the jth of these interest periods starting with December 1, 1995 right after the withdrawal is made is given by

$$V(j,k) = A_0 \cdot \left(1 + \frac{.06}{k}\right)^{5k+(j-1)} - \frac{1000}{k} \cdot \sum_{i=0}^{j-1} \left(1 + \frac{.06}{k}\right)^i$$

$$= A_0 \cdot \left(1 + \frac{.06}{k}\right)^{5k+(j-1)} - \frac{1000}{k} \cdot \frac{1 - (1 + \frac{.06}{k})^j}{1 - (1 + \frac{.06}{k})}$$

$$= A_0 \cdot \left(1 + \frac{.06}{k}\right)^{5k+(j-1)} - \frac{1000}{.06} \cdot \left[\left(1 + \frac{.06}{k}\right)^j - 1\right] .$$

b. Suppose the account runs out on December 1 m years after December 1, 1995 right after the withdrawal is made. Then:

$$A_0 \cdot \left(1+\frac{.06}{k}\right)^{5k+((mk+1)-1)} = \frac{1000}{.06} \cdot \left[\left(1+\frac{.06}{k}\right)^{mk+1} - 1\right] .$$

$$A_0 \cdot \left(1+\frac{.06}{k}\right)^{mk+5k} = \frac{50000}{3} \cdot \left[\left(1+\frac{.06}{k}\right)^{mk+1} - 1\right] .$$

$$A_0 = \frac{50,000}{3} \cdot \frac{\left(1+\frac{.06}{k}\right)^{mk+1} - 1}{\left(1+\frac{.06}{k}\right)^{mk+5k}} .$$

$$A_0 = \frac{50,000}{3} \cdot \left[\frac{\left(1+\frac{.06}{k}\right)^{mk+1}}{\left(1+\frac{.06}{k}\right)^{mk+5k}} - \frac{1}{\left(1+\frac{.06}{k}\right)^{mk+5k}}\right] .$$

Therefore,

$$A_0 = F(m,k) = \frac{50,000}{3} \cdot \left[\frac{1}{\left(1+\frac{.06}{k}\right)^{5k-1}} - \frac{1}{\left(1+\frac{.06}{k}\right)^{mk+5k}}\right] .$$

c. As $m \to \infty$,

$$A_0 = F(m,k) \to \frac{50,000}{3} \cdot \frac{1}{\left(1+\frac{.06}{k}\right)^{5k-1}}$$
$$= \frac{50,000}{3} \cdot \frac{1}{\left(1+\frac{.06}{k}\right)^{5k}\left(1+\frac{.06}{k}\right)^{-1}} = A(k) .$$

d.

$$\lim_{k\to\infty} A(k) = \frac{50,000}{3} \cdot \frac{1}{(e^{.06})^5 \cdot 1} = \frac{50,000}{3} \cdot \frac{1}{e^{.3}}$$
$$= \$12,346.97 .$$

18. For any nonzero real number x and any nonnegative integer n,

$$\frac{\left|\frac{x^{2(n+1)+1}}{(2(n+1)+1)!}\right|}{\left|\frac{x^{2n+1}}{(2n+1)!}\right|} = \frac{x^2}{(2n+2)\cdot(2n+3)} \to 0 \quad \text{as } n \to \infty .$$

Therefore the power series

$$\sum_{n=0}^{\infty} \frac{x^{2n+1}}{(2n+1)!}$$

converges for all real numbers x. Hence

$$\lim_{n\to\infty} \frac{x^{2n+1}}{(2n+1)!} = 0$$

for all real numbers x. For any real number x and any nonnegative integer n,

$$\left|\cos x - \sum_{k=0}^{n}(-1)^k \cdot \frac{x^{2k}}{(2k)!}\right| \leq \frac{|x^{2n+1}|}{(2n+1)!} \to 0 \quad \text{as} \quad n \to \infty\,.$$

CHAPTER 10

SUMMARY OF CHAPTER 10

1. The concept of a differential equation, solution of a differential equation, general solution of a differential equation, initial value problems.

2. Direction field of a differential equation and its use in observing how graphs of vaious solutions behave; use of the direction field of a differential equation to get a sketch of some solution curves.

3. The differential equation $\frac{dy}{dt} = k(M - y)$ for limited growth, and its general solution $y(t) = M\left(1 - Ce^{-kt}\right)$ where C is a constant; Newton's Law of Cooling as an example of limited growth.

4. The differential equation $\frac{dP}{dt} = kP\left(\frac{M-P}{M}\right)$ for logistic growth, and its general solution $P(t) = \frac{M}{1+Ce^{-kt}}$ where C is a constant.

5. The method of separation of variables for solving a differential equation of the form $\frac{dy}{dx} = \frac{f(x)}{g(y)}$, $g(y) \neq 0$; the Law of Natural Growth $\frac{dP}{dt} = kP$ as an example of a separable differential equation, and its general solution $P = Ce^{kt}$ where C is a constant; the differential equation $\frac{dP}{dt} = kP\left(\frac{M-P}{M}\right)$ for logistic growth as an example of a separable differential equation where M is a constant representing the carrying capacity; generalizing the Law of Natural Growth by a differential equation $\frac{dy}{dt} = k(y - a)$ where a is a constant, and its general solu-

tion $y = a + Ce^{kt}$ where C is a constant; certain mixing problems which lead to separable differential equations; annuity problems involving continuous compounding of interest and either continuous deposits or continuous withdrawals which lead to separable differential equations.

6. Second order differential equations of the type

$$\frac{d^2y}{dt^2} + a\frac{dy}{dt} + by = 0$$

where a and b are constants; the characteristic polynomial $r^2 + ar + b$; the result that if the characteristic polynomial has the two distinct real roots r_1 and r_2, then the general solution of the differential equation is given by $y(t) = Ae^{r_1 t} + Be^{r_2 t}$ where A and B are constants; the result that if the characteristic polynomial has only the single real repeated root r_1, then the general solution of the differential equation is given by $y(t) = Ae^{r_1 t} + Bte^{r_1 t}$ where A and B are constants; the result that if the characteristic polynomial has the two nonreal roots $r_1 = \alpha + \beta_i$ and $r_2 = \alpha - \beta_i$ where $\beta \neq 0$, then the general solution of the differential equation is given by $y(t) = e^{\alpha t}(A \sin \beta t + B \cos \beta t)$ where A and B are constants.

7. Use of two initial conditions to determine a unique solution of a second order differential equation.

8. Modelling oscillatory motion by a second order differential equation; using second order differential equations to study predator-prey systems.

9. Use of the techniques for qualitative theory shown in Section 10.5 to infer properties of solutions of an autonomous differential equation $y' = f(y)$ from the equation itself.

10. Euler's Method for approximating the solution to an initial value problem of the form

$$\frac{dy}{dt} = f(t, y), \qquad y(a) = y_0 \,.$$

Selected Solutions to Exercise Set 10.1

1. $y = Ce^{\pi x} \implies \frac{dy}{dx} = \pi Ce^{\pi x} = \pi y.$

3. $y = A\sin 3x + B\cos 3x.$

   $\frac{dy}{dx} = 3A\cos 3x - 3B\sin 3x.$

   $\frac{d^2y}{dx^2} = -9A\sin 3x - 9B\cos 3x = -9y.$

   Therefore $\frac{d^2y}{dx^2} + 9y = 0.$

5. $y = \frac{1}{2} + C\ e^{-4t}.$

   $\frac{dy}{dt} = -4C\ e^{-4t} = -4\left(y - \frac{1}{2}\right) = 2 - 4y.$

7. $y = 22 - C\ e^{-10x}.$

   $\frac{dy}{dx} = 10C\ e^{-10x} = 10 \cdot (22 - y).$

9. $f(t) = Ae^{-2t} + Bte^{-2t}.$

   $f'(t) = -2Ae^{-2t} - 2Bte^{-2t} + Be^{-2t}.$

   $$f''(t) = 4Ae^{-2t} - 2Be^{-2t} + 4Bte^{-2t} - 2Be^{-2t}$$
   $$= (4A - 4B)e^{-2t} + 4Bte^{-2t}.$$

   $$f''(t) + 4f'(t) + 4f(t) = (4A - 4B)e^{-2t} + 4Bte^{-2t}$$
   $$+4(-2Ae^{-2t} + Be^{-2t} - 2Bte^{-2t}) + 4f(t)$$
   $$= -4Ae^{-2t} - 4Bte^{-2t} + 4f(t)$$
   $$= -4(Ae^{-2t} + Bte^{-2t}) + 4f(t)$$
   $$= -4f(t) + 4f(t) = 0.$$

11. $\frac{dy}{dx} = 9 - x.$

$\int dy = \int (9-x)dx \implies y = 9x - \frac{x^2}{2} + C.$

13. $\frac{dy}{dt} = t\ e^t.$

$\int dy = \int te^t\ dt \implies y = t\ e^t - \int e^t\ dt$

$\implies y = t \cdot e^t - e^t + C.$

15. $\frac{d^2y}{dt^2} = \frac{t+1}{\sqrt{t}}, \quad t > 0.$

$\int \frac{d^2y}{dt^2} \cdot dt = \int \frac{t+1}{\sqrt{t}}\ dt \implies \int d\left(\frac{dy}{dt}\right) = \int \frac{t+1}{\sqrt{t}}\ dt$

$\implies \frac{dy}{dt} = \int \left(t^{1/2} + t^{-1/2}\right) dt = \frac{2}{3}t^{3/2} + 2t^{1/2} + C_1.$

$\therefore\ \frac{dy}{dt} = \frac{2}{3}t^{3/2} + 2t^{1/2} + C_1.$

$\int dy = \int \left(\frac{2}{3}t^{3/2} + 2t^{1/2} + C_1\right) dt$

$\implies y = \frac{4}{15}t^{5/2} + \frac{4}{3}t^{3/2} + C_1 t + C_2.$

17. $\frac{dy}{dx} = x\sqrt{x^2+1}.$

$y = \int x\sqrt{x^2+1}\ dx = \int \frac{1}{2}\sqrt{t}\ dt$, where $t = x^2 + 1.$

$\therefore\ y = \frac{1}{3}t^{3/2} + C = \frac{1}{3}\left(x^2+1\right)^{3/2} + C.$

Since $y(0) = 1, \quad y(0) = \frac{1}{3} + C = 1 \implies C = \frac{2}{3}.$

Thus, $y = \frac{1}{3}\left(x^2+1\right)^{3/2} + \frac{2}{3}.$

19. $\frac{dy}{dx} = 2 - 4y.$

By Exercise 5, $y = \frac{1}{2} + C\ e^{-4t}.$

From $y(0) = 3, \quad y(0) = \frac{1}{2} + C = 3 \implies C = \frac{5}{2}.$

Thus, $y = \frac{1}{2} + \frac{5}{2}e^{-4t}.$

21. $\frac{dy}{dx} = -y$.

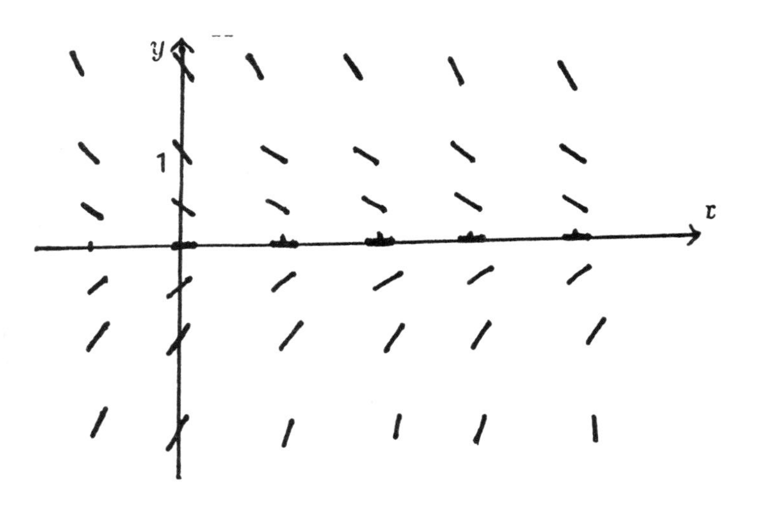

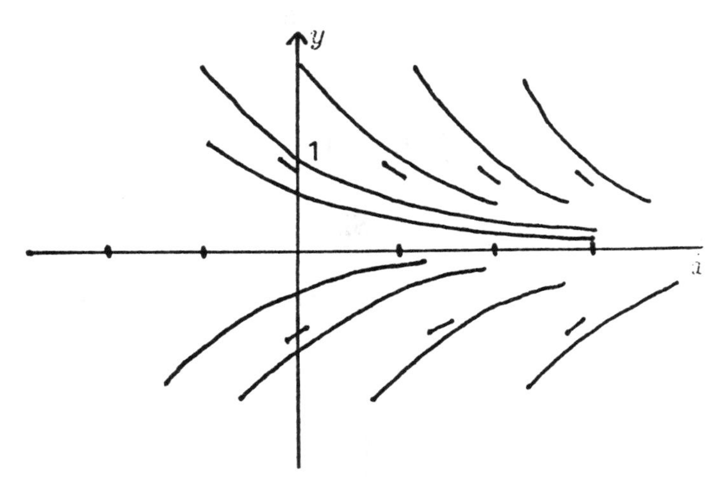

Direction field for

$$\frac{dy}{dx} = -y.$$

Solution curves associated

with direction field.

23. $\frac{dy}{dx} = y(y-3)$.

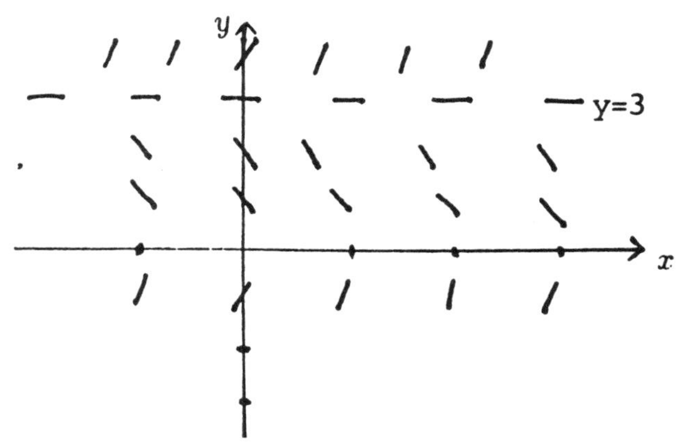

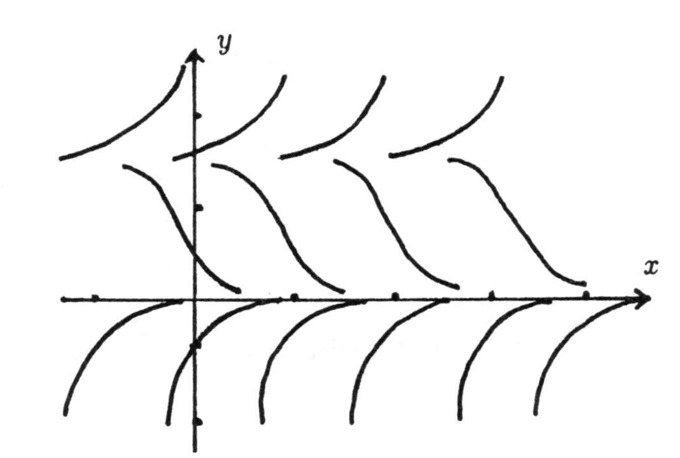

Direction field for

$$\frac{dy}{dx} = y(y-3).$$

Solution curves associated

with direction field.

25. a. $\begin{cases} \frac{dV}{dt} = \sqrt{t+1} \\ V(0) = 5. \end{cases}$

b. $V = \int \sqrt{t+1}\, dt = \frac{2}{3}(t+1)^{3/2} + C.$

Since $V(0) = \frac{2}{3} + C = 5, \quad C = 5 - \frac{2}{3} = \frac{13}{3}.$

$\therefore\ V(t) = \frac{2}{3}(t+1)^{3/2} + \frac{13}{3}.$

27. a. $\begin{cases} U'(x) = 6 - 2x \\ U(0) = 0. \end{cases}$

b. $U(x) = 6x - x^2 + C.$

$U(0) = 0 = C.$

$U(x) = 6x - x^2.$

c. $U'(x) < 0$ for $x > 3$.

Therefore $U(x)$ decreases as x increases for $x > 3$.

Hence, the answer is no.

Selected Solutions to Exercise Set 10.2

1. $\frac{dy}{dx} = k \cdot (M - y)$ where $k = 1$, $M = 1$.

$y = M \cdot \left(1 - Ce^{-kx}\right) \implies y = 1 - Ce^{-x}.$

3. $\frac{dy}{dt} = 10 - 2y = 2 \cdot (5 - y)$. Here $\frac{dy}{dt} = k \cdot (M - y)$ where $k = 2$, $M = 5$.

$y = M \cdot \left(1 - Ce^{-kt}\right) \implies 5 \cdot \left(1 - Ce^{-2t}\right).$

$y(0) = 1 \implies 1 = 5 \cdot \left(1 - Ce^{-2\cdot(0)}\right) \implies .2 = 1 - C \implies C = .8.$

$y = 5 \cdot \left(1 - 0.8 \cdot e^{-2t}\right).$

5. $\frac{dy}{dt} = ky \cdot \left(\frac{M-y}{M}\right)$ where $k = 10$, $M = 50$.

$y = \frac{M}{1+Ce^{-kt}} \implies y = \frac{50}{1+Ce^{-10t}}.$

7. $P(t) = 100\left(1 - e^{\frac{\ln 0.6}{2}t}\right).$

$P(5) = 100\left(1 - e^{\frac{\ln 0.6}{2}\cdot 5}\right) = 100\left(1 - e^{2.5\cdot\ln 0.6}\right)$

$\simeq 100(1 - 0.279) = 72.1\%.$

9. $300 = \frac{600}{1+59\cdot e^{\left(\frac{1}{5}\ln\frac{19}{59}\right)t}}$

$\implies 1 + 59 \cdot e^{\left(\frac{1}{5}\ln\frac{19}{59}\right)t} = 2$

$\implies 59 \cdot e^{\left(\frac{1}{5}\ln\frac{19}{59}\right)t} = 1$

$\implies e^{\left(\frac{1}{5}\ln\frac{19}{59}\right)t} = \frac{1}{59}$

$\implies \left(\frac{1}{5}\ln\frac{19}{59}\right)t = -\ln 59$

$\implies t = -\frac{5\ln 59}{\ln\frac{19}{59}} \doteqdot 17.993.$

$\therefore$ After 18 weeks the pond will contain 300 fish.

11. $N(x) = 10{,}000(1 - e^{-0.002x}).$

$N'(x) = -10{,}000 \cdot e^{-0.002x} \cdot (-0.002) = 20 \cdot e^{-0.002x}.$

At $x = 0$, sales increase most rapidly.

13. $N(t) = \frac{80{,}000}{1+200\cdot e^{-0.2t}}.$

a. $N(6) = \frac{80{,}000}{1+200\cdot e^{-1.2}} \doteqdot 1306.4.$

$\therefore$ 1306 subscribers after 6 months.

b. $\lim_{t\to\infty} N(t) = \lim_{t\to\infty} \frac{80{,}000}{1+200\cdot e^{-0.2t}} = 80{,}000$ subscribers.

15. $N(t) = \frac{400}{1+100e^{-t}}$.

$$N'(t) = \frac{-400(-100e^{-t})}{(1+100e^{-t})^2} = \frac{40000e^{-t}}{(1+100e^{-t})^2}, \quad t > 0.$$

$$N''(t) = \frac{-40000e^{-t}(1+100e^{-t})^2 - 40000e^{-t}\left(2(1+100e^{-t})\cdot(-100\cdot e^{-t})\right)}{(1+100e^{-t})^4}$$

$$= \frac{-40000e^{-t}(1+100e^{-t})+40000\cdot 200\cdot e^{-2t}}{(1+100e^{-t})^3}$$

$$= \frac{40000e^{-t}(-1-100e^{-t}+200e^{-t})}{(1+100e^{-t})^3}$$

$$= \frac{40000e^{-t}(-1+100\cdot e^{-t})}{(1+100e^{-t})^3}.$$

$N''(t) = 0$ for $100 \cdot e^{-t} = 1 \implies e^{-t} = \frac{1}{100} \implies t = -\ln\frac{1}{100} = \ln 100 \doteqdot 4.6$.

Since $N''(t) > 0$ on $(0, \ln 100)$ and $N''(t) < 0$ on $(\ln 100, \infty)$, $N'(t)$ has a maximum at $t = \ln 100$.

Thus the influenza is spreading the greatest after 4.6 days.

17. $N(t) = 30(1 - e^{-0.05t})$.

a. $N(20) = 30(1 - e^{-1}) \doteqdot 18.96 \approx 19$.

$\therefore$ 19 words can be mastered in 20 minutes.

b. $20 = 30\left(1 - e^{-0.05t}\right)$. $\quad 1 - e^{-0.05t} = \frac{2}{3}$.

$e^{-0.05t} = \frac{1}{3} \implies -0.05t = -\ln 3 \implies t = 20 \cdot \ln 3 \doteqdot 21.97$.

$\therefore$ A student has mastered 20 words after 22 minutes.

19. a. $N(t) = \frac{2000}{1+Ce^{-kt}}$.

$N(0) = 500 \implies 500 = \frac{2000}{1+Ce^{-k(0)}} \implies 1 + C = 4 \implies C = 3$.

$N(t) = \frac{2000}{1+3e^{-kt}}$.

$N(2) = 800 \implies 800 = \frac{2000}{1+3e^{-k(2)}} \implies 1 + 3e^{-2k} = 2.5 \implies 3e^{-2k} = 1.5$

$\implies e^{-2k} = .5 \implies -2k = \ln .5 \implies 2k = \ln 2 \implies k = \frac{\ln 2}{2}$.

$N(t) = \frac{2000}{1+3e^{-\left(\frac{\ln 2}{2}\right)t}}.$

b. $N(6) = \frac{2000}{1+3e^{-\left(\frac{\ln 2}{2}\right)\cdot 6}} = \frac{2000}{1+3e^{-3\ln 2}} \approx 1455.$

21. $\frac{dN}{dt} = kN \cdot \left(\frac{M-N}{M}\right) \implies N = \frac{M}{1+Ce^{-kt}}.$

Here $M = 10,000$, $C = 999$, $k = 5$.

$\frac{dN}{dt} = 5N \cdot \left(\frac{10,000-N}{10,000}\right).$

$N(0) = \frac{10,000}{1+999} = 10.$

23. $P(t) = \frac{M}{1+C\cdot e^{-kt}}.$

$$\frac{dP}{dt} = \frac{-M\cdot C\cdot e^{-kt}\cdot(-k)}{(1+C\cdot e^{-kt})^2} = \frac{M\cdot k\cdot C\cdot e^{-kt}}{(1+C\cdot e^{-kt})^2}$$

$$= P(t) \cdot \frac{k\cdot C\cdot e^{-kt}}{(1+C\cdot e^{-kt})} = k \cdot P \cdot \frac{1+C\cdot e^{-kt}-1}{1+C\cdot e^{-kt}}$$

$$= k \cdot P \cdot \left(1 - \frac{1}{1+C\cdot e^{-kt}}\right) = k \cdot P \cdot \left(1 - \frac{P}{M}\right) = k \cdot P \cdot \left(\frac{M-P}{M}\right).$$

Selected Solutions to Exercise Set 10.3

1. $\frac{dy}{dx} = \frac{y}{x}, \quad x \neq 0.$

$\frac{dy}{y} = \frac{dx}{x} \implies \int \frac{dy}{y} = \int \frac{dx}{x} \implies \ln|y| = \ln|x| + C_1$

$\ln|y| = \ln|x| + \ln e^{C_1} = \ln\left(e^{C_1}|x|\right).$

$\therefore\ y = e^{\hat{C}}x = Cx.$

3. $\frac{dy}{dx} = -4xy^2. \qquad \frac{dy}{y^2} = -4x\ dx$

$\int \frac{1}{y^2}\ dy = -\int 4x\ dx \implies -\frac{1}{y} = -2x^2 + C_1.$

$\therefore\ y = \frac{1}{2x^2+C}.$

5. $\frac{dy}{dx} = e^{x-y}$. $\quad \frac{dy}{dx} = \frac{e^x}{e^y} \implies e^y\ dy = e^x\ dx$.

$\int e^y\ dy = \int e^x\ dx \implies e^y = e^x + C$.

$\therefore\ y = \ln(e^x + C)$.

7. $e^y \frac{dy}{dt} - t^2 = 0$. $\quad e^y\ dy = t^2\ dt \implies \int e^y\ dy = \int t^2\ dt$.

$e^y = \frac{1}{3}t^3 + C$.

$\therefore\ y = \ln\left(\frac{1}{3}t^3 + C\right)$.

9. $\frac{dy}{dx} = (1+x)(1+y)$.

$\frac{dy}{1+y} = (1+x)dx \implies \int \frac{dy}{1+y} = \int (1+x)dx$.

$\ln|1+y| = x + \frac{x^2}{2} + C_1 \implies (1+y) = C \cdot e^{x+\frac{x^2}{2}}$.

$\therefore\ y = C \cdot e^{x+\frac{x^2}{2}} - 1$.

11. $y' = 5y \implies \frac{1}{y}\,dy = 5\,dx$.

$\ln|y| = 5x + C_1 \implies y = Ce^{5x}$.

$y(1) = \pi e^5 \implies \pi e^5 = Ce^5 \implies C = \pi$.

$y = \pi e^{5x}$.

13. $\frac{dy}{dx} = xy \implies \frac{1}{y}\,dy = x\,dx \implies \ln|y| = \frac{1}{2}x^2 + C_1 \implies y = Ce^{\frac{1}{2}x^2}$.

$y(2) = 7 \implies 7 = Ce^{\frac{1}{2}\cdot(2)^2} \implies Ce^2 = 7 \implies C = 7e^{-2}$.

$y = 7e^{-2} \cdot e^{\frac{1}{2}x^2} = 7e^{\frac{1}{2}x^2-2}$.

15. $\frac{dy}{dx} = y\cos x, \quad y(0) = \pi$.

$\frac{1}{y}\ dy = \cos x\ dx \implies \int \frac{1}{y}\ dy = \int \cos x\ dx$.

$\ln|y| = \sin x + C_1 \implies y = C \cdot e^{\sin x}$.

From $y(0) = \pi$, $\pi = C$. Thus, $y = \pi \cdot e^{\sin x}$.

17. $y' = 2(y-1), \quad y(0) = 1.$

$\frac{dy}{y-1} = 2\ dx \implies \int \frac{1}{y-1}\ dy = \int 2\ dx \implies \ln|y-1| = 2x + C_1$

$\implies y - 1 = C \cdot e^{2x} \implies y = 1 + Ce^{2x}.$

From $y(0) = 1, \quad 1 = 1 + C \implies C = 0$. Thus, $y \equiv 1$.

19. $y' - 3y = 6, \quad y(0) = 2.$

$y' = 3(y+2) \implies \frac{dy}{y+2} = 3 \cdot dx \implies \ln|y+2| = 3x + C_1.$

$y + 2 = C \cdot e^{3x} \implies y = -2 + C \cdot e^{3x}.$

From $y(0) = 2, \quad 2 = -2 + C \implies C = 4$. Thus, $y = -2 + 4e^{3x}$.

21. From Example 4, $k = 2$, and $K = 100$, the solution $P(t) = \frac{100}{1+M \cdot e^{-2t}}$. The carrying capacity is 100.

23. $\frac{dU}{dt} = kU \quad (k < 0).$

$\frac{dU}{U} = k\ dt \implies \ln|U| = kt + C_1 \implies U = Ce^{kt}.$

Here $U(0) = C \implies U = U(0) \cdot e^{kt}.$

Since $U(4.5) = \frac{1}{2}U(0)$,

$\frac{1}{2} = e^{4.5k} \implies 4.5k = \ln\frac{1}{2}$

$\therefore\ k = -\frac{\ln 2}{4.5} \approx -0.154.$

25. a. $\frac{dp}{dt} = -\frac{3}{200}p, \quad p(0) = 200(0.5) = 100.$

b. $\frac{dp}{p} = -\frac{3}{200}dt.$

$\ln|p| = -\frac{3}{200}t + C_1 \implies p = C \cdot e^{-0.015t}.$

Since $p(0) = 100, \quad C = 100.\ \therefore p(t) = 100 \cdot e^{-0.015t}$

So $p(20) = 100 \cdot e^{-0.3}.$

c. $\lim_{t\to\infty} p(t) = \lim_{t\to\infty}(100 \cdot e^{-0.015t}) = 0.$

27. a. $\frac{dP}{dt} = 0.1P + 100, \quad P(0) = 100.$

b. Solving the above differential equation for P,

$$\ln|0.1P + 100| = \frac{1}{10}t + C_1$$

$$0.1P + 100 = C_2 \cdot e^{0.1t}$$

$$0.1P = C_2 \cdot e^{0.1t} - 100$$

$$\therefore P(t) = C \cdot e^{0.1t} - 1000.$$

Since $P(0) = 100, \quad 100 = C - 1000 \implies C = 1100.$

Hence $P(t) = 1100 \cdot e^{0.1t} - 1000.$

$P(7) = 1100 \cdot e^{0.7} - 1000 = \$1215.13.$

29. $P(t) = \frac{1}{1+M \cdot e^{-kt}}.$

Since $P(0) = \frac{100}{100{,}000} = \frac{1}{1000}, \quad \frac{1}{1+M} = \frac{1}{1000} \implies M = 999.$

Since $P(2) = \frac{500}{100{,}000} = \frac{1}{200}, \quad \frac{1}{200} = \frac{1}{1+999 \cdot e^{-2k}}$

$\implies e^{-2k} = \frac{199}{999}.$

$P(5) = \frac{1}{1+999 \cdot e^{-5k}} = \frac{1}{1+999 \cdot (e^{-2k})^{5/2}}$

$= \frac{1}{1+999 \cdot (\frac{199}{999})^{5/2}} \doteqdot 0.053498.$

Hence the number of people who know the rumor is $100{,}000 \cdot 0.053498 \doteqdot 5350.$

31. $\frac{dy}{dt} = k(y - 30)$, where $y(t)$ is the temperature in t minutes.

So $y(t) = 30 + C \cdot e^{kt}.$

Since $y(0) = 100, \quad C = 70.$

$\therefore \; y(t) = 30 + 70 \cdot e^{kt}.$

Since $y(3) = 80, \quad y(3) = 30 + 70 \cdot e^{3k} = 80,$

$70 \cdot e^{3k} = 50 \implies e^{3k} = \frac{5}{7}.$

Then, $y(10) = 30 + 70 \cdot e^{10k} = 30 + 70 \cdot (e^{3k})^{10/3}$

$= 30 + 70 \cdot \left(\frac{5}{7}\right)^{10/3} = 52.8°$ C.

33. $\frac{dy}{dt} = 60 - \frac{y}{500} \cdot 10$, where $y(t)$ is the number of pounds of salt after t minutes.

$\frac{dy}{dt} = 60 - \frac{y}{50} = 60 - 0.02y = -0.02 \cdot (y - 3000)$

$\Longrightarrow\ y(t) = 3000 + C \cdot e^{-0.02t}$.

Since $y(0) = 1000$, $\quad -2000 = C$.

$\therefore\ y(t) = 3000 - 2000 \cdot e^{-0.02t}$.

Hence the concentration of salt after t minutes is $\frac{y(t)}{500} = 6 - 4 \cdot e^{-0.02t}$.

35. Let A be the value of the annuity after t years.

$\frac{dA}{dt} = 0.1 \cdot A - 2000 \Longrightarrow \frac{dA}{dt} = 0.1(A - 20000)$.

$\Longrightarrow\ A(t) = 20000 + C \cdot e^{0.1t}$.

Since $A(0) = 10000$, $\quad C = -10000$.

$\therefore\ A(t) = 10000\left(2 - e^{0.1t}\right)$.

From $A(t) = 0$, $\quad 0 = 2 - e^{0.1t} \Longrightarrow e^{0.1t} = 2$

$\Longrightarrow\ 0.1t = \ln 2\ \therefore\ t = 10 \cdot \ln 2 \doteqdot 6.9315$ years.

37. $\frac{dy}{ds} = k \cdot \frac{y}{s}$. $\quad \frac{dy}{y} = \frac{k}{s}\,ds$.

$\ln|y| = k\ln|s| + C_1\ \therefore\ y = C \cdot e^{k\ln s} = C \cdot s^k$.

Selected Solutions to Exercise Set 10.4

1. $\frac{d^2y}{dt^2} + 5\frac{dy}{dt} + 6y = 0$.

$r^2 + 5r + 6 = 0 \Longrightarrow (r+2)(r+3) = 0 \Longrightarrow r_1 = -2, \quad r_2 = -3$.

The solution is $y = Ae^{-2t} + Be^{-3t}$.

3. $y'' - y = 0$.

$r^2 - 1 = 0 \Longrightarrow r_1 = 1, \quad r_2 = -1$.

General solution is $y = Ae^{-t} + Be^{t}$.

5. $\frac{d^2y}{dt^2} + 3\frac{dy}{dt} - 10y = 0$.

$r^2 + 3r - 10 = 0 \Longrightarrow (r+5)(r-2) = 0 \Longrightarrow r_1 = -5, \quad r_2 = 2$.

General solution is $y = Ae^{2t} + Be^{-5t}$.

7. $y'' - \pi y = 0$.

$r^2 - \pi = 0 \Longrightarrow r_1 = \sqrt{\pi}, \quad r_2 = -\sqrt{\pi}$.

General solution is $y = Ae^{\sqrt{\pi}t} + Be^{-\sqrt{\pi}t}$.

9. $y'' + 2y' + y = 0.$

$r^2 + 2r + 1 = 0 \implies (r+1)^2 = 0 \implies r_1 = r_2 = -1.$

General solution is $y = Ae^{-t} + Bte^{-t}$.

11. $\frac{d^2y}{dt^2} + 6\frac{dy}{dt} + 9y = 0.$

$r^2 + 6r + 9 = 0 \implies (r+3)^2 = 0 \implies r_1 = r_2 = -3.$

General solution is $y = Ae^{-3t} + Bte^{-3t}$.

13. $y'' - 2Cy' + C^2y = 0.$

$r^2 - 2Cr + C^2 = 0 \implies (r-C)^2 = 0 \implies r_1 = r_2 = C.$

The general solution is $y = Ae^{Ct} + Bte^{Ct}$.

15. $\frac{d^2y}{dt^2} + 4y = 0.$

$r^2 + 4 = 0 \implies r = \pm 2i.$

General solution is $y = A\sin 2t + B\cos 2t$.

17. $y'' = -7y \implies y'' + 7y = 0.$

$r^2 + 7 = 0 \implies r = \pm\sqrt{7}\,i.$

The general solution is $y = A\cos\left(\sqrt{7}\,t\right) + B\sin\left(\sqrt{7}\,t\right)$.

19. $y'' - 4y' + 13y = 0.$

$r^2 - 4r + 13 = 0 \implies r = \frac{4 \pm \sqrt{4^2 - 4(13)}}{2} = 2 \pm \frac{\sqrt{-36}}{2} = 2 \pm 3i.$

General solution is $y = Ae^{2t}\sin 3t + Be^{2t}\cos 3t$.

21. $y'' + y' + \frac{y}{2} = 0.$

$r^2 + r + \frac{1}{2} = 0 \implies r = \frac{-1 \pm \sqrt{1^2 - 4\left(\frac{1}{2}\right)}}{2} = -\frac{1}{2} \pm \frac{i}{2}.$

General solution is $y = Ae^{-t/2}\sin\frac{1}{2}t + Be^{-t/2}\cos\frac{1}{2}t$.

23. $y'' - y' - 6y = 0.$

$r^2 - r - 6 = 0 \implies (r-3)(r+2) = 0 \implies r_1 = 3, \quad r_2 = -2.$ Then

$y = Ae^{3t} + Be^{-2t} \implies y' = 3Ae^{3t} - 2Be^{-2t}.$

$$y(0) = 0, \quad y'(0) = 5 \implies \begin{cases} A + B = 0 & \textbf{(1)} \\ 3A - 2B = 5 & \textbf{(2)} \end{cases}$$

From **(1)** $A = -B$; then **(2)** becomes $3A - 2(-A) = 5 \implies A = 1, \quad B = -1.$

Then the solution is $y = e^{3t} - e^{-2t}$.

25. $y'' + 2y' + y = 0.$

$r^2 + 2r + 1 = 0 \implies r = -1.$

$y = Ae^{-t} + Bte^{-t}.$

$y' = -Ae^{-t} + Be^{-t} - Bte^{-t} = -Ae^{-t} + Be^{-t}(1-t).$

$y(0) = 3, \quad y'(0) = 1 \implies \begin{cases} A = 3 \\ -A + B = 1 \end{cases} \implies A = 3, \quad B = 4.$

The solution is $y = 3e^{-t} + 4te^{-t}$.

27. $y'' + 9y = 0. \qquad r^2 + 9 = 0 \implies r = \pm 3i.$

$y = A\sin 3t + B\cos 3t.$

$y' = 3A\cos 3t - 3B\sin 3t.$

$y(0) = 3, \quad y'(0) = -3 \implies \begin{cases} B = 3 \\ 3A = -3 \end{cases} \implies A = -1, \quad B = 3.$

Thus, the solution is $y = -\sin 3t + 3\cos 3t$.

29. $\frac{dy}{dt} = 4x, \qquad a = 4.$

$\frac{dx}{dt} = -y, \qquad b = 1.$

General solution, from (24) and (25) in Example 8, is

$x(t) = A\sin\sqrt{ab}\,t + B\cos\sqrt{ab}\,t$

$y(t) = -\sqrt{\frac{a}{b}}\left[A\cos\sqrt{ab}\,t - B\sin\sqrt{ab}\,t\right]$

$\implies x(t) = A\sin 2t + B\cos 2t.$

$y(t) = -2A\cos 2t + 2B\sin 2t.$

31. From Example 9, the general solution for the predator-prey system is

$x(t) = 20 + A\sin\sqrt{2}\,t + B\cos\sqrt{2}\,t$

$y(t) = 10 - \sqrt{2}\left(A\cos\sqrt{2}\,t - B\sin\sqrt{2}\,t\right).$

Now $x(0) = 100, \quad y(0) = 40.$

$\implies \begin{cases} 20 + B = 100 \\ 10 - \sqrt{2}\,A = 40 \end{cases} \implies A = -\frac{30}{\sqrt{2}} = -15\sqrt{2}, \quad B = 80.$

Thus the solution is

$x(t) = 20 - 15\sqrt{2}\sin\sqrt{2}\,t + 80\cos\sqrt{2}\,t$

$y(t) = 10 + 30\cos\sqrt{2}\,t + 80\sqrt{2}\sin\sqrt{2}\,t.$

33. In Exercise 32, if $x(0) = 25, \quad y(0) = 16$:
$$\begin{cases} 25 + B = 25 \\ 16 - \frac{1}{\sqrt{2}} A = 16 \end{cases} \implies A = B = 0.$$
The solution then is $x(t) = 25, \quad y(t) = 16$.

35. $y = te^{-2t}. \quad y' = e^{-2t} - 2te^{-2t}.$

$$y'' = -2e^{-2t} - 2e^{-2t} + 4te^{-2t} = -4e^{-2t} + 4te^{-2t}.$$

$$y'' + 4y' + 4y = (-4e^{-2t} + 4te^{-2t}) + 4(e^{-2t} - 2te^{-2t}) + 4te^{-2t}$$
$$= -4e^{-2t} + 8te^{-2t} + 4e^{-2t} - 8te^{-2t} = 0.$$

Thus $y = te^{-2t}$ is a solution to $y'' + 4y' + 4y = 0$.

Selected Solutions to Exercise Set 10.5

1. $y' = y(3 - y)$.

a. From $y(3 - y) = 0$, there are two constant solutions $y = 0$ and $y = 3$.

b.

Range of y	Sign of y'	Behavior of Solution
$y < 0$	Negative	decreasing
$0 < y < 3$	Positive	increasing
$y > 3$	Negative	decreasing

c. $y'' = \frac{d}{dt}(y') = \frac{d}{dt}(y(3 - y)) = \frac{d}{dt}(3y - y^2) = (3 - 2y) \cdot y'$
$= (3 - 2y) \cdot y \cdot (3 - y)$.

Range of y	Sign of y''	Concavity
$y < 0$	Negative	Down
$0 < y < \frac{3}{2}$	Positive	Up
$\frac{3}{2} < y < 3$	Negative	Down
$y > 3$	Positive	Up

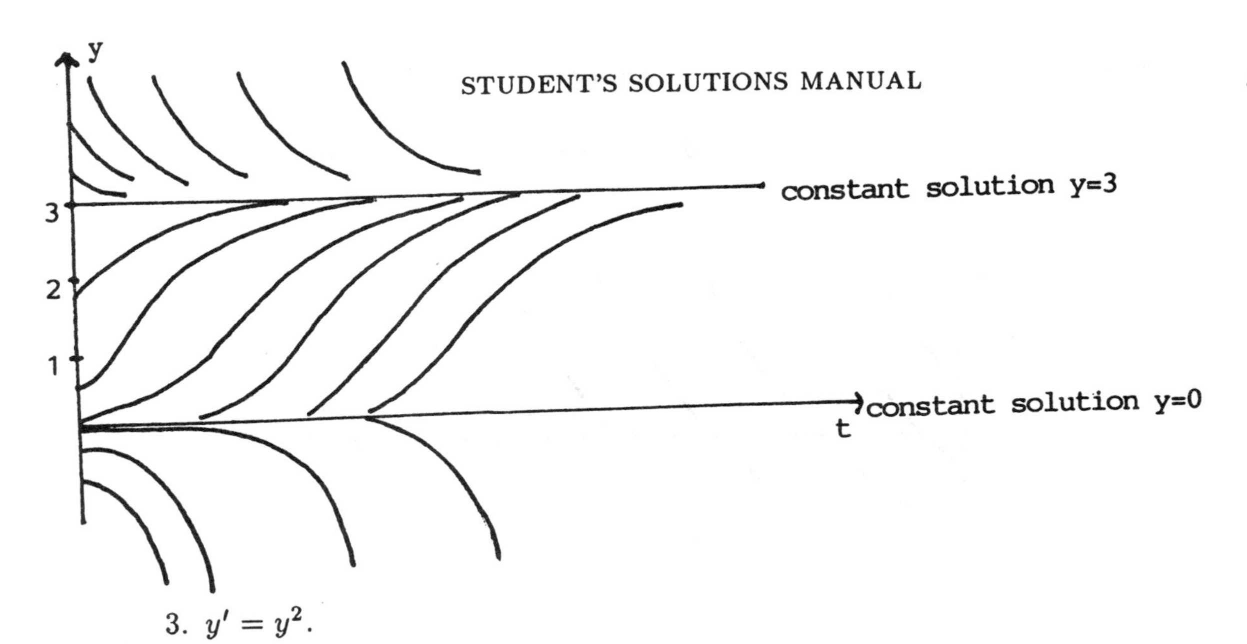

3. $y' = y^2$.

 a. $y = 0$ is a constant solution.

 b. For all $y \neq 0$, y is increasing.

 c. $y'' = 2y \cdot y' = 2y^3$.

Range of y	Sign of y''	Concavity
$y < 0$	Negative	Down
$y > 0$	Positive	Up

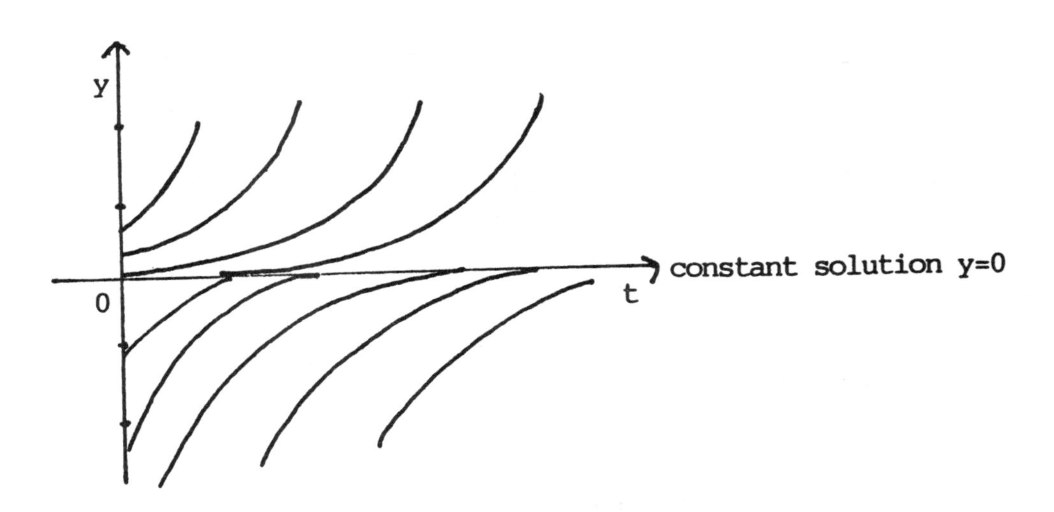

5. $y' = e^y$.

 a. Since $e^y > 0$, there is no constant solution.

 b. y is increasing for all y.

c. $y'' = e^y \cdot y' = e^{2y} > 0$, so y is concave up for all y.

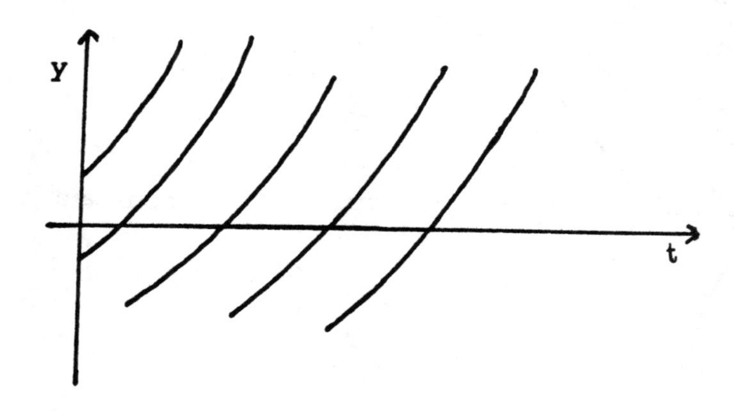

7. $y' = \cos y$.

a. $\cos y = 0$ for $y = \frac{(2n-1)\pi}{2}$, $\quad n = 0, \pm 1, \pm 2, \ldots$.
Thus $y = \left(n - \frac{1}{2}\right)\pi$, $(n = 0, \pm 1, \pm 2, \ldots)$, are constant solutions.

b.

Range of y	Sign of y'	Behavior of Solution
$-\frac{\pi}{2} < y < \frac{\pi}{2}$	Positive	increasing
$\frac{\pi}{2} < y < \frac{3\pi}{2}$	Negative	decreasing
$\left(2n - \frac{1}{2}\right)\pi < y < \left(2n + \frac{1}{2}\right)\pi$	Positive	increasing
$\left(2n - \frac{3}{2}\right)\pi < y < \left(2n - \frac{1}{2}\right)\pi$	Negative	decreasing

c. $y'' = (-\sin y) \cdot y' = -(\sin y)(\cos y) = -\frac{1}{2}\sin 2y$.
$y'' = 0$ for $2y = n\pi \implies y = \frac{n}{2}\pi, \quad n = 0, \pm 1, \pm 2, \ldots$.

Range of y	Sign of y''	Concavity
$0 < y < \frac{\pi}{2}$	Negative	Down
$\frac{\pi}{2} < y < \pi$	Positive	Up
$n\pi < y < \frac{2n+1}{2}\pi$	Negative	Down
$\frac{2n-1}{2}\pi < y < n\pi$	Positive	Up

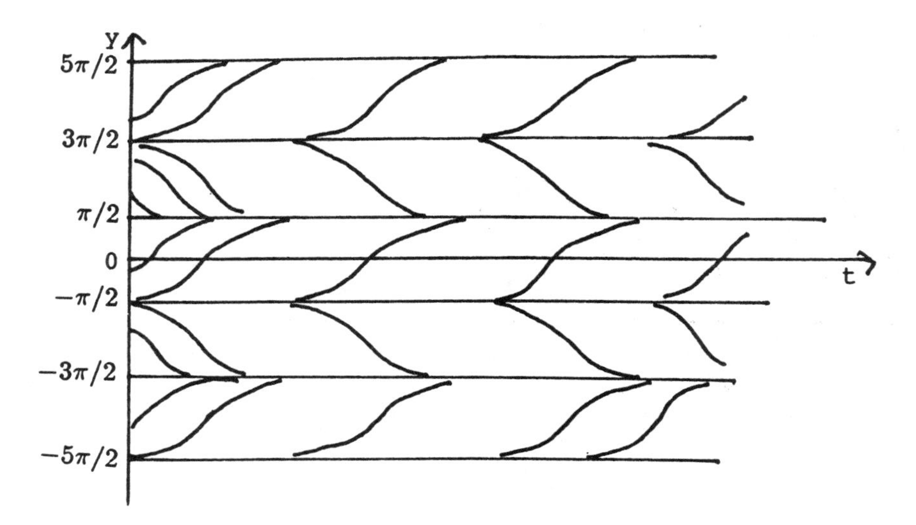

9. $y' = y(1-y)(2-y)$.

a. The constant solutions are $y = 0$, $y = 1$, and $y = 2$.

b.

Range of y	Sign of y'	Behavior of y
$y < 0$	Negative	decreasing
$0 < y < 1$	Positive	increasing
$1 < y < 2$	Negative	decreasing
$y > 2$	Positive	increasing

c. $y'' = [(1-y)(2-y) - y(1-y) - y(2-y)]\, y'.$
$= (3y^2 - 6y + 2)(1-y)(2-y)y.$
$y'' = 0$ for $y = 0, 1, 2$ and $1 \pm \frac{\sqrt{3}}{3}$.

Range of y	Sign of y''	Concavity
$y < 0$	Negative	Down
$0 < y < 1 - \frac{\sqrt{3}}{3}$	Positive	Up
$1 - \frac{\sqrt{3}}{3} < y < 1$	Negative	Down
$1 < y < 1 + \frac{\sqrt{3}}{3}$	Positive	Up
$1 + \frac{\sqrt{3}}{3} < y < 2$	Negative	Down
$y > 2$	Positive	Up

d. 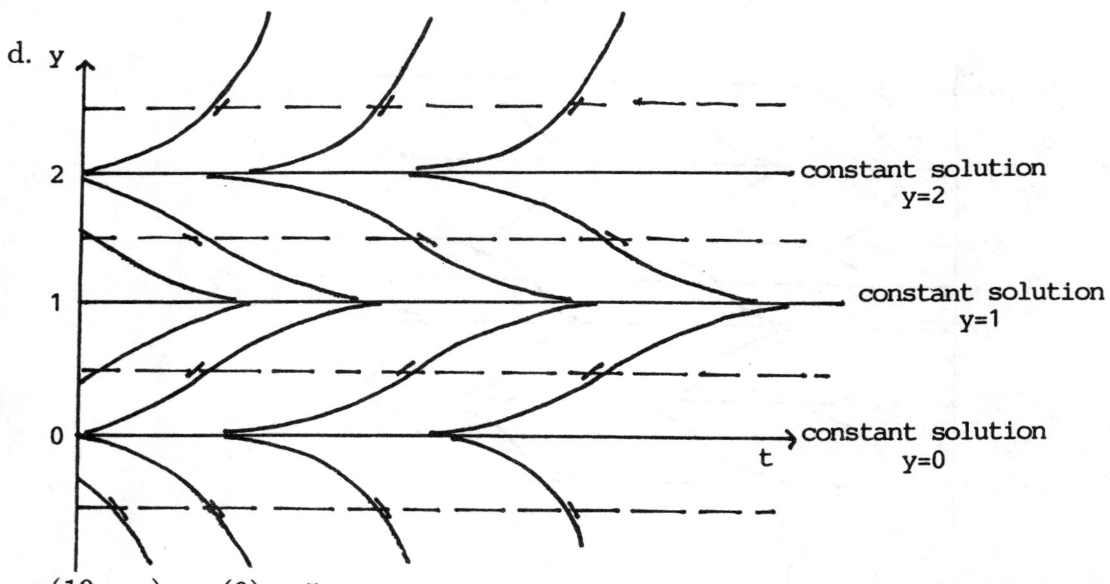

11. $y' = y(10 - y), \quad y(0) = 5.$

a. Because the initial condition $y(0) = 5$ is between $y = 0$ and $y = 10$, the solution $y(t)$ of the initial value problem is such that $0 < y(t) < 10$ for all $t \geq 0$.

b. Because $y'(t) > 0$ for all $t > 0$, it follows that $y(t)$ increases as t increases for all $t \geq 0$, and $5 < y(t) < 10$ for all $t > 0$.

c. $y'' = y' \cdot (10 - y) + y \cdot (-y') = y' \cdot (10 - 2y) = y(10 - y)(10 - 2y)$.
Because $y'' < 0$ for all $5 < y < 10$, the graph of $y(t)$ is concave down for all $t > 0$.

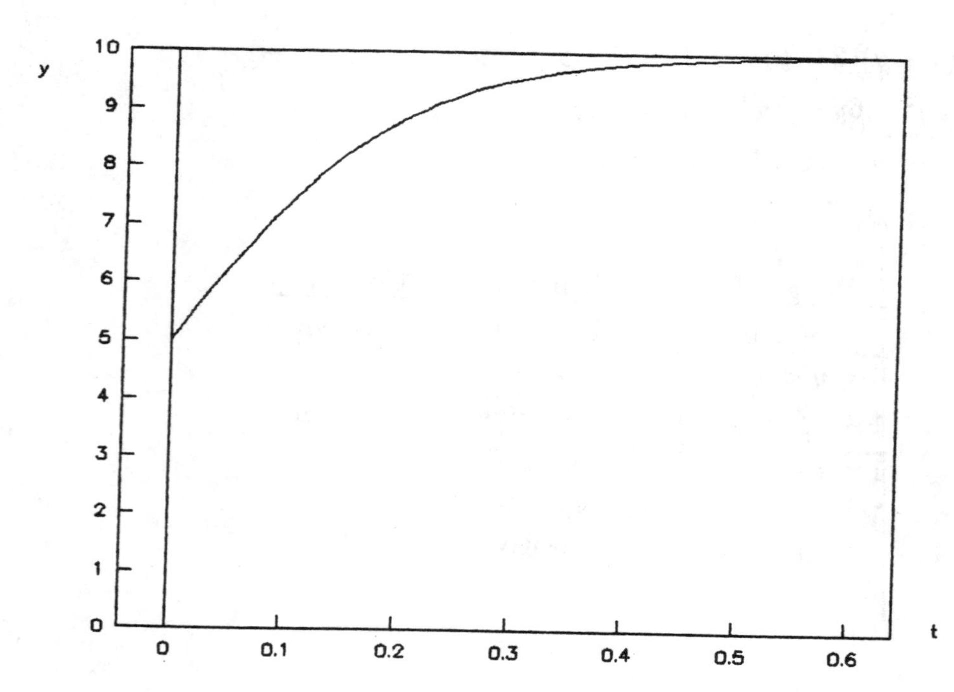

13. $y' = \sin y, \quad y(0) = -\frac{\pi}{4}$.

Since $\sin y = 0$ for $y = n\pi$ for n any integer, and since $-\pi < y(0) < 0$, the solution $y(t)$ to the initial value problem is such that $-\pi < y(t) < 0$ for all $t \geq 0$. Since $y'(t) < 0$ for all $t > 0$, $y(t)$ decreases as t increases for all $t \geq 0$, and therefore $-\pi < y(t) \leq -\frac{\pi}{4}$ for all $t \geq 0$. Since $y''(t) = \cos y \cdot y' = \cos y \cdot \sin y = \frac{1}{2}\sin 2y$, $\quad y'' = 0$ for $y = -\frac{\pi}{2}$, and $y'' > 0$ for $-\pi < y < -\frac{\pi}{2}$ and $y'' < 0$ for

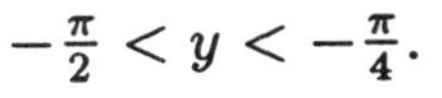
$-\frac{\pi}{2} < y < -\frac{\pi}{4}$.

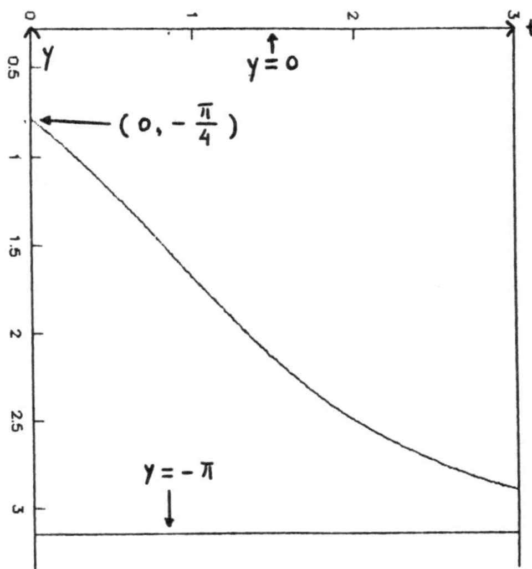

15. $y' = y(1 - y^2), \quad y(0) = -2$.

Since $y = 0, -1$ and 1 are the constant solutions of $y' = y(1-y^2)$, and $y(0) = -2 < -1$, the initial value problem has the solution $y(t)$ such that $y(t) < -1$ for all $t > 0$. Since $y'(t) > 0$ for all $t > 0$, $y(t)$ increases for all $t > 0$. Thus $-2 < y(t) < -1$ for all $t > 0$. Since $y'' = (1 - y^2 - 2y^2)y' = y(1-y^2)(1-3y^2)$, $y'' < 0$ for all $t > 0$. So the graph of $y(t)$ is concave down for all $t > 0$.

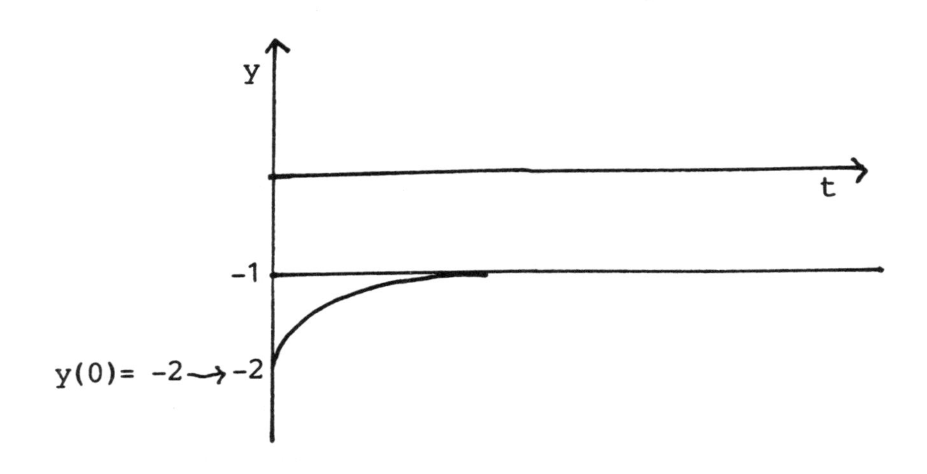

17. $y' = y^2, \quad y(0) = -1.$

Since $y^2 = 0$ for $y = 0$ and $y(0) = -1 < 0$, the solution $y(t)$ of this initial value problem has values with $y(t) < 0$ for all $t > 0$. Here, $y(t)$ is increasing for all $t > 0$. Since $y'' = 2yy' = 2y^3 < 0$ for all $t > 0$, the graph of $y(t)$ is concave down for all $t > 0$.

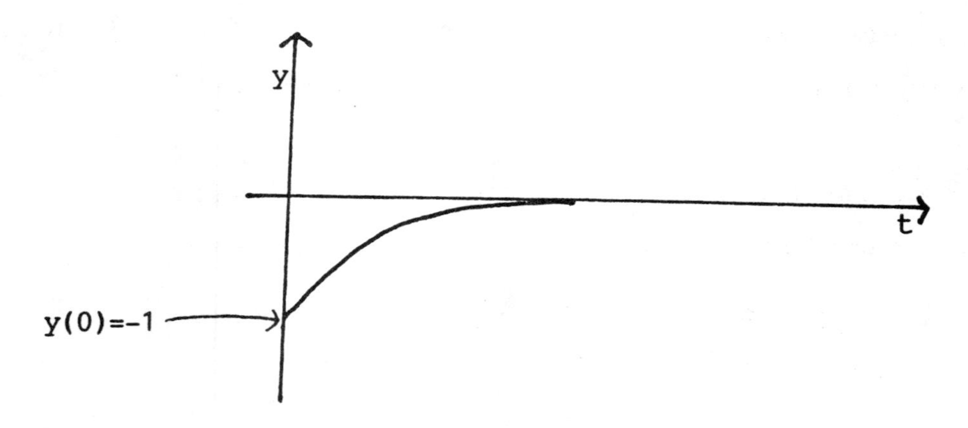

19. $y' = (y+3)(y-2), \quad y(0) = 1.$

Since $(y+3)(y-2) = 0$ for $y = -3$ and 2, and $-3 < y(0) = 1 < 2$, the solution $y(t)$ of this problem has values with $-3 < y(t) < 2$ for all $t > 0$. Since $y'(t) < 0$ for all $t > 0$, $y(t)$ is decreasing for all $t > 0$.

$$y'' = [(y-2) + (y+3)]\, y' = (2y+1)(y+3)(y-2).$$

Range of y	Sign of y''	Concavity
$-3 < y < -\frac{1}{2}$	Positive	Up
$-\frac{1}{2} < y < 1$	Negative	Down

21. The values of y_1 and y_2 become closer together as t increases. Note that $y' = y(1-y)$ has two constant solutions $y = 0$ and $y = 1$, and $0 < y_1(0) = \frac{1}{2}$, $y_2(0) = \frac{3}{4} < 1$. Thus solutions y_1 and y_2 are in the same strip $0 < y < 1$ in which, by property A-111, any non-constant solution approaches the constant solution $y = 1$ as t increases. Hence the values of y_1 and y_2 become closer together as t increases.

23. The solution y with $y(0) = \frac{3}{2}$ decreases for all $t > 0$ because $1 < y(t) < 2$ for all $t > 0$ so that $y'(t) < 0$ for all $t > 0$.

25. The differential equation has three constant solutions $y = 0$, $y = 1$, and $y = 2$. We now determine whether all solutions in each strip decrease or increase.

Range of y	Sign of y'	Conclusion
$(-\infty, 0)$	Negative	decreasing
$(0, 1)$	Positive	increasing
$(1, 2)$	Negative	decreasing
$(2, \infty)$	Positive	increasing

From the above information, the only constant solution $y = c$ which has the described property is $y = 1$. Thus $c = 1$.

27. $y' = 30y(60 - 4y)$ has constant solutions $y = 0$ and $y = 15$.

Range of y	Sign of y'	Conclusion
$(-\infty, 0)$	Negative	decreasing
$(0, 15)$	Positive	increasing
$(15, \infty)$	Negative	decreasing

The constant solution to which all nearby solutions are asymptotic is $y = 15$.

29. Since the solution with $y(0) = 10$ is decreasing for all $t > 0$, we must have $y'(0) < 0 \implies y'(0) = k(10)(20 - 10) = 100k < 0 \implies k < 0$.

Selected Solutions to Exercise Set 10.6

1. $y' = 2y, \quad y(0) = 1.$

   $h = 0.25$ on $[0, 1]$: $t_0 = 0$, $t_1 = \frac{1}{4}$, $t_2 = \frac{1}{2}$, $t_3 = \frac{3}{4}$, $t_4 = 1$.

By Euler's Method the approximation iteration scheme is

$$y_{i+1} = y_i + f(t_i, y_i)h \qquad (i = 0, 1, 2, 3) \quad \text{with } y_0 = y(0) = 1$$

and $f(t, y) = 2y$. Thus

$$y_1 = y_0 + 2y_0 h = 1 + 2(1)\left(\frac{1}{4}\right) = 1.5$$

$$y_2 = y_1 + 2y_1 h = 1.5 + 2(1.5)\left(\frac{1}{4}\right) = 2.25$$

$$y_3 = y_2 + 2y_2 h = 2.25 + 2(2.25)\left(\frac{1}{4}\right) = 3.375$$

$$y_4 = y_3 + 2y_3 h = 3.375 + 2(3.375)\left(\frac{1}{4}\right) = 5.0625$$

3. $\begin{array}{l} y' + 2y = 4 \\ y(0) = 1 \end{array} \implies \begin{array}{l} y' = 4 - 2y \\ y(0) = 1 \end{array}$.

The approximations with $f(t, y) = 4 - 2y$ and $y_0 = y(0) = 1$ are

$$y_1 = y_0 + (4 - 2y_0)h = 1 + (4 - 2(1))\left(\frac{1}{4}\right) = 1.5$$

$$y_2 = y_1 + (4 - 2y_1)h = 1.5 + (4 - 2(1.5))\left(\frac{1}{4}\right) = 1.75$$

$$y_3 = y_2 + (4 - 2y_2)h = 1.75 + (4 - 2(1.75))\left(\frac{1}{4}\right) = 1.875$$

$$y_4 = y_3 + (4 - 2y_3)h = 1.875 + (4 - 2(1.875))\left(\frac{1}{4}\right) = 1.9375.$$

5. $\frac{dy}{dx} = x(y + 1), \quad y(0) = 1.$

The approximations on $[0, 1]$ with $f(x, y) = x(y + 1)$, $h = \frac{1}{4}$, and $y_0 = y(0) = 1$ are:

$$y_1 = y_0 + x_0(y_0 + 1)h = 1 + 0(1 + 1)\left(\frac{1}{4}\right) = 1$$

$$y_2 = y_1 + x_1(y_1 + 1)h = 1 + \frac{1}{4}(1 + 1)\frac{1}{4} = 1.125$$

$$y_3 = y_2 + x_2(y_2 + 1)h = 1.125 + \left(\frac{1}{2}\right)(1.125 + 1)\left(\frac{1}{4}\right) = 1.390625$$

$$y_4 = y_3 + x_3(y_3 + 1)h = 1.390625 + \left(\frac{3}{4}\right)(1.390625 + 1)\left(\frac{1}{4}\right) \approx 1.838867.$$

7. $\begin{array}{l} \frac{dy}{dt} + 2ty = t \\ y(0) = \frac{3}{2} \end{array} \Longrightarrow \begin{array}{l} \frac{dy}{dt} = t(1-2y) \\ y(0) = \frac{3}{2} \end{array}.$

The approximations on [0,1] with $f(t,y) = t(1-2y)$, $h = \frac{1}{4}$, and $y_0 = y(0) = \frac{3}{2}$ are:

$$y_1 = y_0 + t_0(1-2y_0)h = \frac{3}{2} + 0\left(1 - 2\left(\frac{3}{2}\right)\right)\left(\frac{1}{4}\right) = 1.5$$

$$y_2 = y_1 + t_1(1-2y_1)h = 1.5 + \left(\frac{1}{4}\right)(1-2(1.5))\left(\frac{1}{4}\right) = 1.375$$

$$y_3 = y_2 + t_2(1-2y_2)h = 1.375 + \left(\frac{1}{2}\right)(1-2(1.375))\left(\frac{1}{4}\right) = 1.15625$$

$$y_4 = y_3 + t_3(1-2y_3)h = 1.15625 + \left(\frac{3}{4}\right)(1-2(1.15625))\left(\frac{1}{4}\right) = 0.910156.$$

9. $\frac{dy}{dt} = 2y - 2t, \quad y(0) = \frac{3}{2}.$

The approximations on [0,1] with $f(t,y) = 2(y-t)$, $h = \frac{1}{4}$ and $y_0 = y(0) = \frac{3}{2}$ are:

$$y_1 = y_0 + 2(y_0 - t_0)h = \frac{3}{2} + 2\left(\frac{3}{2} - 0\right)\left(\frac{1}{4}\right) = 2.25$$

$$y_2 = y_1 + 2(y_1 - t_1)h = 2.25 + 2(2.25 - 0.25)\left(\frac{1}{4}\right) = 3.25$$

$$y_3 = y_2 + 2(y_2 - t_2)h = 3.25 + 2(3.25 - 0.5)\left(\frac{1}{4}\right) = 4.625$$

$$y_4 = y_3 + 2(y_3 - t_3)h = 4.625 + 2(4.625 - 0.75)\left(\frac{1}{4}\right) = 6.5625.$$

11. The initial value problem in Exercise 3 is

$$\begin{array}{l} y' + 2y = 4 \\ y(0) = 1 \end{array} \Longrightarrow \begin{array}{l} y' = -2(y-2) \\ y(0) = 1 \end{array}.$$

Using equations (14) and (15) of Section 10.3, we have the general solution for the differential equation with $k = -2$ and $a = 2$.

$$y = 2 + Ce^{-2t}$$

Now $y(0) = 1 \Longrightarrow 1 = 2 + Ce^{-2(0)} \Longrightarrow C = -1.$

Thus the actual solution of this initial value problem is

$$y = 2 - e^{-2t}.$$

Now compare the actual values of $y(t)$ with the values from Euler's method. The results are as follows. (See Exercise 3.)

t_j	0	0.25	0.5	0.75	1.0
y_j	1.0	1.5	1.75	1.875	1.9375
$y(t_j)$	1.0	1.393469	1.632121	1.776870	1.864665

Selected Solutions to the Review Exercises - Chapter 10

1. $\frac{dx}{dt} = x + 3.$

   $\frac{dx}{x+3} = dt \implies \int \frac{1}{x+3}\,dx = \int dt \implies \ln|x+3| = t + C_1 \implies$

   $x + 3 = Ce^t \implies x = -3 + Ce^t.$

3. $\frac{dy}{dx} = \frac{\sec^2\sqrt{x}\tan\sqrt{x}}{\sqrt{x}}.$

   Let $u = \tan\sqrt{x}$. Then $du = \sec^2\sqrt{x}\cdot\frac{1}{2\sqrt{x}}\,dx \implies \frac{\sec^2\sqrt{x}}{\sqrt{x}}\,dx = 2\,du.$

   $\int \frac{\sec^2\sqrt{x}\tan\sqrt{x}}{\sqrt{x}}\,dx = \int u(2\,du) = u^2 + C = \tan^2\sqrt{x} + C.$

   $y = \tan^2\sqrt{x} + C.$

5. $\frac{dy}{dt} = 3ty^2.$

   $\int \frac{1}{y^2}dy = \int 3t\,dt \implies -\frac{1}{y} = \frac{3}{2}t^2 + C_1$

   $\implies y = \frac{-1}{\frac{3}{2}t^2+C_1} = -\frac{2}{3t^2+C}$ where $C = 2C_1$.

7. $\frac{dy}{dt} = (1+t)(2+y).$

   $\int \frac{1}{2+y}\,dy = \int(1+t)dt \implies \ln|2+y| = t + \frac{1}{2}t^2 + C_1$

   $\implies 2 + y = C\cdot e^{\frac{1}{2}t^2+t} \implies y = Ce^{\frac{1}{2}t^2+t} - 2.$

9. $y' + 2y = 4.$

   $\frac{dy}{dt} = -2(y-2) \implies \int \frac{1}{y-2}\,dy = \int(-2)dt \implies \ln|y-2| = -2t + C_1$

   $\implies y - 2 = Ce^{-2t} \implies y = 2 + Ce^{-2t}.$

11. $y' + y\cos t = 0.$

    $y' = -y\cos t \implies \int \frac{1}{y}\,dy = \int(-\cos t)dt$

    $\implies \ln|y| = -\sin t + C_1 \implies y = Ce^{-\sin t}.$

13. $(t-1)y' = t+1 \implies dy = \frac{t+1}{t-1}\,dt \implies dy = \frac{(t-1)+2}{t-1}\,dt \implies dy = \left(1 + \frac{2}{t-1}\right)dt.$

    $y = t + 2\ln|t-1| + C.$

15. $\frac{dy}{dx} = y(2-y)$.

$\int \frac{1}{y(2-y)}\,dy = \int\,dx \implies \frac{1}{2}\int\left(\frac{1}{y}+\frac{1}{2-y}\right)dy = x + C_1$

$\implies \ln|y| - \ln|2-y| = 2x + C_2 \implies \ln\left|\frac{y}{2-y}\right| = 2x + C_2$

$\implies \frac{y}{2-y} = C_3e^{2x} \implies y = 2C_3e^{2x} - yC_3e^{2x} \implies y = \frac{2C_3e^{2x}}{1+C_3e^{2x}}$

$= \frac{2}{\frac{1}{C_3}e^{-2x}+1} = \frac{2}{1+Ce^{-2x}}$.

17. $\frac{dy}{dt} = \frac{t+3}{y} \implies y\,dy = (t+3)dt$.

$\frac{1}{2}y^2 = \frac{1}{2}t^2 + 3t + C_1 \implies y^2 = t^2 + 6t + C$.

$y = \pm\sqrt{t^2+6t+C}$.

19. $y'' - 5y' - 6y = 0$.

The characteristic equation is

$(r^2 - 5r - 6) = 0 \implies (r+1)(r-6) = 0 \implies r_1 = -1,\ r_2 = 6$.

General solution is $y = Ae^{-t} + Be^{6t}$.

21. $\frac{d^2y}{dx^2} - 5\frac{dy}{dx} - 14y = 0$.

$r^2 - 5r - 14 = 0 \implies (r-7)(r+2) = 0 \implies r_1 = 7,\ r_2 = -2$.

General solution is $y = Ae^{7x} + Be^{-2x}$.

23. $y'' - 8y' + 16y = 0$.

The characteristic equation is

$(r^2 - 8r + 16) = 0 \implies (r-4)^2 = 0 \implies r = 4$ is a double root.

General solution is $y = Ae^{4t} + Bte^{4t}$.

25. $\frac{dy}{dx} = \frac{x}{\sqrt{1+x^2}}, \quad y(0) = 3$.

$\int dy = \int \frac{x}{\sqrt{1+x^2}}\,dx \implies y = \frac{1}{2}\int\frac{1}{\sqrt{u}}\,du$ where $u = 1 + x^2$

$\implies y = \sqrt{u} + C = \sqrt{1+x^2} + C$.

$y(0) = 3 \implies 3 = 1 + C \implies C = 2$.

The solution is $y = 2 + \sqrt{1+x^2}$.

27. $y' = 4y, \quad y(0) = 3$.

$\frac{dy}{y} = 4\,dt \implies \ln|y| = 4t + C_1 \implies y = Ce^{4t}$.

$y(0) = 3 \implies 3 = Ce^{4(0)} \implies C = 3$.

The solution is $y = 3e^{4t}$.

29. $\frac{dy}{dx} = y\sin x, \quad y(0) = \pi.$

$\int \frac{dy}{y} = \int \sin x \, dx \implies \ln|y| = -\cos x + C_1 \implies y = Ce^{-\cos x}.$

$y(0) = \pi \implies \pi = Ce^{-\cos 0} \implies \pi = C \cdot e^{-1} \implies C = \pi e.$

So the solution is $y = \pi e \cdot e^{-\cos x} = \pi e^{1-\cos x}.$

31. $y' = 10(5-t), \quad y(0) = 2.$

$y = 10\int(5-t)dt = 10\left(5t - \frac{1}{2}t^2\right) + C.$ $y(0) = 2 \implies 2 = C.$

So the solution is $y = 50t - 5t^2 + 2.$

33. $\frac{d^2y}{dx^2} + 16y = 0, \quad y(0) = 5, \quad y'(0) = 0.$

$r^2 + 16 = 0 \implies r = \pm 4i \implies y = A\sin 4x + B\cos 4x.$

$y(0) = 5 \implies B = 5; \quad y'(0) = 0 \implies 4A = 0 \implies A = 0.$

So the solution is $y = 5\cos 4x.$

35. $y'' - 6y' + 9y = 0, \quad y(0) = 0, \quad y'(0) = 6.$

$r^2 - 6r + 9 = 0 \implies (r-3)^2 = 0 \implies r = 3$ is a double root.

So $y = Ae^{3t} + Bte^{3t}.$ $\quad y(0) = 0 \implies A = 0.$

$y' = (Bte^{3t})' = B(e^{3t} + 3te^{3t}). \quad y'(0) = 6 \implies B = 6.$

So the solution is $y = 6te^{3t}.$

37. $\frac{dy}{dt} = 2y(4-y).$

(1) $y = 0$ and $y = 4$ are the constant solutions

(2)

Range of y	Sign of y'	Conclusion
$y < 0$	negative	decreasing
$0 < y < 4$	positive	increasing
$y > 4$	negative	decreasing

(3) $y'' = 2[4 - y - y]\,y' = 8(2-y)y(4-y).$

Range of y	Sign of y''	Concavity
$y < 0$	negative	down
$0 < y < 2$	positive	up
$2 < y < 4$	negative	down
$y > 4$	positive	up

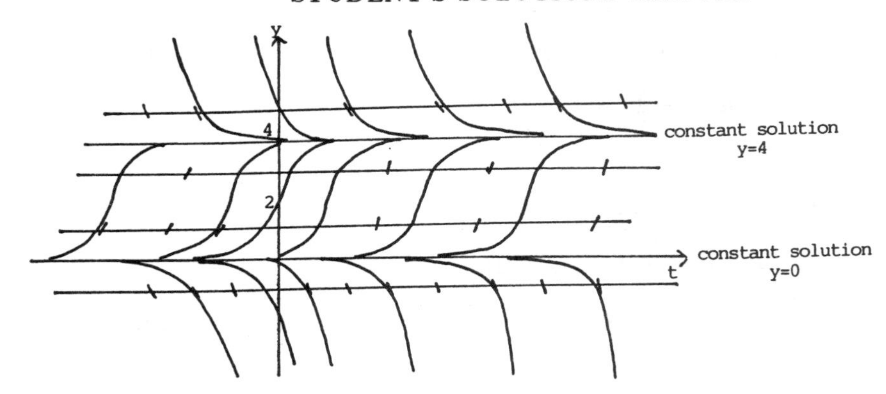

39. $\frac{dy}{dt} = y + 3$.

(1) $y = -3$ is the constant solution.

(2) For $y > -3$, $y' > 0$, so y increases.

For $y < -3$, $y' < 0$, so y decreases.

(3) $y'' = y' = y + 3$.

For $y > -3$, $y'' > 0$, so the graph of y is concave up.

For $y < -3$, $y'' < 0$, so the graph of y is concave down.

(4)

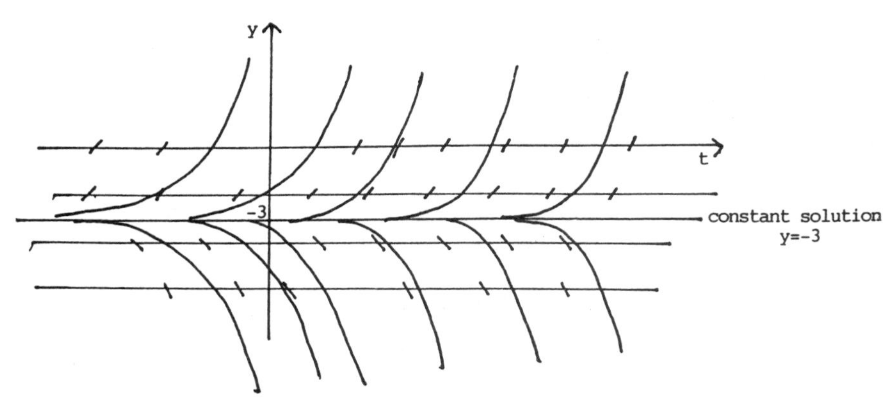

41. $V'(t) = \sqrt{36 + t}$. $V(0) = 200,000$.

$V(t) = \int \sqrt{36+t}\, dt = \frac{2}{3}(36+t)^{3/2} + C$.

$V(0) = 200,000 \implies \frac{2}{3}36^{3/2} + C = 200,000$

$\implies 144 + C = 200,000 \implies C = 199,856$.

So $V(t) = \frac{2}{3}(t+36)^{3/2} + 199,856$ dollars.

43. (a) $y = Ce^{2t}$. $y' = 2Ce^{2t} = 2y$, so $y' = 2y$.

(b) $y = C_1e^t + C_2e^{-t}$.

$y' = C_1e^t - C_2e^{-t}$, $y'' = C_1e^t + C_2e^{-t} = y$. So $y'' - y = 0$.

45. $\frac{du}{dt} = -\frac{u}{5}(1)$, $u(0) = 5$. Here $u =$ the number of gallons of antifreeze.

$\int \frac{du}{u} = \int \left(-\frac{dt}{5}\right) \implies \ln u = -\frac{1}{5}t + C_1$.

$u = Ce^{-t/5}$.

$u(0) = 5 \implies C = 5$. So $u = 5e^{-t/5}$.

The concentration of antifreeze after t minutes is $e^{-t/5}$.

47. Denote by $y(t)$ the temperature of the drink t minutes after removing it. Thus

$y'(t) = K(80 - y(t))$ with $y(0) = 40°\text{F}$.

$\int \frac{dy}{80-y} = Kt + C_1 \implies -\ln(80 - y) = Kt + C_1$.

$80 - y = Ce^{-Kt} \implies y = 80 - Ce^{-Kt}$.

$y(0) = 40 \implies 40 = 80 - C \implies C = 40$.

So $y(t) = 40(2 - e^{-Kt})$.

From $y(5) = 50 \implies 50 = 40(2 - e^{-5K})$

$\implies e^{-5K} = 2 - \frac{5}{4} = \frac{3}{4} \implies e^{-K} = \left(\frac{3}{4}\right)^{1/5}$.

Thus $y(t) = 40(2 - e^{-Kt}) = 40\left[2 - \left(\frac{3}{4}\right)^{(1/5)t}\right]$.

So $y(10) = 40\left[2 - \left(\frac{3}{4}\right)^{\frac{1}{5}(10)}\right] = 40\left[2 - \frac{9}{16}\right] = 57.5°\text{F}$.

49. $\frac{du}{ds} = \frac{1}{s+1}$, $u(0) = 5$.

$u(s) = \ln(s + 1) + C$.

$u(0) = 5 \implies 5 = \ln 1 + C \implies C = 5$.

$u(s) = \ln(s + 1) + 5$.

Practice Problems for Chapter 10

1. Find the general solution of each of the following differential equations.

   a. $\frac{dy}{dt} = \frac{t-1}{t(t-2)}$ b. $2xy\frac{dy}{dx} = 3(y^2+1)$

   c. $(3x+1) + e^{x+y}y' = 0$ d. $\frac{dy}{dx} = -\frac{x}{y^2 \sec^2 x}$

   e. $y^3 + \sqrt{1+x^2}y' = 0$ f. $xy + \ln y \cdot y' = 0$

   g. $y'' - 2y' - 8y = 0$ h. $y'' + y' + y = 0$

   i. $y'' + y' - y = 0$ j. $y'' = 4y' - 4y$

2. Find the solution for each of the following initial value problems.

   a. $\frac{dy}{dx} = x^2 \cdot \sqrt{1-x^3}$, $y(0) = -1$ b. $(x+1)y' + x^2y = 0$, $y(1) = 1$

   c. $(\sin x^2)y' = xy\cos x^2$, $y\left(\sqrt{\frac{\pi}{6}}\right) = 3$ d. $x(3-x)y' - y = 0$, $y(1) = 2$

   e. $\frac{d^2y}{dx^2} + 4y = 0$, $y(0) = 1$, $y'(0) = -1$ f. $y'' - 6y' + 5y = 0$, $y(0) = 3$, $y'(0) = 7$

3. Sketch the direction field and several solution curves for each of the following differential equations from the qualitative information obtained from the equation itself.

   a. $\frac{dy}{dx} = y(3-2y)$ b. $\frac{dy}{dx} = y^2 - 1$

   c. $\frac{dy}{dx} = x + y$ d. $\frac{dy}{dx} = y - 2$

4. Suppose that the acceleration of a particle moving on the x-axis is given by

$$\frac{d^2x}{dt^2} = 12t - 18,$$

and that when $t = 0$, $\frac{dx}{dt} = 12$ and $x = 3$. Find the motion and the total distance traveled by the particle during the first three units of time.

5. The angular velocity $\frac{d\theta}{dt}$ of a particle moving on a circular path is inversely proportional (with proportionality factor k) to the position angle θ. Find θ if $\theta = 3$ when $t = 0$. As t increases, does the particle approach a limiting position on the circle?

6. An object initially at 100°F is placed in air at 60°F. If the temperature of the object drops to 90°F in the first 20 minutes of cooling, find the time required to reach a. 80°F b. 61°F.

7. How long will it take an initial sum of $1000 at 5 percent interest compounded continuously to increase to $3000 if additional deposits of $500 per year are made in a continual manner?

8. The population of the United States grows in according to the logistic growth law

$$\frac{dP}{dt} = kP\left(\frac{M-P}{M}\right).$$

   a. Find $P(t)$ given that: in 1790, $P = 4 \cdot 10^6$; in 1850, $P = 23 \cdot 10^6$; in 1910, $P = 92 \cdot 10^6$.

   b. Find the population for 1960.

9. Suppose that a tank contains water contaminated by some substance thoroughly mixed in the water. In order to reduce the concentration of the contaminant, pure water is poured into the tank and the mixture spills out and is drained away as fast as the pure water enters. Assume that the fluid is always thoroughly mixed. If the original mixture contained 10 pounds of the contaminant in 1000 gallons of water and if pure water enters the tank at the rate of 6 gallons per minute, how long will it take to reduce the amount of contaminant present

   a. to 5 pounds? b. to 1 pound?

10. In problem 9, use Euler's method to approximate $P(1)$, $P(2)$, and $P(3)$ where $P(t)$ is the amount of contaminant present in the fluid in the tank t minutes after mixing begins.

11. Find the general solution of the differential equation

$$\frac{dy}{dt} = y \cdot (2 - y^2).$$

Find and sketch the particular solution such that $y(0) = 0.2$.

12. Find the general solution of the differential equation

$$\frac{dy}{dt} = (y - 1) \cdot (y - 4).$$

Find and sketch the particular solution such that $y(0) = 2$.

13. Suppose you put A_0 dollars into an account on December 1, 1990. Assume a constant annual interest rate of 6 percent with continuous compounding. Beginning on December 1, 1995 you withdraw money continuously from the account at the rate of \$1,000 per year. Determine A_0 so that the account never runs out.

14. Find the general solution of the predator-prey system

$$\begin{cases} \frac{dx}{dt} = -2.5 \cdot (y - 4.375) \\ \frac{dy}{dt} = .875 \cdot (x - 6.25). \end{cases}$$

Find the particular solution of this system such that $x(0) = 5.75$ and $y(0) = 5.5$, and sketch the trajectory of this solution which is the graph in the (x, y)-plane with x and y obtained here as functions of t.

15. Solve Problem 25 in Exercise Set 5.7 by solving a differential equation and an initial condition involving the number of dollars $A(t)$ the state has in an account after t years for paying off Mrs. Brown's winnings.

Solutions to Practice Problems for Chapter 10

1. a. $\frac{dy}{dt} = \frac{t-1}{t(t-2)}$.

$\int dy = \int \frac{t-1}{t(t-2)}\,dt \implies y = \int \frac{t-1}{t^2-2t}\,dt.$

Let $u = t^2 - 2t$. Then $du = (2t-2)dt = 2(t-1)dt$.

So $y = \int \frac{1}{u} \cdot \frac{1}{2}\,du = \frac{1}{2}\ln|u| + C = \frac{1}{2}\ln|t(t-2)| + C$.

b. $2xy\frac{dy}{dx} = 3(y^2+1)$.

$\frac{y}{y^2+1}\,dy = \frac{3}{2x}\,dx \implies \int \frac{y}{y^2+1}\,dy = \int \frac{3}{2x}\,dx$

$\implies \int \frac{1}{2} \cdot \frac{1}{u}\,du = \frac{3}{2}\ln|x| + C_1$ where $u = y^2+1$

$\implies \frac{1}{2}\ln|u| + C_2 = \frac{3}{2}\ln|x| + C_1$

$\implies \ln\sqrt{|u|} = \ln\sqrt{|x|^3} + C_3$

$\implies \sqrt{|u|} = C_4\sqrt{|x|^3} \implies u = Cx^3$

$\implies y^2 + 1 = Cx^3$.

c. $(3x+1) + e^{x+y}y' = 0$.

$e^x \cdot e^y\frac{dy}{dx} = -(3x+1) \implies \int e^y dy = -\int \frac{(3x+1)}{e^x}\,dx$.

$e^y = -\int (3x+1)e^{-x}dx$

Let $u = 3x+1$, $dv = e^{-x}dx$. Then $du = 3dx$, $v = -e^{-x}$.

Then $e^y = -\left[-(3x+1)e^{-x} + \int 3e^{-x}dx\right]$

$= (3x+1)e^{-x} + 3e^{-x} + C = e^{-x}(3x+4) + C$.

So $y = \ln\left[e^{-x}(3x+4) + C\right]$.

d. $\frac{dy}{dx} = -\frac{x}{y^2\sec^2 x}$.

$\int y^2 dy = -\int \frac{x}{\sec^2 x}\,dx$.

$\frac{1}{3}y^3 = -\int x\cos^2 x\,dx$.

Let $u = x$, $dv = \cos^2 x\,dx$.

Then $du = dx$, $v = \int \cos^2 x\,dx = \int \frac{1+\cos 2x}{2}\,dx$

$= \int \left(\frac{1}{2} + \frac{1}{2}\cos 2x\right)dx = \frac{1}{2}x + \frac{1}{4}\sin 2x$.

$$\text{So } \frac{1}{3}y^3 = -\left[x\left(\frac{1}{2}x + \frac{1}{4}\sin 2x\right) - \int\left(\frac{1}{2}x + \frac{1}{4}\sin 2x\right)dx\right]$$
$$= -\frac{1}{2}x^2 - \frac{1}{4}x\sin 2x + \frac{1}{4}x^2 - \frac{1}{8}\cos 2x + C_1$$
$$= -\frac{1}{4}x^2 - \frac{1}{4}x\sin 2x - \frac{1}{8}\cos 2x + C_1.$$

$$y^3 = -\frac{3}{8}(2x^2 + 2x\sin 2x + \cos 2x + C).$$
$$y = -\frac{\sqrt[3]{3}}{2}(2x^2 + 2x\sin 2x + \cos 2x + C)^{1/3}.$$

e. $y^3 + \sqrt{1+x^2}y' = 0.$

$\sqrt{1+x^2}\,\frac{dy}{dx} = -y^3 \implies y^{-3}dy = -\frac{1}{\sqrt{1+x^2}}\,dx.$

$\int y^{-3}dy = -\int \frac{1}{\sqrt{1+x^2}}\,dx.$

From the table of integrals we have then

$$-\frac{1}{2}y^{-2} = -\ln\left|x + \sqrt{1+x^2}\right| + C_1$$
$$y^{-2} = 2\ln\left|x + \sqrt{1+x^2}\right| + C$$
$$y^{-2} = \ln\left(x + \sqrt{1+x^2}\right)^2 + C$$
$$y = \pm\left(\ln\left(x + \sqrt{1+x^2}\right)^2 + C\right)^{-1/2}$$

f. $xy + \ln y \cdot y' = 0.$

$\ln y \cdot \frac{dy}{dx} = -xy \implies \int \frac{\ln y}{y}\,dy = -\int x\,dx.$

$\int u\,du = -\frac{1}{2}x^2 + C_1$ where $u = \ln y$.

$\frac{1}{2}u^2 = -\frac{1}{2}x^2 + C_1 \implies (\ln y)^2 = -x^2 + C$

$\implies \ln y = \pm\sqrt{C - x^2}$

$\implies y = e^{\sqrt{C-x^2}}$ or $y = e^{-\sqrt{C-x^2}}$.

g. $y'' - 2y' - 8y = 0.$

Here the characteristic polynomial is $r^2 - 2r - 8 = 0$.

$$(r-4)(r+2) = 0 \implies r = 4, -2.$$

So $y = Ae^{4t} + Be^{-2t}$.

h. $y'' + y' + y = 0$.

Here the characteristic polynomial is $r^2 + r + 1 = 0$.

$$r = \frac{-1 \pm \sqrt{1-4}}{2} \implies r_1 = -\frac{1}{2} + \frac{\sqrt{3}}{2}i, \quad r_2 = -\frac{1}{2} - \frac{\sqrt{3}}{2}i.$$

The general solution of the differential equation is then

$$\begin{aligned} y &= Ae^{-\frac{1}{2}t} \sin \frac{\sqrt{3}}{2}t + Be^{-\frac{1}{2}t} \cos \frac{\sqrt{3}}{2}t \\ &= e^{-\frac{1}{2}t} \left(A \sin \frac{\sqrt{3}}{2}t + B \cos \frac{\sqrt{3}}{2} t \right). \end{aligned}$$

i. $y'' + y' - y = 0$.

Here the characteristic polynomial is $r^2 + r - 1 = 0$.

$$r_{1,2} = \frac{-1 \pm \sqrt{1+4}}{2} = -\frac{1}{2} \pm \frac{\sqrt{5}}{2}.$$

So the differential equation has the general solution

$$y = Ae^{\left(-\frac{1}{2}+\frac{\sqrt{5}}{2}\right)t} + Be^{\left(-\frac{1}{2}-\frac{\sqrt{5}}{2}\right)t}.$$

j. $y'' = 4y' - 4y$.

$y'' - 4y' + 4y = 0$.

Set the characteristic polynomial $r^2 - 4r + 4 = 0$.

$$(r-2)^2 = 0 \implies r_1 = r_2 = 2.$$

The general solution of the differential equation is then

$$y = Ae^{2t} + Bte^{2t}.$$

2. a. $\frac{dy}{dx} = x^2 \cdot \sqrt{1-x^3}$, $y(0) = -1$.

$y = \int x^2\sqrt{1-x^3}\ dx = -\frac{1}{3} \cdot \frac{2}{3}(1-x^3)^{3/2} + C$
$= -\frac{2}{9}(1-x^3)^{3/2} + C.$
$y(0) = -1 \implies -\frac{2}{9} + C = -1 \implies C = -1 + \frac{2}{9} = -\frac{7}{9}.$
The solution of the initial value problem is then

$$y = -\frac{2}{9}\left(1-x^3\right)^{3/2} - \frac{7}{9}.$$

b. $(x+1)y' + x^2 y = 0,\ \ y(1) = 1.$
$\frac{dy}{dx} = -\frac{x^2 y}{x+1} \implies \int \frac{1}{y}\, dy = -\int \frac{x^2}{x+1}\, dx.$
$\ln|y| = -\int \frac{x^2-1+1}{x+1}\, dx$
$= -\int \left(\frac{(x^2-1)}{x+1} + \frac{1}{x+1}\right) dx$
$= -\int \left(\frac{(x+1)(x-1)}{(x+1)} + \frac{1}{x+1}\right) dx$
$= -\int (x-1)dx - \int \frac{1}{x+1}dx$
$= -\frac{1}{2}x^2 + x - \ln|x+1| + C_1.$
$y = Ce^{\left[-\frac{1}{2}x^2+x-\ln|x+1|\right]} = Ce^{\left(-\frac{1}{2}x^2+x\right)} \cdot e^{\ln\left(\frac{1}{x+1}\right)}$
$= Ce^{\left(-\frac{1}{2}x^2+x\right)} \cdot \frac{1}{x+1}.$
$y(1) = 1 \implies 1 = Ce^{\frac{1}{2}} \cdot \frac{1}{2} \implies C = 2e^{-\frac{1}{2}}.$
$y = 2e^{-\frac{1}{2}} \cdot e^{\left(-\frac{1}{2}x^2+x\right)} \cdot \frac{1}{x+1} = \frac{2e^{\left(-\frac{1}{2}x^2+x-\frac{1}{2}\right)}}{x+1}.$

c. $(\sin x^2)y' = xy\cos x^2$, $y\left(\sqrt{\frac{\pi}{6}}\right) = 3.$
$\int \frac{dy}{y} = \int x\ \frac{\cos x^2}{\sin x^2}\ dx.$
$\ln|y| = \int x\cot x^2 dx$
$= \frac{1}{2}\int \cot x^2 d(x^2)$
$= \frac{1}{2}\ln|\sin x^2| + c_1 = \ln\sqrt{|\sin x^2|} + C_1.$

$y = C \cdot \sin x^2.$
$y\left(\sqrt{\frac{\pi}{6}}\right) = 3 \implies C\sin\frac{\pi}{6} = 3 \implies C \cdot \frac{1}{2} = 3 \implies C = 6.$
So $y = 6\sin x^2$.

d. $x(3-x)y' - y = 0$, $y(1) = 2$.

$\int \frac{dy}{y} = \int \frac{dx}{x(3-x)}$.

$\ln|y| = \int \frac{1}{3} \cdot \left(\frac{1}{x} + \frac{1}{3-x}\right) dx$

$= \frac{1}{3}\left(\ln|x| - \ln|3-x|\right) + C_1 = \ln\left|\frac{x}{3-x}\right|^{1/3} + C_1$.

$y = C\sqrt[3]{\frac{x}{3-x}}$.

$y(1) = 2 \implies C \cdot \sqrt[3]{\frac{1}{2}} = 2 \implies C = 2 \cdot \sqrt[3]{2}$.

So $y = 2 \cdot \sqrt[3]{2} \cdot \sqrt[3]{\frac{x}{3-x}} = 2 \cdot \sqrt[3]{\frac{2x}{3-x}}$.

e. $\frac{d^2y}{dx^2} + 4y = 0$, $y(0) = 1$, $y'(0) = -1$.

From $r^2 + 4 = 0 \implies r = \pm 2i$.

The general solution is

$$y = A\cos 2x + B\sin 2x.$$

From $y(0) = 1$, $A = 1$. $y' = -2A\sin 2x + 2B\cos 2x$.

From $y'(0) = -1$, $2B = -1 \implies B = -\frac{1}{2}$.

Hence $y = \cos 2x - \frac{1}{2}\sin 2x$.

f. $y'' - 6y' + 5y = 0$, $y(0) = 3$, $y'(0) = 7$.

From $r^2 - 6r + 5 = 0$, $r = 1$ or 5.

The general solution is

$$y = Ae^t + Be^{5t} \implies y' = Ae^t + 5Be^{5t}.$$

From $y(0) = 3$ and $y'(0) = 7$,

$$A + B = 3 \text{ and } A + 5B = 7$$
$$\implies A = 2,\ B = 1.$$

Hence $y = 2e^t + e^{5t}$.

3. a. $\frac{dy}{dx} = y(3-2y)$. $y = 0$ and $y = \frac{3}{2}$ are constant solutions.

$$\frac{d^2y}{dx^2} = y'(3-2y) + y(-2y') = y'(3-4y)$$
$$= (3-4y)(3-2y)y.$$

Sign of $\frac{dy}{dx}$	
$y > \frac{3}{2}$	$-$
$y = \frac{3}{2}$	0
$0 < y < \frac{3}{2}$	$+$
$y = 0$	0
$y < 0$	$-$

Sign of $\frac{d^2y}{dx^2}$	
$y > \frac{3}{2}$	$+$
$y = \frac{3}{2}$	0
$\frac{3}{4} < y < \frac{3}{2}$	$-$
$y = \frac{3}{4}$	0
$0 < y < \frac{3}{4}$	$+$
$y = 0$	0
$y < 0$	$-$

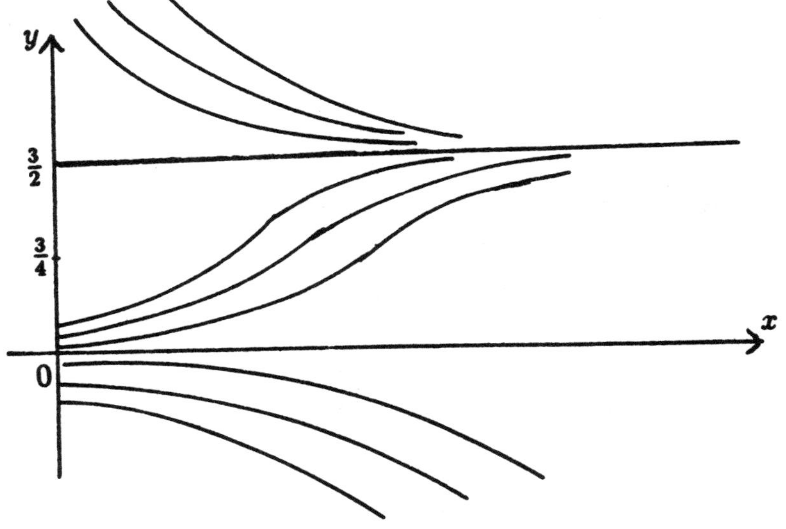

b. $\frac{dy}{dx} = y^2 - 1 = (y+1)(y-1)$. $y = 1$ and $y = -1$ are constant solutions.

$$\frac{d^2y}{dx^2} = 2y \cdot y' = 2y(y+1)(y-1).$$

Sign of $\frac{dy}{dx}$	
$y > 1$	$+$
$y = 1$	0
$-1 < y < 1$	$-$
$y = -1$	0
$y < -1$	$+$

Sign of $\frac{d^2y}{dx^2}$	
$y > 1$	$+$
$y = 1$	0
$0 < y < 1$	$-$
$y = 0$	0
$-1 < y < 0$	$+$
$y = -1$	0
$y < -1$	$-$

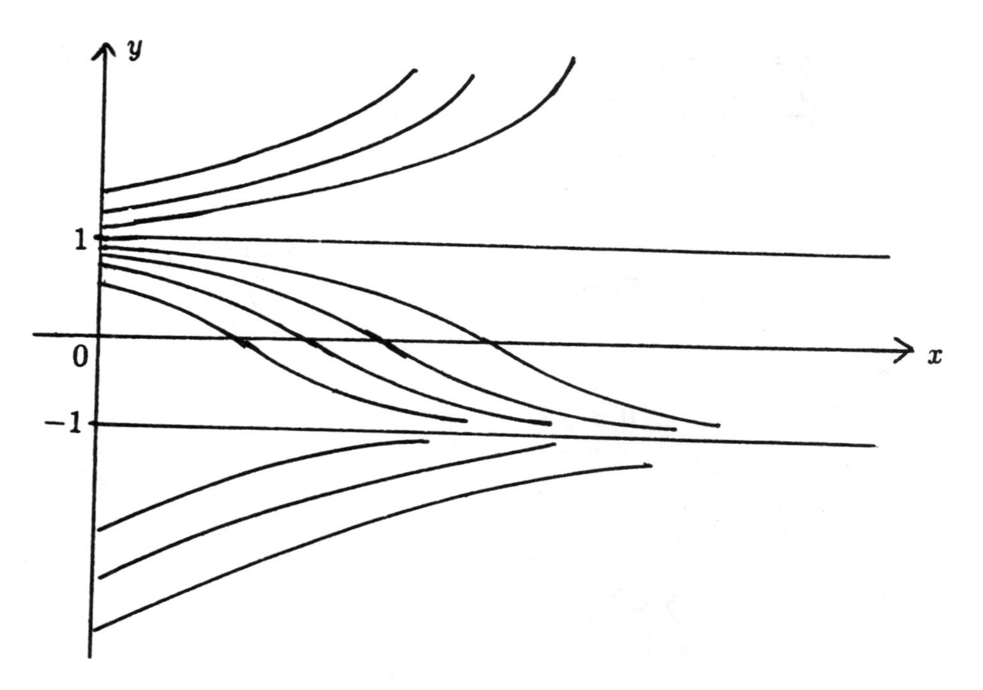

c. $\frac{dy}{dx} = x + y$. $\frac{d^2y}{dx^2} = x + y + 1$.

$\frac{dy}{dx} = 0$ for $y = -x$.

$\frac{d^2y}{dx^2} = 0$ for $y = -x - 1$.

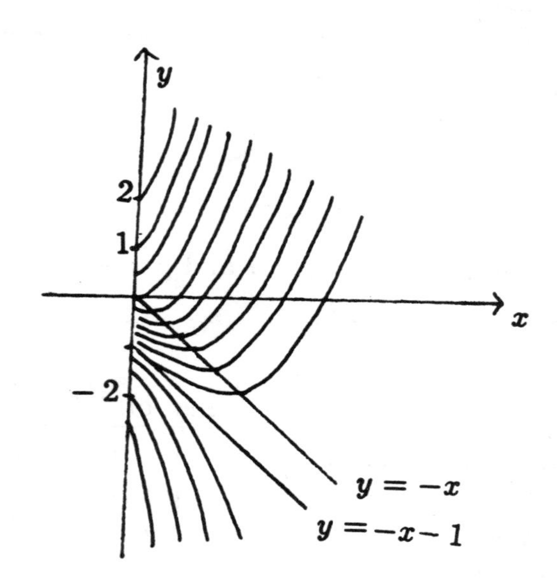

d. $\frac{dy}{dx} = y - 2$. $y = 2$ is a constant solution and $\frac{d^2y}{dx^2} = y - 2$.

Sign of $\frac{dy}{dx}$	
$y > 2$	$+$
$y = 2$	0
$y < 2$	$-$

Sign of $\frac{d^2y}{dx^2}$	
$y > 2$	$+$
$y = 2$	0
$y < -2$	$-$

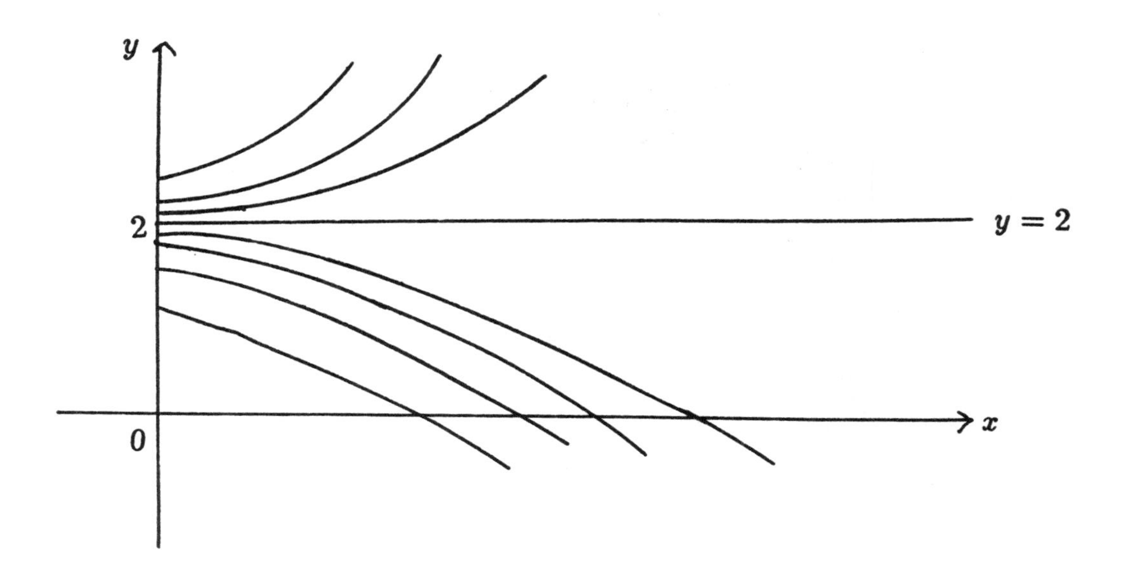

4. $\frac{d^2x}{dt^2} = 12t - 18,$

$\frac{dx}{dt} = 6t^2 - 18t + C_1 \cdot \left.\frac{dx}{dt}\right]_{t=0} = 12 \implies C_1 = 12.$

$x = 2t^3 - 9t^2 + 12t + C_2 \cdot x]_{t=0} = 3 \implies C_2 = 3.$

$x(t) = 2t^3 - 9t^2 + 12t + 3.$

$\frac{dx}{dt} = 6(t^2 - 3t + 2) = 6(t-1)(t-2).$

$\frac{dx}{dt} = 0$ for $t = 1, \quad t = 2.$

Distance traveled $= |x(1) - x(0)| + |x(2) - x(1)| + |x(3) - x(2)|$

$= |8 - 3| + |7 - 8| + |12 - 7| = 5 + 1 + 5 = 11.$

5. $\frac{d\theta}{dt} = \frac{k}{\theta} \implies \theta\, d\theta = kdt \implies \frac{\theta^2}{2} = kt + C.$

$\therefore \theta^2 = 2kt + C.$

Since $\theta(0) = 3$, $C = 9$. Hence $\theta^2 = 2kt + 9$.

$\lim_{t\to\infty} \theta(t) = \lim_{t\to\infty} \sqrt{2kt + 9} = \infty.$

6. $\frac{dy}{dt} = k(60 - y)$

$\Longrightarrow \; y(t) = 60(1 - ce^{-kt}).$

Since $y(0) = 100$ and $y(20) = 90$, one obtains

$$y(0) = 60(1 - C) = 100 \Longrightarrow C = -\frac{2}{3},$$

$$y(20) = 60(1 - Ce^{-20k}) = 90 \Longrightarrow 1 + \frac{2}{3}e^{-20k} = \frac{3}{2}$$

$$\Longrightarrow k = \frac{1}{20} \ln \frac{4}{3}.$$

Therefore, $y(t) = 60\left(1 + \frac{2}{3}e^{\frac{1}{20}(\ln \frac{3}{4})t}\right).$

a. $80 = 60\left(1 + \frac{2}{3}e^{\frac{1}{20}(\ln \frac{3}{4})t}\right)$

$\frac{4}{3} = 1 + \frac{2}{3}e^{\frac{1}{20}(\ln \frac{3}{4})t}$

$\frac{1}{3} = \frac{2}{3}e^{\frac{1}{20}(\ln \frac{3}{4})t}$

$\frac{1}{2} = e^{\frac{1}{20}(\ln \frac{3}{4})t} \Longrightarrow -\ln 2 = \frac{1}{20}\left(\ln \frac{3}{4}\right)t$

$-20 \cdot \ln 2 = \left(\ln \frac{3}{4}\right) \cdot t$

$\therefore t = (-20 \cdot \ln 2) \,/\, \left(\ln \frac{3}{4}\right) \doteq 48.19$ minutes.

b. $61 = 60\left(1 + \frac{2}{3}e^{\frac{1}{20}(\ln \frac{3}{4})t}\right)$

$\frac{61}{60} = 1 + \frac{2}{3}e^{\frac{1}{20}(\ln \frac{3}{4})t}$

$\frac{1}{60} = \frac{2}{3}e^{\frac{1}{20}(\ln \frac{3}{4})t}$

$\frac{1}{40} = e^{\frac{1}{20}(\ln \frac{3}{4})t}$

$-\ln 40 = \frac{1}{20}\left(\ln \frac{3}{4}\right)t$

$\therefore t = \frac{20 \cdot \ln 40}{\ln \frac{4}{3}} \doteq 256.46$ minutes.

7. Let V be the value of the investment after t years.

$\frac{dV}{dt} = 0.05V + 500.$

$\frac{dV}{0.05V + 500} = dt.$

$20 \cdot \ln(0.05V + 500) = t + C_1 \Longrightarrow \ln(0.05V + 500) = \frac{t}{20} + C_2$

$0.05V + 500 = C \cdot e^{t/20} \Longrightarrow 0.05V = C \cdot e^{t/20} - 500.$

$\therefore V = C \cdot e^{t/20} - 10000.$

Since $V(0) = 1000$, $C = 11000$. $\therefore V = 11000 \cdot e^{t/20} - 10000.$

$3000 = 11000 \cdot e^{t/20} - 10000 \Longrightarrow 3 = 11 \cdot e^{t/20} - 10 \Longrightarrow$

$11 \cdot e^{t/20} = 13 \Longrightarrow e^{t/20} = \frac{13}{11} \Longrightarrow t = 20 \cdot \ln \frac{13}{11} \doteq 3.34$ years.

8. From $\frac{dP}{dt} = k \cdot P\left(\frac{M-P}{M}\right)$, $P(t) = \frac{M}{1+Ce^{-kt}}$.

a. $4 \cdot 10^6 = \frac{M}{1+C}$ **(1)**

$23 \cdot 10^6 = \frac{M}{1+Ce^{-60k}}$ **(2)**

$92 \cdot 10^6 = \frac{M}{1+Ce^{-120k}}$ **(3)**

Dividing **(3)** by **(2)**: $4 = \frac{1+Ce^{-60k}}{1+Ce^{-120k}}$ **(4)**

Dividing **(2)** by **(1)**: $\frac{23}{4} = \frac{1+C}{1+Ce^{-60k}}$ **(5)**

From **(4)**: $4 + 4Ce^{-120k} = 1 + 1Ce^{-60k}$

$$3 = Ce^{-60k}(1 - 4e^{-60k})$$

$$Ce^{-60k} = \frac{3}{1-4e^{-60k}}.$$

$$\therefore C = \frac{3}{1e^{-60k} - 4e^{-120k}} \quad \textbf{(6)}$$

From **(5)** and **(6)**:

$$\frac{23}{4} = \frac{e^{-60k} - 4e^{-120k} + 3}{e^{-60k} - 4e^{-120k} + 3e^{-60k}}$$

$$\frac{23}{4} = \frac{e^{-60k} - 4e^{-120k} + 3}{4e^{-60k} - 4e^{-120k}}$$

Put $A = e^{-60k}$ and $A \neq 1$.

$$19A^2 - 22A + 3 = 0$$

$$(19A - 3) \cdot (A - 1) = 0$$

$$A = \frac{3}{19}.$$

For $A = \frac{3}{19} \implies \frac{3}{19} = e^{-60k} \implies -60k = \ln\frac{3}{19}$.

$\therefore k = \frac{1}{60}\ln\frac{19}{3}$.

Substitute k into **(5)**:

$$\frac{23}{4} = \frac{1+C}{1+C\cdot\frac{3}{19}} \implies 23 + \frac{69}{19}C = 4 + 4C \implies 19 = \frac{7}{19}C$$

$$\therefore C = \frac{361}{7}.$$

$$4 \cdot 10^6 = \frac{M}{1+C} \implies M = 4 \cdot 10^6 \left(1 + \frac{361}{7}\right) \doteq 210.3 \times 10^6.$$

$$\therefore P(t) = \frac{210.3 \times 10^6}{1 + \frac{361}{7} \cdot e^{\left(-\frac{1}{60} \ln \frac{19}{3} \cdot t\right)}}.$$

b. $P(170) = \frac{210.3 \times 10^6}{1 + \frac{361}{7} \cdot e^{-\frac{1}{60}\left(\ln \frac{19}{3}\right) \cdot 170}} \doteq 164.8 \times 10^6.$

9. Let x = number of pounds of contaminant present after t minutes.

a. $\frac{dx}{dt} = \frac{-x \cdot (6)}{1000}$.

$\frac{dx}{dt} = \frac{-3x}{500} \implies \int \frac{dx}{x} = \int \frac{-3}{500} dt \implies$

$\ln x + C_1 = \frac{-3t}{500} + C_2 \implies x = e^{\frac{-3t}{500}} \cdot C.$

$x(0) = 10 \implies C = 10.$

$x = 10 \cdot e^{\frac{-3t}{500}} \implies$

$\ln x = \ln 10 - \frac{3t}{500} \implies \frac{3t}{500} = \ln 10 - \ln x \implies$

$\frac{3t}{500} = \ln\left(\frac{10}{x}\right) \implies 3t = 500 \ln\left(\frac{10}{x}\right) \implies t = \frac{500 \ln\left(\frac{10}{x}\right)}{3}$

So for 5 pounds,

$t(5) = \frac{500 \ln\left(\frac{10}{5}\right)}{3} = \frac{500 \ln(2)}{3} = 115.5$ min.

b. For 1 pound,

$t(1) = \frac{500 \ln 10}{3} = 383.8$ min.

10. $P'(t) = -\frac{3}{500} \cdot P(t).$

As in Problem 9, let $x = P(t)$.

$\frac{dx}{dt} = -\frac{3}{500} \cdot x.$

Let $f(t, x) = -\frac{3}{500} x.$

$x_{j+1} = x_j + f(t_j, x_j) \cdot 1$ for $j = 0, 1, 2.$

$t_0 = 0,\ t_1 = 1,\ t_2 = 2,\ t_3 = 3.$

$x_0 = 10 = P(0).$

$x_1 = x_0 + f(t_0, x_0) \cdot 1 = 10 - \frac{3}{500} \cdot 10 = 9.94.$

$x_2 = x_1 + f(t_1, x_1) \cdot 1 = 9.94 - \frac{3}{500} \cdot 9.94 = 9.88036.$

$x_3 = x_2 + f(t_2, x_2) \cdot 1 = 9.88036 - \frac{3}{500} \cdot 9.88036 = 9.82107784.$

11. $\frac{dy}{dt} = y \cdot (2 - y^2)$.

$\frac{dy}{y(\sqrt{2}-y)\cdot(\sqrt{2}+y)} = dt.$

$\frac{1}{y(\sqrt{2}-y)\cdot(\sqrt{2}+y)} = \frac{A}{y} + \frac{B}{\sqrt{2}-y} + \frac{D}{\sqrt{2}+y}$

$1 = A\left(\sqrt{2}-y\right)\left(\sqrt{2}+y\right) + By\left(\sqrt{2}+y\right) + Dy\left(\sqrt{2}-y\right).$

$y = 0\colon 1 = A\left(\sqrt{2}\right)\left(\sqrt{2}\right) \implies A = \frac{1}{2}.$

$y = \sqrt{2}\colon 1 = B\sqrt{2}\left(\sqrt{2}+\sqrt{2}\right) \implies B = \frac{1}{4}.$

$y = -\sqrt{2}\colon 1 = D \cdot \left(-\sqrt{2}\right) \cdot \left(\sqrt{2}+\sqrt{2}\right) \implies D = -\frac{1}{4}.$

$\frac{1}{4} \cdot \left(\frac{2}{y} + \frac{1}{\sqrt{2}-y} - \frac{1}{\sqrt{2}+y}\right) dy = dt.$

$\frac{1}{4} \cdot \left(\ln y^2 - \ln\left|\sqrt{2}-y\right| - \ln\left|\sqrt{2}+y\right|\right) = t + C_1.$

$\frac{1}{4} \cdot \left(\ln y^2 - \ln\left|2-y^2\right|\right) = t + C_1.$

$\frac{1}{4} \ln\left|\frac{y^2}{2-y^2}\right| = t + C_1.$

$\ln\left|\frac{y^2}{2-y^2}\right| = 4t + C_2.$

$\frac{y^2}{2-y^2} = C_3 e^{4t}.$

$y^2 = C_3 e^{4t} \cdot (2 - y^2).$

$y^2 = 2C_3 e^{4t} - C_3 e^{4t} y^2.$

$y^2\left(1 + C_3 e^{4t}\right) = 2C_3 e^{4t}.$

$y^2 = \frac{2C_3 e^{4t}}{1+C_3 e^{4t}} = \frac{2}{\frac{1}{C_3} e^{-4t}+1}.$

$y^2 = \frac{2}{1+Ce^{-4t}}$ for $\left(1 + Ce^{-4t}\right) > 0.$

$y = y_1(t) = \frac{\sqrt{2}}{\sqrt{1+Ce^{-4t}}}, \qquad y = y_2(t) = \frac{-\sqrt{2}}{\sqrt{1+Ce^{-4t}}}.$

$y_1(0) = .2$:

$.2 = \frac{\sqrt{2}}{\sqrt{1+C}}.$

$\sqrt{1+C} = \frac{\sqrt{2}}{.2}.$

$1 + C = \frac{2}{.04} \implies 1 + C = 50.$

$C = 49.$

$y = y_1(t) = \frac{\sqrt{2}}{\sqrt{1+49e^{-4t}}}.$

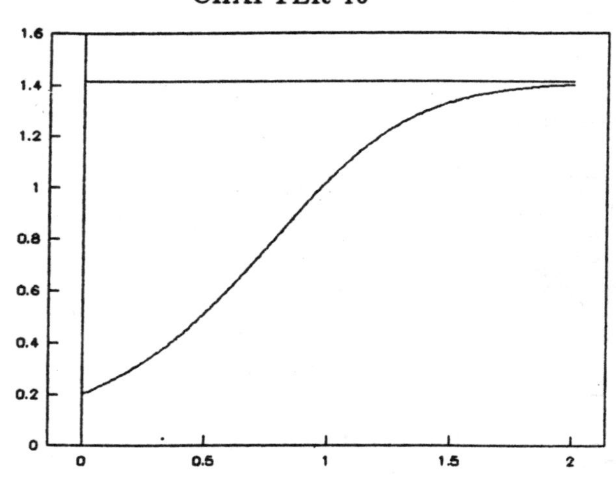

12. $\frac{dy}{dt} = (y-1)\cdot(y-4)$.

$\frac{dy}{(y-1)\cdot(y-4)} = dt$.

$\frac{1}{(y-1)\cdot(y-4)} = \frac{A}{y-1} + \frac{B}{y-4}$.

$1 = A\cdot(y-4) + B\cdot(y-1)$.

$y = 1$: $1 = A\cdot(1-4) \implies A = -\frac{1}{3}$.

$y = 4$: $1 = B\cdot(4-1)\cdot B = \frac{1}{3}$.

$\frac{1}{3}\cdot\left(-\frac{1}{y-1} + \frac{1}{y-4}\right) dy = dt$.

$\frac{1}{3}\cdot(-\ln|y-1| + \ln|y-4|) = t + C_1$.

$\ln\left|\frac{y-4}{y-1}\right| = 3t + C_2$.

$\frac{y-4}{y-1} = C_3e^{3t}$.

$y - 4 = C_3e^{3t}\cdot(y-1)$.

$\left(1 - C_3e^{3t}\right)\cdot y = 4 - C_3e^{3t}$.

$y = \frac{4-C_3e^{3t}}{1-C_3e^{3t}}$.

$y = \frac{1+\frac{4}{-C_3}e^{-3t}}{1+\frac{1}{-C_3}e^{-3t}}$.

$y = \frac{1+4Ce^{-3t}}{1+Ce^{-3t}}$.

$y(0) = 2$:

$2 = \frac{1+4C}{1+C}$.

$2\cdot(1+C) = 1 + 4C$.

$2 + 2C = 1 + 4C$.

$2C = 1.$

$C = \frac{1}{2}.$

$y(t) = \frac{1+2e^{-3t}}{1+\frac{1}{2}e^{-3t}}.$

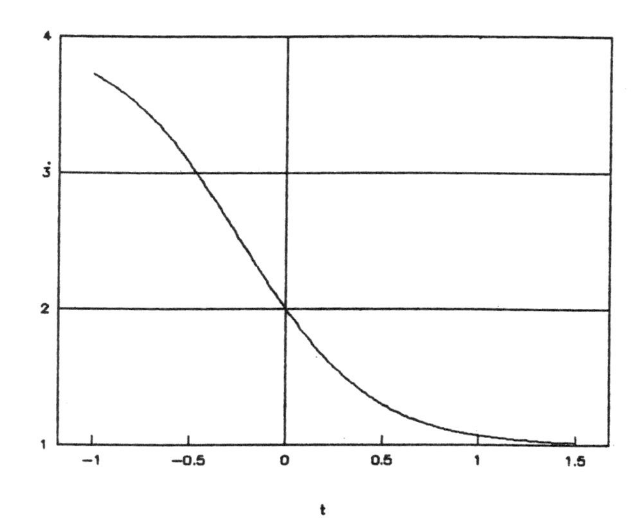

13. Let $V(t)$ denote the value of the account in dollars t years after December 1, 1995. Note that $V(0) = Ae^{.06 \cdot 5} = A_0e^{.3}$. The differential equation here is

$$\frac{dV}{dt} = .06V - 1000.$$

$$\frac{dV}{.06V - 1000} = dt.$$

$$\frac{1}{.06} \cdot \ln(.06V - 1000) = t + C_1.$$

$$\ln(.06V - 1000) = .06t + C_2.$$

$$.06V - 1000 = C_3e^{.06t}.$$

$$.06V = C_3e^{.06t} + 1000.$$

$$V = C_4e^{.06t} + \frac{50,000}{3}.$$

$$V(0) = A_0e^{.3} \implies A_0e^{.3} = C_4 + \frac{50,000}{3}.$$

$$C_4 = A_0e^{.3} - \frac{50,000}{3}.$$

Thus

$$V = A_0e^{.3+.05t} - \frac{50,000}{3} \cdot (e^{.06t} - 1).$$

If $V(T) = 0$, then

$$A_0 e^{.3+.06T} = \frac{50,000}{3} \cdot (e^{.06T} - 1)$$
$$A_0 = \frac{50,000}{3} \cdot \frac{e^{.06T} - 1}{e^{.3+.06T}}$$
$$A_0 = \frac{50,000}{3 \cdot e^{.3}} \cdot \left(1 - e^{-.06T}\right).$$

Note that

$$\lim_{T \to \infty} \frac{50,000}{3 \cdot e^{.3}} \cdot \left(1 - e^{-.06T}\right) = \frac{50,000}{3 \cdot e^{.3}} = \$12,346.97.$$

Therefore you should take $A_0 \geq \$12,346.97$.

14. We use the predator-prey equations (22), (23), (24), (25) in Section 10.4. Here $a = .875$, $b = 2.5$, $\bar{x} = 6.25$, $\bar{y} = 4.375$. The general solution is

$$\begin{cases} x(t) = 6.25 + A \cdot \sin\left(\sqrt{2.1875}\, t\right) + B \cos\left(\sqrt{2.1875}\, t\right) \\ y(t) = 4.375 - \sqrt{.35} \cdot \left[A \cdot \cos\left(\sqrt{2.1875}\, t\right) \right. \\ \qquad \left. - B \cdot \sin\left(\sqrt{2.1875}\, t\right)\right]. \end{cases}$$

If $x(0) = 5.75$ and $y(0) = 5.5$:

$$5.75 = 6.25 + B \implies B = -0.5.$$

$$5.5 = 4.375 - \sqrt{.35}\, A \implies A = -1.9016.$$

$$\begin{cases} x(t) = 6.25 - 1.9016 \sin\left(\sqrt{2.1875}\, t\right) - 0.5 \cos\left(\sqrt{2.1875}\, t\right) \\ y(t) = 4.375 + 1.125 \cos\left(\sqrt{2.1875}\, t\right) - 0.2958 \sin\left(\sqrt{2.1875}\, t\right). \end{cases}$$

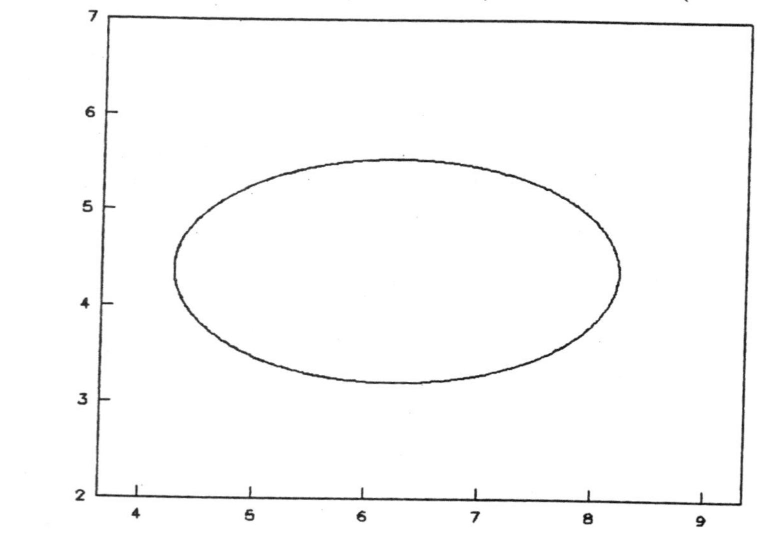

15. Let $A(t)$ denote the number of dollars the state has in an account after t years for paying off Mrs. Brown's winnings. Let k denote the number of dollars per year she will receive in a continuous manner over 20 years.

$$\frac{dA}{dt} = -k + .08A.$$

$$\frac{dA}{-k + .08A} = dt.$$

$$\frac{1}{.08} \cdot \ln(-k + .08A) = t + C_1.$$

$$\ln(-k + .08A) = .08t + C_2.$$

$$-k + .08A = C_3 e^{.08t}.$$

$$.08A = k + C_3 e^{.08t}.$$

$$A = \frac{k}{.08} + Ce^{.08t}.$$

$$A(0) = 1,000,000 \implies 1,000,000 = \frac{k}{.08} + C.$$

$$C = 1,000,000 - \frac{k}{.08}.$$

$$A(t) = \frac{k}{.08} \cdot (1 - e^{.08t}) + 1,000,000e^{.08t}.$$

$A(20) = 0$:

$$\frac{k}{.08} \cdot (e^{1.6} - 1) = 1,000,000e^{1.6}.$$

$$k = 1,000,000 \cdot .08 \cdot \frac{e^{1.6}}{e^{1.6} - 1}.$$

$$k = 1,000,000 \cdot .08 \cdot \frac{1}{1 - e^{-1.6}}.$$

$$k = \$100,237.63.$$

CHAPTER 11

SUMMARY OF CHAPTER 11

1. Outcomes of an experiment, sample space for the experiment, event for the experiment, random variable, discrete random variable, probability distribution for a discrete random variable referred to as a discrete probability distribution.

2. For a discrete probability distribution: rules for probabilities, probability for events, probability histograms, expected value, variance, and standard deviation.

3. For a continuous random variable: probability density function, properties of a probability density function, distribution function, relationship between a probability density function f and the associated distribution function F, expected value, variance, and standard deviation.

4. For the normal random variable X with probability density function

$$f(x) = \frac{1}{\sqrt{2\pi} \cdot \sigma}\, e^{-\frac{1}{2}\left(\frac{x-\mu}{\sigma}\right)^2},$$

   where $\sigma > 0$, referred to as an $N(\mu, \sigma)$ random variable: μ is the expected value (mean), σ is the standard deviation, the vertical line $x = \mu$ is the axis of symmetry for the graph of f, the graph of $y = f(x)$ has inflection points for $x = \mu - \sigma$ and $x = \mu + \sigma$.

5. The standard normal random variable $N(0, 1)$

6. Calculating probabilities for $N(\mu, \sigma)$ random variables by first converting such variables to standard normal variables.

7. For the exponential random variable X with parameter α having probability density function

$$f(x) = \begin{cases} \alpha e^{-\alpha x}, & x \geq 0 \\ 0, & x < 0 \end{cases}$$

where $\alpha > 0$: the expected value is $\frac{1}{\alpha}$, the variance is $\frac{1}{\alpha^2}$.

Selected Solutions to Exercise Set 11.1

1. a. $Pr(E_1) = Pr(1) + Pr(3) = \frac{1}{3} + \frac{1}{6} = \frac{1}{2}$.

   b. $Pr(E_2) = Pr(1) + Pr(4) + Pr(5) = \frac{1}{3} + \frac{1}{12} + \frac{1}{12} = \frac{1}{2}$.

   c. $Pr(E_3) = Pr(2) + Pr(3) + Pr(5) = \frac{1}{3} + \frac{1}{6} + \frac{1}{12} = \frac{7}{12}$.

3.

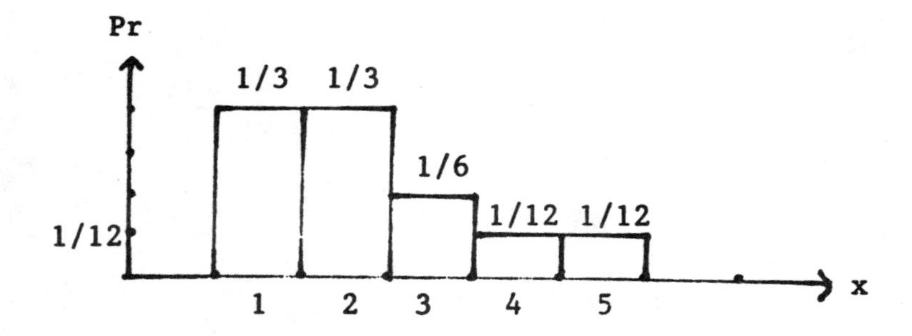

5. a. $Pr(X = 3) + Pr(X = 5) + Pr(X = 9)$

   $= \frac{2}{36} + \frac{4}{36} + \frac{4}{36} = \frac{10}{36} = \frac{5}{18}$.

   b. $Pr(X = 8) + Pr(X = 9) + Pr(X = 10) + Pr(X = 11) + Pr(X = 12)$

   $= \frac{5}{36} + \frac{4}{36} + \frac{3}{36} + \frac{2}{36} + \frac{1}{36} = \frac{15}{36} = \frac{5}{12}$.

   c. $1 - Pr(X = 6) = 1 - \frac{5}{36} = \frac{31}{36}$.

7. a. $Pr(\text{red}) = \frac{3}{5}$, (b) $Pr(\text{black}) = \frac{2}{5}$.

9. a. $\frac{8}{15}$ b. $\frac{7}{15}$ c. $\frac{3}{15} = \frac{1}{5}$.

11. The area of the target is $12^2\pi = 144\pi$ and the area of the bull's eye is $3^2\pi = 9\pi$. So the probability that it hits the bull's eye is $\frac{9\pi}{144\pi} = \frac{1}{16}$.

13. a. $E(X) = x_0p_0 + x_1p_1 + x_2p_2 + x_3p_3$

    $= 0 \cdot 0.2 + 1 \cdot 0.2 + 2 \cdot 0.3 + 3 \cdot 0.3 = 0.2 + 0.6 + 0.9 = 1.7$.

    b. $\text{Var}(X) = 0.2(0 - 1.7)^2 + 0.2(1 - 1.7)^2 + 0.3(2 - 1.7)^2 + 0.3(3 - 1.7)^2$

    $= 0.2 \cdot 2.89 + 0.2 \cdot 0.49 + 0.3 \cdot 0.09 + 0.3 \cdot 1.69$

    $= 0.578 + 0.098 + 0.027 + 0.507 = 1.21$.

    $\sigma = \sqrt{\text{Var}(X)} = 1.1$.

15. $E(X) = 0.2 \cdot 10 + 0.1 \cdot 100 + 0.7 \cdot 0 = 2 + 10 = \$12.$

17. $E(X) = 0.005 \cdot 100{,}000 + 0.025 \cdot 50{,}000 + 0.97 \cdot 0 = 500 + 1250 = \$1750.$

Selected Solutions to Exercise Set 11.2

1. a. continuous b. discrete c. continuous d. continuous

   e. continuous f. discrete g. discrete

3. a. $\int_0^2 f(x)\,dx = \int_0^2 \frac{3}{16}\sqrt{x}\,dx = \frac{3}{16} \cdot \frac{2}{3}x^{3/2}\Big]_0^2 = \frac{1}{8} \cdot 2^{3/2} = \frac{\sqrt{2}}{4}.$

   b. $\int_2^4 f(x)dx = \int_2^4 \frac{3}{16}\sqrt{x}\,dx = \frac{1}{8}x^{3/2}\Big]_2^4$

   $= \frac{1}{8}\left(4^{3/2} - 2^{3/2}\right) = \frac{1}{8}\left(\sqrt{64} - \sqrt{8}\right) = \frac{1}{8}\left(8 - 2\sqrt{2}\right) \approx 0.6464$

   c. $\int_0^4 f(x)dx = \int_0^4 \frac{3}{16}\sqrt{x}\,dx = \frac{1}{8}x^{3/2}\Big]_0^4 = 1.$

5. a. $\int_0^3 f(x)\,dx = \int_0^3 \frac{x}{9}\,dx = \frac{x^2}{18}\Big]_0^3 = \frac{9}{18} = \frac{1}{2}.$

   b. $\int_1^5 f(x)dx = \int_1^3 \frac{x}{9}\,dx + \int_3^5 \left(\frac{2}{3} - \frac{1}{9}x\right)dx$

   $= \frac{x^2}{18}\Big]_1^3 + \left(\frac{2}{3}x - \frac{1}{18}x^2\right)\Big]_3^5$

   $= \frac{1}{2} - \frac{1}{18} + \frac{10}{3} - \frac{25}{18} - 2 + \frac{1}{2}$

   $= -1 + \frac{10}{3} - \frac{26}{18} = \frac{7}{3} - \frac{26}{18} = \frac{16}{18} = \frac{8}{9}.$

   c. $\int_2^6 f(x)dx = \int_2^3 \frac{x}{9}\,dx + \int_3^6 \left(\frac{2}{3} - \frac{1}{9}x\right)dx$

   $= \frac{x^2}{18}\Big]_2^3 + \left(\frac{2}{3}x - \frac{1}{18}x^2\right)\Big]_3^6$

   $= \frac{5}{18} + 2 - \frac{27}{18} = 2 - \frac{22}{18} = \frac{14}{18} = \frac{7}{9}.$

7. $\int_0^{\ln 2} a \cdot e^{-x}\,dx = 1 \implies a\left(-e^{-x}\right)\Big]_0^{\ln 2} = 1$

   $\implies a\left(-\frac{1}{2} + 1\right) = 1 \implies a = 2.$

9. $E(X) = \int_0^4 x \cdot f(x)dx = \int_0^4 \frac{x^2}{8}\, dx = \left.\frac{x^3}{24}\right]_0^4 = \frac{64}{24} = \frac{8}{3}.$

11. $\mathrm{Var}(X) = \int_0^5 (x - E(X))^2 f(x)dx$

$= \int_0^5 (x - 2.5)^2 \cdot \frac{1}{5}\, dx = \frac{25}{12}.$

13. $\mathrm{Var}(X) = \int_0^4 (x-2)^2 \cdot \frac{3}{32}x(4-x)dx$

$= \frac{3}{32}\int_0^4 (x-2)^2(4x - x^2)dx$

$= \frac{3}{32}\left.\left(-\frac{x^5}{5} + 2x^4 - \frac{20}{3}x^3 + 8x^2\right)\right]_0^4$

$= \frac{3}{32}\left(-\frac{1024}{5} + 512 - \frac{1280}{3} + 128\right)$

$= \frac{3}{32}\left(640 - \frac{9472}{15}\right) = \frac{3}{32} \cdot \frac{128}{15} = \frac{4}{5}.$

15. a. $Pr\{0 \le X \le \frac{1}{2}\} = F\left(\frac{1}{2}\right) - F(0) = 3 \cdot \frac{1}{4} - 2 \cdot \frac{1}{8} - 0 = \frac{1}{2}.$

b. $Pr\{\frac{1}{2} \le X \le 1\} = F(1) - F\left(\frac{1}{2}\right) = 1 - \frac{1}{2} = \frac{1}{2}.$

c. $f(x) = F'(x) = 6x - 6x^2.$

17. $F(x) = \int f(x)dx = \int \frac{3}{32}x(4-x)dx$

$= \frac{3}{32}\left(2x^2 - \frac{1}{3}x^3\right) + C = \frac{3}{16}x^2 - \frac{1}{32}x^3 + C.$

Since $F(0) = 0$, $C = 0$. $\therefore F(x) = \frac{3}{16}x^2 - \frac{1}{32}x^3.$

19. a. $\int_0^2 \frac{5}{4}\left(\frac{1}{1+x}\right)^2 dx = \left.\frac{5}{4}\left(-\frac{1}{1+x}\right)\right]_0^2 = \frac{5}{4}\left(-\frac{1}{3} + 1\right) = \frac{5}{6}.$

b. $\int_1^4 \frac{5}{4}\left(\frac{1}{1+x}\right)^2 dx = \left.\frac{5}{4}\left(-\frac{1}{1+x}\right)\right]_1^4 = \frac{3}{8}.$

21. a. $\int_0^\infty \frac{k}{(1+t)^2}dt = 1 \implies k\int_0^\infty (1+t)^{-2}dt = 1 \implies k\left[-(1+t)^{-1}\right]_0^\infty = 1$

$\implies k(0+1) = 1.$ $\therefore k = 1.$

b. $Pr\{0 \le X \le 9\} = \int_0^9 \frac{1}{(1+t)^2}\, dt = \left.-(1+t)^{-1}\right]_0^9 = -\frac{1}{10} + 1 = \frac{9}{10}.$

Selected Solutions to Exercise Set 11.3

1. 0.9772

3. $Pr\{1.2 \leq Z \leq 2.8\} = Pr\{Z \leq 2.8\} - Pr\{Z \leq 1.2\} = 0.9974 - 0.8849 = 0.1125$

5. $Pr\{-1 \leq Z \leq 2\} = Pr\{Z \leq 2\} - Pr\{Z \leq -1\} = Pr\{Z \leq 2\} - Pr\{Z \geq 1\}$
$$= Pr\{Z \leq 2\} - (1 - Pr\{Z \leq 1\}) = 0.9772 - (1 - 0.8413)$$
$$= 0.9772 - 0.1587 = 0.8185$$

7. $Pr\{-1.6 \leq Z \leq 1.55\} = Pr\{Z \leq 1.55\} - Pr\{Z \leq -1.6\}$
$$= Pr\{Z \leq 1.55\} - Pr\{Z \geq 1.6\}$$
$$= Pr\{Z \leq 1.55\} - (1 - Pr\{Z \leq 1.6\})$$
$$= 0.9394 - (1 - 0.9452) = 0.9394 - 0.0548 = 0.8846.$$

9. $Z = \frac{X-\mu}{\sigma} = \frac{X-4}{2}$.
When $X = 7$, $Z = \frac{7-4}{2} = \frac{3}{2}$.
$Pr\{X \leq 7\} = Pr\{Z \leq 1.5\} = 0.9332.$

11. $A = \frac{1-2}{2} = -0.5$, $B = \frac{7-2}{2} = 2.5.$
So $Pr\{1 \leq X \leq 7\} = Pr\{-0.5 \leq Z \leq 2.5\} = Pr\{Z \leq 2.5\} - Pr\{Z \geq 0.5\}$
$$= 0.9938 - (1 - 0.6915) = 0.9938 - 0.3085 = 0.6853.$$

13. $A = \frac{35-50}{10} = -1.5$, $B = \frac{65-50}{10} = 1.5.$
So $Pr\{35 \leq X \leq 65\} = Pr\{-1.5 \leq Z \leq 1.5\}$
$$= Pr\{Z \leq 1.5\} - (1 - Pr\{Z \leq 1.5\})$$
$$= 0.9332 - (1 - 0.9332) = 0.9332 - 0.0668 = 0.8664.$$

15. $A = \frac{-5-(-1)}{4} = -1$, $B = \frac{1-(-1)}{4} = 0.5.$
So $Pr\{-5 \leq X \leq 1\} = Pr\{-1 \leq Z \leq 0.5\}$
$$= Pr\{Z \leq 0.5\} - (1 - Pr\{Z \leq 1\})$$
$$= 0.6915 - (1 - 0.8413) = 0.6915 - 0.1587 = 0.5328.$$

17. a. $A = \frac{72-70}{2} = 1$, so $Pr\{Z \geq 1\} = 1 - Pr\{Z \leq 1\} = 1 - 0.8413 = 0.1587$.

b. $A = \frac{66-70}{2} = -2, \quad B = \frac{74-70}{2} = 2,$

so $Pr\{-2 \leq Z \leq 2\} = 0.9772 - (1 - 0.9772) = 0.9544$

19. $A = \frac{15-10}{3} = \frac{5}{3} = 1.67.$

$Pr\{Z \geq 1.67\} = 1 - Pr\{Z \leq 1.67\} = 1 - 0.9525 = 0.0475.$

21. $A = \frac{5-6}{1.5} = -0.67, \quad B = \frac{7-6}{1.5} = 0.67.$

$Pr\{-0.67 \leq Z \leq 0.67\} = 0.7486 - (1 - 0.7486) = 0.4972.$

23. $A = \frac{42000-30000}{8000} = \frac{12000}{8000} = 1.5.$

$Pr\{Z > 1.5\} = 1 - Pr\{Z \leq 1.5\} = 1 - 0.9332 = 0.0668.$

25.
$$\begin{aligned}
E(Z) &= \int_{-\infty}^{\infty} \frac{1}{\sqrt{2\pi}} z e^{-\frac{1}{2}z^2}\, dz \\
&= \lim_{t\to-\infty} \int_t^0 \frac{1}{\sqrt{2\pi}} z e^{-\frac{1}{2}z^2}\, dz + \lim_{t\to\infty} \int_0^t \frac{1}{\sqrt{2\pi}} z e^{-\frac{1}{2}z^2}\, dz \\
&= \lim_{t\to-\infty} \int_{\frac{1}{2}t^2}^0 \frac{1}{\sqrt{2\pi}} e^{-u}\, du + \lim_{t\to\infty} \int_0^{\frac{1}{2}t^2} \frac{1}{\sqrt{2\pi}} e^{-u}\, du \\
&= \lim_{t\to-\infty} \left(-\frac{1}{\sqrt{2\pi}} e^{-u}\right)\Big]_{\frac{1}{2}t^2}^0 + \lim_{t\to\infty} \left(-\frac{1}{\sqrt{2\pi}} e^{-u}\right)\Big]_0^{\frac{1}{2}t^2} \\
&= \lim_{t\to-\infty} \left(-\frac{1}{\sqrt{2\pi}} + \frac{1}{\sqrt{2\pi}} e^{-\frac{1}{2}t^2}\right) + \lim_{t\to\infty} \left(-\frac{1}{\sqrt{2\pi}} e^{-\frac{1}{2}t^2} + \frac{1}{\sqrt{2\pi}}\right) \\
&= -\frac{1}{\sqrt{2\pi}} + \frac{1}{\sqrt{2\pi}} = 0.
\end{aligned}$$

Selected Solutions to Exercise Set 11.4

1. $$\int_0^6 \alpha e^{-\alpha x}\, dx = \left(-e^{-\alpha x}\right)\Big]_0^6 = -e^{-(0.05)(6)} + e^0 = -e^{-0.3} + 1$$

$$= 0.2592.$$

3. $$\int_2^6 \alpha e^{-\alpha x}\, dx = \left(-e^{-\alpha x}\right)\Big]_2^6 = -e^{-(0.25)(6)} + e^{-(0.25)(2)}$$

$$= -e^{-1.5} + e^{-0.5} = 0.3834.$$

5. $\int_2^5 \alpha e^{-\alpha x}\,dx = (-e^{-\alpha x})]_2^5 = -e^{-(0.50)(5)} + e^{-(0.50)(2)}$

$= -e^{-2.5} + e^{-1} = 0.2858.$

7. $Pr\{0 \le X \le 6\} = 1 - e^{-0.1\cdot 6} = 1 - e^{-0.6} = 0.4512$

9. $Pr\{5 \le X \le 20\} = Pr\{X \le 20\} - Pr\{X \le 5\} = e^{-0.02\cdot 5} - e^{-0.02\cdot 20}$

$= e^{-0.1} - e^{-0.4}$

$= 0.9048 - 0.6703 = 0.2345.$

11. $\frac{1}{\alpha} = 20 \implies \alpha = \frac{1}{20} = 0.05.$

$Pr\{5 \le X \le 30\} = Pr\{X \le 30\} - Pr\{X \le 5\} = e^{-0.05\cdot 5} - e^{-0.05\cdot 30}$

$= e^{-0.25} - e^{-1.5} = 0.7788 - 0.2231 = 0.5557.$

13. $\frac{1}{\alpha} = 1 \implies \alpha = 1.$

15. $\frac{1}{\alpha} = 4 \implies \alpha = \frac{1}{4}.$

$Pr\{X \le 2\} = 1 - e^{-\frac{1}{4}\cdot 2} = 1 - e^{-0.5} = 1 - 0.6065 = 0.3935.$

17. $E(X) = \frac{1}{\alpha}.$

$Pr\{X > E(X)\} = Pr\{X > \frac{1}{\alpha}\} = 1 - Pr\{X \le \frac{1}{\alpha}\}$

$= 1 - \left(1 - e^{-\alpha\cdot\frac{1}{\alpha}}\right) = e^{-1} = 0.3679.$

The answer is independent of α since $Pr\{X > E(X)\} = e^{-1}$ for any α.

19. $\mathrm{Var}(X) = \int_0^\infty (x - E(X))^2 \cdot \alpha \cdot e^{-\alpha x}\,dx = \int_0^\infty \left(x - \frac{1}{\alpha}\right)^2 \cdot \alpha \cdot e^{-\alpha x}\,dx$

$= \int_0^\infty \left(x^2 - \frac{2}{\alpha}x + \frac{1}{\alpha^2}\right) \cdot \alpha \cdot e^{-\alpha x}\,dx$

$= \alpha \int_0^\infty x^2 e^{-\alpha x}\,dx - 2\int_0^\infty x e^{-\alpha x}\,dx + \int_0^\infty \frac{1}{\alpha} e^{-\alpha x}\,dx$

$= \alpha \lim_{t\to\infty} \left(-\frac{x^2}{\alpha}e^{-\alpha x} - \frac{2x}{\alpha^2}e^{-\alpha x} - \frac{2}{\alpha^3}e^{-\alpha x}\right)\Big]_0^t$

$-2\lim_{t\to\infty} \left(-\frac{x}{\alpha}e^{-\alpha x} - \frac{1}{\alpha^2}e^{-\alpha x}\right)]_0^t + \lim_{t\to\infty} \left(-\frac{1}{\alpha^2}e^{-\alpha x}\right)]_0^t$

$= \frac{2}{\alpha^2} - 2\cdot\frac{1}{\alpha^2} + \frac{1}{\alpha^2} = \frac{1}{\alpha^2}.$

Selected Solutions to the Review Exercises for Chapter 11

1. a. $0.1 + 0.1 + 0.3 = 0.5$

   b. $0.1 + 0.2 + 0.3 = 0.6$

   c. $0.1 + 0.1 + 0.2 = 0.4$

3. $\mathrm{Var}(X) = E\left([X - E(X)]^2\right)$

$$= (1-3.6)^2 \cdot 0.1 + (2-3.6)^2 \cdot 0.1 + (3-3.6)^2 \cdot 0.2 + (4-3.6)^2 \cdot 0.3 + (5-3.6)^2 \cdot 0.3$$

$$= 0.676 + 0.256 + 0.072 + 0.048 + 0.588 = 1.640.$$

5.

x_i	\$490	−\$10
p_i	$\frac{1}{2000}$	$\frac{1999}{2000}$

$E(X) = 490 \cdot \frac{1}{2000} - \frac{1999}{2000} \cdot 10 = \frac{-19500}{2000} = -\$9.75.$

7. $\int_0^5 kx\ dx = 1 \implies \left.\frac{kx^2}{2}\right]_0^5 = 1 \implies \frac{25}{2}k = 1 \therefore k = \frac{2}{25}.$

9. a. $\int_0^2 k(4 - x^2)dx = 1 \implies k\left(4x - \frac{1}{3}x^3\right)\Big]_0^2 = 1$

$\implies k\left(8 - \frac{8}{3}\right) = 1 \therefore k = \frac{3}{16}.$

   b. $\int_0^1 \frac{3}{16}(4 - x^2)dx = \frac{3}{16}\left(4x - \frac{1}{3}x^3\right)\Big]_0^1 = \frac{3}{16}\left(\frac{11}{3}\right) = \frac{11}{16}.$

   c. $\int_1^2 \frac{3}{16}(4 - x^2)dx = \frac{3}{16}\left(4x - \frac{1}{3}x^3\right)\Big]_1^2$

$$= \frac{3}{16}\left(4 - \frac{7}{3}\right) = \frac{3}{16} \cdot \frac{5}{3} = \frac{5}{16}.$$

11. $\mathrm{Var}(X) = \int_0^5 \left(x - \frac{10}{3}\right)^2 \cdot \frac{2}{25} \cdot x\ dx$

$$= \frac{2}{25}\int_0^5 \left(x^3 - \frac{20}{3}x^2 + \frac{100}{9}x\right) dx$$

$$= \frac{2}{25}\left(\frac{x^4}{4} - \frac{20}{9}x^3 + \frac{50}{9}x^2\right)\Bigg]_0^5$$

$$= \frac{2}{25}\left(\frac{625}{4} - \frac{20 \cdot 125}{9} + \frac{50 \cdot 25}{9}\right) = \frac{25}{2} - \frac{200}{9} + \frac{100}{9}$$

$$= \frac{25}{2} - \frac{100}{9} = \frac{225-200}{18} = \frac{25}{18}.$$

13. $Pr\{Z \leq -0.5\} = Pr\{Z \geq 0.5\} = 1 - Pr\{Z \leq 0.5\} = 1 - 0.6915 = 0.3085.$

15. $Pr\{Z \geq -0.6\} = Pr\{Z \leq 0.6\} = 0.7257.$

17. $A = \frac{7-10}{2} = -1.5.$

$Pr\{X \geq 7\} = Pr\{Z \geq -1.5\} = Pr\{Z \leq 1.5\} = 0.9332.$

19. $A = \frac{-11-2}{6} = -2.17, \qquad B = \frac{13-2}{6} = 1.83.$

$$\begin{aligned} Pr\{-11 \leq X \leq 13\} &= Pr\{-2.17 \leq Z \leq 1.83\} \\ &= Pr\{Z \leq 1.83\} - Pr\{Z \leq -2.17\} \\ &= Pr\{Z \leq 1.83\} - (1 - Pr\{Z \leq 2.17\}) \\ &= 0.9664 - (1 - 0.9850) = 0.9514. \end{aligned}$$

21. a. $A = \frac{110-100}{10} = 1.$

$Pr\{X \geq 110\} = Pr\{Z \geq 1\} = 1 - Pr\{Z \leq 1\} = 1 - 0.8413 = 0.1587.$

b. $A = \frac{120-100}{10} = 2.$

$Pr\{X \geq 120\} = Pr\{Z \geq 2\} = 1 - Pr\{Z \leq 2\} = 1 - 0.9772 = 0.0228.$

23. $\mu = 16$ and $\sigma = 3.$

$$\begin{aligned} \Longrightarrow Pr\{X < 10\} &= Pr\{Z < \tfrac{10-16}{3}\} = Pr\{Z < -2\} \\ &= 1 - Pr\{Z < 2\} = 1 - 0.9772 = 0.0228. \end{aligned}$$

25. $\int\limits_4^\infty 0.2e^{-0.2x}dx = 0.2\left(\frac{e^{-0.2x}}{-0.2}\right)\Big]_4^\infty = e^{-0.8} = 0.4493.$

27. $\int\limits_{12}^\infty 0.1e^{-0.1x}dx = \left(-e^{-0.1x}\right)\Big]_{12}^\infty = e^{-1.2} = 0.3012.$

29. Since $\alpha = \frac{1}{4}$,

$$\begin{aligned} Pr\{2 \leq X \leq 6\} &= \int_2^6 \frac{1}{4}e^{-\frac{1}{4}x}dx \\ &= e^{-\frac{1}{2}} - e^{-\frac{3}{2}} = 0.6065 - 0.2231 \\ &= 0.3834. \end{aligned}$$

31. $\int\limits_0^5 \alpha e^{-\alpha x}dx = 1 - e^{-5\alpha} = 0.221$

$\Longrightarrow e^{-5\alpha} = 0.779 \Longrightarrow -5\alpha = \ln 0.779.$

$\Longrightarrow \alpha = -\frac{\ln 0.779}{5} \qquad \therefore\ E(X) = \frac{1}{\alpha} = -\frac{5}{\ln 0.779} = 20.020482.$

Practice Problems for Chapter 11

1. A card is drawn from an ordinary deck of 52 cards. If it is a face card, what is the probability that:

   a. it is a king? b. it is a diamond?

   c. it is either a king or a diamond?

2. An experiment consists of selecting a number at random from the set of numbers $\{1, 2, 3, 4, 5, 6, 7, 8, 9\}$. Find the probability that the number selected is:

   a. less than 5 b. even

   c. less than 7 and odd

3. For the probability distribution in the table below, find

   a. the expected value $E(X)$

   b. the variance, $\text{Var}(X)$, and the standard deviation σ.

x_i	0	1	2	3
p_i	0.2	0.3	0.3	0.2

4. An experiment has the probability density function $f(x) = 6(x - x^2)$ and outcomes lying between 0 and 1. Determine the probability that an outcome:

   a. lies between $\frac{1}{4}$ and $\frac{1}{2}$ b. lies betwen 0 and $\frac{1}{3}$

   c. is at least $\frac{1}{4}$ d. is at most $\frac{3}{4}$

5. The probability density function of a random variable X is

$$f(x) = 1 - \frac{1}{\sqrt{x}} \quad \text{for } 1 \leq x \leq 4.$$

   Find the following probabilities

   a. $Pr(X \geq 3)$ b. $Pr(X \leq 2)$ c. $Pr(2 \leq X \leq 3)$

6. Suppose that the outcomes X of an experiment are exponentially distributed, with density function $f(x) = 0.20e^{-0.20x}$. Find:

   a. $Pr(1 \leq X \leq 2)$

   b. $Pr(X \leq 3)$

   c. $Pr(X \geq 4)$

   d. the expected value of X.

7. Suppose that the average lifetime of an electronic component is 60 months, and the lifetimes are exponentially distributed. Find the probability that a component lasts for more than 20 months.

8. Suppose that the lifetimes of a certain type of light bulb are normally distributed with $\mu = 1200$ hours and $\sigma = 160$. Find the probability that a light bulb will burn out in less than 800 hours.

9. The price of an item in dollars is a continuous random variable with a probability density function of

$$f(x) = 2 \quad \text{for } 1.25 \leq x \leq 1.75.$$

   Find the probability that the price will be less than \$1.45.

10. A game consists of tossing a fair coin 3 times. You pay \$4 to play this game. If the coin comes up heads on exactly 2 of your tosses, you receive a prize of \$6. If the coin comes up heads on all of your tosses, you receive a prize of \$10. Otherwise, you lose your \$4. Find the expected value of the game for you.

11. The average monthly income of the trainees of a consulting firm is \$1000 with a standard deviation of \$150. Find the probability that an individual trainee earns less than \$600 per month assuming that the incomes are normally distributed.

12. a. Find the expected value and the variance of the random variable X in Problem 5.

    b. Find the distribution function F associated with the random variable X in Problem 5.

13. You pay a gambling house 15 cents to play a game which consists of your drawing at random one card from a standard deck of 52 cards. If you draw a heart other than the ace of hearts, the house pays you back 40 cents. If you draw an ace other than the ace of hearts, the house pays you back 50 cents. If you draw the ace of hearts, the house pays you back 90 cents. If you draw neither a heart nor an ace, you lose your 15 cents. Find the expected value of the game for you.

14. Your insurance company sells \$10,000 term policies insuring 25-year old persons against death for one year. The probability of a 25-year old person dying during the life of the policy is .002. Determine how much you must charge for such a policy in order to come out ahead in the long run.

15. Let θ be a positive real number, and let X be the continuous random variable having the probability density function f such that

$$f(x) = \begin{cases} \frac{1}{\theta} & \text{if } 0 \leq x \leq \theta \\ 0 & \text{if } x < 0 \text{ or } x > \theta. \end{cases}$$

Find the expected value and the variance of X.

Solutions to Practice Problems for Chapter 11

1. a. Probability that it is a king is $\frac{\frac{4}{52}}{\frac{12}{52}} = \frac{1}{3}$.

   b. Probability that it is a diamond is $\frac{\frac{3}{52}}{\frac{12}{52}} = \frac{1}{4}$.

   c. Probability that it is either a king or a diamond is $\frac{\frac{6}{52}}{\frac{12}{52}} = \frac{1}{2}$.

2. a. Probability of less than 5 $= \frac{4}{9}$.

   b. Probability of even number $= \frac{4}{9}$.

   c. Probability of less than 7 and odd $= \frac{3}{9}$.

3. a. The expected value $E(X) = 0(0.2) + 1(0.3) + 2(0.3) + 3(0.2) = 1.5$.

   b. Variance
$$\begin{aligned}\text{Var}(X) &= (0-1.5)^2(0.2) + (1-1.5)^2(0.3) + (2-1.5)^2(0.3) \\ &\quad +(3-1.5)^2(0.2) \\ &= 0.45 + 0.075 + 0.075 + 0.45 = 1.05.\end{aligned}$$

   Standard deviation $\sigma = \sqrt{1.05} = 1.0247$.

4. a.
$$\begin{aligned}Pr\left(\tfrac{1}{4} \le X \le \tfrac{1}{2}\right) &= \int_{1/4}^{1/2} 6(x - x^2)dx \\ &= (3x^2 - 2x^3)\Big]_{1/4}^{1/2} \\ &= \left\{3\left(\tfrac{1}{2}\right)^2 - 2\left(\tfrac{1}{2}\right)^3\right\} - \left\{3\left(\tfrac{1}{4}\right)^2 - 2\left(\tfrac{1}{4}\right)^3\right\} \\ &= \tfrac{3}{4} - \tfrac{2}{8} - \tfrac{3}{16} + \tfrac{2}{64} = \tfrac{22}{64} = \tfrac{11}{32}.\end{aligned}$$

   b.
$$\begin{aligned}Pr\left(0 \le X \le \tfrac{1}{3}\right) &= \int_{0}^{1/3} 6(x - x^2)dx \\ &= (3x^2 - 2x^3)\Big]_{0}^{1/3} \\ &= 3\left(\tfrac{1}{3}\right)^2 - 2\left(\tfrac{1}{3}\right)^3 - 0 = \tfrac{3}{9} - \tfrac{2}{27} = \tfrac{7}{27}.\end{aligned}$$

   c.
$$\begin{aligned}Pr\left(X \ge \tfrac{1}{4}\right) &= \int_{1/4}^{1} 6(x - x^2)dx \\ &= (3x^2 - 2x^3)\Big]_{1/4}^{1} \\ &= \{3(1)^2 - 2(1)^3\} - \left\{3\left(\tfrac{1}{4}\right)^2 - 2\left(\tfrac{1}{4}\right)^3\right\} \\ &= 3 - 2 - \tfrac{3}{16} + \tfrac{2}{64} = \tfrac{54}{64} = \tfrac{27}{32}.\end{aligned}$$

d. $Pr\left(X \le \frac{3}{4}\right) = \int_0^{3/4} 6(x - x^2)dx$

$= \left(3x^2 - 2x^3\right)\Big]_0^{3/4}$

$= \left\{3\left(\frac{3}{4}\right)^2 - 2\left(\frac{3}{4}\right)^3\right\} - 0$

$= \frac{27}{16} - \frac{54}{64} = \frac{54}{64} = \frac{27}{32}.$

5. a. $Pr(X \ge 3) = \int_3^4 \left(1 - \frac{1}{\sqrt{x}}\right) dx$

$= \left(x - 2x^{1/2}\right)\Big]_3^4$

$= \left\{4 - 2(4)^{1/2}\right\} - \left\{3 - 2(3)^{1/2}\right\}$

$= 0 - 3 + 3.4641 = 0.4641.$

b. $Pr(X \le 2) = \int_1^2 \left(1 - \frac{1}{\sqrt{x}}\right) dx$

$= \left(x - 2x^{1/2}\right)\Big]_1^2$

$= \left\{2 - 2(2)^{1/2}\right\} - \left\{1 - 2(1)^{1/2}\right\} = 2 - 2.8284 - 1 + 2 = 0.1716.$

c. $Pr(2 \le X \le 3) = \int_2^3 \left(1 - \frac{1}{\sqrt{x}}\right) dx$

$= \left(x - 2x^{1/2}\right)\Big]_2^3$

$= \left\{3 - 2(3)^{1/2}\right\} - \left\{2 - 2(2)^{1/2}\right\}$

$= 3 - 3.4641 - 2 + 2.8284 = 0.3643.$

6. a. $Pr(1 \le X \le 2) = \int_1^2 0.20e^{-0.20x} dx$

$= -e^{-0.20x}\Big]_1^2$

$= -e^{-0.4} + e^{-0.2} = 0.14841.$

b. $Pr(X \le 3) = \int_0^3 0.20e^{-0.20x} dx$

$= -e^{-0.20x}\Big]_0^3$

$= -e^{-0.6} + e^0 = -0.5488 + 1 = 0.4512.$

c. $Pr(X \ge 4) = \int_4^\infty 0.20e^{-0.20x} dx$

$= -e^{-0.20x}\Big]_4^\infty$

$= -e^{-\infty} + e^{-0.8} = 0.4493.$

d. The expected value of $X = E(X) = \frac{1}{\alpha} = \frac{1}{0.20} = 5.$

7. Since $E(X) = \frac{1}{\alpha} = 60 \implies \alpha = \frac{1}{60}$,

$$Pr(X > 20) = 1 - Pr(X \leq 20) = 1 - \left(1 - e^{-\left(\frac{1}{60}\right)\cdot 20}\right)$$
$$= e^{-1/3} = 0.7165.$$

8. $B = \frac{800-1200}{160} = -\frac{400}{160} = -2.5.$

$Pr(X < 800) = Pr(Z < -2.5) = 0.0062.$

9. $$Pr(X < 1.45) = \int_{1.25}^{1.45} (2)dx = (2x)]_{1.25}^{1.45}$$
$$= 2(1.45) - 2(1.25) = 0.4.$$

10. $$E(X) = \tfrac{3}{8}(6-4) + \tfrac{1}{8}(10-4) + \tfrac{4}{8}(-4)$$
$$= \tfrac{3}{4} + \tfrac{3}{4} - \tfrac{4}{2} = -\tfrac{1}{2} \text{ or } -50 \text{ cents.}$$

11. $B = \frac{600-1000}{150} = \frac{-400}{150} = -\frac{8}{3}$

$Pr(X < 600) = Pr\left(Z < -\frac{8}{3}\right) = 0.004.$

12. a. Expected value for X:

$$\int_1^4 \left(1 - \frac{1}{\sqrt{x}}\right) x\, dx = \int_1^4 (x - x^{1/2})dx = \left(\frac{1}{2}x^2 - \frac{2}{3}x^{3/2}\right)\Bigg]_1^4$$
$$= \left(8 - \frac{16}{3}\right) - \left(\frac{1}{2} - \frac{2}{3}\right) = \frac{8}{3} - \left(-\frac{1}{6}\right) = \frac{17}{6}.$$

Variance for X:

$$\int_1^4 \left(1 - \frac{17}{6}\right)^2 \cdot \left(1 - \frac{1}{\sqrt{x}}\right) dx$$
$$= \int_1^4 \left(x^2 - \frac{17}{3}x + \frac{289}{36}\right) \cdot \left(1 - x^{-1/2}\right) dx$$
$$= \int_1^4 \left(x^2 - \frac{17}{3}x + \frac{289}{36} - x^{3/2} + \frac{17}{3}x^{1/2} - \frac{289}{36}x^{-1/2}\right) dx$$

$$= \left(\frac{1}{3}x^3 - \frac{17}{6}x^2 + \frac{289}{36}x - \frac{2}{5}x^{5/2} + \frac{34}{9}x^{3/2} - \frac{289}{18}x^{1/2}\right)\Bigg]_1^4$$

$$= \left(\frac{64}{3} - \frac{136}{3} + \frac{1156}{36} - \frac{64}{5} + \frac{272}{9} - \frac{289}{9}\right)$$

$$- \left(\frac{1}{3} - \frac{17}{6} + \frac{289}{36} - \frac{2}{5} + \frac{34}{9} - \frac{289}{18}\right)$$

$$= \frac{3840 - 8160 + 5780 - 2304 + 5440 - 5780}{180}$$

$$- \frac{60 - 510 + 1445 - 72 + 680 - 2890}{180}$$

$$= -\frac{1184}{180} - \frac{-1287}{180} = \frac{103}{180} = 0.5722.$$

b. $F(x) = \int f(x)dx = \int \left(1 - x^{-1/2}\right) dx = x - 2x^{1/2} + C.$

$F(1) = 0 \implies C = 1.$

$F(x) = x - 2x^{1/2} + 1.$

13. Let X be the random variable representing your net gain in cents corresponding to the card you draw. These net gains are as follows.

$x_1 = 40 - 15 = 25$ cents in case you draw a heart other than the ace of hearts.

$x_2 = 50 - 15 = 35$ cents in case you draw an ace other than the ace of hearts.

$x_3 = 90 - 15 = 75$ cents in case you draw the ace of hearts.

$x_4 = -15$ cents in case you draw neither a heart nor an ace.

The probabilities associated with x_1, x_2, x_3, x_4 are, respectively, $p_1 = \frac{12}{52}$, $p_2 = \frac{3}{52}$, $p_3 = \frac{1}{52}$, $p_4 = \frac{36}{52}$. The expected value of the game for you is

$$E(X) = x_1p_1 + x_2p_2 + x_3p_3 + x_4p_4$$

$$= 25 \cdot \frac{12}{52} + 35 \cdot \frac{3}{52} + 75 \cdot \frac{1}{52} + (-15) \cdot \frac{36}{52}$$

$$= \frac{-60}{52} = -1.1538 \text{ cents.}$$

14. Let X be the random variable representing the net gain for your insurance company depending on whether the insured 25-year old person dies or does not die during the life of the policy. Let a be the amount in dollars your insurance company charges for the policy. Then the net gains for your insurance company are as follows.

$x_1 = (-10,000 + a)$ dollars in case the insured person dies during the life of the policy.

$x_2 = a$ dollars in case the insured person does not die during the life of the policy.

The probabilities associated with x_1, x_2 are, respectively, $p_1 = .002$, $p_2 = .998$. The expected value of the policy for your insurance company is

$$E(X) = x_1p_1 + x_2p_2 = (-10,000 + a) \cdot .002 + a \cdot .998 = -20 + a.$$

Thus your insurance company must charge more than \$20 for the policy in order to come out ahead in the long run.

15. $E(X) = \int_0^\theta \frac{1}{\theta} x\, dx = \frac{1}{\theta} \cdot \frac{1}{2}x^2\Big]_{x=0}^{x=\theta} = \frac{\theta}{2}.$

$V(X) = \int_0^\theta \left(x - \frac{\theta}{2}\right)^2 \cdot \frac{1}{\theta}\, dx = \frac{1}{\theta} \cdot \frac{1}{3} \cdot \left(x - \frac{\theta}{2}\right)^3\Big]_{x=0}^{x=\theta}$

$= \frac{1}{3\theta} \cdot \left(\left(\frac{\theta}{2}\right)^3 - \left(-\frac{\theta}{2}\right)^3\right) = \frac{1}{3\theta} \cdot 2 \cdot \frac{\theta^3}{8} = \frac{\theta^2}{12}.$